MATERIALS SCIENCE RESEARCH TRENDS

MATERIALS SCIENCE RESEARCH TRENDS

LAWRENCE V. OLIVANTE
Editor

Nova Science Publishers, Inc.
New York

For permission to use material from this book please contact us:
Telephone 631-231-7269; Fax 631-231-8175
Web Site: http://www.novapublishers.com

Library of Congress Cataloging-in-Publication Data

Materials science research trends / Lawrence V. Olivante, editor.
p. cm.
Includes index.
ISBN-13: 978-1-60021-654-1 (hardcover)
ISBN-10: 1-60021-654-4 (hardcover)
1. Materials science. I. Olivante, Lawrence V.
TA403.M34717 2006
620.1'1--dc22

2007011010

Published by Nova Science Publishers, Inc. ✢ New York

CONTENTS

PREFACE

Materials science includes those parts of chemistry and physics that deal with the properties of materials. It encompasses four classes of materials, the study of each of which may be considered a separate field: metals; ceramics; polymers and composites. Materials science is often referred to as materials science and engineering because it has many applications. Industrial applications of materials science include processing techniques (casting, rolling, welding, ion implantation, crystal growth, thin-film deposition, sintering, glassblowing, etc.), analytical techniques (electron microscopy, x-ray diffraction, calorimetry, nuclear microscopy (HEFIB) etc.), materials design, and cost/benefit tradeoffs in industrial production of materials. This new book presents new leading-edge research in the field.

Chapter 1 - Copper based delafossite transparent semiconducting oxide thin films have recently gained tremendous interest in the field of optoelectronic technology, after the discovery of p-type conductivity in a transparent thin film of copper aluminum oxide ($CuAlO_2$). Most of the well-known and widely used transparent conducting oxide thin films such as ZnO, SnO_2, ITO etc. and their doped versions are n-type material, but corresponding p-type transparent conducting oxides were surprisingly missing for a long time until the fabrication of above-mentioned p-$CuAlO_2$ thin film have been published (*Nature* 1997, 389, 939). This has opened up a new field in opto-electronics device technology, the so-called "Transparent Electronics", where a combination of the two types of transparent conducting oxides in the form of a p-n junction could lead to a 'functional' window, which transmits visible portion of solar radiation yet generates electricity by the absorption of UV part of it. Non-stoichiometric and doped versions of various new types of p-type transparent conducting oxides with improved optical and electrical properties have been synthesized in the last few years in this direction. Wide range of deposition techniques have been adopted to prepare the films. But fabrication of device quality films by cost-effective deposition techniques such as sputtering, chemical vapor deposition, wet-chemical dip-coating technique etc. are the need of the hour for large-scale production of these films for diverse device applications. Here the authors have discussed the fabrication and opto-electrical characterization of p-$CuAlO_{2+x}$ thin films by cost-effective and scaleable deposition routes such as sputtering and wet-chemical dip-coating technique. The authors have also discussed briefly some of the new developments in the field of p-type transparent conducting oxide thin film technology and an up-to-date and comprehensive description of different Cu-based p-type transparent conducting oxide thin films is presented. Also the origin of p-type conductivity in these transparent oxides has been dealt with considerable attention. Fabrication of all-transparent junctions is also discussed which is most important in the development of 'Transparent Electronics'. Field emission properties of thin films are currently of much interest due to the potential application in field emission displays (FEDs), which are considered to be strong candidate for low-power panel applications. The low-threshold field emission properties of wide-bandgap $CuAlO_2$ thin films have been investigated for its potential applications in FED technology. The films showed considerable low turn-on field. This finding might open up a new direction in the field-emission technology, and a new group of materials (such as, different transparent conducting oxides) might become a promising candidate for low-threshold field emitter. Also, recently, the research on nanostructured materials generates great interest in the scientific community and offers tremendous opportunities in the field of science and technology. Here, the authors have also discussed in brief, the formation of nanocrystalline p-$CuAlO_2$ films, which may open up an extremely important and interesting field of research for the fabrication of all-transparent nano-active devices. This will not only give a new dimension in the field of 'Transparent Electronics', but new avenues may open up in the nanoparticle research keeping an eye on its tremendous applications in optoelectronics technology.

Chapter 2 - Plastic deformation in a copper crystal is modeled using three dimensional atomistic simulations. The primary objective is to gain fundamental insight into the deformation features in face-centered-cubic materials in the form of a nanowire under tensile loading. An initial defect is utilized in the molecular statics model to trigger plasticity in a controlled manner. A parametric study is then performed by varying the atomic interaction range for the Morse interatomic potential used in the model. The simulation parameters are employed such that dislocation slip behavior and/or phase transformation can be observed without the influence of an unstable surface state of the specimen. The authors focus on tensile loading along a low-symmetry orientation where single slip prevails upon yielding. When the interaction distance is small, slip is seen to be the dominant deformation mechanism. A slight increase in the interaction range results in phase transition from the FCC structure to a BCC structure. Re-orientation of the BCC lattice also occurs at later stages of the deformation via a twinning operation. The phase transition mechanism is further enhanced if the nanowire is attached to a flat substrate parallel to the initial close-packed plane. The mechanisms of dislocation evolution, phase transformation, and crystal re-orientation features are discussed.

Chapter 3 - Progress in materials science and engineering is closely related to material characterization. Materials performance is highly dependent on its microstructure. Microstructural characterization has long surpassed the optical microscopy era. Advanced techniques including scanning electron microscopy (SEM) and transmission electron microscopy (TEM) have been well integrated into routine characterization excises. Other microscopy techniques like electron probe microanalyzer, Auger, X-ray photon spectroscopy (XPS) and secondary ion mass spectroscopy (SIMS) are also well recognized in the past years. In recent years, the focused ion beam (FIB) microscope has gradually evolved into an important microstructure characterization instrument. The combination of high-resolution imaging and stress-free site-specific cross sectioning provides valuable microstructure information both at the specimen surface and beneath. In addition, FIB techniques are often the preferred method to prepare TEM specimens, which, in many circumstances, are impossible to make by any other conventional methods. In this chapter, various FIB microscopy applications in microstructural characterizations will be discussed using practical examples in the authors recent research.

Chapter 4 - To fabricate high-quality functional thin films at very high deposition rates on large-sized substrates, the authors have proposed an atmospheric-pressure plasma chemical vapor deposition (AP-PCVD) technique. In the AP-PCVD process, stable glow plasma of gas mixtures containing carrier gases and source gases is generated at atmospheric pressure, and is effectively used to deposit thin films. Since the partial pressure of source gases can be high, the deposition rate is significantly increased. In the AP-PCVD system, combination of the rotary electrode and 150-MHz very high frequency (VHF) power supply makes it possible not only to stably generate high-density atmospheric-pressure plasma but also to suppress ion impingement upon the film surface. The AP-PCVD system equips a gas circulation system connected with the reaction chamber for efficiently collecting and removing particles that float around the plasma region. By virtue of these noble characteristics of the system, it has become possible to fabricate high quality films at extremely high deposition rates.

In this article, the basic concept and principle of the AP-PCVD technique are described first. Then, some of the fundamental research results on the property of atmospheric-pressure plasma and the elemental technologies for the AP-PCVD system are given. To evaluate the performance of the AP-PCVD system, the authors have deposited silicon (Si) films using silane (SiH_4) diluted with hydrogen (H_2) and helium. The deposition rate, morphology, and structural and electrical properties of the deposited Si films are discussed as functions of the deposition parameters, such as VHF power, SiH_4 and H_2 concentrations, and substrate temperature. The results show that homogeneous amorphous Si films having smooth surface and cross-sectional morphology can be successfully formed at unprecedented high rates. When the ratio of H_2 to SiH_4 and/or the substrate temperature is increased, polycrystalline and single crystalline films grow on a variety of substrate materials, such as Si and SiO_2, even at temperatures lower than in conventional deposition techniques. It is shown that the VHF power is a very important deposition parameter, which dominates the dissociation of SiH_4 molecules and the structural relaxation of a growing film. Note that the plasma gas temperature, including rotational and vibrational temperatures of molecules, and high-density atomic hydrogen in the atmospheric-pressure plasma can supply considerable physical and chemical energies to the film-growing surface, enhancing the film-forming reactions even at low temperatures.

Chapter 5 - In aluminum alloys one of the most pervasive and important impurity elements is iron, stemming from the impurities in bauxite ores and the contamination of ferrous metals such as melting tools. Since iron has a very low solid solubility in aluminum (max. 0.05%), almost all iron in aluminum alloys is present in the form of second intermetallic phases. One of the most common Fe-rich intermetallics that form in cast and wrought aluminum alloys upon solidification is the β-Al_5FeSi phase. This phase has long been thought to be brittle and responsible for the inferior mechanical properties (in particular ductility) of aluminum cast alloys. The commonly accepted method to ameliorate the harmful influence of iron is the addition of one or more corrective elements. Such additions generally convert the β-Fe platelets into α-Fe dendrites. Various studies have been carried out by researchers on the modification of β-Al_5FeSi intermetallics in aluminum alloys using Mn, Cr, Co, Mg, Sr, Li and Be. The relative effectiveness of these elements is collected and compared in the present review. The mechanisms for the action of the chemical modifiers are critically reviewed particularly in the light of the modern theory of their nucleation on oxide films present in aluminum melts, probably in large populations. The new insights into the Fe-rich phase in aluminum alloys will aid in better understanding the role of iron in aluminum alloys.

Chapter 6 - Superselection rules induced by the interaction with a mass zero Boson field are investigated for a class of exactly soluble Hamiltonian models. The calculations apply as well to discrete as to continuous superselection rules. The initial state (reference state) of the Boson field is either a normal state or a KMS state. The superselection sectors emerge if and only if the Boson field is infrared divergent, i. e. the bare photon number diverges and the ground state of the Boson field disappears in the continuum. The time scale of the decoherence depends on the strength of the infrared contributions of the interaction and on properties of the initial state of the Boson system. These results are first derived for a Hamiltonian with conservation laws. But in the most general case the Hamiltonian includes an additional scattering potential, and the only conserved quantity is the energy of the total system. The superselection sectors are stable against the perturbation by the scattering processes.

Chapter 7 - The microstructure evolution of ultra-thin Al films deposited on Si and SiO_2 substrates using molecular beam epitaxy (MBE) and, the effect of microstructure on electronic properties has been studied. First, the authors present a literature review on the "microstructure formation phenomena" and "structure zone model" for metallic films and, various existing theoretical models to explain electronic transport in these films. The authors present a systematic study on the evolution of microstructure in ultra-thin Al films on Si as a function of: (i) Film thickness: film thickness is varied between 10 and 200 nm, while keeping deposition temperature to a fix value; (ii) Deposition temperature: films are insitu deposited at different temperature between 25 and 600°C, while keeping thickness fixed; (iii) Post-annealing: annealing the room temperature deposited at higher temperature under UHV conditions.

The results reveal that in-situ deposited films grow in a columnar structure, forming a random 2D network of islands. The low temperature electrical transport in these films could not be accounted by the existing theoretical models. The authors have found that the charge conduction is governed by 2D variable range hopping mechanism. The coalescence of columnar Al islands is found to take place at a critical thickness, and this thickness is found to anomalously increase with increasing deposition temperature and the authors have proposed an explanation for this phenomenon. Post-annealing of films leads to the normal and abnormal growth, owing to the grain boundary migration. On SiO_2 substrates, the Al film picks up oxygen during in-situ deposition at elevated temperature as well as during post-annealing process, leading to the formation of Al_2O_3 at the grain boundaries.

Chapter 8 - Combustion synthesis is a novel processing technique in which the compacted powders are first ignited by an external heating source to induce the chemical reaction inside the heated materials. Propagation of a combustion front during Ni-Al unstable combustion synthesis often extinguishes in the half way, due to the lower exothermic heat of the metallic reactions. To facilitate the combustion front to propagate completely, the reaction is always ignited again during the experimental demonstration. In this numerical study, the different second ignition positions in the combusted region, the reacting region, and the pre-heating region as well as the different second ignition times before and after the stop of the first combustion front are chosen to study the effect of the second ignition. The second ignition position and time are found to influence the subsequent temperature profiles. The stable propagation is observed as the reaction is ignited again in the reacting region. When the reaction is ignited secondly in the combusted region or the pre-heating region, part of the specimens cannot be synthesized at the theoretical combustion temperature due to low combustion temperature. In addition, the combustion temperature may be significantly enhanced for some area, and results in heterogeneous microstructure. Delay of the second ignition time is also found to increase the initial propagation velocity of the new combustion front. From the results generated in this study, the process map of double ignitions is established. The process map provides appropriate double-ignition circumstances to propagate the combustion front completely and achieve homogeneous microstructure product.

EXPERT COMMENTARY

In: Materials Science Research Trends
Editor: Lawrence V. Olivante, pp. 3-14
ISBN: 978-1-60021-654-1

EFFECT OF AGING TREATMENTS ON SEVERELY DEFORMED MICROSTRUCTURE OF DIFFERENT AL-MG-SI ALLOYS

Emanuela Cerri[1*], Paola Leo[1] and H. J.Roven[2]
[1]Dept. of Ingegneria dell'Innovazione, University of Lecce, via Arnesano, 73100-Lecce, Italy
[2]NTNU, Inst. For Material Technology, Alfred Getz vei 2, 7491-Trondheim, Norway

Abstract

Equal channel angular pressing (ECAP) is a processing method in which a metal is subjected to an intense plastic straining through simple shear without any corresponding change in the cross –sectional dimension of the sample. The main purpose of ECAP is to obtain ultrafine grained materials. The influence of severe plastic deformation induced by ECAP on microstructural modification and aging effect was studied in as-cast and extruded Al-Mg-Si aluminum alloys. The microstructure of the alloys in different heat treated and deformed state was characterised by X-Rays diffraction, polarised light microscopy and scanning electron microscopy. The effect of post ECAP aging was investigated on samples after different number of pressings by hardness and electrical conductivity measurements. At higher aging temperature (170 and 190°C) the alloys showed an increasing softening with time due to recovery or/and grain coarsening effect. At the lower aging temperature, the hardness remains almost constant due to enhanced precipitation hardening effect. The solution treatment prior to ECAP enhances the post ECAP hardness values even if the general trend is similar to the untreated alloys.

Keywords: ECAP, aging, Al-Mg-Si alloys

[*] E-mail address: emanuela.cerri@unile.it (Emanuela Cerri). Corresponding author. Tel.: +39 0832 297324

Introduction

Equal channel angular pressing (ECAP) is a processing method in which a metal is subjected to an intense plastic straining through simple shear without any corresponding change in the cross–sectional dimension of the sample [1]. The main purpose of ECAP is to obtain ultrafine grained materials. Fine grains are beneficial in viewpoints of increased strength, according to Hall Petch relation [2,3] and improved superplasticity, according to constitutive equation [4,5,6].

Al-Mg-Si alloys are widely used in automotive and aerospace industries as a result of their good physical and chemical properties such as corrosion, formability, weldability [7] and because they are age hardenable to develop adeguate strength [8]. Moreover Sc and Zr addition play a critical role in these alloys by providing precipitates which impede grain growth at elevate temperatures (superplasticity) [9].

The aim of the present study is to understand the influence of severe plastic deformation on aging treatments performed on 6082 aluminium alloys. Two of them are extruded and contains 0.1% Zr (one also 0.1% Sc) (%in wt.). The addition of Zr let the precipitation of Al_3Zr from the melt during solidification. The addition of Zr and Sc makes the alloy able to form $Al_3(Sc,Zr)$ particles always during solidification. The effect in both cases is to obtain a very refined cast structure because these particles act as crystallisation nuclei. In the second case, the grain refining effect is higher, because it is reduced the necessary Sc content for getting critical size of Al_3Sc as a crystallisation nuclei [10]. The as-cast alloy contains higher quantity of Mg, Si and Mn (almost double) and so a higher potential of aging.

Experimental Procedures

Three different 6082 aluminum alloys were processed by ECAP. The chemical compositions are reported in Table 1. The 6082Zr and the 6082ZrSc were supplied as extruded bars of 12 mm in diameter, while the AA6082 was in the as cast state.

Table 1. Chemical composition of the 6082 alloys

	Fe	Si	Mg	Mn	Zr	Sc	Ti	Cr	Al
6082Zr	0.16	0.51	0.34	0.014	0.1				bal
6082ZrSc	0.16	0.51	0.34	0.014	0.1	0.1			bal.
AA6082	0.19	0.98	0.64	0.51			0.012	0.0037	bal

ECAP was conducted using a die with an internal angle (Φ) of 90 deg and a curvature angle (Ψ) of 35 deg. For this design, it has been shown that the effective strain occurred on a single pass through the die is close to 1 [11] . Molybdenum bisulfide (MoS_2) was used as lubrificant. Rods with diameter of 10mm and length of 100mm were cut from the extruded bars, while billets of 100mm in length and a square section of 20x20 mm^2 were machined from the cast alloy. They were pressed through the dies at room temperature. Repetitive pressing were conducted on each sample according to route B_c (extruded) or route A (as cast). After ECAP, specimens cut from the rods were statically aged at temperature ranging from Room Temperature (RT) to 190°C. Static aging was performed on the as received samples to verify the aging potential of the materials. Microhardness (HV 0.5), hardness and electrical conductivity measurements were carried out on cross sections to evaluate the effect of heat treatment on both ECAP processed and aged samples.

Microstructural observations were conducted by polarized light. Samples were ground according to standard method and then electropolished (80ml perchloric acid, 120ml distilled water, 800ml ethanol, 20V) and anodyzed (5% HBF_4 in distilled water, 20V). Grain size measurements were carried out on the cross section of the extruded samples. Scanning electron microscopy was performed on the as received and severely deformed structures by a FEG-SEM. Images were obtained by channelling contrast. The samples were suitable for observations only after electropolishing.

X-Ray diffraction measurements were also performed on aged samples and processed samples to complete the microstructural investigations (Cu K_α radiation, 45KV, 40mmA).

Results and Discussion

The microstructure of the as-received extruded bars is shown in Fig. 1. The 6082Zr aluminium alloy is illustrated in Fig. 1a showing equiaxed grains of an average size of (50 ±10) μm, while Fig. 1b reports the microstructure of the 6082ZrSc alloy showing a finer grain size with an average of (5±1) μm. Elongated rod-shaped intermetallics of 2-3 μm in length are present in the extruded samples identified as Mg_2Si [12]. Moreover, X-rays diffractometry (Fig. 1c) shows the presence of Al-Sc and Al-Zr type particles, AlMnSi and and AlFeSi based intermetallics. The 6082Zr alloy contains the same kind of particles with the exception of the AlSc phases. The microstructure of the cast alloy is presented in Fig. 2a showing a very coarse grains (300-400 μm) surrounded by intermetallic particles (fig. 2b). EDS identified AlFeSi and AlMnFe particles on the grain boundaries as well as Mg_2Si phase (fig. 2c and 2d).

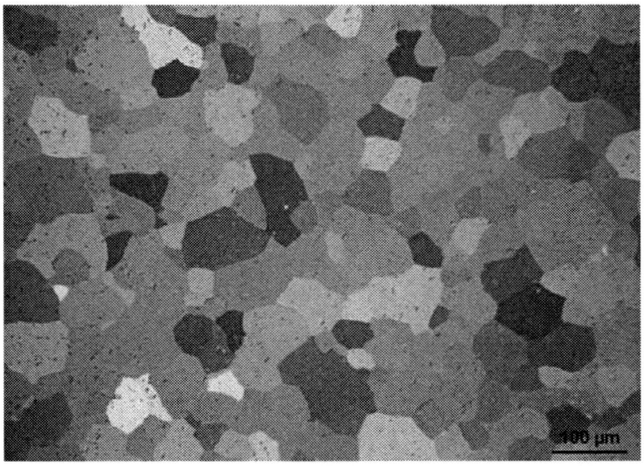

a)

Figure 1. Continued on next page.

b)

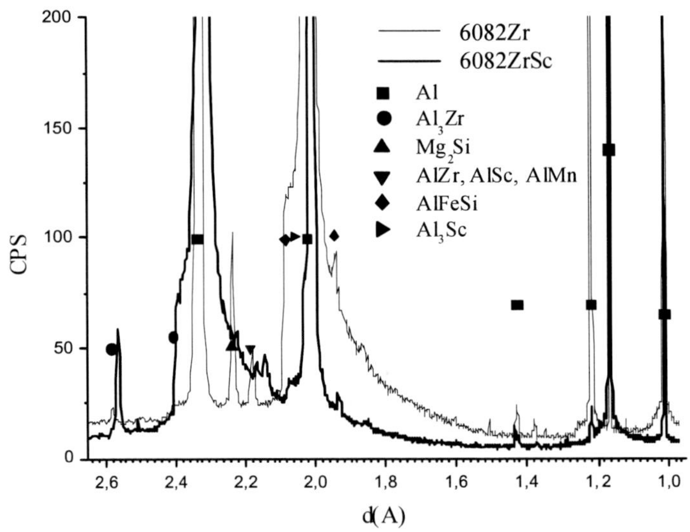

c)

Figure 1. Microstructure of the as-received a) 6082Zr and b) 6082ZrSc alloys. c) X-rays spectra.

After 4 pass (via route B_c), the grains have been severely deformed and the heavily strained microstructure is no more resolved by light microscopy. Only elongated parallel bands are visible (Fig. 3a). SEM observations performed by channelling contrast, revealed the presence of very fine grains, with an average size of 0.3-0.4 μm (Fig. 3b). The microstructure has been refined by a factor of 10^2 after 4 passes via route Bc in the 6082Zr alloy. In the 6082ZrSc, the grain size is very similar to the 6082Zr (approximately 0.3-0.4 μm) but the refining factor is 10 (Fig.4). In both alloys there are many grains which are well defined, indicating that the microstructures of the two ECA-pressed alloys are in equilibrium state.

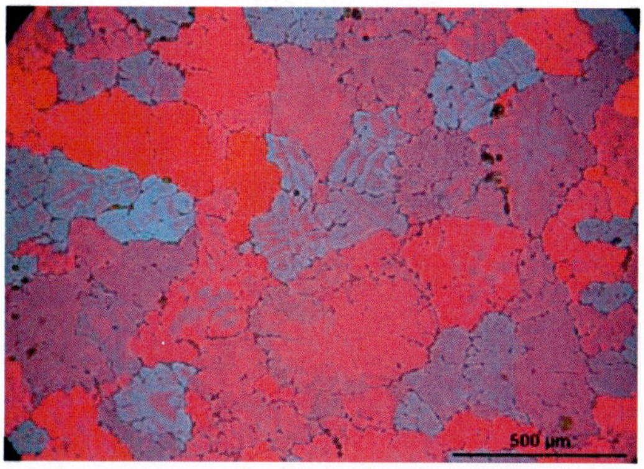

a)

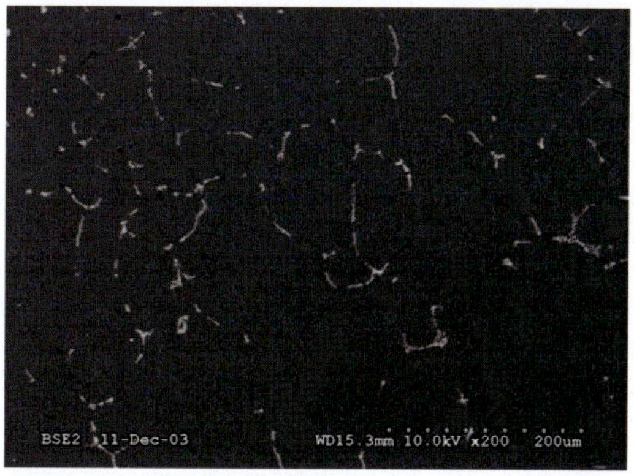

b)

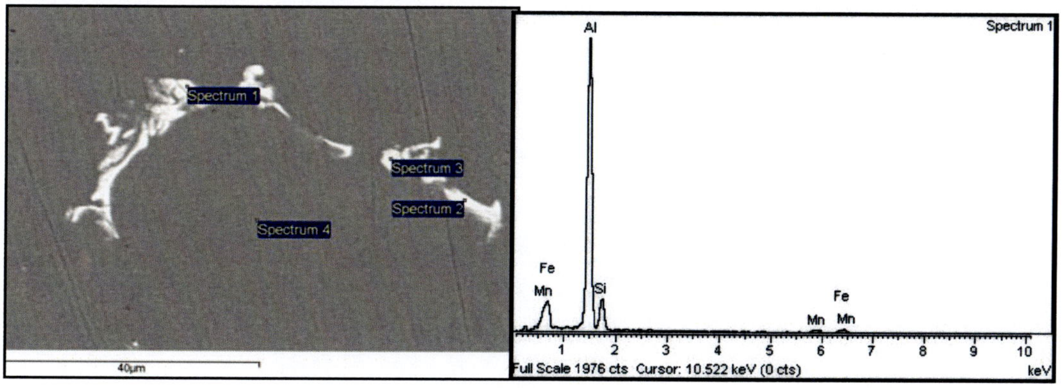

c)

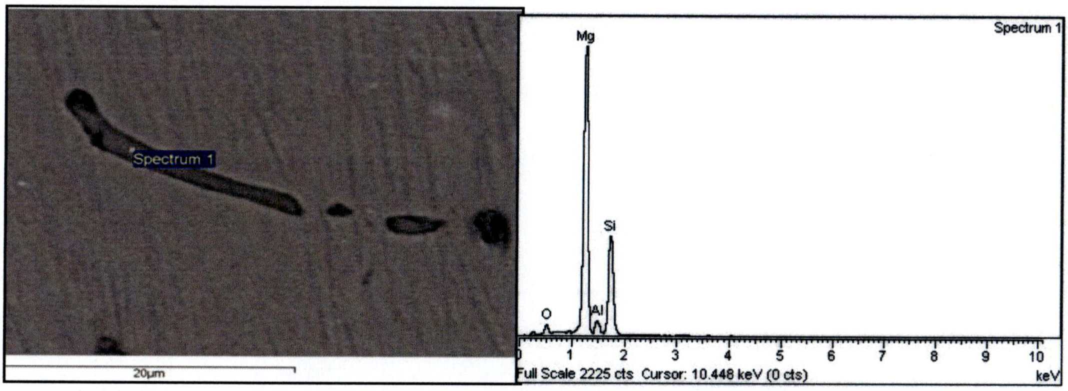

d)

Figure 2. Microstructure of the AA6082. a) anodized structure showing grain size, b)SEM image showing intermetallic particles, c) and d) EDS of the intermetallic phase surrounding the grains.

Fig. 5 shows the microstructure of the AA6082 after different ECA pressings. Fig. 5a shows the anodized grains still well defined and distinguishable with deformation bands visible inside. The intermetallic particles have been refined during ECAP due to the fracture phenomena and their distribution is reported in Fig. 5b. The image is taken in the plane perpendicular to the ECA extrusion direction.

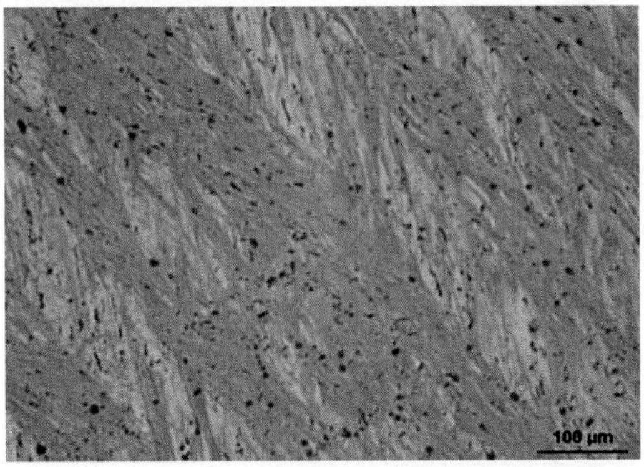

a)

b)

Figure 3. a) Anodized microstructure of 6082Zr after ecap (N=4) and b) electron backscattered channelling contrast images showing the grain size.

Figure 4. Electron backscattered channelling contrast images of the 6082ZrSc showing the grain size after 4 ECA pressing (route Bc).

a) N=1

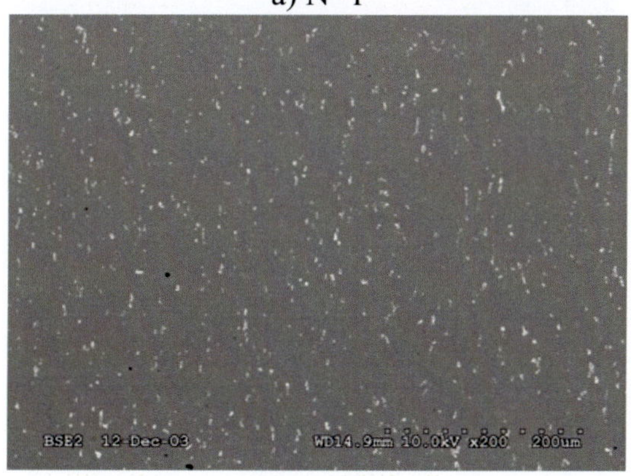

b) N=7

Figure 5. Microstructure of AA6082 after a) one pass and b) intermetallic particle evolution after 7 pressings (route A)

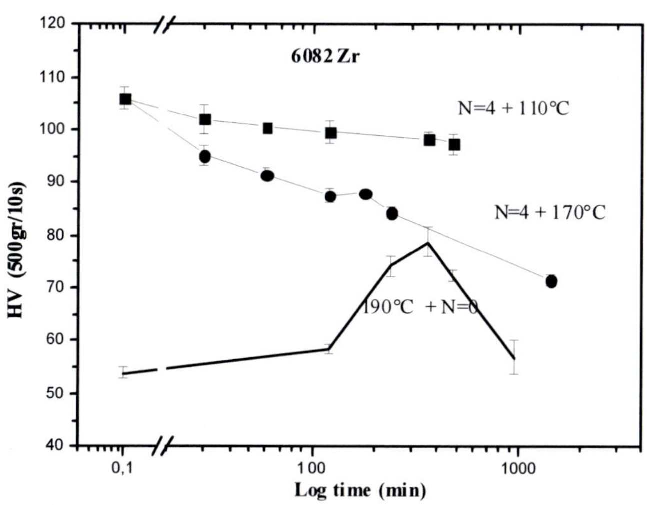

a)

Figure 6. Continued on next page.

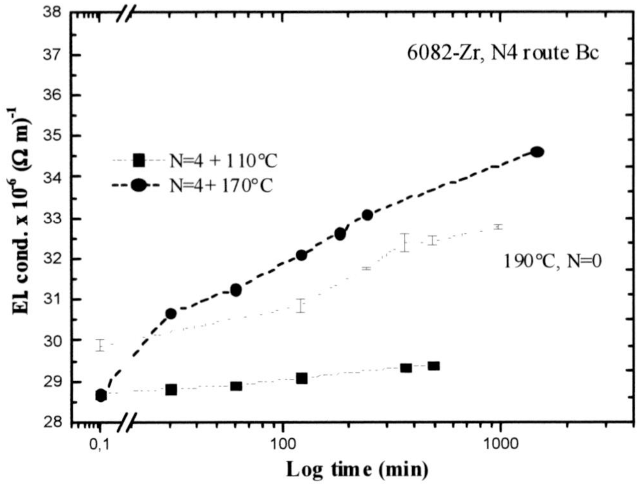

b)

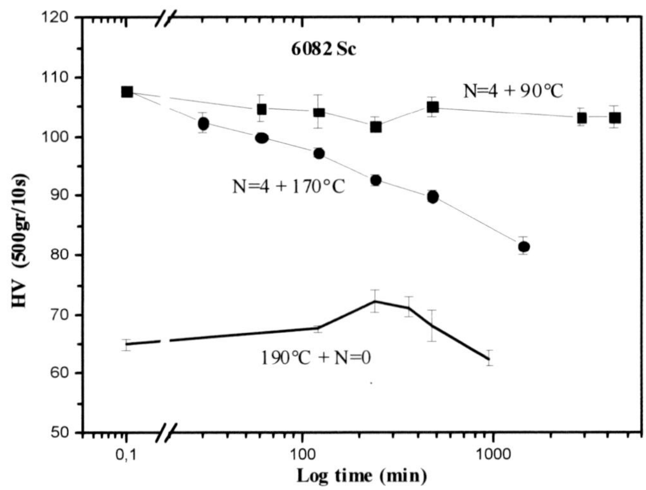

c)

Figure 6. Vickers hardness a) and electrical conductivity measurements b) as a function of time at different temperatures for the 6082Zr alloy before and after ECA pressing. c) Hardness evolution for the 6082ZrSc alloy.

The Vickers hardnesses measured on the plane perpendicular to the longitudinal axis of the ECA-pressed samples after different aging conditions are plotted in Fig. 6 and 7. Fig. 6(a) shows the aging curves at 190°C of the as-extruded samples and the post-ECA aging at 110 and 170°C for the Zr containing alloy, while Fig. 6(b) illustrates the plot of the electrical conductivity versus time. A significant increase in hardness occurs after 4 pressings respect to the as-extruded state (54±2HV) up to 106HV. This value is much higher than the static aging peak at 190°C (79±2HV). The large increase in hardness during ECAP of the extruded material (almost doubled after 4 passes) can be attributed to the considerable substructure refinement which occurs during intensive plastic deformation [13,14]. During post ECAP aging, the pressed materials exhibits a decrease in hardness with time. The hardness of the post ECAP aged sample at 170°C for 8 hours results comparable with the static peak-aging value for the 6082Zr alloy. The decreasing of hardness with time depends on the recovery process and/or recrystallisation that maybe have occurred during annealing of the severely strained microstructure. Precipitation may also have occurred during post ECAP aging treatments. In fact, as the precipitation process occurs in the static case (Fig. 6a), it is reasonable to suppose that precipitation may occur in samples with a high density of dislocations as ECA pressed specimens. If the aging is performed at a relatively high temperature like 170°C, the effect of recovery overwhelms the hardening associated to precipitation, leading to a decreasing stage in the hardness curves.

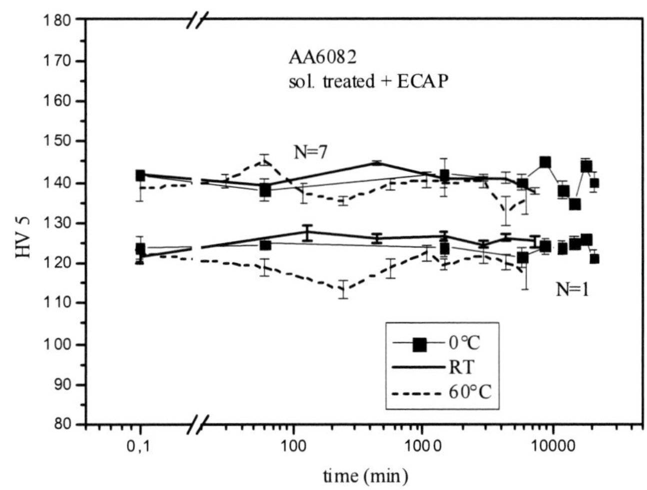

a)

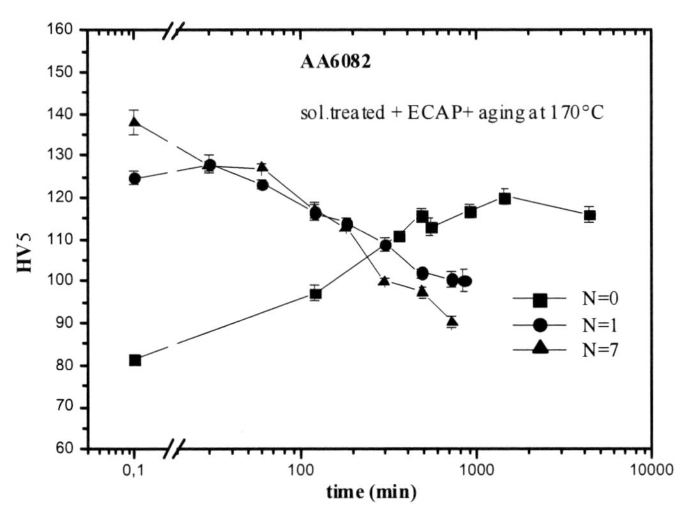

b)

Figure 7. Continued on next page.

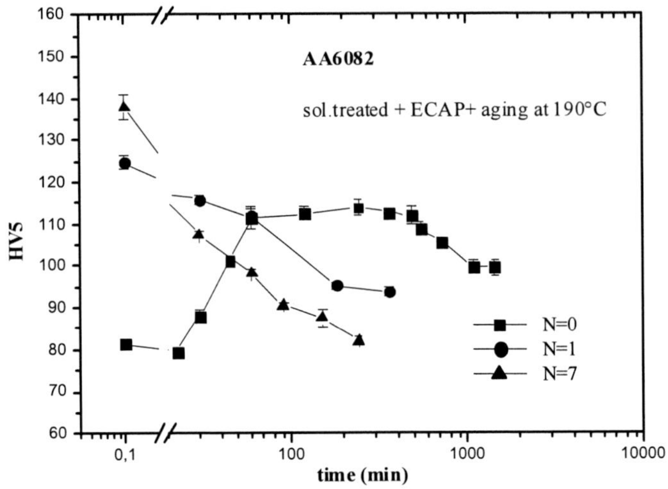

c)

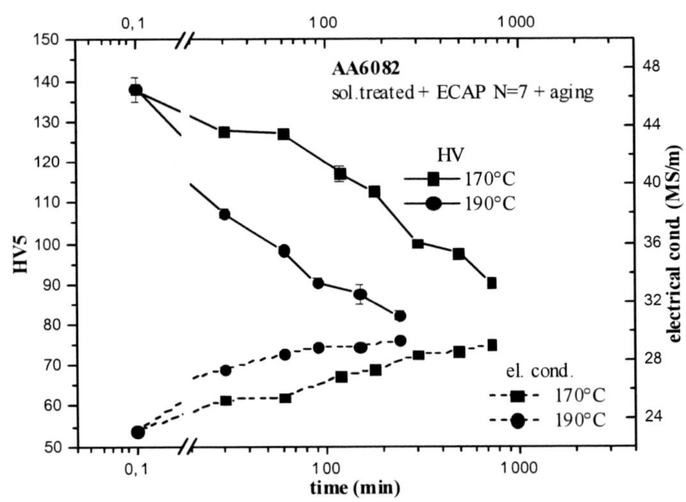

d)

Figure 7. Hardness evolution for the AA6082 alloy a) at low temperatures, b) at 170°C, c) at 190°C and d) comparison at high T after 7 passes.

In order to reduce the effect of recovery process during post ECAP treatments, an aging was performed at a lower temperature, 110°C. A decrease in hardness is still present, but the entity is very low compared to the former case (less than 10% in 8hours). This observations confirms that the softening is reduced at this temperature. The electrical conductivity (Fig. 6b) remains almost constant at the lower temperature, because of the contrasting phenomena, while at 170°C and 190°C increases with time. At 170°C (postECAP aging) the increment should be addressed to microstructural recovery, while at 190°C (static aging) the increase is due to precipitation.

Fig. 6c shows the Vickers hardness measured on the as-extruded samples of the 6082ZrSc alloy during static aging at 190°C and post ECAP aging at 170°C and 90°C. A significant increase in hardness occurs after 4 pressings respect to the as-extruded state (65±2HV) up to 108HV. This value is 50% higher than the static aging peak at 190°C. Even in the present case, the pressed materials exhibits a decrease in hardness with time. At 90°C the hardness remains constant during time due to the reduced effect of recovery at this low temperature.

X-ray diffraction analysis performed on post ECAP aged 6082Zr samples (110°C, 15h and 170°C,24h) [15] shows the presence of Mg_2Si particles at the highest temperature. Nevertheless, recovery phenomena produces a decrease in hardness with time shown at 170°C (Fig. 6a) [13,14,16]. In fact, the increased diffusion and the strong stress field induced by the significant amount of dislocation density during ECAP pressing, may influence the kinetics of precipitation and morphology of particles, leading to the development of incoherent interface and to negative contribution to hardness.

Fig. 7 illustrates the response of the solution treated AA6082 to post ECAP heat treatment performed at low and high aging temperatures. The number of pressing was increased to 7 to verify the enhancement of hardness at low temperatures. The results show that the hardness remains constant with time during aging at 0°, RT and 60°C and it increases with the number of extrusions (Fig. 7a). If the post ECAP aging temperature is increased to 170 or 190°C (Fig. 7b and 7c respectively), the measured values decreases with time and equals the static peak hardness after 1-2 h. At high aging temperatures, the hardness decreases with increased number of extrusions. This behaviour is completely reversed respect to the low temperatures. The exposure at high aging temperatures enhances the dislocation mobility and recovery/recrystallization phenomena as well as precipitation, inducing a higher rate of softening in the 7 passes sample respect to the 1 pass sample. Fig. 7d is a comparison of post ECAP aging results after 7 passes showing the softening effect due to the higher temperature of aging.

Conclusion

In the present study, the effect of post ECAP aging on hardness has been studied on three 6082 aluminium alloys supplied as cast or extruded. In all the alloys, the post ECAP aging curves shows a decreasing values of hardness with time at the higher temperature, while a reduction of aging temperature to 100°C or less, enhances the precipitation hardening contribution over the recovery and/or grain coarsening effect. In any case, the ECAP process substantially increases the hardness of the alloys respect to their peak values obtained without any previous deformation (almost doubled). The presence of precipitate during post ECAP aging has been confirmed by X-rays analysis, even though their nature is not well defined. The solution treatment prior to deformation is more effective in increasing hardness during ECAP respect to the alloys with no heat treatments. The effect of post ECAP heat treatment is similar in all the alloys investigated, independently from the state of the material prior to ECAP.

References

[1] M.Furukawa, Z.Horita, T.G.Langdon, *J. of Mat. Sci.* 2001, 36, 2835-2843

[2] E.O.Hall, *Proc. Roy. Soc.* 1951, B64 747

[3] N.J.Petch, *J. Iron Steel Inst.,* 1953, 174, 25

[4] A. Ball, M.M. Hutchinson, *Met. Sci. J.* , 1969, 3

[5] T.G.Langdon, *Acta Metall. Mater.* 1994, 42, 2437

[6] R.Z. Valiev, R.K. Islamgaliev, I.V. Alexandrov, *Progr. in Mat. Sci.,* 2000, 45, 156.

[7] C.D.Maioara,S.J. Andersen, J.Jansen, H.W.Zandbergen, *Acta Mater.* 2001, 49, 321-328

[8] G.A. Edwaeds, K.Stiller, G.L.Dunlop, M.J.Couper, *Acta Mater.* 1998, 46, 3893-3904

[9] S.Lee, A.Utsunomiya, H.Akamatsu, K.Neishi, M.Furukawa, Z.Horita, T.G.Langdon, *Acta Mater.* 2002, 50, 553–564

[10] Z.Yin, Q.Pan, Y.Zhang, F.Jiang, *Mater. Scie. Eng. A,* 2000, 280, 151-155

[11] Y. Iwahashi, Z. Horita, M. Nemoto and T. G. Langdon, *Acta Materialia,* 1997, 45, 4733-4741

[12] A.K. Asby, L. Edwards, J.W.Martin, *Mat. Sci & Tech.*, 1986, 2, 363-367

[13] J.K.Kim, W.J.Kim, T.Y.Park, S.I.Hong, D.I.Kim, Y.S.Kim,J.D.Lee, *Metall. and Mat. Trans.A*, 2002, 33, 3155-3164

[14] J.K.Kim, H.G. Jeong, S.I.Hong, Y.S.Kim, W.J.Kim, *Scripta Mater.* 2001, 45, 901-907

[15] P. Leo, E. Cerri, *Mater. Sci.Eng.A* , 2005, 410-411, 226-229

[16] M. Murayama, Z.Horita, K.Hono, *Acta Materialia* , 2001, 49, 21-29

RESEARCH AND REVIEW STUDIES

In: Materials Science Research Trends
Editor: Lawrence V. Olivante, pp. 17-132
ISBN: 978-1-60021-654-1

Chapter 1

P-TYPE TRANSPARENT SEMICONDUCTING DELAFOSSITE $CuAlO_{2+x}$ THIN FILM: PROMISING MATERIAL FOR OPTOELECTRONIC DEVICES AND FIELD-EMISSION DISPLAYS

***Arghya N. Banerjee*[1,a] *and Kalyan K. Chattopadhyay*[2,b]**
[1]Nevada Nanotechnology Center, Department of Electrical and Computer Engineering, University of Nevada, Las Vegas, Nevada-89154, US.
[2]Thin Film and Nanoscience Laboratory, Department of Physics, Jadavpur University, Kolkata-700032, India.

Abstract

Copper based delafossite transparent semiconducting oxide thin films have recently gained tremendous interest in the field of optoelectronic technology, after the discovery of p-type conductivity in a transparent thin film of copper aluminum oxide ($CuAlO_2$). Most of the well-known and widely used transparent conducting oxide thin films such as ZnO, SnO_2, ITO etc. and their doped versions are n-type material, but corresponding p-type transparent conducting oxides were surprisingly missing for a long time until the fabrication of above-mentioned p-$CuAlO_2$ thin film have been published (*Nature* 1997, 389, 939). This has opened up a new field in opto-electronics device technology, the so-called "Transparent Electronics", where a combination of the two types of transparent conducting oxides in the form of a p-n junction could lead to a 'functional' window, which transmits visible portion of solar radiation yet generates electricity by the absorption of UV part of it. Non-stoichiometric and doped versions of various new types of p-type transparent conducting oxides with improved optical and electrical properties have been synthesized in the last few years in this direction. Wide range of deposition techniques have been adopted to prepare the films. But fabrication of device quality films by cost-effective deposition techniques such as sputtering, chemical vapor deposition, wet-chemical dip-coating technique etc. are the need of the hour for large-scale production of these films for diverse device applications. Here we have discussed the fabrication and opto-electrical characterization of p-$CuAlO_{2+x}$ thin films by cost-effective and

[a] E-mail address: banerjee_arghya@hotmail.com;arghya@egr.unlv.edu (Arghya N. Banerjee, corresponding author)
[b] E-mail address: kalyan_chattopadhyay@yahoo.com (Kalyan K Chattopadhyay)

scaleable deposition routes such as sputtering and wet-chemical dip-coating technique. We have also discussed briefly some of the new developments in the field of p-type transparent conducting oxide thin film technology and an up-to-date and comprehensive description of different Cu-based p-type transparent conducting oxide thin films is presented. Also the origin of p-type conductivity in these transparent oxides has been dealt with considerable attention. Fabrication of all-transparent junctions is also discussed which is most important in the development of 'Transparent Electronics'. Field emission properties of thin films are currently of much interest due to the potential application in field emission displays (FEDs), which are considered to be strong candidate for low-power panel applications. The low-threshold field emission properties of wide-bandgap $CuAlO_2$ thin films have been investigated for its potential applications in FED technology. The films showed considerable low turn-on field. This finding might open up a new direction in the field-emission technology, and a new group of materials (such as, different transparent conducting oxides) might become a promising candidate for low-threshold field emitter. Also, recently, the research on nanostructured materials generates great interest in the scientific community and offers tremendous opportunities in the field of science and technology. Here, we have also discussed in brief, the formation of nanocrystalline p-$CuAlO_2$ films, which may open up an extremely important and interesting field of research for the fabrication of all-transparent nano-active devices. This will not only give a new dimension in the field of 'Transparent Electronics', but new avenues may open up in the nanoparticle research keeping an eye on its tremendous applications in optoelectronics technology.

1. Introduction

In the last century, scientists have made rapid and significant advances in the field of semiconductor physics. Semiconducting materials have been the subjects of great interest due to their numerous practical applications and also they provide fundamental insights into the electronic processes involved. Thus material processing has similarly become an increasingly important research field. Many new materials and devices, which possess specific properties for special purposes, have now become available, but material limitations are often the major deterrent to the achievement of new technological advances. Material scientists are now particularly interested in developing materials which maintain their required properties in extreme environment. In general it is the aim of the material scientists to find ways of improving qualities and increasing productivity, whilst reducing the manufacturing cost.

One of the most important fields of interest in materials science is the fundamental aspects and applications of semiconducting transparent films, which are more popularly known as "Transparent Conducting Oxides" (TCO) in opto-electronic device technology. The characteristics of such films are high room-temperature electrical conductivity (~ 10^3 S cm^{-1} or more) and high optical transparency (more than 80 %) in the visible region. TCOs are well-known and widely used for a long time in opto-electronics industries as well as in research fields. After the first report of transparent conducting cadmium oxide (CdO) thin film by Badekar [1] in 1907, extensive works have been done in the field of TCO technology to prepare new types of TCOs with wide ranging applications [2 – 12]. Some of these well-known and widely used TCOs include In_2O_3: Sn/F/Sb/Pb, ZnO: In/Al/F/B/Ga, Cd_2SnO_4, SnO_2: Sb/F etc. as well as some new TCOs such as $CdIn_2O_4$: Sn, $CdSb_2O_6$: Y, $GaInO_3$: Ge/Sn, $AgInO_2$: Sn, $MgIn_2O_4$, $In_4Sn_3O_{12}$, Zn_2SnO_4, $ZnSnO_3$, $Zn_2In_2O_5$, $ZnGa_2O_4$ etc. [13 – 43].

Technologically, these TCOs are being used extensively in various fields, which include solar cells, flat panel displays (FPD), low-emissivity ("low-e") windows, electromagnetic shielding of cathode-ray tubes in video display terminals, electrochromic (EC) materials in rear-view mirrors of automobiles, EC-windows for privacy (so-called "smart windows"), oven windows, touch-sensitive control panels, defrosting windows in refrigerators and airplanes, invisible security circuits, gas sensors, biosensors, organic light emitting diodes (OLED), polymer light emitting diodes (PLED), antistatic coatings, cold heat mirrors, etc. [1 – 4, 8 - 9, 13, 44 – 54]. Also some new applications of TCOs have been proposed recently such as holographic recording medium, high-refractive index waveguide overlays for sensors and telecommunication applications, write-once read-many-times memory chips (WORM), electronic ink etc. [55 – 58]. And lastly, the low-temperature deposition of TCOs onto poly(ethylene terephthalate) (PET), polyamides and other polymer substrates in roll-coating processes for touch-screen and infra-red reflector applications are the recent challenges for the TCO industries [59 – 61].

Possibility of the above-mentioned novel applications of TCOs is based on the fact that the electronic band gap of a TCO is higher than 3.1 eV (corresponding to the energy of a 400 nm blue photon). So visible photons (having energy between 2.1 to 3.1 eV) cannot excite electrons from valence band (VB) to the conduction band (CB) and hence are transmitted through it, whereas they have enough energy to excite electrons from donor level to CB (for n-type TCO) or holes from acceptor level to VB (for p-type TCO). And these acceptor or donor levels are created in the TCOs by introducing non-stoichiometry and (or) appropriate dopants in a controlled manner. A schematic representation of the bandgap designing for transparent conductors is shown in Fig. 1(a).

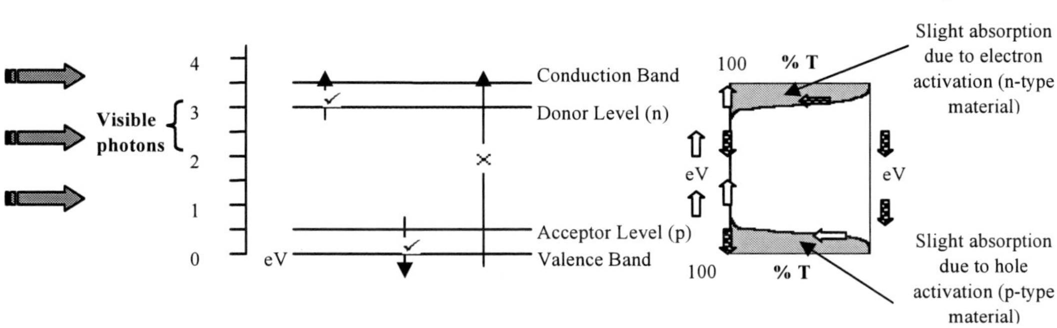

Figure 1(a). Bandgap designing for transparent conductors. Visible photons (2.1 eV to 3.1 eV) do not have enough energy to excite electrons from valence band to conduction band, but have enough energy to excite holes (for p-type) from acceptor level to VB or electrons (for n-type) from donor level to CB. Right hand side shows the transmittance of the TCO with respect to incident radiation. The arrow '⇨' indicates the transmittance graph for a p-type TCO, where a slight absorption can be observed (indicated by shaded part) at low energy region, due to the activation of holes from acceptor level to VB. Similarly, the arrow '➡' indicates the same for n-type TCO, where slight absorption at low energy region takes place due to electron activation from donor level to CB.

Although the TCOs have vast range of applications as mentioned above, very little work have been done on the active device fabrication using TCOs [62, 63]. This is because most of the aforementioned TCOs are n-type semiconductors. But the corresponding p-type transparent conducting oxides (p-TCO), which are essential for junctional devices, were surprisingly missing in thin film form for a long time, until in 1997, Kawazoe and co-authors reported the p-type conductivity in a highly transparent thin film of copper aluminum oxide ($CuAlO_{2+x}$) [64]. This has opened up a new field in opto-electronics device technology, the so-called "Transparent Electronics" or "Invisible Electronics" [65], where a combination of the two types of TCOs in the form of a p-n junction could lead to a 'functional' window, which transmits visible portion of solar radiation yet generates electricity by the absorption of UV part [64]. It must be mentioned here that the first report of semi-transparent p-type conducting thin film of nickel oxide was published in 1993 by Sato et al [66]. They observed only 40 % transmittance of the NiO films in the visible region and when they tried to fabricate an all-TCO p-i-n diode of the form p-NiO/i-NiO/i-ZnO/n-ZnO, the visible transmittance further reduced to almost 20 %. Although this low transmittance was not favorable for superior device applications, but still this report was an important milestone in the field of "Transparent Electronics" and in the development of TCO technology.

Now for diverse device applications, it is utmost important to prepare various new types of p-TCOs with superior optical and electrical characteristics, at least comparable to the existing, widely used n-TCOs, which are having transparency above 80 % in the visible region and conductivity about 1000 S cm^{-1} or more. Intense works have been done for the last few years in this direction to fabricate new p-TCOs by various deposition techniques. Also quite a number of works have been carried out for proper understanding of the structural, optical and electrical characteristics of p-TCOs. As this is an emerging field in TCO technology, preparation of new materials as well as existing materials with new deposition techniques is the need of the hour.

Copper aluminum oxide ($CuAlO_2$) is the first and the most important p-TCO material reported in thin film form [64], which has reasonable optical and electrical properties for diverse device applications. The reported visible transparency of this material is around 80 % with a direct bandgap value of 3.5 eV, whereas the room temperature conductivity (σ_{RT}) is 0.34 S cm^{-1} with a carrier concentration ~ 3.0 x 10^{19} cm^{-3} [67]. Although the reported transparency is quite high but the hole concentration is one to two orders of magnitude lower than the corresponding well-known and widely used n-TCO thin films e.g. ITO, ZnO, SnO_2 etc. Therefore, as far as technological aspects are concerned, improvement in the electrical characteristics of this material is the need of the hour alongwith the reproducibility with the required opto-electrical properties.

The electrical, optical, structural as well as morphological properties and hence device performance of $CuAlO_2$ thin films are highly correlated with the deposition techniques. The growth parameters, especially deposition atmosphere, substrate temperature, post-deposition annealing of the films etc., control the properties of $CuAlO_2$ thin films to a large extent. Defect chemistry plays a major role in the enhancement of the p-type conductivity of the films. So a systematic study of the effect of different growth parameters on the characteristics of the films is needed for improved material synthesis. Also low-cost processes to deposit device quality $CuAlO_2$ and similar types of p-TCO thin films are the most important issue for large-scale production of these films for diverse device applications. And most importantly, transparent junction fabrication with superior opto-electronic properties will be the next significant step towards the realization of "Transparent Electronics".

Nanocrystalline $CuAlO_2$ thin films: After the pioneering works of Efros and Efros [68] and Brus [69] on the size-quantization effect in semiconductor nanoparticles, the research on nanostructured materials generates great interest in the scientific community and offers tremendous opportunities in science and technology because of new properties exhibited by these materials and challenging problems thrown up for providing theoretical concepts in physics associated with it [70-72]. Infact, both the natural as well as the artificial world can now be categorized in two regimes: micro regime and nano regime. Starting from a human hair to DNA structure – the nature evolves itself from micro to nano scale structures. Similarly, man-made world is now shifting its attention from micro devices to nano materials. Fig. 1(b) schematically represents the broad spectrum of the micro and nano regime, indicating how natural and artificial world evolve into smaller domain.

Optical properties of nanocrystals are markedly related to their size and surface chemistry and drastically differ from those of bulk materials. Preparation and study of high quality quantum dots [73], nanobelts [74] and nanowires [75] have been reported widely. These achievements in the last few years have focused nanoparticle research on their applications in electrical and optoelectronics devices [76-77].

Syntheses and characterizations of nanostructured n-TCOs are very important and well-established field in nanotechnology and still growing in stature. Therefore, the formation of nanocrystalline p-type counterpart may open up an extremely important and interesting field of research for the fabrication of all-transparent nanoactive devices. This will not only give a new dimension in the field of ''Transparent Electronics'', but new avenues may open up in the nanoparticle research keeping an eye on its tremendous applications in optoelectronics technology.

Field-emission displays: Low-macroscopic field (LMF) emission of electrons from the surface of a thin film to the vacuum in the presence of a macroscopic electric field (mean field between the parallel plates in a capacitor configuration) is currently of much interest due to the potential application in cold cathode devices. Also field emission displays (FEDs) are considered to be strong candidate for low-power panel application because of its very thin profile, high production efficiency, fast response, high brightness, wide operating temperature, possible expansion of size and last but not the least, high picture quality at a lower cost [78]. Spindt tip cathodes made up of materials with high work function, e.g. Mo, W, Si etc. are used in typical FEDs. For extraction of electrons from these cathodes, sharp tips with radii as low as 20 nm were constructed, which enhances the macroscopic field at the emitter-tip and supplies the necessary barrier-field (also called local-field at the emitter-tip) to produce the field-emission-tunneling. These emitted electrons are then allowed to collide with fluorescent material applied to the cathode, thus emitting light. A schematic diagram of the light emitting principle of the FED system is shown in Fig. 1(c). While the cathode of a CRT uses a point electron source, an FED uses a surface electron source. 6-inch color FED panels have already been manufactured, and research and development on 10-inch FEDs is proceeding very rapidly. When compared with TFT LCDs, FEDs offer a superior viewing angle (160 degrees both vertically and horizontally) and are several microseconds quicker in response speed. In the last decade, low-macroscopic field emission from carbon based films like diamond, diamond like carbon (DLC), amorphous carbon (a: C) etc. [79, 80], made them strong candidate materials for FEDs.

It was found that the materials with wide bandgap (such as diamond) have low or negative electron affinity, which, in turn, enhances the low-macroscopic field emission properties of diamond films [81]. Also p-type semiconducting diamond film showed low-threshold field emission properties [82].

$CuAlO_2$ - being a p-type wide bandgap semiconducting material, can become a candidate material for potential field-emitters. Infact, we have first reported the low macroscopic field emission, at a relatively lower threshold, from $CuAlO_2$ thin film, deposited on glass substrate.

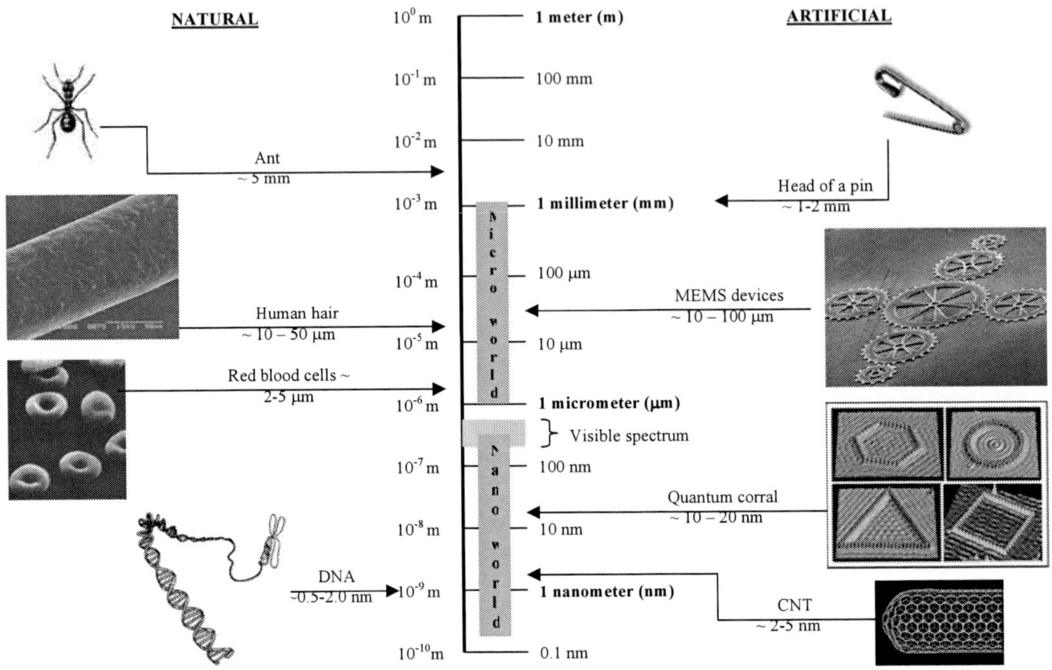

Figure 1(b). Description of nano regime. From Forbes-Nanotech Report.

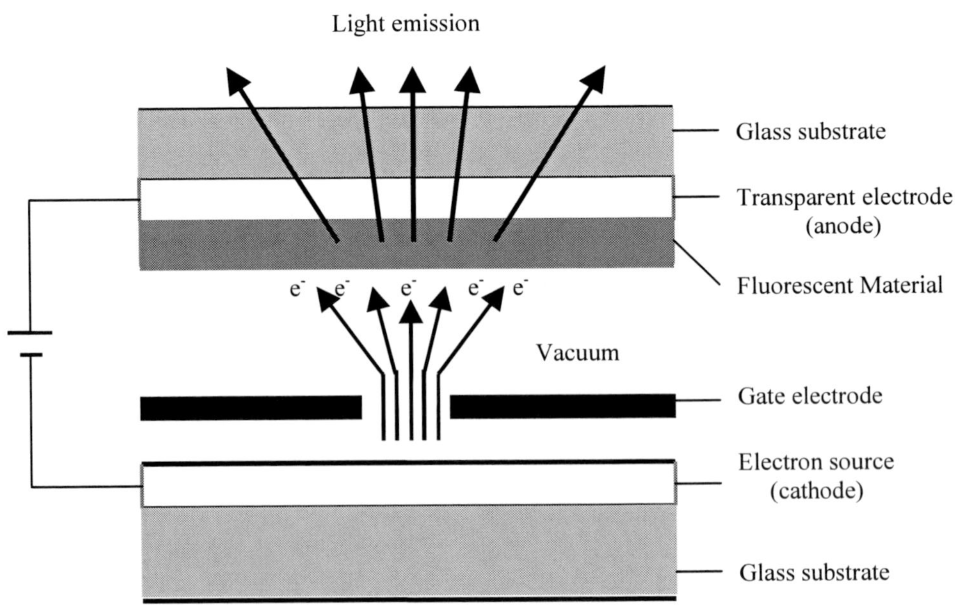

Figure 1(c). Schematic diagram of the light emitting principle of the FED system.

The emission properties have been studied for different anode-sample spacing. The threshold field and approximate local work function are calculated and we have tried to explain the emission mechanism therefrom. As mentioned above, $CuAlO_2$ is a transparent p-type semiconducting material, which has excellent potential to be used in opto-electronics device technology [64, 83]. Its field emission properties have given an additional impetus on the properties of this technologically important material and may open up a new window in the field-emission technology with a new group of materials other than carbon-based films like amorphous carbon (a: C), diamond like carbon (DLC), diamond, carbon nano-tubes (CNT), silicon carbide (Si: C) nano-rods etc. [84-87].

2. Brief Review of Past Work

As p-TCO technology is an important and emerging field of research, hence a systematic review of the major developments in p-TCO materials with regard to the different deposition techniques and properties of the films so obtained are the needs of the hour. A brief and partial review on this field had been reported previously by Tate and co-authors [89] as well as by Nagarajan and co-authors [89-90]. Also Norton [91] presented a detailed review on the synthesis and properties of oxide thin films, which briefly includes the importance of p-TCO too. We have recently published a detailed and up-to-date review on the recent developments in this interesting and challenging field of p-type transparent conducting oxides [92]. Fig. 2(a) describes different p-TCO materials fabricated so far by various groups around the globe. Here we have tried to give a comprehensive picture of the developments in various Cu-based p-TCO thin films, starting with $CuAlO_2$, which is the first and most important material in this family. We have also discussed, in details, the fabrication of various transparent junctions, which is the most important aspect of the "Invisible Electronics". Also a brief review on the recent activities on nanostructured p-TCO thin films have been presented, which is an extremely important field of research for the development of nano-active devices. And lastly, field-emission properties of various wide bandgap materials and their applications in FED technology have been presented.

2.1. Copper Based p-TCO Films

2.1.1. Delafossite Films

Delafossite materials have the chemical formula $M^IM^{III}O_2$, where M^I: monovalent cations such as Cu^+, Ag^+, Pd^+, Pt^+ etc. and M^{III}: trivalent cations such as Al^{+3}, Ga^{+3}, In^{+3}, Cr^{+3}, Fe^{+3}, Co^{+3}, Y^{+3}, La^{+3}, Sc^{+3} etc. Amongst them, those materials having d^{10} orbital (Cu, Ag) show p-type semiconducting behavior whereas those with d^9 orbital (Pt, Pd) show metallic conductivity [93-95]. For p-TCO technology, materials from the former group show the required properties for possible device applications. The first and the most important material in this group is the copper aluminum oxide ($CuAlO_2$). Although this material is known to exist for nearly 50 years [96] and back in 1984 its p-type conductivity was first reported by Benko and Koffyberg [97], but Kawazoe and co-authors [64] first prepared it in transparent thin film form for possible applications in p-TCO technology. The structural properties of this material were extensively studied by Ishiguro and co-authors [98-100]. The structure is shown in Fig. 2(b) and described in details later. It belongs to $R\bar{3}m$ (D_{3d}) space group with rhombohedral crystal structure [95]. The crystal data of $CuAlO_2$ is given in Table 1. Other p-TCO thin films belonging to this group are copper gallium oxide ($CuGaO_2$) and copper indium oxide ($CuInO_2$) [101-103]. The lattice parameters of these materials were reported in various literatures [94, 104-105]. Also, the band structures of these materials were calculated by Yanagi et al. [67], Robertson et al. [106] and Ingram et al. [107] in details. Doped versions of some similar types of p-TCO thin films have also been reported which include iron doped copper gallium oxide ($CuGaO_2$: Fe), calcium doped copper indium oxide ($CuInO_2$: Ca), magnesium doped copper scandium oxide ($CuScO_2$: Mg), magnesium doped copper chromium oxide ($CuCrO_2$: Mg), calcium doped copper yttrium oxide ($CuYO_2$: Ca) etc. [88, 102-103, 108-110]. Crystallographic data as well as band structure calculations of these materials had also been reported in various literatures [104, 111-112]. Preparation of some other highly resistive (~ 10^6 Ω-cm) new delafossite materials such as $CuFe_{1-x}V_xO_2$ (x = 0.5), $CuNi_{1-x}Sb_xO_2$, $CuZn_{1-x}Sb_xO_2$, $CuCo_{1-x}Sb_xO_2$, $CuMg_{1-x}Sb_xO_2$, $CuMn_{1-x}Sb_xO_2$ (x = 0.33) in powder form had been reported by Nagarajan et al. [89-90] (but no thin film preparation of these materials has been reported so far). Preparation of 10 % Sn doped $CuNi_{1-x}Sb_xO_2$ thin film has been reported by the same group [88, 90], having reasonable visible transparency (60 %) and conductivity (5 x 10^{-2} S cm^{-1}). The electrical and optical properties of these films are described in Table 2.

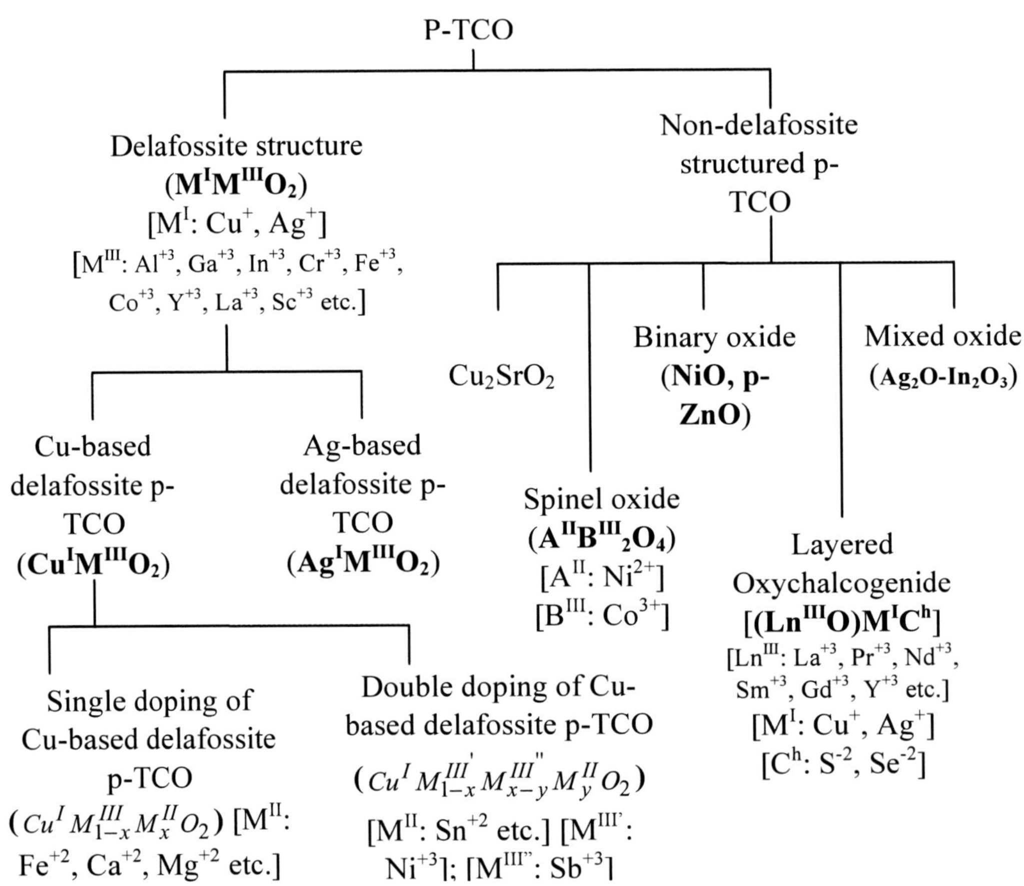

Figure 2(a). Chart of various p-TCO materials reported so far. Here the doped versions of Cu-based delafossite p-TCOs have been mentioned only.

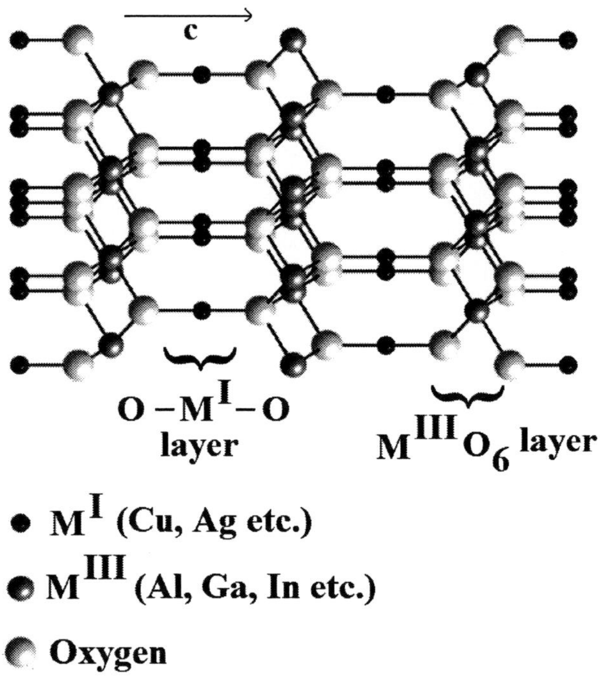

Figure 2(b). Delafossite Crystal Structure.

Table 1. Crystal data for $CuAlO_2$ [98, 104, 106, 113]

System - Rhombohedral

Space Group – $R\bar{3}m$ (D_{3d})

a = 2.858 Å = b, c = 16.958 Å

$\alpha = 28.1^O = \beta$

Cu-O = 1.86 Å, Al-O = 1.91 Å, Cu-Cu = 2.86 Å

Table 2. Delafossite p-TCO thin films with different doping concentrations and their respective opto-electrical parameters.

Material	Dopant	% dop-ing	Average film thickness (nm)	T (%)	$E_{g\text{-direct}}$ (eV)	σ_{RT} (S cm^{-1})	S_{RT} ($\mu V K^{-1}$)	Ref.
$CuAlO_2$	undoped	---	230	70	3.5	0.34	+ 214	67
$CuGaO_2$	undoped	---	500	80	3.6	0.063	+ 560	101
$CuGa_{1-x}Fe_xO_2$	Fe	0.5	150	60	3.4	1.0	+ 500	88
$CuIn_{1-x}Ca_xO_2$	Ca	0.07	170	70	~ 3.9	0.028	+ 480	103
$CuCrO_2$	undoped	---	250	40	~ 3.1	1.0	---	109
$CuCr_{1-x}Mg_xO_2$	Mg	0.5	270	50	3.1	220.0	+ 150	89, 109
$CuYO_2$	undoped	---	200	60	~ 3.5	0.025	---	89, 110

Table 2. Continued

Material	Dopant	% dop-ing	Average film thickness (nm)	T (%)	$E_{g\text{-direct}}$ (eV)	σ_{RT} (S cm^{-1})	S_{RT} ($\mu V\ K^{-1}$)	Ref.
$CuY_{1-x}Ca_xO_2$	Ca	0.01-0.02	240	50	3.5	1.05	+ 275	89, 110
$CuScO_2$*	undoped	---	110	40	~ 3.3	30.0	---	89, 108
$CuSc_{1-x}Mg_xO_2$[#]	Mg	0.05	220 - 250	80	3.3 -3.6	~ 0.07	---	88, 114
				60	-do-	~ 0.1	---	
				25	-do-	~ 0.8	---	
				15	-do-	~ 20.0	---	
$CuNi_{1-x}Sb_{x-y}Sn_yO_2$	Ni	0.66	~ 200	60	3.4	0.05	+ 250	88
	Sb	0.30						
	Sn	0.033						

* Maximum of 25 % oxygen was intercalated.

[#] The variation of transparency of the films at the expense of conductivity was due to a variation of oxygen pressure from 3 Torr (for most transparent film) to 15,000 Torr (for least transparent film). Also according to Ref. [115] the doping concentration of Mg was 1 %.

2.1.2. Nondelafossite Films

Cu_2SrO_2: Besides delafossite films, another Cu-based p-TCO thin film in the form of Cu_2SrO_2 has been synthesized by Kudo et al. [116]. The crystallographic data and band structure calculations were done by Teske et al. [117], Boudin et al. [118] and Robertson et al. [106]. Undoped and 3 % K doped sintered discs and films were prepared by Kudo et al [116]. The transparency remains almost same (~ 70 % to 75 %) for both types of films whereas the conductivity increased slightly from 3.9 x 10^{-3} S cm^{-1} to 4.8 x 10^{-2} S cm^{-1}.

Layered oxychalcogenide films: Layered-structure oxychalcogenide films of the form (LaO)CuC^h (C^h = Chalcogenides e.g. S, Se) [115, 119] showed high optical transparency and reasonable p-type conductivity to become promising material for “Transparent Electronics”. Although this material was first prepared almost two decades ago by Palazzi [120] and its p-type conductivity was reported more than a decade ago [121, 122], but Ueda, Hiramatsu and co-authors first prepared it in transparent thin film form to extend its application into p-TCO technology [115, 119]. Also this material shows room-temperature band edge emission under UV-excitation, extending its application in light emitting devices (LEDs) and similar fields [123-129]. Crystallographic parameters of these materials were extensively studied by Palazzi [120] as well as by others [121, 130-131]. Also band structure calculations were done by Inoue et al. [132]. Different physical properties of the material were also studied by various groups [133-135].

Non-oxide Cu-based transparent semiconductors: There are reports on the fabrication of non-oxide p-type transparent conductors like Cu_2BaS_2, CuBaSF [136-138] etc. Park and co-authors [136] synthesized α-Cu_2BaS_2 thin film, which crystallizes at low temperature in orthorhombic structure [139]. They obtained a visible transmittance of 70 % for a 430 nm thick film with a rather low bandgap of 2.3 eV. The room-temperature conductivity was reported as 17 S cm^{-1} with a Hall mobility of 3.5 cm^2 V^{-1} S^{-1}. Later the same group reported the preparation of undoped and K doped CuBaSF pellets and thin films [138]. The transmittance of the undoped film was ~ 85 % in the visible region with an estimated direct bandgap value of 3.2 eV. A decrease in the transmittance with increase in the K-dopant was observed for the doped films. Room-temperature conductivity of the polycrystalline thin film was obtained as 1.0 S cm^{-1}. Although these materials cannot be classified as p-TCO, but still they have scientific importance in the field of "Transparent Electronics".

2.1.3. Deposition Techniques

Growth technique of thin films plays the most significant role on the properties of the films. Different deposition routes yield films with diverse structural, optical and electrical properties. Even for the same deposition technique, slight variation in the deposition parameters produce films with different properties. So it is very important to have a comparative study on the properties of various films produced by different deposition routes. Detailed description of different deposition techniques along-with their schematic diagrams and related parameters are reported in various literatures [5, 8, 140-143].

$CuAlO_2$ thin films prepared by various techniques include pulsed laser deposition (PLD) [64, 67, 144, 145], R. F. magnetron sputtering [145], R. F. magnetron reactive co-sputtering of Cu and Al metal targets [146], D. C. reactive sputtering of facing targets of Cu and Al metals and a rotating substrate [147], pulsed magnetron sputtering [148], chemical vapor deposition (CVD) [149-151], e-beam evaporation [152], wet-chemical solution growth technique [153], hydrothermal cation exchange reaction followed by spin-on technique [154], rapid thermal annealing [155], spray technique [156], sol-gel technique [157] etc. Also ion exchange method is used to prepare $CuAlO_2$ powder from $LiAlO_2$ [158]. Although no film preparation was reported by this method but this process may become an important target preparation procedure for PLD or sputtering. Similarly, hydrothermal process [159] had been adopted to synthesize $CuAlO_2$ and Ga doped $CuAlO_2$ solid solutions. Previously, we have reported the syntheses of phase pure p-$CuAlO_2$ thin films by D. C. sputtering of sintered pellet of copper aluminum oxide [160] as well as reactive sputtering of a mixture of Cu and Al metal target pellets in oxygen diluted Ar atmosphere [161]. Wet-chemical deposition of highly oriented $CuAlO_2$ thin film has also been carried out by us [92, 162], which showed very good optical properties.

Also, based on ab initio electronic structure calculations, new methods have been proposed by Yoshida and co-authors [163-164] to fabricate high-conducting p-$CuAlO_2$. They proposed that in thermal non-equilibrium PLD or molecular beam epitaxy (MBE) crystal growth techniques, induction of high concentration of Cu vacancies, to form impurity band, by reducing the Cu vapor pressure during deposition might enhance the p-type conductivity in the material. On the other hand, doping of Mg or Be at Al-sites to form acceptor levels by decreasing the Al vapor pressure and increasing the Cu vapor pressure during low temperature PLD, MBE or MOCVD process might also increase the p-type conductivity of the material. Optical and Electrical properties of $CuAlO_2$ thin films synthesized by various growth techniques are furnished in Table 3.

Table 3. Optical and Electrical properties of $CuAlO_2$ thin films synthesized by various growth techniques.

Growth Technique	Thickness (nm)	Avg. Visible Transmittance (%)	Room-Temp. Conductivity ($S\ cm^{-1}$)	E_a (meV)	Carrier density (cm^{-3})	Ref.	Remarks
PLD	500	70	0.095	~ 200	1.3×10^{17}	64	---
PLD	230	80	0.34	220	2.7×10^{19}	67	Films were post-annealed in O_2 atmosphere (1.3 Pa)
R. F. Sputtering	180	85	---	---	---	145	Preliminary Hall and TEP measurements confirmed p-type conductivity
R. F. Magnetron Reactive Co-Sputtering	250	20 – 80	---	---	---	146	Small amount of CuO was present in the film
Pulsed magnetron sputtering	20	85-95	0.20	---	---	148	Film deposited directly from a blended Cu_2O/Al_2O_3 powder target.
Reactive D. C. Sputtering	400 – 800	50 – 60	0.01 – 0.1	---	---	147	With facing metal targets and rotating substrate. Films were annealed at 1050 ^{O}C in N_2 atmosphere.
MOCVD	250	40	2.0	120	2.6×10^{19}	150	The films were a mixture of $CuAlO_2$, Cu_2O and $CuAl_2O_4$.
PE-MOCVD	120	40	17.08	32	1.17×10^{20}	149	For those samples annealed in air for 5 min (at 350 ^{O}C)
E-beam evaporation	---	50 - 85	1.0	---	10^{18}-10^{20}	152	Hole concentration decreases with increasing water vapor pressure
Dip-coating	1000	---	5×10^{-3}	---	---	153	Results given for films deposited via Nitrate route.
RTA	360	60	0.57	---	---	155	RTA was performed over 1000 ^{o}C.
Hydrothermal cation exchange	420	60	2.4	140	5.4×10^{18}	154	Film is nanocrystalline in nature with grain size around 14-16 nm
Spray pyrolyses	1000	30 - 70	---	---	---	156	Transmittance increases as Cu:Al ratio approaches to 1.0
Sol-Gel synthesis	1100	---	0.004	---	---	157	High resistivity is due to porous structure of the film.

2.2. Transparent Junctions

Junctional devices fabricated by both n and p types of TCO thin films are the key structure for "Invisible Electronics" [65]. The simplest of them is the p-n junction diodes with rectifying properties. The importance of these types of devices lies in the fact that 'functional windows' can be fabricated by these devices, which would transmit the visible solar radiation but absorb the UV part [64]. Therefore, simultaneously these devices can act as 'UV-shields' as well as 'electricity generators' by the UV absorption. A schematic diagram of all-TCO diode is shown in Fig. 2(c). Fabrication of a number of all-TCO junctional devices have been reported, which include both p-n and p-i-n homo-junctions and hetero-junctions as well as transparent field effect transistors (TFET) .

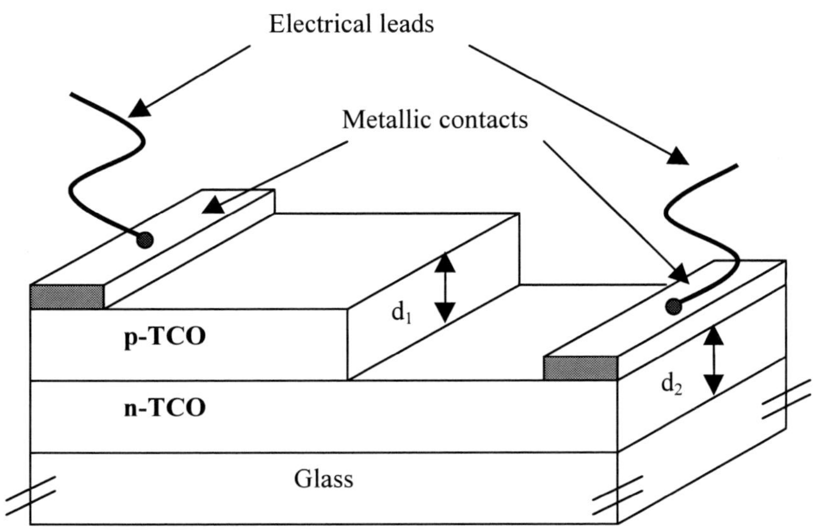

Figure 2(c). Schematic diagram of an all-TCO p-n junction diode on glass substrate.

Heterojunction: The first all-TCO diodes were reported by Sato and co-authors [66]. They fabricated a semi-transparent thin film p-i-n structure consisting of p-NiO / i-NiO / i-ZnO / n-ZnO: Al. The rectifying properties of the structure confirmed the formation of the junction. Similarly, fabrication of all-TCO p-n hetero-junction thin film diode of the form p-$SrCu_2O_2$ / n-ZnO was reported by Kudo and co-authors [165]. The same group also reported UV emission from a p-n hetero-junction diode composed of p-$SrCu_2O_2$ / n-ZnO after current injection through it [166-169]. P-i-n hetero-junction in the form of p-$SrCu_2O_2$: K / i-ZnO / n-ZnO was also constructed by this group [168]. Similarly p-i-n hetero-junction of the form p-$CuYO_2$: Ca / i-ZnO / n-ITO was fabricated by Hoffman et al [170]. Jayaraj and co-authors [110, 171] fabricated p-n hetero-junction using p-$CuY_{1-x}Ca_xO_2$ ($x = 0.01$-0.02) / n-$Zn_{1-x}Al_xO$ ($x = 0.02$) structure. Tonooka and co-authors [172] reported the fabrication of n-ZnO/p-$CuAlO_2$ diode structure with rectifying characteristics and observed a photovoltaic effect (as large as 80 mV) under illumination of blue radiation. Although the performance of the diode was restricted by the low crystallinity of the $CuAlO_2$ layer but the forward-to-reverse current ratio showed a moderate value of 90 between –1.5 to +1.5 volt. Also the transparency of the structure was 40 % to 70 % in the visible region.

Homojunction: Besides hetero-junctions, fabrications of p-n homo-junctions were also reported by few authors. Importance of the homo-junctions lies in the fact that lattice matching is supposed to be automatic during the formation of diodes. First all-delafossite p-n homo-junction diode was fabricated by Yanagi and co-authors [103] of the form YSZ (111) / ITO / p-$CuInO_2$: Ca / n-$CuInO_2$: Sn / ITO. Similarly all-ZnO p-n homo-junctions were reported by Hwang and co-authors [173], Tüzemen and co-authors [174] and Aoki and co-authors [175]. Hwang et al [173] fabricated n-ZnO: Al / p-ZnO: As and observed rectifying characteristics with a turn-on voltage around 2.5 volt. Tüzemen et al [174] reported intrinsic p and n-type ZnO homo-junction, prepared by reactive R. F. magnetron sputtering. The n-type and p-type conductivities were obtained by varying O_2 partial pressure in the Ar + O_2 sputtering atmosphere. It is worthwhile to be noted that the first report on p-type conductivity in intrinsic ZnO was published by Butkhuzi and co-authors [176], where post-annealing of the as-grown material in atomic oxygen atmosphere was performed to achieve intrinsic p-type conductivity. Aoki et al [175] fabricated p-ZnO: P / n-ZnO homo-junction and observed rectifying I-V characteristics. If the carrier concentration can be increased by optimizing the deposition parameters then these all-ZnO diode structures may open up a new horizon in the field of "Transparent Electronics". Parameters of deferent all-transparent diodes are compared in Table 4(a).

TFET: Another important area in the field of "Transparent Electronics" is the fabrication of transparent field-effect transistors (TFET) [177]. A schematic diagram of top-Gated TFET is shown in Fig. 2(d). Prins and co-authors [178, 179] reported the fabrication of ferroelectric TFETs, based on transparent SnO_2: Sb thin films. They have observed the field-effect mobility around 10 $cm^2V^{-1}s^{-1}$, with an on/off current ratio ~ 10^4. Later various groups [180-181] reported the fabrication of ZnO based TFETs with reasonable device properties. Hoffman et al [180] reported 75 % visible transparency in their ZnO-TFETs with mobility and on/off ratio around 2.5 $cm^2V^{-1}s^{-1}$ and 10^7 respectively. Masuda et al [181] observed these values around 1.0 $cm^2V^{-1}s^{-1}$ and 10^5 respectively, with optical transmittance more than 80 % in the visible region. Similarly Carcia et al [182] obtained these values around 2.0 $cm^2V^{-1}s^{-1}$, 10^6 and >80 % respectively for their ZnO-TFETs. Recently, Nomura, Ohta and co-authors [183, 184] reported the successful fabrication of high mobility top-gate TFETs based on single crystalline transparent $InGaO_3(ZnO)_5$ thin film. The device shows the mobility as high as 80 $cm^2 V^{-1}s^{-1}$ with on/off current ratio ~ 10^6 and more than 80 % transparency in the visible and near infra-red region. The deposition techniques for the fabrication of these TFETs include pulsed laser deposition (PLD) [178, 179, 181], ion beam sputtering [180], r.f. magnetron sputtering [181], reactive solid-phase epitaxy [183] etc. The deposition routes and various parameters of different TFETs are furnished in Table 4(b). These reports provide a significant step towards the realization of "Invisible Electronics".

Table 4(a). Parameters of deferent all-transparent diodes

Diode structure		n-ZnO/ p-$SrCu_2O_2$	n-ZnO/ p-$SrCu_2O_2$:K	n-ZnO:Al/ p-$CuYO_2$:Ca	n-ZnO/ p-$CuAlO_2$	n-$CuInO_2$:Sn / p-$CuInO_2$:Ca	n-ZnO: Al/ p-ZnO: As	n-ZnO/ p-ZnO
Thickness (nm)	p-layer	300	200	300	400	400	1500-2000	5000
	n-layer	300 – 1000	200	250	400	400	600	5000
Carrier concentra-- tion (cm^{-3})	p-layer	10^{17}	~10^{18}	---	---	---	---	~ 5 x 10^{15}
	n-layer	5 x 10^{18}	~10^{18}	---	---	---	---	~ 6 x 10^{15}
Substrate		Glass	YSZ (111)	Glass	Glass	YSZ (111)	GaAs (001)	Si (100)
Deposition technique	p-layer	Reactive co-evaporation in O_2 atmosphere	PLD	Reactive co-evaporation in O_2 atmosphere	PLD	PLD	R. F. Magnetron sputtering	R. F. Magnetron reactive sputtering
	n-layer	Magnetron sputtering	PLD	R. F. Magnetron sputtering	PLD	PLD	R. F. Magnetron sputtering	R. F. Magnetron reactive sputtering
Electrodes	p-side	ITO	Ni	In	ITO	ITO	In	Au/Al
	n-side	n^+-ZnO	ITO	ITO	n^+ - ZnO	ITO	In	Au/Al
Turn-on voltage (V)		~ 0.5	~ 1.0	0.4 – 0.8	0.4 – 1.0	1.8	~ 2.5	~ 1.0
Reference		165	166	110	172	103	173	174

Table 4(b). Parameters of various TFETs

Active Channel	Gate Insulator	Gate Electrode	Substrate used	Mobility ($cm^2V^{-1}s^{-1}$)	On/ off ratio	Visible Transparency (%)	Ref
SnO_2: Sb (Thickness ~ 110 nm) (Deposition technique: PLD)	$PbZr_{0.2}Ti_{0.8}O_3$ (Thickness ~ 160 nm) (Deposition technique: PLD)	$SrRuO_3$ (Thickness ~ 140 nm) (Deposition technique: PLD)	$SrTiO_3$ (100)	10.0	~ 10^4	Transparent, as seen in the figure provided, but no numerical data given.	178* 179
ZnO (Thickness ~ 100 nm) (Deposition technique: PLD)	$Al_2O_3 + TiO_2$ # (Thickness ~ 220 nm) (Deposition technique: ALD)	ITO (Thickness ~ 200 nm) (Deposition technique: Sputtering)	Glass	2.5	~ 10^7	75	180
ZnO (Thickness ~ 140 nm) (Deposition technique: PLD)	$SiO_2 + SiN_x$ † (Thickness ~ 250 & 50 nm respectively) (Deposition technique: PECVD)	ITO (Thickness ~ 100 nm) (Deposition technique: e-beam evaporation)	Glass	1.0	10^5	80	181
$InGaO_3(ZnO)_5$ ‡ (Thickness ~ 120 nm) (Deposition technique: PLD)	a-HfO_2 (Thickness ~ 80 nm) (Deposition technique: PLD)	ITO (Thickness ~ 30 nm) (Deposition technique: PLD)	YSZ (111)	80.0	~ 10^6	80	183‡‡

* Due to the presence of ferroelectric insulator $PbZr_{0.2}Ti_{0.8}O_3$, the device showed intrinsic memory function.

$Al_2O_3 + TiO_2$ is an alternative layers of Al_2O_3 & TiO_2.

† This TFET has a double layer Gate insulator.

‡ Single-crystalline $InGaO_3(ZnO)_5$ is used as active channel layer.

‡‡ The device has a top gate structure.

Table 5(a). Electro-optical properties of nanostructured $CuAlO_2$ thin films synthesized by various processes.

Process	Avg. particle size (nm)	Band -gap (eV)	Room-temp. conductivity ($S\ cm^{-1}$)	Carrier concentra-tion (cm^{-3})	Ref	Remarks
MO-CVD	10	3.75	2.0	1.8×10^{19}	150	The film contains nanocrystalline phases of $CuAlO_2$ and Cu_2O.
Spin-on technique	10	3.75	2.4	5.4×10^{18}	154	Initially $CuAlO_2$ nanocrystalline powder was prepared by hydrothermal cation exchange reaction between $NaAlO_2$ and CuCl. Then the powder was dispersed in alcohol and deposited as thin film.
Sputtering	~10	3.94	---	---	185	Deposition time was varied to decrease the particle size. With decrease in the particle size, an increase in the bandgap is observed due to quantum confinement effect. Also room-temp. photoluminescence properties were observed first time in these nanocrystalline $CuAlO_2$ thin films.

Table 5(b). Progress in the development of F-N theory and F-N equation.

Theory	Author(s)	Year	Assumptions and remarks	Ref.
'Original' F-N theory	R. H. Fowler and L. Nordheim	1928	The authors treated the effect as wave-mechanical tunneling through a triangular potential barrier. They have carried out an exact solution of the Schrödinger equation for a simple triangular barrier.	209
'Standard' F-N theory	E. L. Murphy and R. H. Good Jr.	1956	These authors used more realistic barrier and introduced 'exchange-and-correlation interaction' between the emitted electron and the surface, into the original F-N theory.	210
'Modified' standard F-N theory	H. A. Schwettman, J. P. Turneaure, and R.F.Waites	1974	These authors introduced the local field enhancement factor 'β'. β was initially postulated to arise from the geometrical irregularities on the emitting surface.	211
Latham's model for 'field enhancement'	R. V. Latham	1983	This model introduced field enhancement due to semiconducting or insulating materials on the emitting metal surface.	212
'Generalized' F-N equation	R. G. Forbes	1999	This equation combined various models for F-N theory and depended on the particular assumption(s) and approximation(s) made. The proper choice of the 'generalized correction factors' in the generalized F-N equation would lead to the required F-N equation.	214
'ENH' theory	R. G. Forbes	2001	The emitting films are assumed to be 'electrically nanostructured heterogeneous' (ENH) materials, where internal nanostructure creates geometrical field enhancement inside and at the film-vacuum interface. Thus the macroscopic field is enhanced by a factor,β, to produce the required barrier-field for electron tunneling.	213

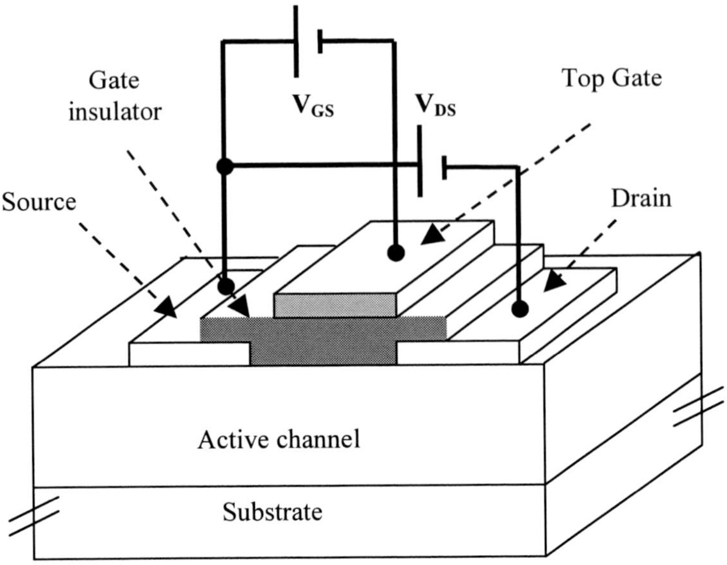

Figure 2(d). Schematic diagram of a top-gated TFET structure. After [183].

2.3. Nanostructured p-$CuAlO_2$ Thin Films

As far as syntheses of nanocrystalline $CuAlO_2$ thin film is concerned, Gong and co-authors [150] first reported the preparation of phase impure copper aluminum oxide films by chemical vapor deposition (CVD) method, which contain nanocrystalline phases of $CuAlO_2$ and Cu_2O. They have used metalorganic precursors $Cu(acac)_2$ and $Al(acac)_3$ (acac = acetylacetonate) as the source material. The crystallite size was found to be below 10 nm with an optical bandgap of 3.75 eV. The carrier concentration was $\sim 10^{19}$ cm^{-3} [150]. Also later, Gao and co-authors [154] reported the synthesis of phase pure nanocrystalline $CuAlO_2$ thin film by spin-on technique. Initially $CuAlO_2$ nanocrystalline powder was prepared by hydrothermal cation exchange reaction between $NaAlO_2$ and CuCl. Then the powder was dispersed in alcohol and deposited as thin film on glass substrates by spin-on technique [154]. The average grain size obtained by this group was around 10 nm with an optical bandgap around 3.75 eV. The room temperature conductivity was found to be 2.4 S cm^{-1} with a hole concentration around 10^{18} cm^{-3}. We have also reported the synthesis of $CuAlO_2$ nanoparticles by D. C. sputtering technique from a sintered disk of copper aluminum oxide [185]. The particle size was found to be as low as 10 nm. We have observed an increase in the particle size with an increase in the deposition time. Also an increase in the bandgap from 3.60 to 3.94 was observed with the decrease in the particle size. And this bandgap enhancement is attributed to the quantum confinement effect as often found in semiconductor nanocrystals. Various opto-electronic properties of nanocrystalline $CuAlO_2$ thin film are furnished in Table 5(a). We have also observed for the first time some photoluminescence properties of nanocrystalline $CuAlO_2$ thin films and tried to explain it with existing theories [185]. Photoluminescence properties of p-type transparent semiconducting layered oxysulphide thin films of LaO(CuS) have been reported previously by Ueda and co-authors [115]. Also, as far as luminescence properties of copper based delafosite oxide materials are concerned, Jacob and co-authors [186] reported the luminescence properties of $CuLaO_2$ and $CuYO_2$ pellets.

2.4. Wide Bandgap Field-Emitters

It was found that the materials with wide bandgap (such as diamond) have low or negative electron affinity, which, in turn, enhances the low-macroscopic field emission properties of diamond films [81]. Also p-type semiconducting diamond film showed low-threshold field emission properties [82]. As far as the field emission from other wide bandgap thin films is concerned, GaN and AlN showed very good field emission properties. Polycrystalline GaN films [187] as well as nanostructured GaN, in the form of nanorods [188], nanobelts [189], nanoneedles [190], nanowires [191] etc. showed very low threshold field emission properties. Similarly, oriented AlN [192, 193] and nanostructured AlN [194-197] also showed very good field emission properties.

As far as field emission properties of various TCOs are concerned, Olesik et al. [198] reported the field emission properties of Sb and Sn doped Indium tin oxide (ITO) thin films under an average external field of 5 $V\mu m^{-1}$. Similarly, Baranauskas and co-authors [199] reported the field emission properties of SnO_2 thin films and observed that the deposition temperature had a dramatic influence on the electron emission properties of the material. Also nanostructured ZnO films have recently generated great interest in the field of FED technology due to their superior field emission properties over carbon based field emitters. ZnO nanowires [200-202], nanoneedles [203] etc. have been reported to show excellent low-threshold cold field emission properties. Previously, our group also reported very good field emission properties of ZnO nanowires [204-205] synthesized by catalyst free solution route. The enhanced field emission properties of this material were attributed to the geometrical structure of the nanowires, due to which considerable field enhancement was manifested at the emitter tip to show low-threshold field emission. Also recently, Yang et al. [206] observed that a nanocomposite film composed of ZnO nanowire and amorphous diamond layer showed remarkable enhancement in the field-emission properties over the intrinsic diamond and ZnO films. This is very important in the sense that both diamond, being a robust material, and ZnO, a highly chemically stable and structurally rigid material, a nanocomposite of these two may produce much stable field emitters for diverse applications. As far as the field emission properties of wide bandgap p-$CuAlO_2$ thin film is concerned we have first reported its field emission properties and tried to explain the field emission mechanism therefrom [207, 208].

The emission of electrons from a metal-vacuum interface, in the presence of an external electric field normal to the emitting surface, was initially treated as a quantum mechanical tunneling process by Fowler and Nordheim ('original' F-N theory)[209]. Later, Murphy and Good [210] proposed a more rigorous theory, called 'standard F-N theory', where the 'exchange-and-correlation interaction' between the emitted electron and the surface was included into the original F-N theory [209]. Schwettman et al. [211] further modified the 'standard F-N theory' by introducing the local field enhancement factor 'β'. β was initially postulated to arise from the geometrical irregularities on the emitting surface, but later Latham [212] proposed a model, which introduced field enhancement due to semiconducting or insulating materials on the metal surface. Extending the Latham's model, Forbes [213] tried to explain the low-macroscopic field emission of various films by assuming that the thin films are 'electrically nanostructured heterogeneous' (ENH) materials, where internal nanostructure creates geometrical field enhancement inside the film as well as at the film-vacuum interface. Thus the macroscopic field is enhanced by a factor, which can be related to the above mentioned field-enhancement factor β, to produce the required barrier-field for electron tunneling. In fact, as the original F-N theory (so also F-N equation) [209] is based on certain assumptions, any deviation from one (or more) of these assumptions, leads to some 'specialized versions' of elementary F-N equation. So, Forbes proposed a 'generalized F-N equation' [214], whose form in a given case depends on the particular assumption(s) and approximation(s) made. He showed that the proper choice of the 'generalised correction factors' in the generalized F-N equation, leads to the standard F-N equation, proposed by Murphy and Good [210], which was the first fully satisfactory treatment of standard physical assumptions. The development in the F-N theory is tabulated in Table 5(b). In fact there are many different mechanisms involved as the electrons, in the presence of an external electric field, travel through the bulk of the film to the surface via different interfacial contacts, followed by the emission to vacuum, propagated through the electrode gap and finally reaching the anode. The exact nature of these mechanisms is yet to be explored completely.

3. Origin of P-type Conductivity in P-TCO

Most of the existing TCOs are n-type, whereas it is very difficult to prepare binary metal oxides with p-type conductivity. A possible reason for this has been described by Kawazoe et al. [64, 144], where they argued that this is probably because of the electronic structure of these metal oxides. Strong localization of holes (it can be successfully introduced by intentional substitutional doping or by producing non-stoichiometry within the material) at oxygen 2p levels or an upper edge of the valence band due to high electronegative nature of oxygen, i.e. this localization is due to the ionicity of metallic oxides. O 2p levels are far lower lying than the valence orbit of metallic atoms [215], leading to the formation of deep acceptor level by the holes. In other words, the holes, therefore, have high probability to be localized around the oxygen atoms. Hence these holes require high enough energy to overcome large barrier height in order to migrate within the crystal lattice, resulting in poor conductivity and hole mobility.

A possible solution proposed by Kawazoe and co-authors [144] is to introduce a "degree of covalency" in the metal-oxygen bondings to induce the formation of an extended valence band structure, i.e. the valence band edge should be modified by mixing orbitals of appropriate counter cations that have energy-filled-levels comparable to O 2p level. This would reduce the strong coulombic force by oxygen ions and thereby delocalizing the holes. This is the essential approach to obtain p-TCO, which is called "Chemical Modulation of the Valence Band (CMVB)" [144].

But the next requirement is the choice of appropriate cationic species that will serve for CMVB technique. Investigations showed that the required cationic species are $3d^{10}$-closed shell of Cu^+ ions and $4d^{10}$-closed shell of Ag^+ ions [144, 215]. Although some transition metal cations with open d-shell may fulfill the energy requirement [216] for CMVB technique, but they usually show strong coloration due to d-d transition, which is not expected for transparent materials. Hence focus had been concentrated on the cations mentioned above, with closed ($d^{10}s^0$) electronic configuration. Fig. 3 shows a schematic illustration of CMVB technique. Both of the atomic orbitals are occupied by electron pairs, and the resulting antibonding level becomes the highest occupied level, i.e. the valence band edge.

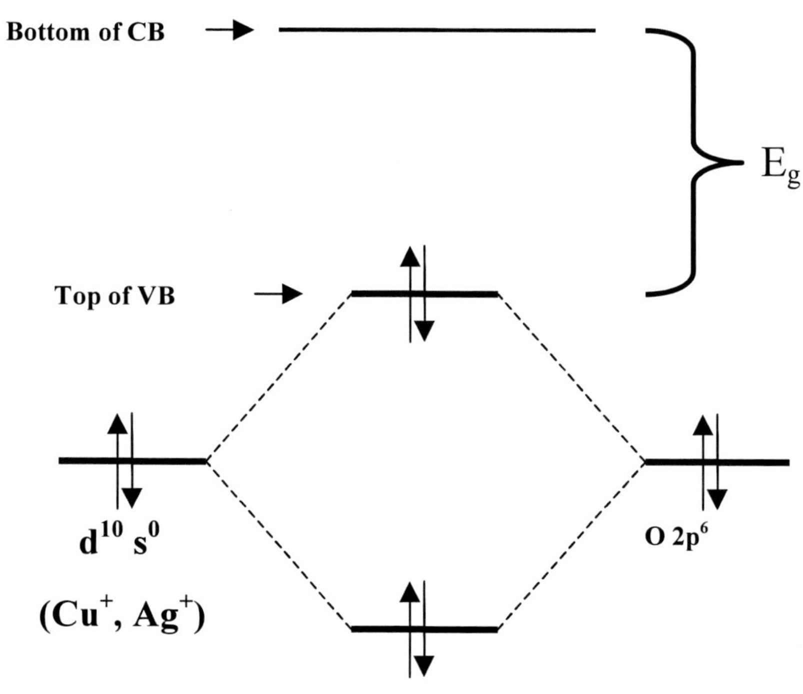

Figure 3. Schematic diagram of CMVB method. Energy levels are to the scale. After Ref. [144].

Next is the structural requirement for designing p-TCO materials. Tetrahedral coordination of oxide ions is advantageous for p-type conductivity, as it acts in reducing the localization behavior of 2p electrons on oxide ions [144]. The valence state of the oxide ions can be expressed as sp^3 in this conformation. Eight electrons (including $2s^2$) on an oxide ion are distributed in the four σ bonds with the coordination cations. This electronic configuration reduces the non-bonding nature of the oxide ions and increases the delocalization of holes at the valence band edge (that is why Cu_2O is a p-type conducting oxide [217-220]). But Cu_2O, although p-type in nature, has rather small bandgap (2.17 eV) [218]. This is probably because of the three-dimensional interactions between $3d^{10}$ electrons of neighboring Cu^+ ions. It is expected that the low-dimensional crystal structure would suppress this interaction [67]. As we are interested in transparent conducting oxides, bandgap of the material (E_g) should be greater than 3.1 eV. Hence enlargement of bandgap would be another structural requirement for designing p-TCO, so that there is no absorption of visible photons. Materials with delafossite crystal structure $M^IM^{III}O_2$ (M^I = Monovalent ions, Cu^+, Ag^+; M^{III} = Trivalent ions, Al^{+3}, Ga^{+3}, In^{+3}, Cr^{+3}, Fe^{+3}, Co^{+3}, Sc^{+3}, Y^{+3} etc.) [93-95] were chosen as the candidates for p-TCOs for several reasons. Firstly, if we investigate the delafossite structure as shown in Fig. 2(a), we see an alternative stacking of M^I and layers of nominal $M^{III}O_2$ composition consisting M^{III}-O_6 octahedra sharing edges. Each M^I atom is linearly coordinated with two oxygen atoms to form a O-M^I-O dumbbell unit placed parallel to the c-axis. O-atoms of O-M^I-O dumbbell link all M^I layers with the $M^{III}O_2$ layers. On the other hand, each oxide ion in the $M^{III}O_2$ layer forms a "pseudo-tetrahedral coordination ($M^{III}{}_3M^IO$)" [144] with the neighboring M^{III} and M^I ions. Hence, as previously mentioned, this electronic configuration reduces the non-bonding nature of the oxide ions and, therefore, delocalizes the holes at the valence band edge. Secondly, this layered structure (O-M^I-O dumbbell layer and $M^{III}O_2$ layer) effectively reduces the dimension of cross-linking of M^I ions and, thus enlarging the bandgap [64]. And finally, another important factor in this structure, is the low coordination number of the M^I ions, due to the large separation from oxygen legands, which is the result of the strong coulombic repulsion between 2p electrons in oxygen legands and M^I d^{10} electrons. This leads to the M^I d^{10} energy levels almost comparable to the O 2p level, resulting in a high degree of mixing of these levels, which is essential for CMVB technique [144].

As the importance of p-TCO lies in the active device fabrication, it is very important to have lattice matching between both p and n-types of TCOs to form p-n homojunctions. Both types of TCOs with delafossite structure may serve this requirement. In this regard, it is also worthwhile to mention that the $M^{III}O_2$ layers of this structure is also important for designing n-TCOs, specially for the cations like Ga^{+3}, In^{+3} in the M^{III} sites with s^0 configuration [144]. Following the above argument, delafossite $AgInO_2$ thin film with n-type semiconductivity had already been established [221].

Non-stoichiometry and doping in p-TCO: The cause of p-type conductivity shown by p-type transparent conducting oxide materials is due to excess oxygen (or metal deficit) within the crystallite sites of the material, i.e. the defect chemistry plays an important role. This deviation from the stoichiometric composition of the components can be induced by regulating the preparation condition of the materials. The defect reaction may be represented by the following equation [222, 223]:

$$O_2(g) \;=\; 2O_O^x + V_{M^I}^- + V_{M^{III}}^{-3} + 4h^+ \qquad (1)$$

where 'O_O' denotes the lattice oxygen, 'V' denotes the vacancies of monovalent cation M^I and trivalent cation M^{III} respectively and 'h' denotes the hole. Superscripts ^{X}, $^-$, and $^+$ denote effective neutral, negative, and positive charge states respectively.

Also, intercalation of excess O^{-2} ions in the interstitial sites may trap electrons, leaving behind empty states in the valence band, which act as holes. The formula for oxygen-excess delafossite films may be written as $M^IM^{III}O_{2+x}$ (M^I = Cu^+, Ag^+ and M^{III} = Al^{+3}, Ga^{+3}, In^{+3}, Y^{+3}, Sc^{+3} cations etc.). The value of x i.e. the percentage of excess oxygen may be as low as 0.001 % in $CuAlO_{2+x}$ thin film [65] to more than 25 % in $CuYO_{2+x}$ polycrystalline powder and $CuScO_{2+x}$ thin films [89, 224-226]. Fig. 4(a), 4(b) and 4(c) show schematic representation of stoichiometric ABO_2 crystal and non-stoichiometric ABO_2 crystal with "excess" oxygen in lattice sites and interstitial sites.

Oxygen intercalation in delafossite p-TCOs only showed a maximum reported conductivity around 3 x 10^1 S cm^{-1} [108]. But this is still quite less than that of commercially available n-TCOs like indium tin oxide (ITO), which is having room temperature conductivity more than 1 x 10^3 S cm^{-1}. So next attention was focused on the substitutional doping of these materials by appropriate dopants to increase the conductivity. Doping of $CuAlO_2$ was first attempted, as it was the first reported material amongst p-TCOs. Several groups theoretically calculated the effects on the electronic behavior of the material due to the presence of various cations in Cu and (or) Al sites. Lalić and co-authors [227, 228] showed that Cd and Zn substitutions on Cu site would produce n-type conductivity in the material, whereas Ni doping in Cu sites would enhance the p-type conductivity of the material. But Cd doping on Al sites would have no effect on the electrical properties of the material. Preparation of a solid solution of gallium doped copper aluminum oxide in the form of $CuAl_{1-x}Ga_xO_2$ ($0 \leq x \leq 0.5$) was reported by Shahriari et al [159]. But no film preparation of this material was reported by them. Also any other experimental data on the doping of $CuAlO_2$ thin film has yet been reported. Heavy doping (~ 50 %) of $CuGaO_2$ by Fe^{+3} in Ga sites has

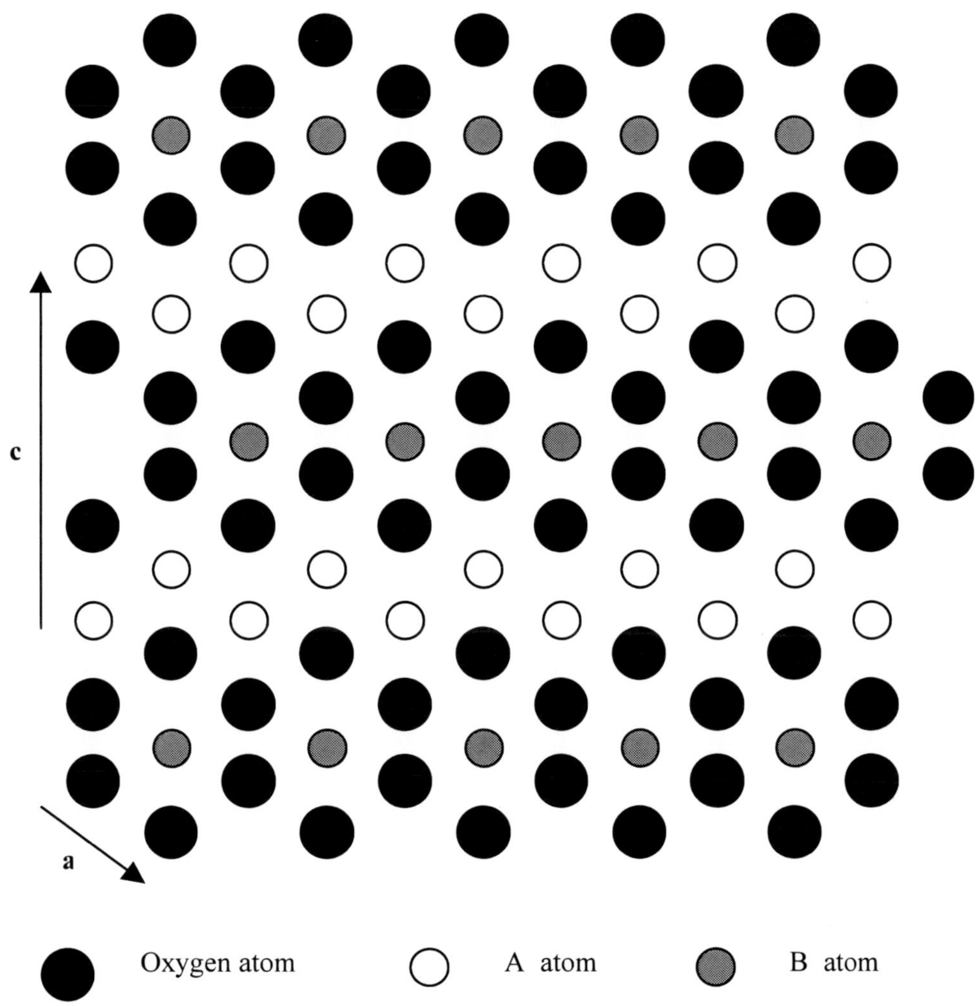

Figure 4(a). Stoichiometric ABO_2 lattice. The diagram is not according to the relative lattice parameters.

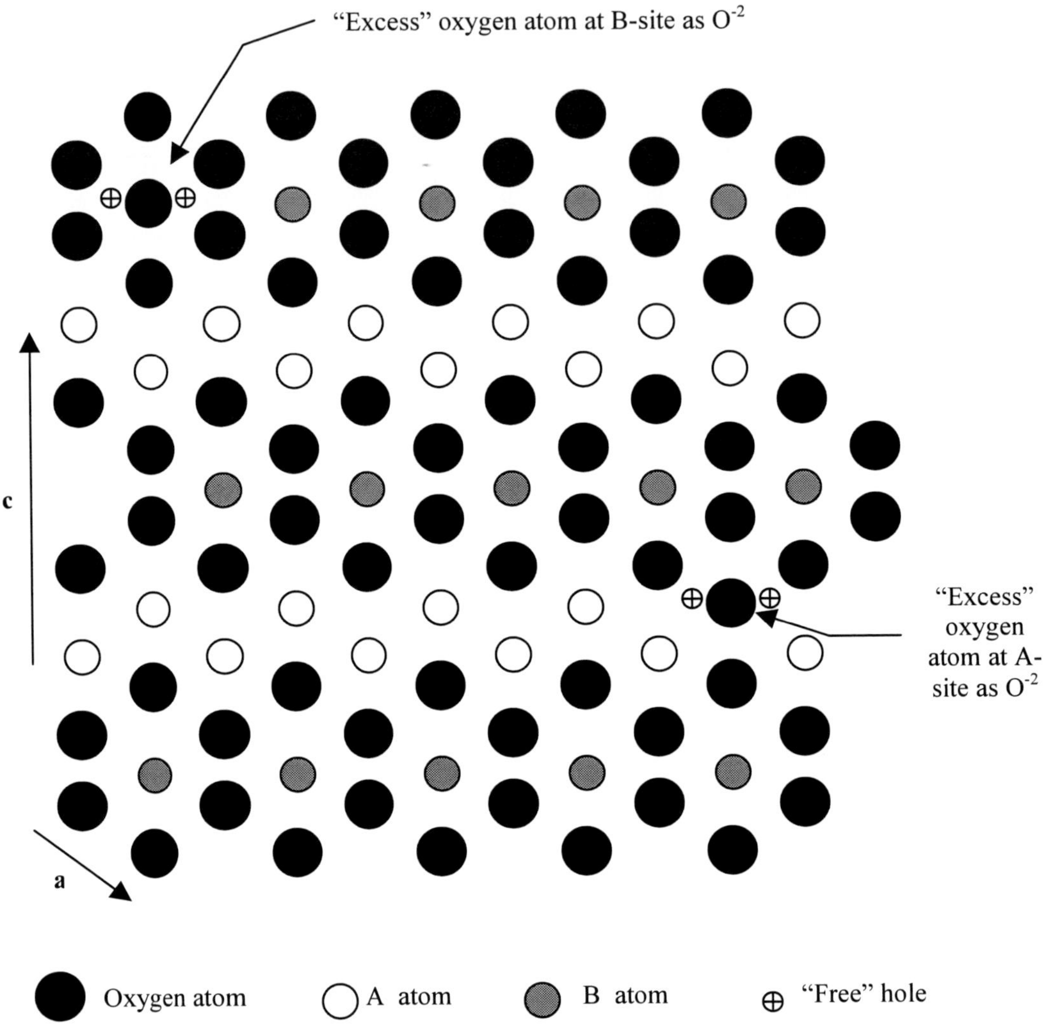

Figure 4(b). Non-stoichiometric ABO_2 structure with "excess" oxygen in lattice sites.

been reported by Tate et al. [88]. Their strategy was to combine high transparency of $CuGaO_2$ thin film (~ 80 % in visible region [101]) with better conductivity (over other Cu and Ag based delafossites [95]) of $CuFeO_2$ pellets (2.0 S cm^{-1} [95, 229]). Both the polycrystalline powder and thin film of $CuGa_{1-x}Fe_xO_2$ ($0 \leq x \leq 1$) have shown p-type conductivity. It was observed that high Fe doping had increased the conductivity of the film from 2 x 10^{-2} S cm^{-1} (for undoped $CuGaO_2$ thin film) to almost 1.0 S cm^{-1} for $CuGa_{1-x}Fe_xO_2$ ($x = 0.5$) thin film, whereas transparency of the films became ~ 60 % in the visible region [88].

Doping of $CuInO_2$, $CuYO_2$, $CuScO_2$, $CuCrO_2$ by divalent cations e.g. Ca^{+2}, Mg^{+2} etc. were reported by various groups [88, 102-103, 108-110]. When a trivalent cation was

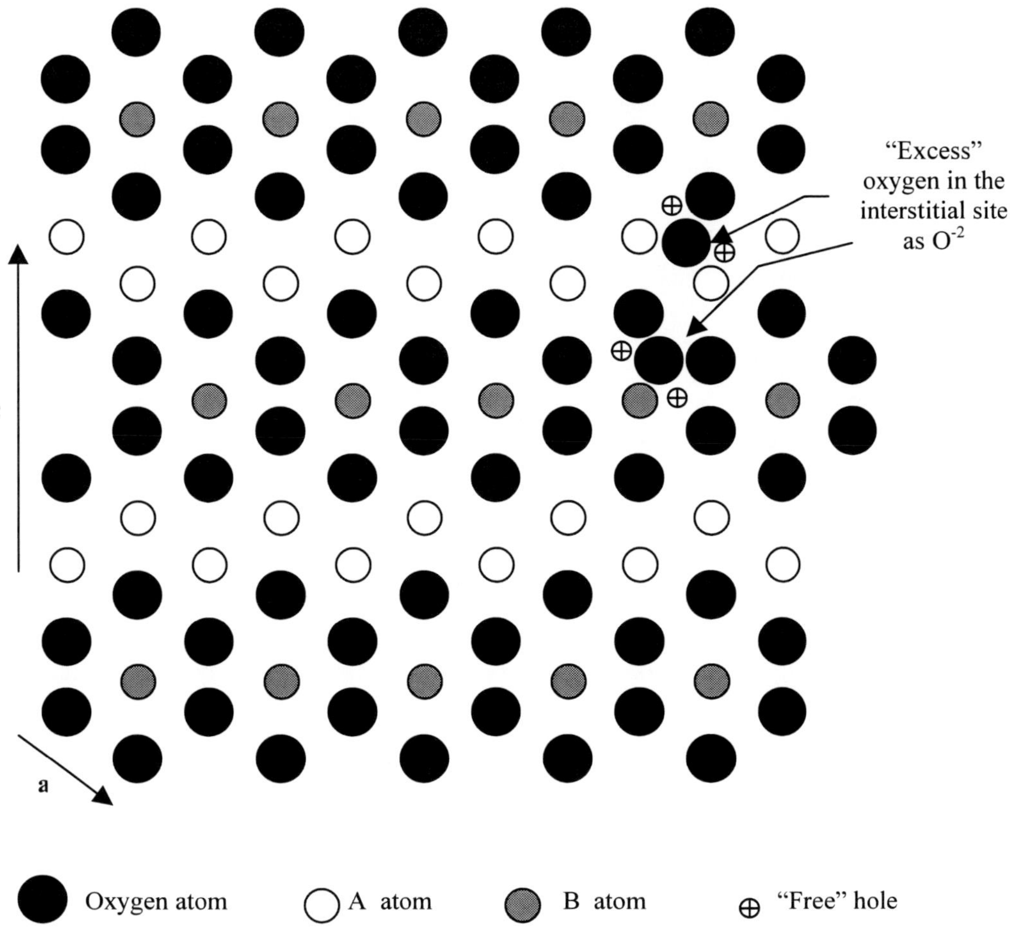

Figure 4(c). Non-stoichiometric ABO_2 lattice with "excess" oxygen at interstitial site.

replaced by a divalent one, one empty state in the valence band was created, which acts as a hole, thus increasing hole conductivity. The method may be described by the following equation:

$$(M^{III^{+3}} + 3e^-)\uparrow \;=\; (M^{II^{+2}} + 2e^-)\downarrow + V^- + h^+ \tag{2}$$

where $M^{III^{+3}}$ and $M^{II^{+2}}$ are trivalent and divalent cations, V^- is a negatively charged vacant state, e^- and h^+ is an electron and "free" hole respectively. The symbols $\uparrow$ and $\downarrow$ denote the replacement of trivalent cation by divalent one in the lattice sites. Such doped delafossite films like $CuCr_{1-x}Mg_xO_2$ (x = 0.05), $CuY_{1-x}Ca_xO_2$ (x = 0.01 – 0.02), $CuSc_{1-x}Mg_xO_2$ (x=0.05) showed better hole conductivity over the corresponding undoped films.

Some Ag based delafossite materials like $AgM^{III}O_2$ ($M^{III} = Sc^{+3}, Cr^{+3}, Ga^{+3}$ etc.) with 5 % Mg doping at M^{III} sites was reported by Nagarajan et al. [89]. The conductivities of these sintered powders were very low (~ 10^{-5} –10^{-4} S cm^{-1}) and also no film preparation of these materials were reported anywhere so far.

There are also reports in the literature about the double substitution of trivalent M^{III} sites by divalent and pentavalent cations e.g. $CuFe_{1-x}V_xO_2$ (x = 0.5), $CuNi_{1-x}Sb_xO_2$, $CuZn_{1-x}Sb_xO_2$, $CuCo_{1-x}Sb_xO_2$, $CuMg_{1-x}Sb_xO_2$, $CuMn_{1-x}Sb_xO_2$ (x = 0.33), $AgNi_{1-x}Sb_xO_2$, $AgZn_{1-x}Sb_xO_2$ (x = 0.33) etc., but all in the form of sintered powder [90, 224]. Also triple substitution of trivalent cation had been reported by Tate and co-authors [88, 90] in the form of $CuNi_{1-x}Sb_xSn_yO_2$ (x = 0.3, y = 0.033). Thin film of this material showed an average of 60 % transmittance with a room temperature conductivity of 5 x 10^{-2} S cm^{-1}.

4. Syntheses of p-$CuAlO_{2+x}$ Thin Films

Copper aluminium oxide ($CuAlO_2$) thin films were prepared by three routes:

(a) Direct current (d.c.) sputtering of a prefabricated $CuAlO_2$ powder pellet,
(b) Reactive d.c. sputtering of a mixture of copper and aluminium metal powder pellet in oxygen-diluted argon atmosphere.
(c) Wet-chemical dip-coating technique from a solution of CuCl and $AlCl_3$ dissolved in HCl.

4.1. Synthesis of $CuAlO_2$ Films by D.C. Sputtering

The d.c. sputtering technique to prepare the film, involved the following three steps:

(i) $CuAlO_2$ powder preparation

Polycrystalline $CuAlO_2$ powder was synthesized by heating stoichiometric mixture of Cu_2O and Al_2O_3 according to the reaction: $Cu_2O + Al_2O_3 = 2CuAlO_2$. At first Cu_2O and Al_2O_3 powder (99.99 %) were taken with Cu / Al atomic ratio 1 : 1 and mixed for 1 hour. Then the mixture was heated in alumina boat at 1100°C for 24 hours. In every 6 hours the mixture was taken out of the furnace after proper cooling, remixed and placed into the furnace at the same temperature. The sintered body was reground and pressed into a pellet by hydrostatic pressure of about 200 kgf / cm^2. These pellets were then placed into a grooved aluminium holder by appropriate arrangement, which was used as the target for sputtering.

(ii) Substrate cleaning

Before placing into the deposition chamber the glass substrates were cleaned at first by mild soap solution, then washed thoroughly in deionized water and also in boiling water. Finally they were ultrasonically cleaned in acetone for 15 minutes. Si substrates were first immersed in 20 % HF solution for 5 minutes for removing surface oxide layers. Then they were cleaned in deionized water and finally with alcohol in an ultrasonic cleaner.

(iii) Film deposition

Our sputtering system consists of a conventional vacuum system, which was evacuated to 10^{-6} mbar by rotary and diffusion pump arrangement. The chamber was back filled with Ar and O_2 (40 vol %) gas mixture. The target was pre-sputtered for 10 minutes to remove contamination, if any, from the surface and then the shutter was displaced to expose the substrates in the sputtering plasma. Si (400) and glass were used as substrates. The target was connected to the negative terminal of high voltage d.c. power supply and the substrate was placed on the ground electrode. Summary of the deposition conditions is shown in Table 6 and a photograph of the d.c. plasma generated during deposition is shown in Fig. 5. After the deposition was over, the films were post-annealed in the same vacuum chamber at 473 K for 30 minutes to 150 minutes (at pressure 0.2 mbar) maintaining the oxygen flow to induce nonstoichiometry in the film, which is an important precondition for enhancing p-type conductivity of the film.

Table 6. Summary of deposition parameters for D. C. sputtered films [160]

Electrode distance	:	1.8 cm
Sputtering Voltage	:	1.1 kV
Current Density	:	10 mA / cm_2
Substrates	:	Si (400), glass
Base pressure	:	10-6 mbar
Sputtering Gasses	:	Ar & O_2 (3 : 2 volume ratio)
Deposition Pressure	:	0.2 mbar
Substrate Temperature	:	453 K
Deposition Time	:	4 hr
Post-annealing time	:	30 to 150 min
Post-annealing temperature	:	473 K
Post-annealing atmosphere	:	O_2 (0.2 mbar)

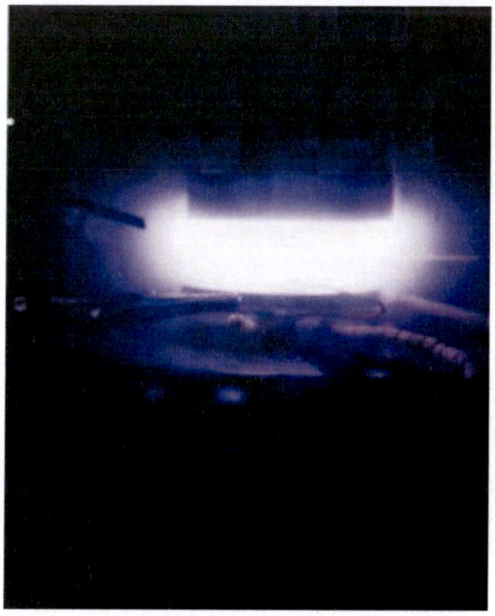

Figure 5. Photograph of D. C. sputtering plasma.

4.2. Synthesis of $CuAlO_2$ Film by Reactive Sputtering

The reactive d.c. sputtering technique also involved three steps:

(a) Target preparation

Firstly, a mixture of ultra pure copper and aluminum powders (99.99 %) were taken with Cu / Al atomic ratio as 1 : 1 and then they were mixed thoroughly for 1½ hour. The mixture was then pelletized into a grooved aluminium holder by hydrostatic pressure of 150 Kgf / cm^2 to use as target for sputtering.

(b) Substrate cleaning

The substrates used, were glass and Si (400). The substrate cleaning procedure was same as that one described in Section 4.1.

(c) Film deposition

Negative terminal of the d.c. generator was connected with the target and the substrates were placed on the grounded electrode. Si (400) and glass were used as substrates for film

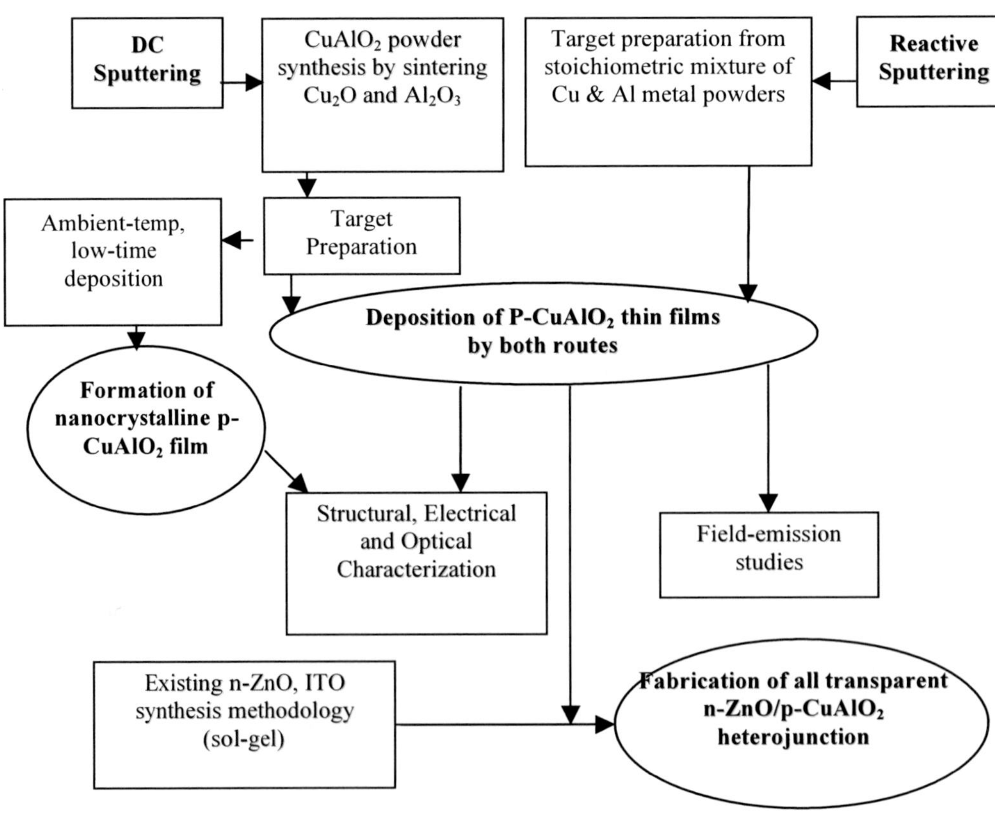

Figure 6(a). Layout of deposition and characterization process.

deposition. Prior to the deposition, the chamber was evacuated by standard rotary and diffusion pumping arrangements to a base pressure of 10^{-6} mbar. Subsequently, the chamber was flushed with Ar several times and then the target was pre-sputtered at 0.05 mbar in Ar atmosphere for 10 minutes to remove contaminations, if any, present on the target surface. The summary of the deposition conditions is shown in Table 7. After every 2 hours of deposition, the films were post annealed in the same vacuum chamber at 493 K for 1 hour.(at pressure 0.2 mbar) maintaining the oxygen flow to induce excess oxygen into the film to increase p-type semiconductivity of the film. A simple flow-chart describing the steps followed for the preparation and characterization of $CuAlO_2$ thin film by D.C. and reactive sputtering is shown in Fig. 6(a).

Table 7. Summary of deposition parameters by reactive D. C. sputtering [161]

Electrode distance	:	1.8 cm
Sputtering Voltage	:	1.0 kV
Current density	:	12 mA / cm^2
Substrates	:	Si (400), glass
Base pressure	:	10^{-6} mbar
Sputtering gasses	:	Ar & O_2 (3 : 2 volume ratio)
Deposition pressure	:	0.2 mbar
Substrate temperature	:	475 K
Deposition Time	:	4 hr
Post-annealing time	:	60 min
Post-annealing temperature	:	493 K
Post-annealing atmosphere	:	O_2 (0.2 mbar)

4.3. Synthesis of $CuAlO_2$ Film by Wet-Chemical Dip-Coating Technique

The wet-chemical synthesis procedure also involved three steps:

(i) Sol preparation

The sol required for deposition of the films was prepared as described in the following steps. Firstly, 2.5 cc of concentrated HCl was added slowly to 0.015 moles of cuprous chloride (CuCl, 99.99%) and the solution was stirred continuously by a magnetic stirrer. During the stirring process, further addition of 4 - 5 drops (0.2 cc) of HCl to the solution was done until all the salts were dissolved into it. On the other hand, another solution was prepared by adding 30 cc of distilled water drop by drop to 0.015 moles of aluminium chloride ($AlCl_3$, 99.9%) to dissolve it completely. Two solutions were then mixed and 50 cc of distilled water was also added to it. The mixed solution was then stirred continuously at an elevated temperature of 85° C for 2 hrs. During the stirring process, 0.002 moles (approx.) of NaOH pellets (99.99%) were added to the solution to control the pH value around 2. In the resulting solution, the concentrations of Cu and Al were calculated to be 0.187 moles / liter each. The solution was then aged for 3 hrs to get the required sol which was used for dip coating process.

(ii) Substrate cleaning

The substrates used, were glass and Si (400). The substrate cleaning procedure was same as that one described in Section 4.1.

(iii) Dip-coating

Substrates were first dipped into and then withdrawn vertically from the solution slowly at the rate of 6 cm / min for 12 to 15 times. Between two successive dipping, the substrate along with the sol was dried at ~ 100° C - 120° C to have quick geletion. After the dipping, withdrawing and drying procedure, the resulting films were annealed at ~ 480° C to 500° C in air for 3 hrs to form the desired copper aluminium oxide thin film. A flow chart of the dip-coating procedure is shown in Fig. 6(b).

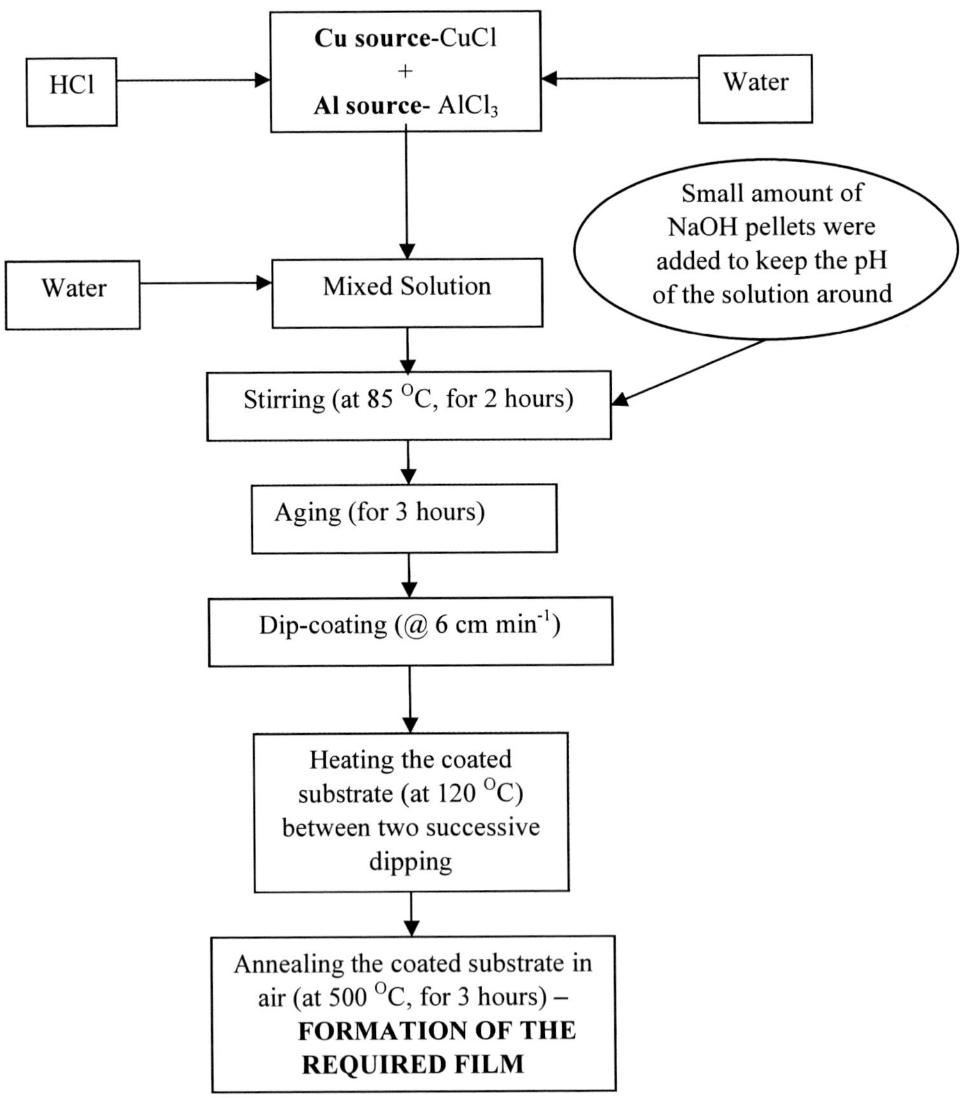

Figure 6(b). Flow Chart of wet-chemical dip-coating process for $CuAlO_2$ thin films.

4.4. Characterizations of the Films

Structural properties of the films were investigated by X-Ray Diffraction (XRD) measurements. The films were deposited on Si (400) and glass substrates. A Bruker X-ray diffractometer (D8, AXS, ADVANCE) was used for recording the diffraction traces of the films in θ - 2θ mode. Germanium (022) monocromator was used for CuK_{α} (1.4506 Å) radiation from a highly stabilized Bruker X-ray generator (K 780). Diffraction traces were recorded at room temperature. Another X-ray diffractometer (Philips PW 1730 / PW 1710, by $CuK\alpha$ line) was also used for structural studies of some of the samples.

Surface morphology and microstructural properties of the films were studied by Scanning Electron Microscopy (SEM) and Transmission Electron Microscopy (TEM) respectively. An SEM (JEOL, JSM 5200) was used to determine the growth and morphology of the samples. The resolution of the instrument was 5.5 nm. It is designed to operate at voltages between 1 to 25 kV (7 steps). The magnification could be varied from 15 X to 200,000 X (25 steps). Probe-current range was 10^{-12} to 10^{-9} A. The instrument can be operated in two types of image modes e.g. Secondary Electron Image (SEI) and Backscattered Electron Image (BEI). The camera was 35 mm, single-lens reflex type (MP 35051, CSI 3) with focal length = 50 mm.

A TEM (HITACHI H 600) was used to study the microstructure of the film. The instrument has a guaranteed resolution of 2.04 Å. But the resolution attained during routine measurement was 8–10 Å. Magnification could be varied from 100 X to 30,000 X with accelerating voltage 25, 50, 75 or 100 kV. Selected area electron diffraction pattern (SAED) could be obtained with diffraction camera length 0.2 to 1.6 m.

For SEM studies, films were deposited on both glass and Si substrates whereas for TEM studies, films were deposited directly on carbon coated Cu-grids. The film thicknesses in this case were maintained between 50 – 100 nm by reducing the deposition time.

Compositions of the films were determined from an Energy Dispersive X-Ray (EDAX, Leica S-440 Oxford ISIS) instrument. The instrument has the capability to detect elements from Boron (5) to Uranium (92).

The optical transmittance (T) and reflectance (R) spectra of the films were measured by UV-Vis-NIR spectrophotometer. A Shimadzu-UV-3101-PC spectrophotometer was used to determine the optical properties. It is a double beam spectrophotometer with integrating sphere attachment for reflectance measurement within the wavelength range of 190 nm to 2600 nm. The attachment is mainly used for measurement of both transmittance as well as diffuse/specular reflectance of the films. The integrating sphere equipped with photomultiplier (UV-Vis region) and PbS cell (NIR region) detectors. Both the optical transmission and reflection spectra of the films deposited on glass substrates were recorded taking similar glass as reference, and hence the spectrum gives transmittance and reflectance of the films only. Another Hitachi (U 3410) spectrophotometer was used to measure transmittance spectra of some of the samples. The wavelength for this instrument could be varied from 180 to 3500 nm.

A Nicolet Magna (IR-750 Series-II) FT-IR was used to obtain different bonding information in the sample. The resolution of the instrument was 4 cm^{-1} with the wavenumber range of 4000 cm^{-1} to 400 cm^{-1}. Number of scan steps was 50.

The sheet resistance and temperature dependence of electrical conductivity of the films were studied by linear four-probe method using Kiethley electrometer (Model- 6514) from

300 to 550 K. All The contacts were made with silver paint, which showed linear I-V characteristic over a wide range of applied voltage. Films were deposited on glass substrates.

Thermoelectric power (TEP) and Hall effect studies were used to determine the type of conduction taking place within the deposited films. For thermoelectric power measurement (temperature variation of Seebeck coefficient), a temperature gradient across the sample was created by keeping one end of the film in a hot-head and the other in a cold-head. The hot-head temperature was varied from room temperature to 460 K, whereas the cold-head was kept at room temperature. And these temperatures of the hot and cold-ends of the film were measured by proper thermocouple arrangements. The thermoemfs generated between the hot and cold ends of the sample, at different hot-end temperatures, were used to determine the Seebeck coefficients (S) of the material. The entire system was kept under vacuum condition.

For room temperature Hall-study, standard van der Pauw method was used, with rectangular van der Pauw configuration. The electrical connections were made at the four corners of the sample. For the measurement of Hall-voltage and related parameters, an electromagnet (Polytronic Corporation, India) with 4 inches pole pieces was used alongwith a stabilized power supply (Current range – 0 to 6 A, Voltage range – 0 to 100 V) to monitor the field strength. The distance between the pole pieces could be varied and for a separation of 3.0 cm of pole pieces, the field strength could be adjusted to a maximum of 10 K Gauss. The field within the measuring system was determined by using Differential Gaussmeter. Flow-diagram of various characterizations done on the $CuAlO_2$ films are furnished in Fig. 6(c), 6(d) and 6(e).

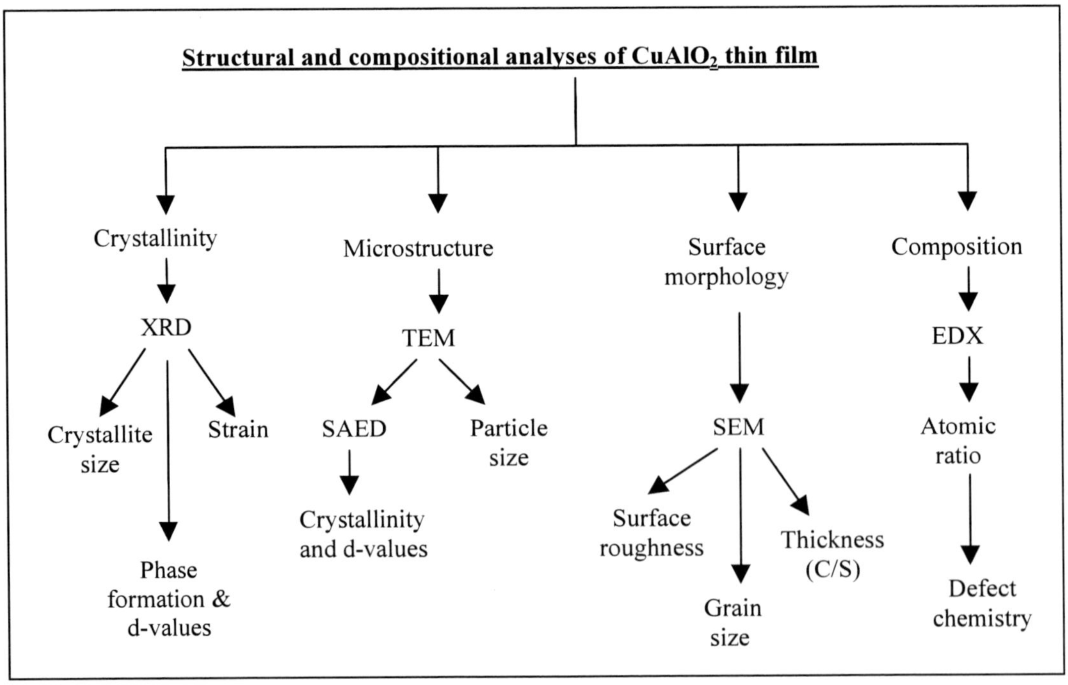

Figure 6(c). Flow-chart of structural characterizations and compositional analyses of $CuAlO_2$ thin film.

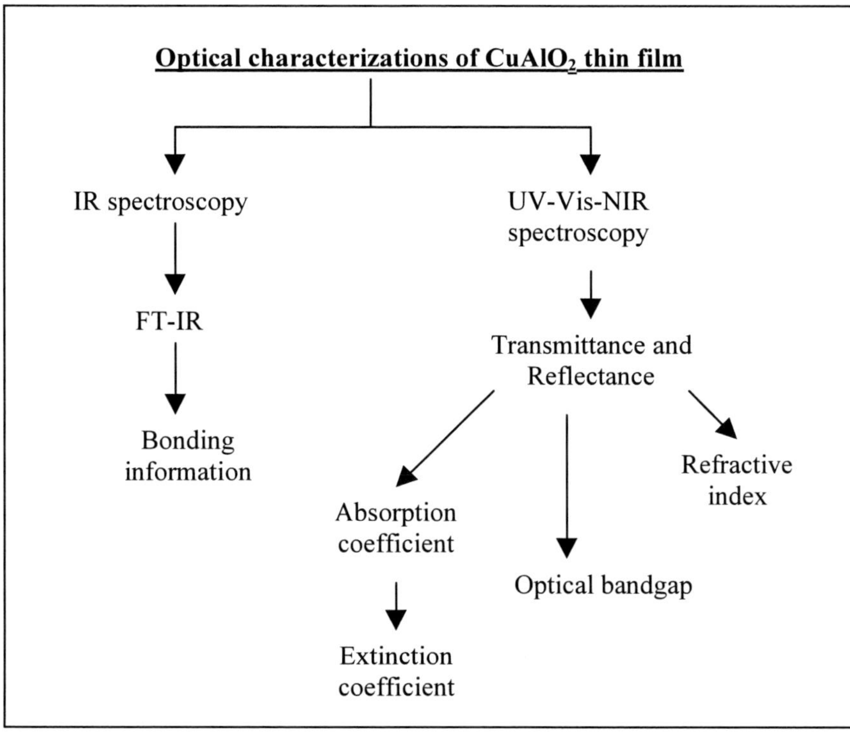

Figure 6(d). Flow-chart of optical characterizations $CuAlO_2$ thin film.

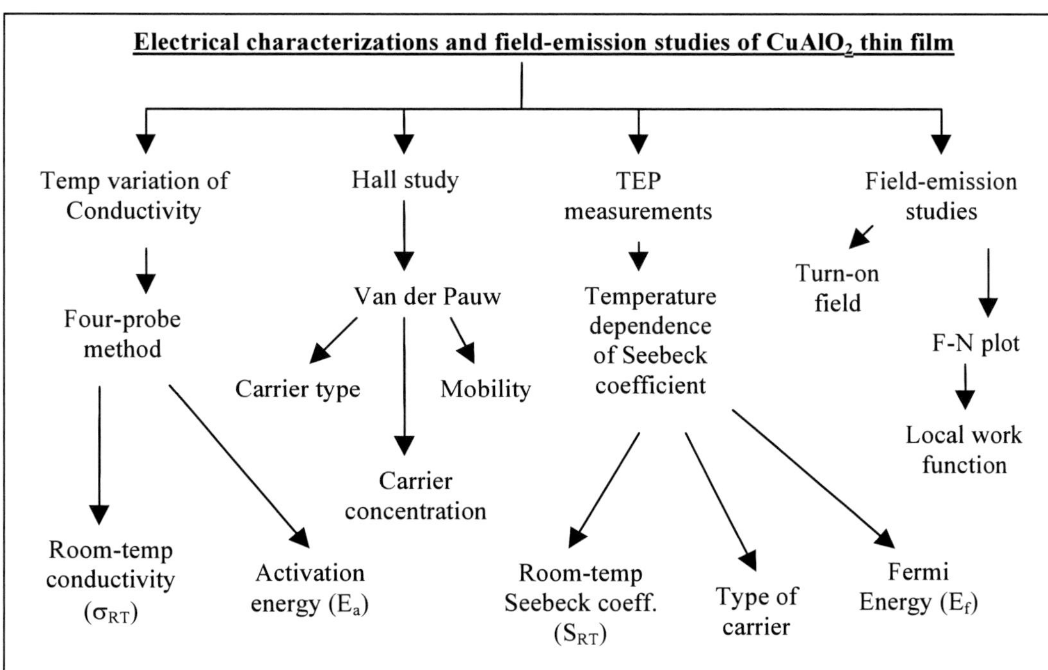

Figure 6(e). Flow-chart for the Electrical and FE characterizations of $CuAlO_2$ thin film.

5. Results and Discussion

5.1. Properties of D. C. Sputtered Films

i) X-ray diffraction studies

X-ray diffraction study in thin film technology is essential to identify proper phase formation of the required polycrystalline films as well as the degree of crystallinity of the materials. In this section we have presented the results of the XRD analyses of sintered $CuAlO_2$ target as well as thin films prepared both by D. C. and reactive D. C. sputtering methods. Also the semiquantitative information of strain and particle size of the films are obtained from the XRD data.

Fig. 7 shows the X-ray diffraction pattern (XRD) of the synthesized $CuAlO_2$ powder, which was used for target preparation. 2θ values for the scanned pattern range from 10 degree to 100 degree. The peaks of the powdered material are identified to originate from (006), (101), (012), (104), (107), (018), (110), $(00\overline{12})$, (116), (202) and (119) reflections. This pattern closely reflects the rhombohedral crystal structure with $R\overline{3}m$ space group [113]. From the XRD pattern it is observed that the target material contains no unreacted species, such as Cu_2O or Al_2O_3 or any other phase of copper aluminium oxide (e.g. $Cu_2Al_2O_4$). The crystallographic data and bond lengths of $CuAlO_2$ are furnished in Table 1.

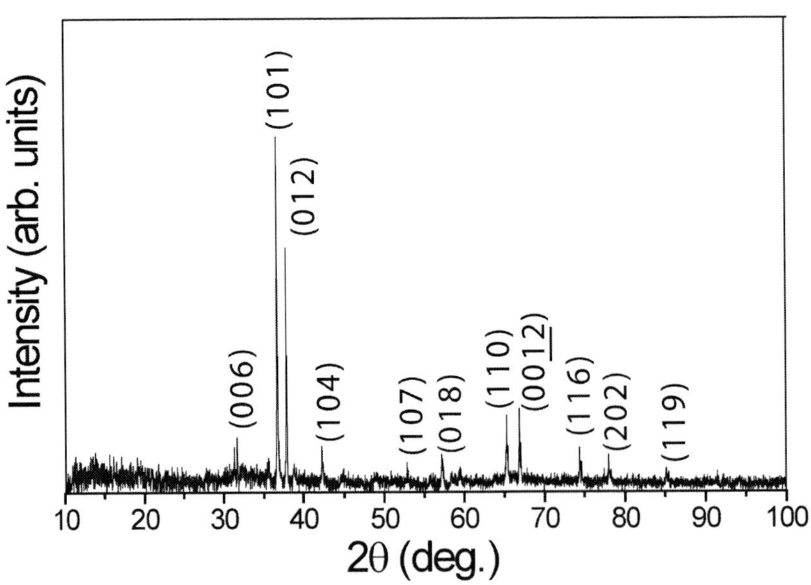

Figure 7. X-ray diffraction pattern of the synthesized $CuAlO_2$ powder.

Fig. 8(a) shows the XRD pattern of the D. C. sputter-deposited $CuAlO_2$ thin film on Si (400) substrate for post-deposition oxygen annealing time (t_a) 60 min. The XRD pattern shows a strong (006) orientation. Two other small peaks e.g. (003) and (018) have also been observed in the pattern. It is worthwhile to mention that, for XRD patterns of $CuAlO_2$ powdered samples, Kawazoe et al [64] and Yanagi et al. [67] previously reported a high (012) orientation, whereas for the thin films deposited on sapphire substrates by PLD method, they observed a strong (006) orientation. For the XRD pattern of our $CuAlO_2$ powder, we observed a maximum intensity at (101) peak as shown in Fig. 7, whereas the $CuAlO_2$ thin films deposited by D. C. sputtering on Si substrate, a strong (006) orientation was observed as reported previously [64, 67]. It is also noteworthy that the $CuAlO_2$ thin films deposited previously by other techniques, such as R. F. sputtering [147], CVD [149-151], wet-chemical method [153] etc., either the crystal quality of the films were not very good or the films were phase impure (i.e. the films contained some amounts of impurity such as CuO, Cu_2O, $Cu_2Al_2O_4$ etc.). This would result in the poor electrical characteristics of those films. But as evidenced from the XRD pattern of our D. C. sputter-deposited $CuAlO_2$ thin films, these films are highly crystalline, and there are no unreacted species and any impurity present in the films.

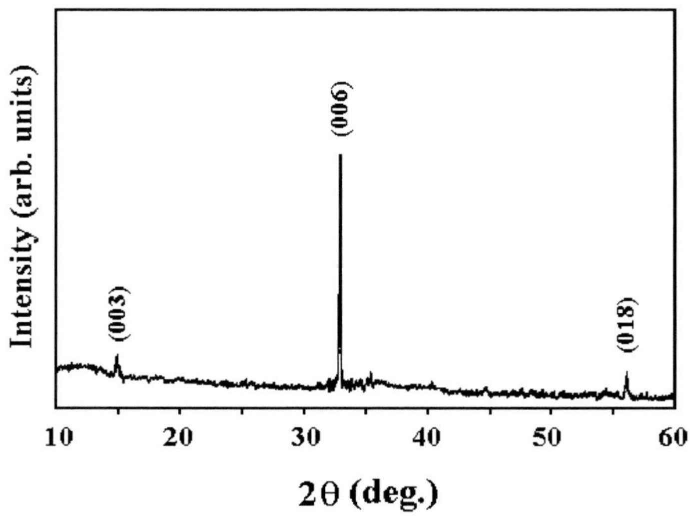

Figure 8(a). XRD pattern of $CuAlO_2$ thin film deposited on Si substrate, with post-annealing time (t_a) 60 min.

For the films deposited at other annealing times (t_a) e.g. 30 min, 90 min, 120 min and 150 min, the XRD patterns show identical peaks and no significant changes have been observed in the intensity of the peaks and, therefore, not shown here. This is probably because, in all cases, the annealing temperature was kept fixed (at 473 K) and the lowest time of annealing (i.e. 30 min) of our films may be sufficient enough to saturate the grain growth at that particular deposition temperature (473 K) and, hence, no further change in the XRD patterns of our films with increase in post-annealing time was observed. This indicates that in our case, post deposition annealing time has no (or almost insignificant) effect on the structural properties of the films. Fig. 8(b) shows the film deposited on glass substrates with 60 min post-annealing time. The figure shows similar peaks as that deposited on Si substrate [Fig. 8(a)], but the intensities of the peaks were slightly lesser and the peak-sizes were slightly broader than that deposited on Si substrates. Table 8 shows the comparison between the theoretical d-values given in JCPDS file and observed d-values obtained from XRD data of sintered $CuAlO_2$ powder (Fig. 7), D. C. sputtered $CuAlO_2$ thin film (Fig. 8).

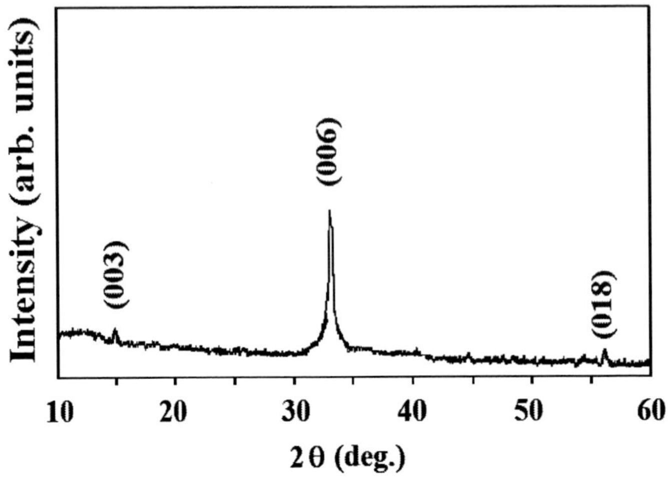

Figure 8(b). XRD pattern of $CuAlO_2$ thin film deposited on Si substrate, with post-annealing time (t_a) 60 min.

Table 8 Comparison between the theoretical d-values, observed d-values of $CuAlO_2$ powder, D. C. sputtered and reactive D. C. sputtered $CuAlO_2$ thin films.

h k l	d-values from JCPDS file card # 35-1401 (d_{JCPDS}) (Å)	Observed d-values for $CuAlO_2$ powder (d_{powder}) (Å)	Observed d-values for $CuAlO_2$ thin film deposited by D. C. sputtering on Si substrate ($d_{DC\text{-}Sputter}$) (Å)
0 0 3	5.610	---	5.700
0 0 6	2.820	2.830	2.800
1 0 1	2.440	2.450	---
0 1 2	2.376	2.378	---
1 0 4	2.133	2.133	---
1 0 7	1.732	1.732	---
0 1 8	1.612	1.611	1.620
1 1 0	1.426	1.428	---
0 0 $\overline{12}$	1.401	1.401	---
1 1 6	1.274	1.275	---
2 0 2	1.225	1.225	---
1 1 9	1.148	1.140	---

The information on particle size of very small crystallites from the measured Full-Widths-at-Half-Maximum (FWHM) of the diffraction peaks can be estimated from the well-known Scherrer formula [230]

$$L = \frac{x\,\lambda}{\beta_1 \cos\theta} \tag{3}$$

where L is the particle size, β_1 is the particle-broadening of diffraction peaks measured at FWHM of the peak at a certain 2θ value, x is a correction factor (= 0.9) and λ is the wavelength of the X-ray used. It is to be mentioned here that when the size of the individual crystallites in a polycrystalline sample is less than 100 nm, the term "particle size" is usually used [230]. Although the grain size of our sample is not clearly determined from SEM micrograph (shown in Ref. [160]), but a rough estimation shows that it may fall within the limit mentioned above and, therefore, we have used the term "particle size" here.

In polycrystalline thin films, due to the interaction between grains of the films as well as that with the substrate, a single grain in the polycrystalline thin film is not free to deform in the same way as an isolated crystal would, if subjected to the same deforming force. As a result of this restraint by its neighbors, a deformed grain in a polycrystalline aggregate usually is in a state of tension or compression. Thus an "internal stress" or "residual stress" is generated within the films. This residual stress produces uniform or non-uniform strain within the film. If the grains are subjected to a uniform tensile strain at right angles to the X-ray reflecting planes, corresponding diffraction peaks shift to the lower angles but do not change otherwise. Similarly for uniform compressive strain, the diffraction peaks shift to the higher angles with no change otherwise. On the other hand, if the strain is non-uniform then the diffraction peak will be broadened, which is called "strain broadening" [230]. The relation between this broadening and the strain can be obtained by differentiating the Bragg's law as follows [230]:

$$\begin{aligned} & 2\Delta d\, Sin\theta \;+\; 2d\, Cos\theta\, \Delta\theta = 0 \\ \Rightarrow\quad & \Delta\theta = -\frac{\Delta d}{d}\tan\theta \\ \Rightarrow\quad & \Delta(2\theta) = -2\,\varepsilon\tan\theta; \qquad [\Delta d / d = \varepsilon] \\ \Rightarrow\quad & \Delta\beta = -2\,\varepsilon\tan\theta; \qquad [\beta = 2\theta] \end{aligned} \tag{4}$$

where $\Delta\beta$ is the extra broadening of the diffraction peaks over and above the instrumental breadth (therefore also called "instrumental broadening"), ε is the strain generated within the films, θ is the Bragg angle. Now the above equation contains both tensile and compressive strain and must be divided by two to obtain maximum tensile strain alone or maximum compressive strain alone, if these two are assumed equal. Hence the equation for strain broadening for only one type of strain will be

$$\Delta\beta = -\,\varepsilon\tan\theta \tag{5}$$

Now if both the effect of "particle-size broadening" and "strain-broadening" is taken into consideration, then the total broadening (β) can be expressed as a linear combination of equations 3 and 5 as follows [231]:

$$\beta = \beta_1 + |\Delta\beta| = \frac{\lambda}{L\,Cos\theta} + \varepsilon\tan\theta$$

$$\Rightarrow\quad \frac{\beta\, Cos\theta}{\lambda} = \frac{1}{L} + \frac{\varepsilon\, Sin\theta}{\lambda} \tag{6}$$

where β is the FWHM of the observed peaks, L is the effective particle size, ε is the effective strain. A plot of $\frac{\beta\ Cos\theta}{\lambda}$ vs. $\frac{Sin\theta}{\lambda}$ will be a straight-line, slope of which will give the estimation of the effective strain, whereas the intercept on $\frac{\beta\ Cos\theta}{\lambda}$ axis will carry the information of the effective particle size. Fig. 9 represents the plot of $\frac{\beta\ Cos\theta}{\lambda}$ vs. $\frac{Sin\theta}{\lambda}$, obtained from the XRD pattern of the $CuAlO_2$ thin film deposited by D. C. sputtering on Si substrate, with t_a = 60 min (shown in Fig. 8(a)). Slope of the graph depicts the strain value as 8.52×10^{-3} and the intercept on y-axis gives the particle size as ~ 26 nm.

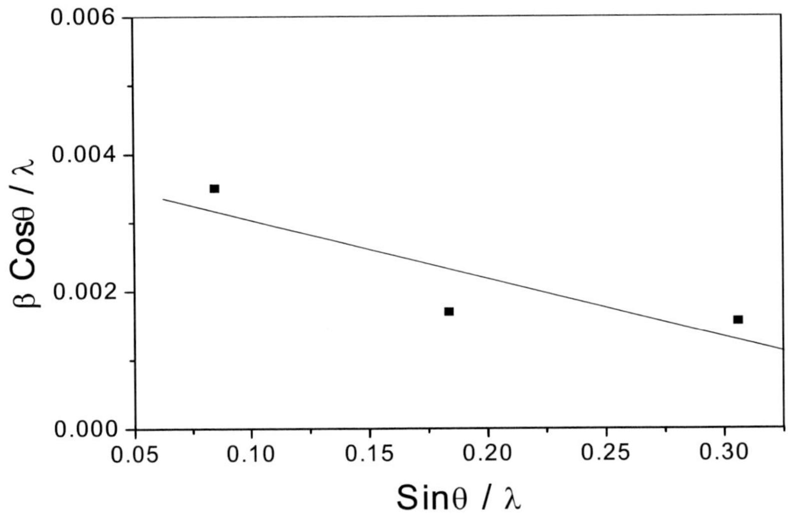

Figure 9. Plot to determine strain & particle size of $CuAlO_2$ thin film deposited by D. C. sputtering, with t_a = 60 min.

(ii) Compositional analyses

Compositional analyses of the D. C. sputtered films deposited with various post-deposition oxygen annealing times (t_a) were done by EDX measurements. Results suggest slight deviation from the stoichiometric composition within the films with increase in post-deposition oxygen-annealing time (t_a). The percentage of excess oxygen within the films ranges from 0.5 at % (for annealing time 30 min) to 10 at % (for the films annealed for 120 min and above) over stoichiometric value. The Cu : Al stoichiometry remained close to 1 : 1 for all the samples (i.e. in the ratio Cu : Al : O = 1 : 1 : 2+x, percentage of x w.r.t. 2 is given here). Previously, Gao and co-authors [154] also observed similar 1:1 atomic ratio of Cu:Al (more precisely 1.06:1.00, and taken as unity within experimental error) in their nanocrystalline $CuAlO_2$ thin film from EDX analysis. We have observed that for the films post-annealed for 30, 60 and 90 min, the percentages of excess oxygen were around 0.5 at %, 2.5 at % and 5 at % respectively over stoichiometric value. Compositional analyses of the films post-annealed for 120 min and above, show percentage of excess oxygen within the films more than 10 at % over stoichiometric value. Table 9 shows the composition of D. C. sputtered $CuAlO_{2+x}$ thin films for different values of t_a and corresponding chemical formula of the material. These "excess" oxygen atoms are supposed to lie in the lattice positions and (or) interstitial positions and produce enhanced p-type conductivity of the films, which will be discussed in details in later chapters.

Table 9. Composition of D. C. sputtered $CuAlO_2$ thin films for different values of t_a.

Post-annealing time (t_a) (min)	Cu / Al ratio	Atomic % of excess oxygen	Chemical formula of the film
30	1	0.5	$CuAlO_{2.01}$
60	1	2.5	$CuAlO_{2.05}$
90	1	5.0	$CuAlO_{2.10}$
120	1	10.0	$CuAlO_{2.20}$
150	1	12.0	$CuAlO_{2.24}$

(iii) FT-IR studies

Fourier Transform Infra-red spectroscopic (FT-IR) analyses of D.C. sputtered $CuAlO_2$ thin films were performed. Films were deposited on Si substrates. Wavenumber varied from 400 cm^{-1} to 4000 cm^{-1}. Fig. 10 represents the FT-IR spectra of the $CuAlO_2$ film deposited by D. C. sputtering technique and post-annealed for 60 min. All bands have been assigned to the absorption peaks of Cu-O, O-Cu-O, Al-O bond vibrations. The broad peak ranging from 500 cm^{-1} to 900 cm^{-1} is actually consisting of a number of peaks, which can be obtained by deconvoluting the peak. The absorption peaks near 550 cm^{-1} and 600 cm^{-1} may be assigned to Cu-O stretching vibration and O-Cu-O antisymmetric vibration respectively. The peak around 600 cm^{-1} originates due to Al-O stretching vibration in AlO_6 octahedra of $CuAlO_2$ structure. Peaks ranging from 700 cm^{-1} to 900 cm^{-1} may be assigned to short Al-O stretching vibrations in distorted AlO_6 octahedra. Peak around 1000 cm^{-1} may be assigned to Si-O-Al vibration, which occurs due to Si substrate used [232, 233]. Peak at 2349 cm^{-1} is a CO_2 peak and the broad peak around 3000 cm^{-1} - 3500 cm^{-1} is due to O-H stretching vibration, which may be incorporated from the atmospheric contaminations. From the literature survey, it becomes clear that there is no reported study on FT-IR of the $CuAlO_2$. So there may remain some unidentified peaks, such as ~ 1633 cm^{-1} in our FT-IR spectra. It must be mentioned here that the assignments of the peaks for different vibrational modes of $CuAlO_2$ is a simplification of the vibrational treatment of different inorganic aluminates as well as copper complexes in organic solvents. A rigorous vibrational treatment of inorganic solids is generally very difficult. Strictly speaking, the different vibrational modes are those of the whole unit cell of

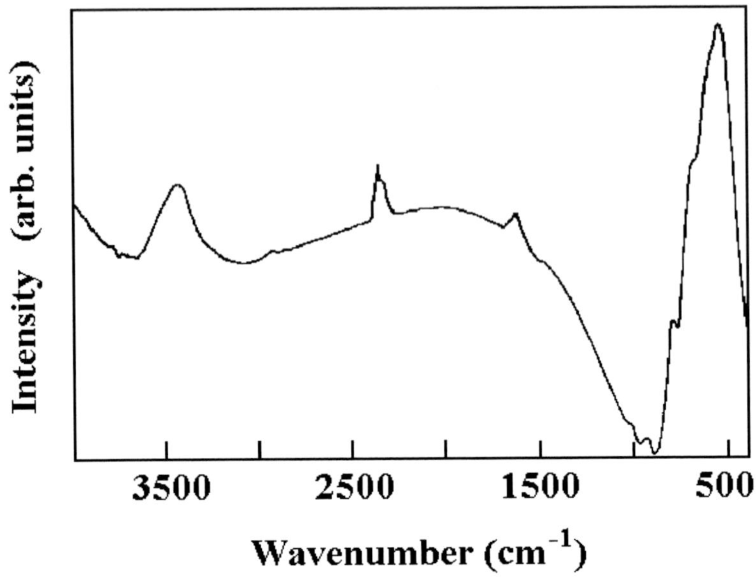

Figure 10 FT-IR spectrum of $CuAlO_2$ thin film deposited by D.C. sputtering on Si substrate.

the crystal and, therefore, the number of fundamental frequencies are quite high and, hence,detailed assignment of the observed frequencies to the vibrational modes is nearly impossible [233]. In such cases, simplified methods have been applied as follows [233-235]: if a solid $A_xB_yO_z$ is constituted of AO_m and BO_n coordinated groups, two extreme cases must be considered:

a) If the AO_m and BO_n groups have different vibrational frequencies, then the vibrational interactions between these groups are weak and, therefore, neglected. The groups are assumed to be vibrating 'independently' [236].
b) If the AO_m and BO_n groups have similar vibrational frequencies, then the vibrational interactions between the groups are very large and the vibrations of those groups are taken as a whole [235].

Between these extreme cases, a number of intermediate cases are characterized by weak or moderate interactions [234]. It is quite evident that the assignment of an absorption peak to a vibrational mode of a given coordinated group is meaningful only if the concept of "independent" vibrations is a good approximation for the group under consideration. Now, as $CuAlO_2$ is a layered-structured material with AlO_6 octahedral layers connected by O-Cu-O dumbbell layers (shown in Fig. 2b), the two layers may be approximated to be vibrating independently. This argument seems reasonable if we theoretically calculate the vibrational frequencies of Cu-O and Al-O bonds. From the equation of simple harmonic oscillator, the frequency of oscillation will be expressed as

$$\nu = \frac{1}{2\pi}\sqrt{\frac{K}{\mu}} \tag{7}$$

where ν is the vibrational frequency, K is the force constant of the bond, μ is the reduced mass. For Cu-O, $\mu_{Cu\text{-}O} = 2.12 \times 10^{-23}$ gm and $K = 2.25 \times 10^{5}$ dynes cm^{-1}. For Al-O, $\mu_{Al\text{-}O} = 1.67 \times 10^{-23}$ gm and $K = 2.60 \times 10^{5}$ dynes cm^{-1} [232]. Calculations show that $\bar{\nu}_{Cu-O} = 546.798\ cm^{-1}$ and $\bar{\nu}_{Al-O} = 662.35\ cm^{-1}$. Although these values will be quite different from the 'actual' values when these bond vibrations will be influenced by the neighboring atoms of a three-dimensional network, but, here we are concerned about the difference between the above-mentioned two values. As these values are fairly different, therefore, our argument of independent vibrations of the two coordinated groups in $CuAlO_2$ is reasonable. And, that is why, we assigned the broad peaks around 400 cm^{-1} to 700 cm^{-1} shown in Fig. 10 to the absorption peaks of Cu-O, O-Cu-O, Al-O bond vibrations.

But, strictly speaking, the concept of 'independent' vibration is an approximate one because there is always a more or less important influence of neighboring groups to a certain bond vibration. Also the vibrational frequencies are influenced by any distortion or deformation of the coordinated groups (which is very frequent in thin films). Another additional effect may be present where the coordinated groups are interlinked by common oxygen atoms (as in our case) to form a chain or sheet or three-dimensional network. This affects the vibrational frequencies of a certain bond vibration. As a consequence, the calculated frequencies and the observed values will be quite different. That is why in our cases we have not assigned a vibrational mode to certain frequency, rather to a range of frequencies.

(iv) UV-Vis-NIR measurements

Optical properties of $CuAlO_2$ thin films are extremely important because of its possible applications in the field of optoelectronics technology. High transparency coupled with high conductivity is the main feature for TCOs as mentioned earlier. Therefore detailed optical characterization and determination of related parameters are the most significant part of the analyses of TCOs. Following this point of view, we have studied the optical properties of $CuAlO_2$ thin films in details. Three types of films with different post-deposition annealing times (t_a = 30, 60 and 90 min) were studied. Fig. 11, 12 & 13 show the transmittance (T) and reflectance (R) spectra of the films with t_a = 60, 90 and 30 min respectively. The films were deposited on glass substrates, taking similar glass as reference. Hence the spectra are for the film only. The thicknesses of all the films were 500 nm. Slight noises present around 800 nm to 900 nm in all the graphs are artifacts of detector crossover.

The transmittance (T), reflectance (R) and absorption coefficient (α) of a specimen is related by the equation [237]

$$T = \frac{(1-R)^2 e^{-\alpha d}}{1-R^2 e^{-2\alpha d}} \qquad (8)$$

where d is the film thickness and here the multiple internal reflections within the film are considered. Now at the region of fundamental absorption, α will be quite high, so also αd. So we can neglect the 2nd term of the denominator of eqn. (8) and rewrite it as [237, 238]

$$T \approx (1-R)^2 e^{-\alpha d} \tag{9}$$

Knowing T, R and d, absorption coefficients can be determined. If R is not known, then from transmittance data of two samples of known thicknesses d_1 & d_2, α can be obtained from the relation [237]

$$\frac{T_1}{T_2} \approx e^{\alpha(d_2 - d_1)} \tag{10}$$

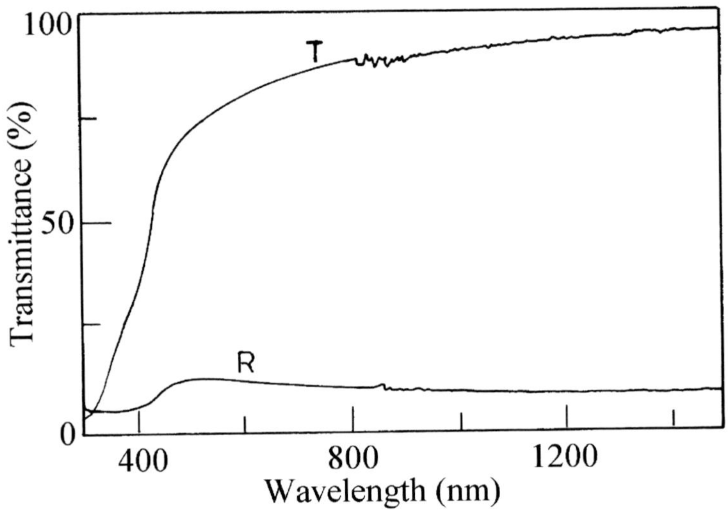

Figure 11. Transmittance (T) and reflectance (R) spectra of $CuAlO_2$ thin film, post annealed for 60 min. The spectral range is from 300 nm to 1500 nm.

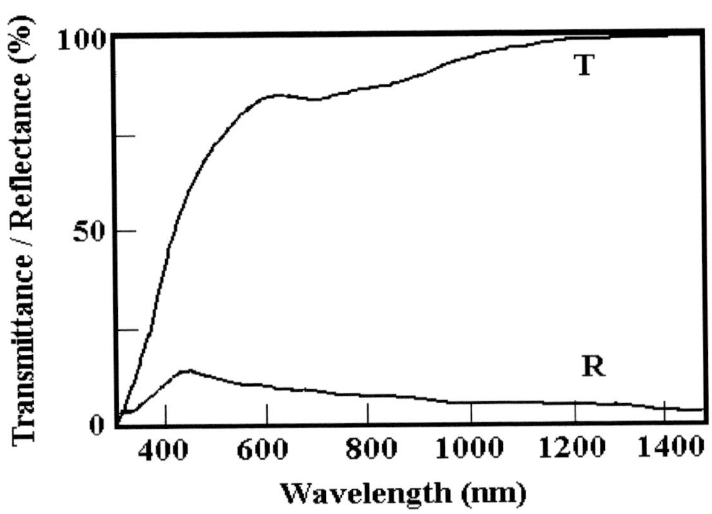

Figure 12. UV-Vis-NIR spectra of D. C. sputtered $CuAlO_2$ thin film with t_a = 90 min.

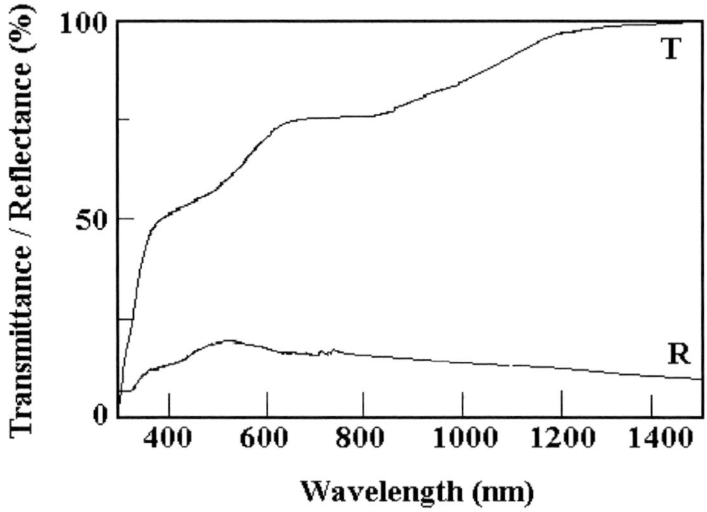

Figure 13. UV-Vis-NIR spectra of D. C. sputtered $CuAlO_2$ thin film with t_a = 30 min.

Beyond the absorption edge if one can observe the interference effect in the transmittance and reflectance spectra due to the multiple internal reflections within the film, then it will be possible to find the refractive index (n) of the material by measuring the wavelengths (λ_1 and λ_2) at two adjacent maxima. The expression will be [237]

$$n = \frac{\lambda_1 \lambda_2}{|\lambda_1 - \lambda_2|} \tag{11}$$

Now, according to the schematic diagram shown in Fig. 14, in the spectral region of fundamental absorption, as a first approximation, T, R and α will be related by the following equation [239] (here, we have neglect the internal multiple reflections for TCOs, unlike Eqs. 8 and 9)

$$T \approx (1-R)e^{-\alpha d} \tag{12}$$

and

$$R = \frac{(n-1)^2 + k^2}{(n+1)^2 + k^2} \tag{13}$$

where n is the refractive index and k is the extinction coefficient, which is related to the wavelength (λ) and absorption coefficient (α) by the following equation:

$$k = \frac{\lambda \alpha}{4\pi} \tag{14}$$

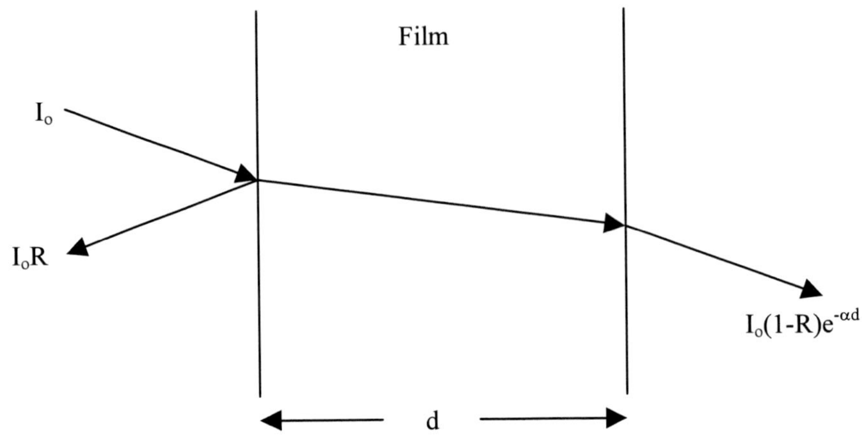

Figure 14. Schematic diagram of incident (I_o), reflected (I_oR) and transmitted [$I_o(1\text{-}R)e^{-\alpha d}$] rays in a thin film of thickness d. Multiple internal reflections are neglected.

Now, for transparent medium (as in our p-$CuAlO_2$ films), $k^2 \ll (n\text{-}1)^2$ and Eq. 13 will be reduced to

$$n = \frac{1+\sqrt{R}}{1-\sqrt{R}} \tag{15}$$

and the absorption coefficients (α) can be calculated by rewriting Eq. 12 as

$$\alpha = \frac{1}{d}\ln[\frac{1-R}{T}] \tag{16}$$

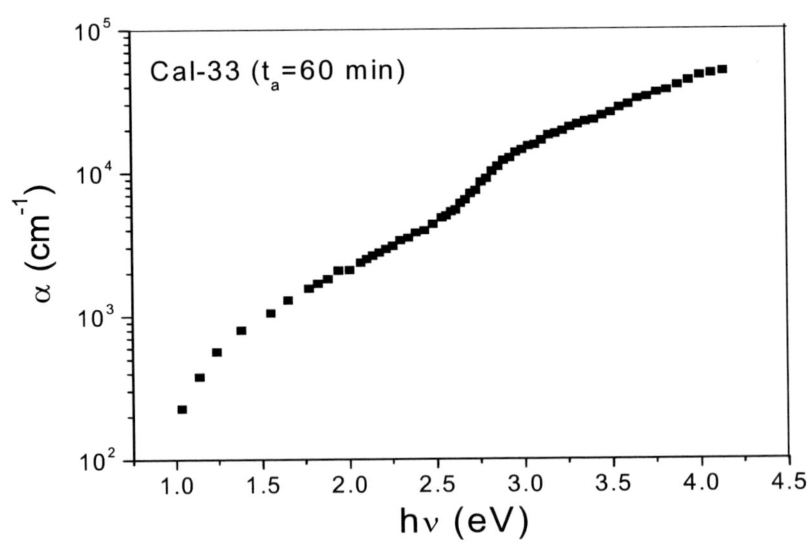

Figure 15(a). Energy dependence of absorption coefficient of $CuAlO_2$ thin film, post-annealed for 60 min.

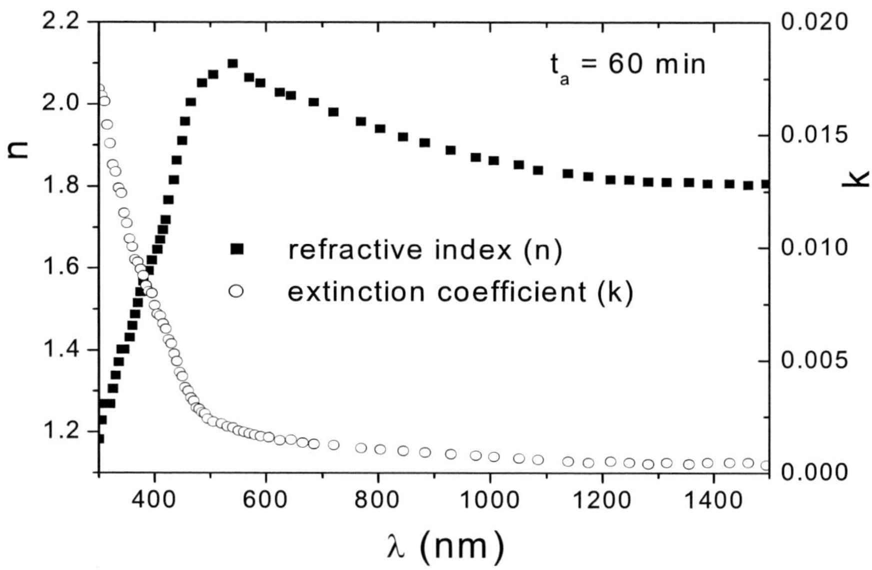

Figure 15(b). Spectral variation of refractive indices (n) and extinction coefficients (k) of $CuAlO_2$ thin film post-annealed for 60 min.

Fig. 15(a) and **(b)** show the spectral variation of α, k and n of $CuAlO_2$ thin film post-annealed for 60 min. It has been observed that the refractive index varies between 1.2 to 2.1 in the wavelength range of 300 nm to 1500 nm. Although there are no reported data on the refractive indices of $CuAlO_2$ thin film, but these data are reasonable when compared with other TCOs such as ITO (1.75 to 2.0 in the wavelength range of 400-1200 nm [240]) and CdO (1.31 to 2.84 in the wavelength range of 500 – 2500 nm [241]).

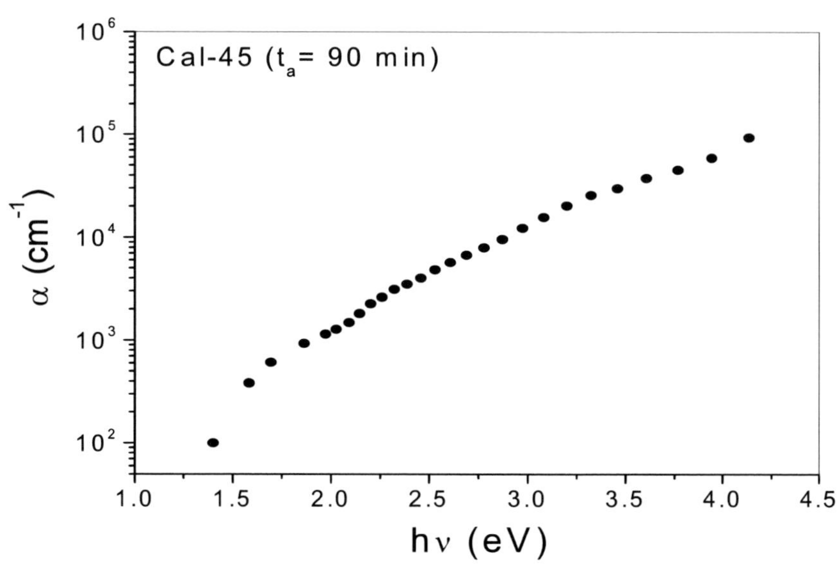

Figure 16(a) Energy dependence of absorption coefficient of $CuAlO_2$ thin film, post-annealed for 90 min.

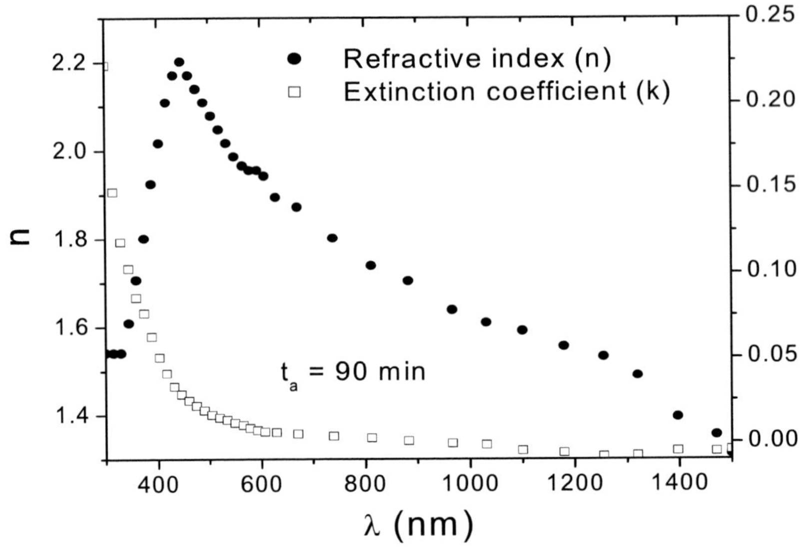

Figure 16(b). Spectral variation of refractive indices (n) and extinction coefficients (k) of $CuAlO_2$ thin film post-annealed for 90 min.

Fig. 16(a) and (b) show the spectral variation of α, k and n of $CuAlO_2$ thin film post-annealed for 90 min, whereas **Fig. 17(a) and (b)** show the same for $CuAlO_2$ film post-annealed for 30 min. In these cases also, we have observed the variation of n between 1.3 and 2.5 within the specified wavelength range. A comparative study of the average visible transmittance, and the ranges of α and n of the three types of films (with different t_a) are furnished in **Table 10**.

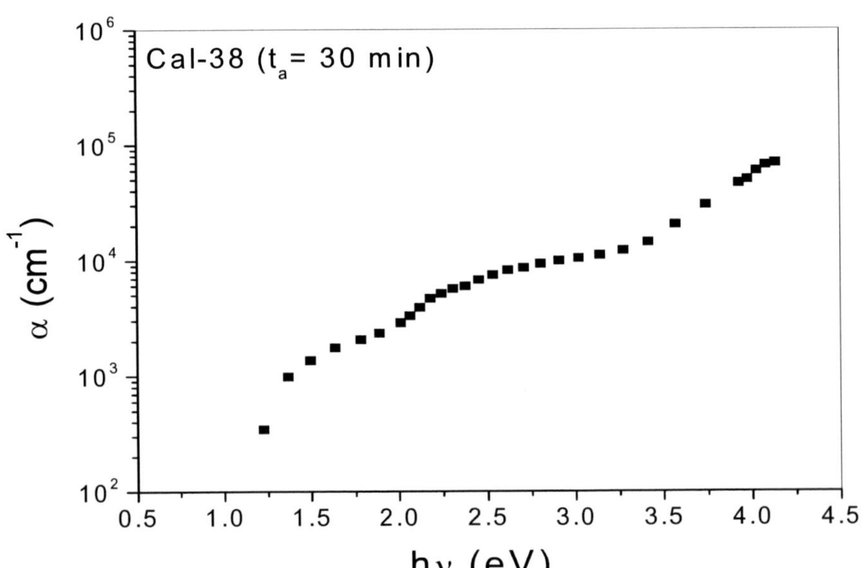

Figure 17(a). Energy dependence of absorption coefficient of $CuAlO_2$ thin film, post-annealed for 30 min.

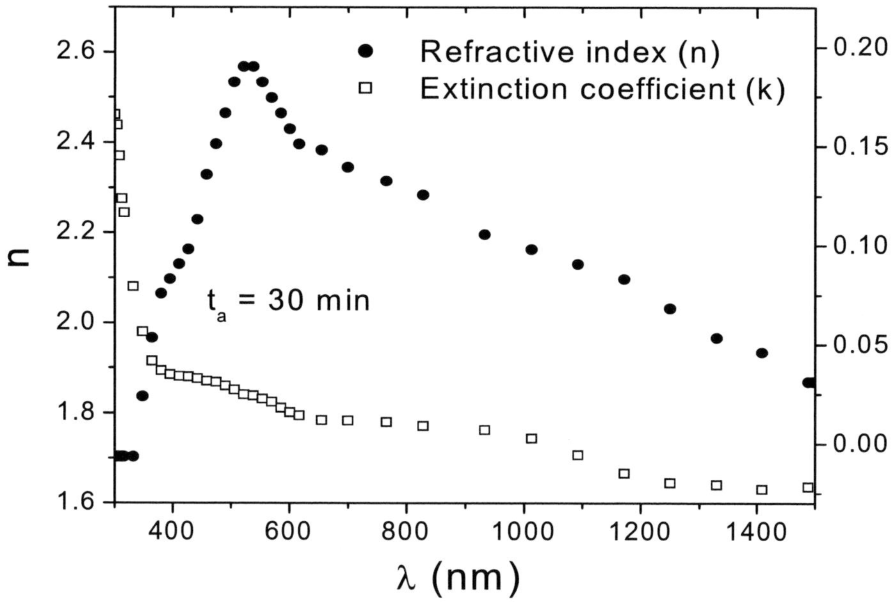

Figure 17(b). Spectral variation of refractive indices (n) and extinction coefficients (k) of $CuAlO_2$ thin film post-annealed for 30 min.

Table 10. Comparison of different optical parameters of three types of $CuAlO_2$ thin films with different post-annealing times (t_a).

Film No.	t_a (min)	Average visible transmittance (%)	Range of refractive indices (n)	Range of absorption coefficient (α) (cm^{-1})	Bandgap (E_g)	
					Direct (eV)	Indirect (eV)
CAL-38	30	65	1.7 – 2.6	3.4×10^2 – 7.0×10^4	3.81	2.8
CAL-33	60	80	1.2 – 2.1	2.3×10^2 – 5.9×10^4	3.7	2.1
CAL-45	90	75	1.3 – 2.2	9.9×10^1 – 9.4×10^4	3.8	2.32

In the range of the onset of absorption edge, the absorption coefficients (α) can be described by the relation for parabolic bands, i.e. [237, 238].

$$(\alpha h\nu)^{1/n} = A(h\nu - E_g) \qquad (17)$$

where E_g is the band gap of the material, the exponent n depends on the type of transition. For direct allowed transition, n=1/2, for indirect allowed transition, n=2, and for direct forbidden transition, n=3/2. The factor A also depends on the type of transition. For direct allowed transition, $A \approx \frac{e^2 \left(2 \frac{m_h^* m_e^*}{m_h^* + m_e^*}\right)^{3/2}}{nch^2 m_e^*}$; for direct forbidden transition, $A = \frac{4}{3} \frac{e^2 \left(\frac{m_h^* m_e^*}{m_h^* + m_e^*}\right)^{5/2}}{nch^2 m_e^* m_h^* h\nu}$; and for indirect allowed transition, $A \propto \frac{(m_h^* m_e^*)^{3/2}}{\pi^4 h^6}$; (n = refractive index of the material, m_h^* & m_e^* are the effective masses of holes and electrons respectively) [242]. To determine the possible transitions, $(\alpha h\nu)^{1/n}$ vs. hν were plotted for different values of n. The $(\alpha h\nu)^2$ vs. hν and $(\alpha h\nu)^{1/n}$ vs. hν plots for three types of films post-annealed for 90 min, 60 min and 30 min are shown in **Fig. 18(a), (b) and (c)** respectively. Extrapolating the linear portion of the graphs to the hν axis we have obtained the direct and indirect band gaps as ~3.8 eV and 2.32 eV for the sample post-annealed for 90 min respectively, ~3.7 eV and 2.1 eV for the sample with t_a = 60 min and ~3.81 eV and 2.8 eV for the sample post-annealed for 30 min respectively (shown in **Table 10**). These values are comparable to those reported previously by Kawazoe et al (3.5 eV)[64] and Yanagi et al (3.5 eV & 1.8 eV) [67] for their pulsed laser deposited $CuAlO_2$ thin films and also fall in the range theoretically calculated by Robertson et al (3.91 eV & 2.1 eV) [106]. Also Stauber et al [145] obtained the direct

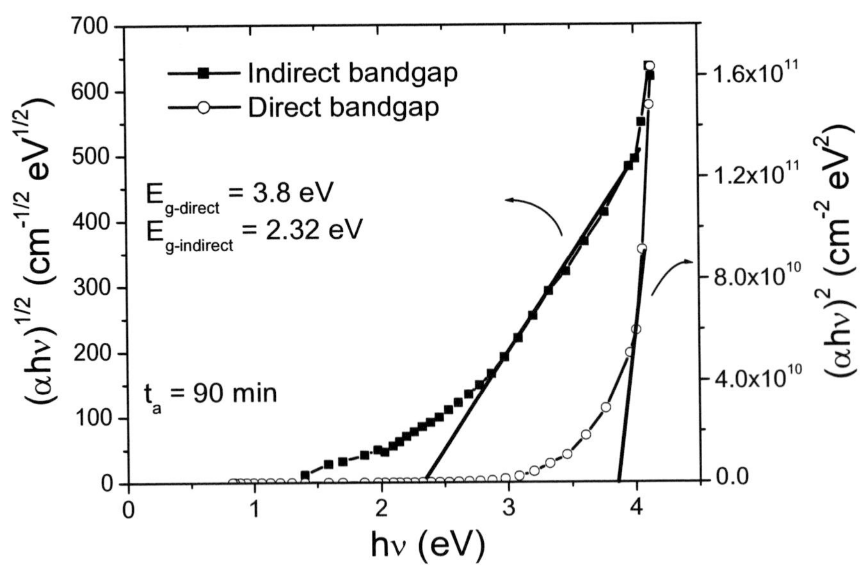

Figure 18(a). Determination of bandgaps of $CuAlO_2$ thin film post-annealed for 90 min.

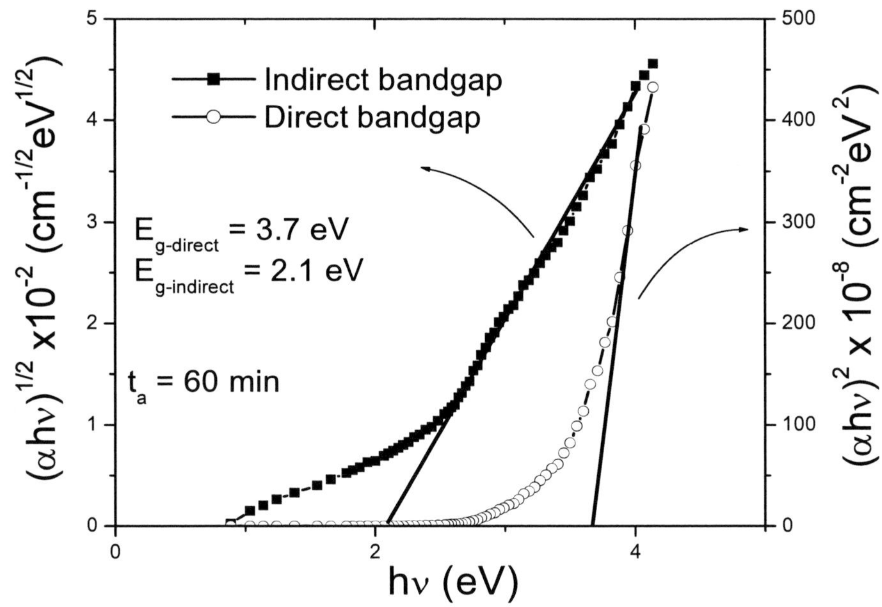

Figure 18(b). Determination of bandgaps of $CuAlO_2$ thin film post-annealed for 60 min.

bandgap of their R. F. sputter deposited $CuAlO_2$ thin film around 3.5 eV whereas Wang and Gong [149] reported the direct bandgap of their plasma enhanced chemical vapor deposited (PECVD) copper aluminum oxide films around 3.6 to 3.75 eV.

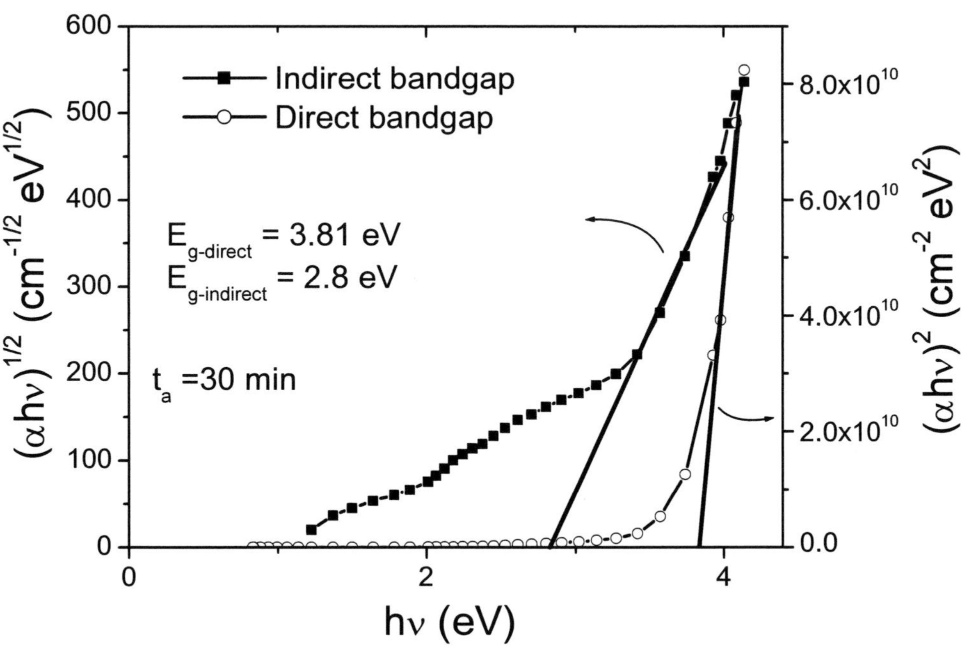

Figure 18(c). Determination of bandgaps of $CuAlO_2$ thin film post-annealed for 30 min.

(v) Electrical properties and Hall studies

Electrical properties of $CuAlO_2$ thin films have been studied by standard four-probe methods. All electrical contacts were made by silver paint, which showed linear I-V characteristics over a wide range of voltages and temperatures. **Fig. 19** shows I-V characteristics of one contact at room temperature indicating ohmic nature of it over the voltage range upto 150 V.

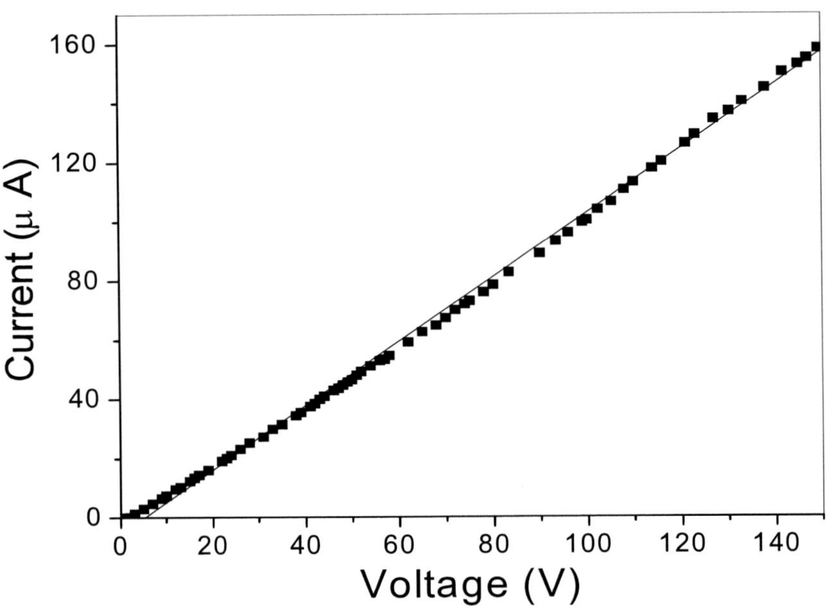

Figure 19. Verification of ohmic nature of the contacts.

The thermally activated conduction of a semiconductor can be given by the relation [142]

$$\sigma = \sigma_o \exp[-\frac{E_a}{kT}] \tag{18}$$

where σ_0 is a temperature independent factor, E_a is the activation energy of the material. For p-type semiconductor (as in our p-$CuAlO_2$ sample), this is the energy difference between the acceptor level and the top of the valence. Therefore, a plot of lnσ vs. 1/T should be a straight-line whose slope would carry the information of the activation energy of the material. We have determined the temperature dependence of conductivity of D. C. sputtered $CuAlO_2$ thin films for several sets of samples having different post-deposition oxygen-annealing times (t_a) ranging from 30 min 150 min respectively and observed any variation in the electrical characteristics of these films and then tried to explain it. **Fig. 20** represents the temperature variation (from 300 K to 575 K) of the conductivity (σ) of the films for t_a = 30, 60, 90, 120 and 150 min. The thickness of the films was around 500 nm estimated from cross-sectional SEM. An increase in the room temperature conductivity (σ_{RT}) was observed with the increase in annealing times (t_a) upto 90 min. (For example, films with t_a = 90, 60 and 30 min, σ_{RT} =

0.39, 0.16 and 0.09 S cm^{-1} respectively). The conductivities range from a minimum of 0.014 S cm^{-1} (for t_a = 150 min) to a maximum of 5.0 S cm^{-1} (for t_a = 90 min) in the above temperature range. Previously, Kawazoe et al. [64] and Yanagi et al. [67] obtained the room-temperature conductivities for their pulsed laser deposited $CuAlO_2$ thin films on sapphire substrates as 0.095 S cm^{-1} and 0.34 S cm^{-1} respectively. These values are quite comparable to our films post-annealed for 30 min and 90 min respectively.

Defect chemistry plays an important role for the increase in p-type conductivity of this material. Metal deficit (or excess oxygen) within the crystallite sites of the material enhances the p-type conductivity. This deviation from the stoichiometric composition of the components can be induced by regulating the post-deposition annealing time (t_a) in oxygen atmosphere. A detailed discussion on nonsoichiometry and defect chemistry is given in **Section 3**. Now re-writing **Eq. 1** for $CuAlO_2$, we get [222, 223]:

$$O_2(g) = 2O_O^X + V_{Cu}^- + V_{Al}^{-3} + 4h^+ \tag{19}$$

where O_O, V_{Cu}, V_{Al} and h denote lattice oxygen, Cu vacancy, Al vacancy and hole respectively. Superscripts ^{X}, $^-$ and $^+$ denote effective neutral, negative and positive charge states respectively.

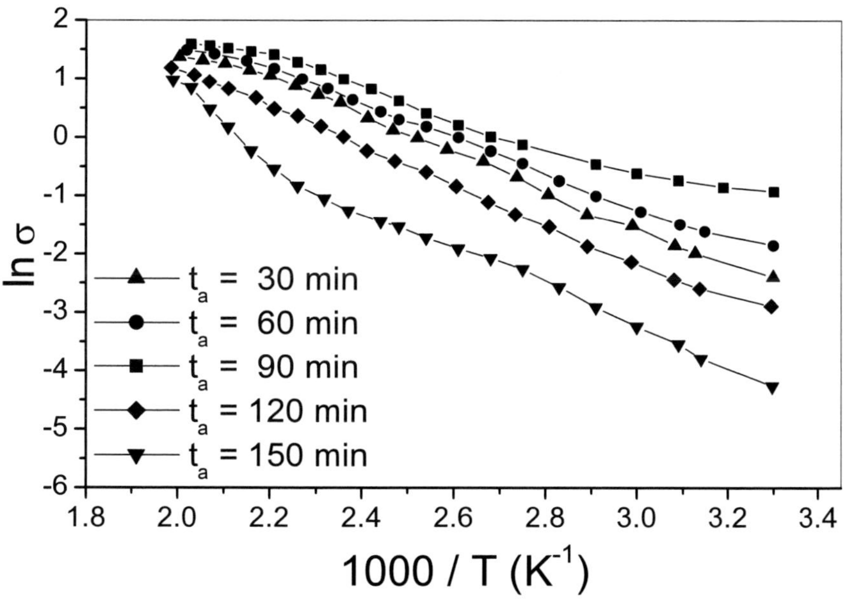

Figure 20. Temperature variation of conductivity of $CuAlO_2$ thin films for five sets of samples post-annealed for 30 min, 60 min, 90 min, 120 min and 150 min.

Composition analyses of the films [as given in the **Section 5.1.(ii)**] showed that for the unannealed films, the composition is almost stoichiometric. But Hot-probe measurement confirmed the p-type nature of these films. Therefore, it can be argued that some amount of excess oxygen may be present in the unannealed films but the amount is so low that it could not be measured within our experimental limit. This argument seems reasonable if we compare our result with the previously reported values where it has been stated that intercalation of oxygen into the $CuAlO_2$ thin film was not easy [109] and Thomas [65] suggested that the chemical formula of this material would be $CuAlO_{2+x}$ with as low as 0.001 at % of excess oxygen (i.e. x = 1/50,000) over stoichiometric value within the film prepared by Kawazoe et al [64]. Later Yanagi et al. [67] performed post-deposition oxygen annealing of the films prepared by the same method as that of Kawazoe and co-authors [64] and observed a significant increase in the carrier conduction within the films. Although they have not reported the composition of the films, but this enhanced p-type conductivity is most probably due to the presence of excess (nonstoichiometric) oxygen within the film, induced due to post-annealing. Similarly, Wang and Gong [149] observed a significant increase in the conductivity of their copper aluminum oxide films after annealing in air. This may be another experimental proof of the suspected p-type conduction caused by excess oxygen. Following this argument we have performed the post-deposition oxygen annealing of our films to induce excess oxygen within the films for getting enhanced p-type conductivity. We have observed that for the films post-annealed for 30, 60 and 90 min, the percentages of excess oxygen were around 0.5 at %, 2.5 at % and 5 at % respectively over stoichiometric value (c.f. **Table 9**). The Cu : Al stoichiometry remained close to 1 : 1 for all the samples. On the other hand, as shown in **Fig. 20**, an increase in the conductivity with t_a has been observed upto 90 min. Although very little, but still, slight increase in the oxygen content within the films leads to an increase in the conductivity of the films, supporting the above reasoning. But we have seen a decrease in the conductivity when the annealing times were 120 min and above (for t_a = 120 min, σ_{RT} = 0.055 S cm^{-1} and for t_a = 150 min, σ_{RT} = 0.014 S cm^{-1}, as shown in **Fig. 20**). Compositional analyses of the films post-annealed for 120 min and above show percentage of excess oxygen within the films more than 10 at % over stoichiometric value (cf. **Table 9**). This suggests that although the presence of excess oxygen within the films (with t_a = 120 min and above) are evidenced but they are not acting favorably to increase the hole conductivity within the films. These excess oxygen atoms, most probably, lay in the grain boundary regions as trap states, which put hindrance in the carrier conduction and, hence, a decrease in the conductivity of these films is observed. On the other hand, for the films post-annealed for 30, 60 and 90 min, show an increase in the conductivity along-with an increase in the excess oxygen content within the films as mentioned earlier. Therefore, in these cases, the excess oxygen atoms may be acting favorably to generate holes within the films. But it must be admitted that the maximum conductivity obtained for our films was not as much as it would have been. So, we suppose that in all cases, whether it is for the films with t_a = 90 min or less (when an increase in σ with t_a was observed) or those with t_a = 120 min or above (when a decrease in σ with t_a was observed), adsorbed oxygen atoms as 'trapped states' in the grain-boundary regions are always present. In the previous section [**Section 5.1(i)**], we have estimated the particle size of our films as 26 nm from XRD data. As the particle size is in nanometer order, a large number of grain-boundaries are present in the films, so also considerable amount of trapped states in these grain-boundary regions are present, which put hindrance in the carrier conduction. But for the films with t_a = 90 min or less, greater proportion of excess oxygen may be acting favorably towards the hole generation and, hence, dominate the grain-boundary scattering. This may be the reason for the increase in the conductivity with t_a in this region. On the other hand, for the films with t_a = 120 min and above, greater proportion of the excess oxygen atoms may be adsorbed in the grain-boundary regions, which probably be correlated with the larger time of exposure of these films in the oxygen atmosphere [243]. Hence, grain-boundary scattering, masks the increase in the conductivity and, therefore, we observe a decrease in σ with t_a in this region. But the exact mechanism is still not fully understood.

The activation energies (E_a), which correspond to the minimum energy required to transfer carriers from acceptor level to the valence band (for p-type materials), have been obtained from the slope of the graphs (**Fig. 20**). The values are 196, 245, 270, 284 and 325 meV for t_a = 90, 60, 30, 120 and 150 min respectively. As expected, the sample with highest conductivity has least E_a value and vice-versa. These activation energy values are comparable to the previously obtained values by Kawazoe et al. (200 meV) [64] and Yanagi et al. (220 meV)[67]).

Hall effect measurements were done for three types of samples with t_a = 30, 60 and 90 min. All the Hall coefficients were positive, which confirms the p-type nature of the samples. The Hall coefficient (R_H) for the sample with t_a = 90 min (σ_{RT} = 0.39 S cm^{-1}), has been obtained as + 4.6 cm^3 C^{-1} corresponding to a hole concentration (n_p) of 1.2 x 10^{18} cm^{-3}. For the sample with t_a = 60 min (σ_{RT} = 0.16 S cm^{-1}), these values are +13.9 cm^3 C^{-1} and 4.5 x 10^{17} cm^{-3} respectively. And for the sample with t_a = 30 min (σ_{RT} = 0.09 S cm^{-1}), R_H and n_p are +22.5 cm^3 C^{-1} and 2.8 x 10^{17} cm^{-3} respectively. For unannealed films as well as for the films post-annealed for 120 min and above, the Hall measurements could not be performed, but the p-type nature of the films was confirmed by Hot-probe method. Maximum carrier concentration obtained by us is one order of magnitude higher than that reported by Kawazoe et al. [64], but still one order less than that of Yanagi et al. [67]. Details of the different electrical parameters of $CuAlO_2$ thin films are furnished in **Table 11** and the variation of these parameters with respect to t_a is shown in **Fig. 21 (a) and (b)**.

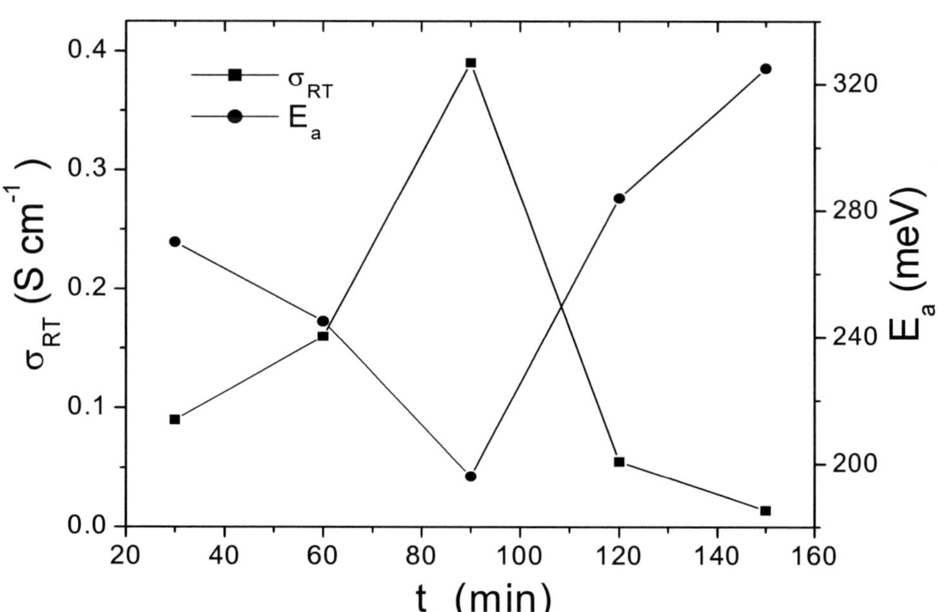

Figure 21(a). Variations of room-temperature conductivity and activation energy of D. C. sputtered $CuAlO_2$ thin films with post-annealing times.

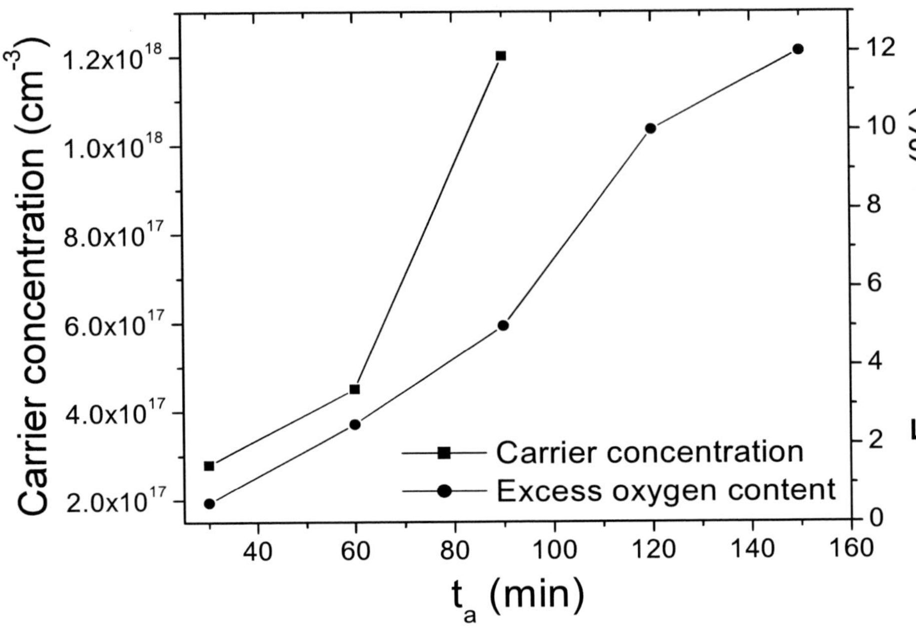

Figure 21(b). Variations of room-temp carrier concentration and excess oxygen content.

Table 11. Different Electrical parameters of $CuAlO_2$ thin films, deposited at different annealing times.

t_a (min)	σ_{RT} (S cm^{-1})	R_H (cm^3 C^{-1})	n_p (cm^{-3})	E_a (meV)	Chemical formula
90	0.39	+4.60	1.2×10^{18}	196	$CuAlO_{2.10}$
60	0.16	+13.9	4.5×10^{17}	245	$CuAlO_{2.05}$
30	0.09	+22.5	2.8×10^{17}	270	$CuAlO_{2.01}$
120	0.055	---	---	284	$CuAlO_{2.20}$
150	0.014	---	---	325	$CuAlO_{2.24}$

(vi) Thermoelectric properties

Thermoelectric properties of D. C. sputtered $CuAlO_2$ thin films have been studied for three types of films having different post-deposition annealing times e.g. 30, 60 and 90 min. The measurements were done from room temperature (300 K) to 550 K. **Fig. 22** shows the temperature dependence of Seebeck coefficients (S) for three types of films. All the Seebeck coefficients are positive in nature, which again confirmed p-type nature of the films. Room temperature Seebeck coefficients (S_{RT}) of the films were obtained as + 230 μV K^{-1} (for $t_a = 90$ min), +141 μV K^{-1} (for $t_a = 60$ min) and +120 μV K^{-1} (for $t_a = 30$ min). As shown in the figure, the Seebeck coefficients initially decrease from room-temperature to around 390 K and then increase to almost +400 μV K^{-1}, for further increase in temperature [244]. Previously, Kawazoe et al. [64] and Yanagi et al. [67] obtained the room temperature Seebeck coefficients for their pulsed laser deposited $CuAlO_2$ thin film as +183 μV K^{-1} and +214 μV K^{-1} respectively, very much comparable to our values. On the other hand, Koumoto et al. [223] determined the Seebeck coefficient of $CuAlO_2$ single crystal as well as polycrystal at 600 K around 180 μV K^{-1} and 150 μV K^{-1} respectively. Also Benko and Koffyberg [97] reported a relatively high value of S_{RT} (670 μV K^{-1}) of $CuAlO_2$ powdered pellets. It has been observed that, in our D. C. sputter-deposited $CuAlO_2$ thin film, S_{RT} increases with the increase in conductivity of the films. This observation is consistent with the Hicks model [245, 246], where the natural superlattice structure was proposed to show high thermoelectric figure of merit (ZT) due to increase in both S and σ according to the following equation:

$$ZT = \frac{S^2 \sigma T}{\kappa} \tag{20}$$

where 'σ' is the electrical conductivity, 'κ' is thermal conductivity and 'S' is the Seebeck coefficient.

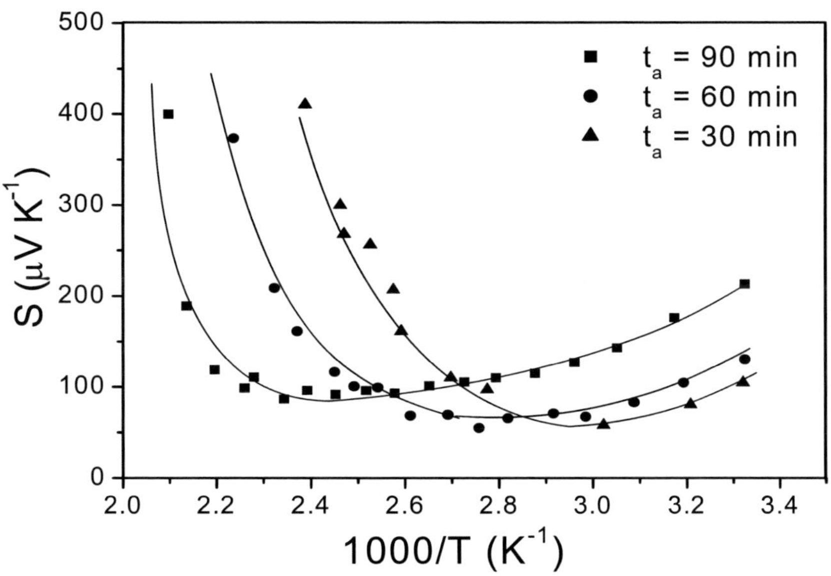

Figure 22. Seebeck coefficient vs.1000 / T of $CuAlO_2$ thin films.

To achieve high ZT, increase in S and (or) σ and decrease in κ are required. But for simple materials, increase in S leads to a decrease in σ. Similarly, an increase in σ is followed by an increase in κ according to Wiedemann-Franz law. So ZT effectively remains more or less constant. To increase Z, various models have been proposed in the last decade. Amongst them, the most exciting proposal by Hick et al. [245, 246] was superlattice quantum-well materials, having an effective two-dimensional density of states for carriers. This density of state is given by $\frac{m}{\pi \hbar^2 a}$, where 'm' is the carrier mass and 'a' is the quantum-well width. These authors assumed infinite potential barrier with zero barrier width and showed a considerable increase in Z. Later, Lin-Chung et al. [247] and Broido et al. [248] included the effects of thermal transport in the finite barrier layers and carrier tunneling between layers in the above model to get a modified Z. Encouraged by these findings, various new materials, having layered structure, have been investigated in the last few years, which include $NaCo_2O_4$ [249], $(ZnO)_5In_2O_3$ [250] and $CuAlO_2$ single crystal [223] etc.

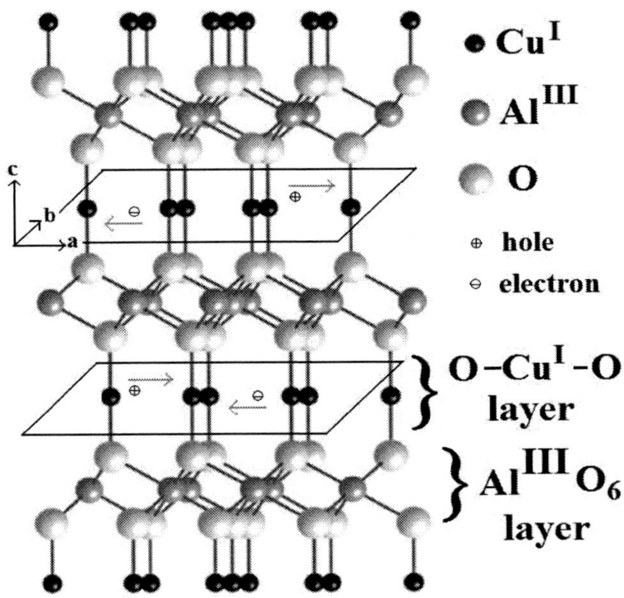

Figure 23. Layered-structure of $CuAlO_2$ showing the carriers confined in the ab-plane.

Structure of $CuAlO_2$ delafossite has been shown in **Fig. 23** and described in details in **Section 3**. The crystal structure is an alternative stacking of Cu^I and layers of nominal AlO_2 composition consisting of $Al\text{-}O_6$ octahedra sharing edges. Each Cu atom is linearly coordinated with two oxygen atoms to form a O-Cu-O dumbbell unit placed parallel to the c-axis. O-atoms of O-Cu-O dumbbell link all Cu layers with the AlO_2 layers [99]. This structure suggests that $CuAlO_2$ has a layered structure where carriers can easily move two-dimensionally along ab-plane than to move across the Al-O insulating layers. In the XRD pattern of our $CuAlO_2$ thin film (shown in **Fig. 8**, section 5.1), we have obtained a strong (006) peak, which is typical of a texture where the c-axis is perpendicular to the substrate (hence parallel to the normal, 'n' to the substrate, i.e. c || n). Now, according to our experimental set up (cf. section 4.4), carriers in the films are expected to move along the ab-plane. Hence the above argument of two-dimensional confinement of carriers along the ab-plane is valid for our films. Although, the reason behind the enhanced thermoelectric properties shown by the materials possessing layered structure, is still not fully understood, but Koumoto et al. [223] suggested that this may be correlated with the low dimensionality of the crystal structure and behavior of electrons and phonons in an anisotropic structural environment. Recently, Wang et al. [251] suggested that spin entropy might be responsible for enhanced thermopower in $Na_xCo_2O_4$ having layered structure [249]. Whether this can be correlated with the good thermoelectric properties of $CuAlO_2$, is a question and intense research is needed in this direction.

The variation of thermo-electric power (S) with temperature is given by [252]:

$$S = \frac{k}{e}\left(A + \frac{\Delta E_f}{kT}\right) \tag{20}$$

with

$$A = \frac{5}{2} - s \tag{21}$$

and

$$\tau = \tau_o e^{-s} \tag{22}$$

where 'k' is the Boltzmann constant, 'e' is the electronic charge, 'ΔE_f' is the energy difference between Fermi level and the upper edge of the valence band, 'τ' is the relaxation time for electron scattering, 's' is a constant, which is different for different scattering mechanism and 'τ_o' is a constant, which is a function of temperature but independent of the electronic charge, e.

From Eq. 20, we can obtain the Fermi level (E_f), from the slope of the S vs. 1/T graph. From the **Fig. 22**, we determined the Fermi energies for three types of samples from the linear portion of the graphs near room temperature, and the values are 130, 151 and 200 meV for t_a = 90, 60 and 30 min respectively. Previously, Benko and Koffyberg [97] have determined the Fermi energy of $CuAlO_2$ powder (σ = 1.69 x 10^{-3} S cm^{-1}) from the thermopower measurement, as 190 meV, which is comparable to our sample having lowest room temperature conductivity (σ_{RT} = 0.09 S cm^{-1}, t_a = 30 min). As previously mentioned, from the slope of the ln σ vs. 1000/T plots (**Fig. 20**), we have obtained the activation energy (E_a) values, which give the estimation of acceptor levels. Comparing these values with the values of Fermi levels, we can say that according to the band picture, Fermi level lies between the upper edge of the valence band and the acceptor level, which is typical of a non-degenerate p-type semiconducting material with acceptors not fully ionized. Hence a continuous increase in conductivity with temperature was observed for all three types of samples. Also it has been observed that the sample with maximum conductivity has its Fermi level nearest to the valence band, which is obvious for a p-type material. The values of various thermoelectric parameters of the D. C. sputtered $CuAlO_2$ thin films are furnished in **Table 12** and a comparison between the activation energy (E_a) and Fermi energy (E_f) is also given in **Table 12**. **Fig. 24** represents the temperature dependence of thermoelectric power factor (σS^2) of $CuAlO_2$ thin film for the temperature range of 300 K to 500 K. The values range from 1.1 x 10^{-7} $Wm^{-1}K^{-2}$ at a temperature around 300 K (for t_a = 30 min) to 7.5 x 10^{-5} $Wm^{-1}K^{-2}$ around 500 K (for t_a = 90 min). Koumoto et al. [223] obtained these values roughly as 1.12 x 10^{-5} $Wm^{-1}K^{-2}$ at 550 K for $CuAlO_2$ single crystal and 7.1 x10^{-6} $Wm^{-1}K^{-2}$ at 700 K for $CuAlO_2$ polycrystal. Also Park et al [253] obtained the power factor for $CuAlO_2$ ceramic as 2 x 10^{-5}

$Wm^{-1}K^{-2}$ at 550 K. These values are comparable to the values reported by us. Also, recently Kurotori and co-authors [254] reported significant increase in the thermoelectric properties of $CuAlO_2$, when doped with Zn and Ca. All these reports suggest that this class of material may become very good candidate of thermoelectric converter, and may bring renaissance in the thermopower industry.

Table 12. Different thermoelectric properties of D. C. sputtered $CuAlO_2$ thin films with different post-deposition oxygen annealing times (t_a).

t_a (min)	S_{RT} ($\mu V K^1$)	E_a (meV)	E_f (meV)	σS^2 ($W m^{-1} K^{-2}$)
90	+230	196	130	2.16×10^{-6}
60	+141	245	151	2.43×10^{-7}
30	+120	270	200	1.10×10^{-7}

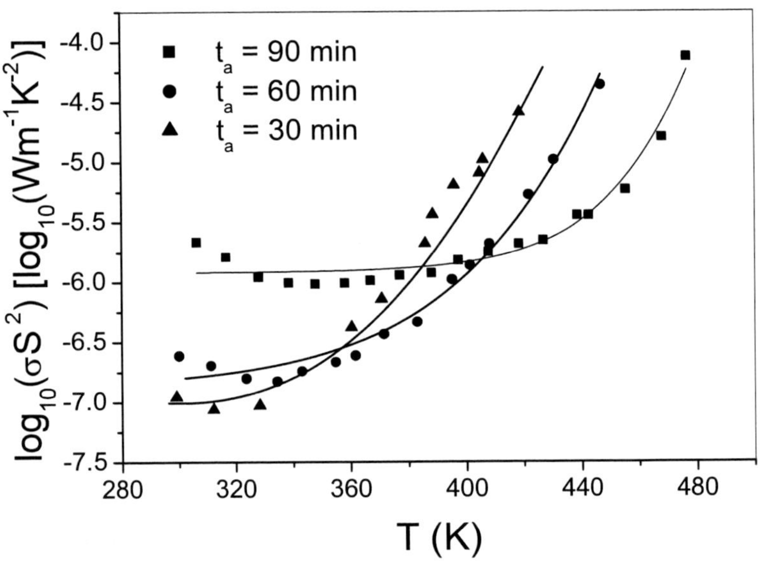

Figure 24. Thermoelectric power factor vs. temperature of $CuAlO_2$ thin films.

5.2. Properties of Reactive Sputtered Films

i) X-ray diffraction studies

We have also prepared the $CuAlO_2$ thin films by reactive D. C. sputtering technique. Details of the experimental conditions are furnished in **Section-4.2.** **Fig. 25(a)** shows the XRD spectrum of the reactive D. C. sputtered $CuAlO_2$ thin film deposited on Si (400) substrate. It shows a strong (006) orientation. Similar orientations was observed by previous workers [64, 67] for their pulsed laser deposited film as well as by ours DC sputter deposited $CuAlO_2$ thin films [160]. Alongwith the above peak, other peaks were also observed in the XRD spectrum, which could be assigned for (003), (101), (012), (104) and (018) reflections of crystalline $CuAlO_2$. Also no peaks corresponding to starting materials e.g. Cu and Al metal powders as well as their oxides, were found in the pattern. This conclusively inferred that the reactants were completely mixed to give the proper phase of the copper aluminium oxide and no residual metal oxides remained in the film. It is to be noted that previously Ong and Gong [146] deposited copper aluminum oxide thin films by R. F. magnetron reactive co-sputtering of Cu and Al metal targets, whereas Tsuboi and co-authors [147] used D. C. reactive sputtering of facing targets of Cu and Al metals and a rotating substrate. But they obtained

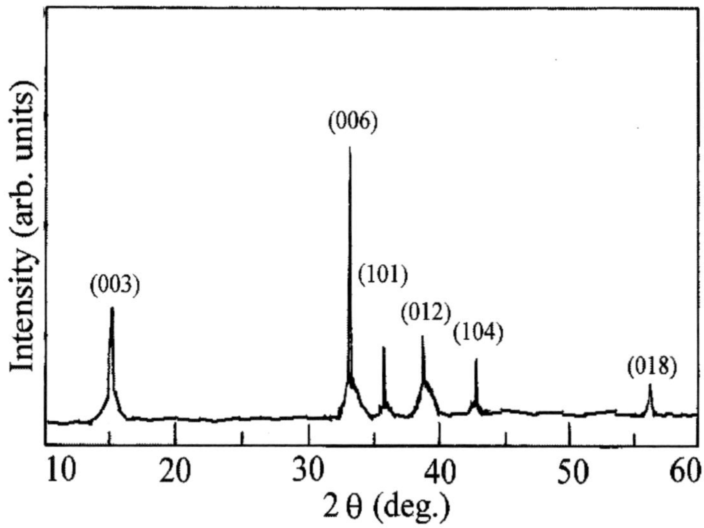

Figure 25(a). XRD pattern of the reactive D. C. sputtered $CuAlO_2$ thin film.

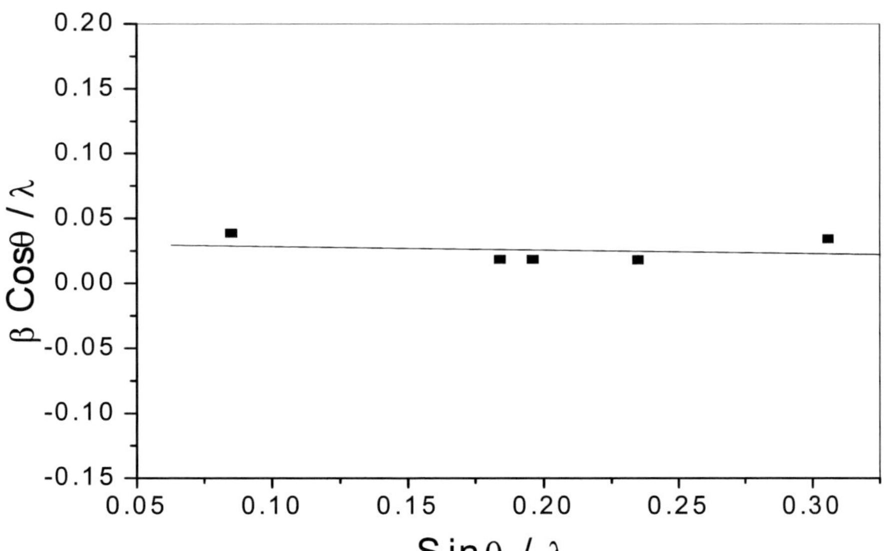

Figure 25(b). Plot to determine strain and particle size of $CuAlO_2$ thin film deposited by reactive D. C. sputtering.

phase impure films. Also the crystallinity of the films was poor. XRD pattern of our reactive D. C. sputtered film shows better crystallinity, resulting in good electrical properties of the films as described in later chapters. **Fig. 25(b)** gives the plot of $\frac{\beta \, Cos\theta}{\lambda}$ vs. $\frac{Sin\theta}{\lambda}$. From the slope and intercept, the strain and particle size were determined according to **Eq. 6**. These values were found to be ~ 32 nm and 2.7 x 10^{-2} respectively.

Table 13 shows the comparison between the theoretical d-values obtained from JCPDS file (d_{JCPDS}) and observed d-values obtained from XRD data of reactive sputtered $CuAlO_2$ thin film ($d_{reactive}$) (**Fig. 25-b**) and also compared with D. C. sputtered $CuAlO_2$ thin film ($d_{DC\text{-}Sputter}$) (**Fig. 8**). **Table 14** compares the effective particle size and effective strain of reactive sputtered and D. C. sputtered (**Fig. 9**) $CuAlO_2$ thin films. All the data furnished in these two tables for D. C. sputtered films have the post-annealing time 60 min. Since the XRD patterns of the other post-annealed (e.g. 30 min, 90 min) D. C. sputtered films have identical peaks as mentioned earlier, all the related parameters will be identical and, therefore, not furnished here.

Table 13. Comparison between the theoretical d-values, observed d-values of $CuAlO_2$ powder, D. C. sputtered and reactive D. C. sputtered $CuAlO_2$ thin films.

h k l	d_{JCPDS} (Å)	$d_{reactive}$ (Å)	$d_{DC\text{-}Sputter}$ (Å)
0 0 3	5.61	5.67	5.70
0 0 6	2.82	2.79	2.80
1 0 1	2.44	2.48	---
0 1 2	2.376	2.374	---
1 0 4	2.133	2.111	---
1 0 7	1.732	---	---
0 1 8	1.612	1.618	1.620

Table 14. Comparison between the effective particle size and effective strain of D. C. sputtered and reactive D. C. sputtered $CuAlO_2$ thin films.

Deposition technique	Effective particle size (nm)	Effective strain
Reactive sputtered thin film	32	2.70 x 10^{-2}
D. C. sputtered thin film	26	8.52 x 10^{-3}

(ii) Compositional analyses

In reactive D. C. sputtering method, films were post-annealed for 60 min and the composition of the film was in the ratio of Cu : Al : O = 1 : 1 : 2.08. Therefore, the chemical formula of the films may be written as $CuAlO_{2.08}$. This means that the percentage of excess oxygen in the reactive D. C. sputtered films is around 4 at %. It has been observed that the percentage of excess oxygen within the reactive sputtered films, is close to that of D. C. sputtered films with t_a = 90 min (cf. **Table 9**). This is probably because of the presence of excess oxygen atmosphere during the reactive sputtering process and higher substrate temperature.

(iii) FT-IR studies

FT-IR spectrum of reactive sputtered films are shown in **Fig. (26)**. As expected, most of the peaks are similar to that obtained in the films deposited by D. C. sputtering (shown in **Fig. 10**). The broad peak around 400 cm^{-1} to 600 cm^{-1} consists of a number of peaks which have been assigned to the absorption peaks of Cu-O, O-Cu-O, Al-O bond vibrations as mentioned in the **Section 5.1 (iii)**. The broad peak around 800 cm^{-1} to 1100 cm^{-1} is actually consisting of two peaks. Deconvoluting it, the two peaks are obtained around 900 cm^{-1} and 1030 cm^{-1}. First one may be assigned to the short Al-O stretching vibrations in distorted AlO_6 octahedra, whereas second one may be assigned to Si-O-Al vibration, which occurs due to Si substrate used. Small hump around 2500 cm^{-1} is a CO_2 peak and the broad peak around 3100 cm^{-1}-3500 cm^{-1} is due to O-H stretching vibration, which may be incorporated from the atmospheric contamination. Similar to the **Fig. 10**, here also, we have got an unidentified peak around 1600 cm^{-1}. Also another peak around 2800 cm^{-1} remained unidentified in the spectrum. As there is no detailed literature on the FT-IR studies of this material, this may become an important field of work in the recent future.

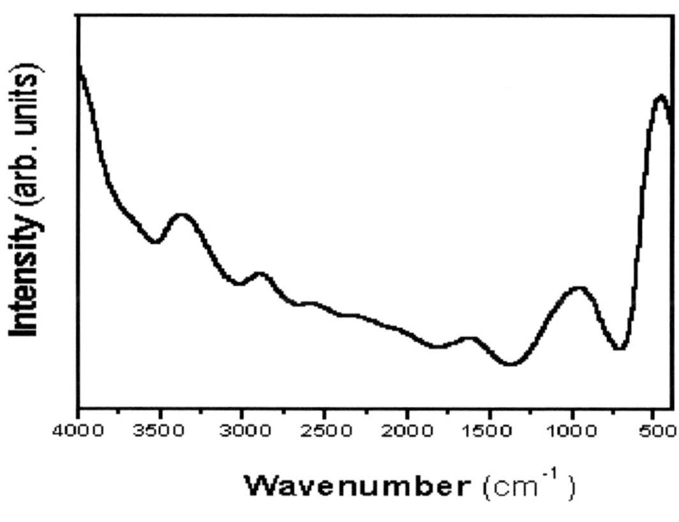

Figure 26. FT-IR spectra of reactive sputtered $CuAlO_2$ thin films.

(iv) UV-Vis-NIR measurements

The optical properties of the reactive D. C. sputtered thin films have also been studied. The films were deposited on glass substrates and the film thicknesses were measured around 500 nm from cross-sectional SEM. **Fig. 27** shows the UV-Vis-NIR spectra of reactive D. C. sputtered $CuAlO_2$ thin film in the wavelength range of 300 nm to 1500 nm. These films were post-annealed for 60 min. The average visible transmittance of the film is found to be ~ 85 - 90 %. We have critically analyzed the variations of transmittance (T) and reflectance (R) in terms of absorption coefficients (α) to derive information on the optical transitions occurring in these films. Now in the fundamental absorption region, the value of α is calculated according to the **Eq. 16**. Also the extinction coefficients (k) and refractive indices (n) are calculated from **Eq. 14** and **Eq. 15** respectively. The spectral variations of α, n and k are shown in the inset of **Fig. 27** and **Fig. 28** respectively. For the determination of the bandgaps

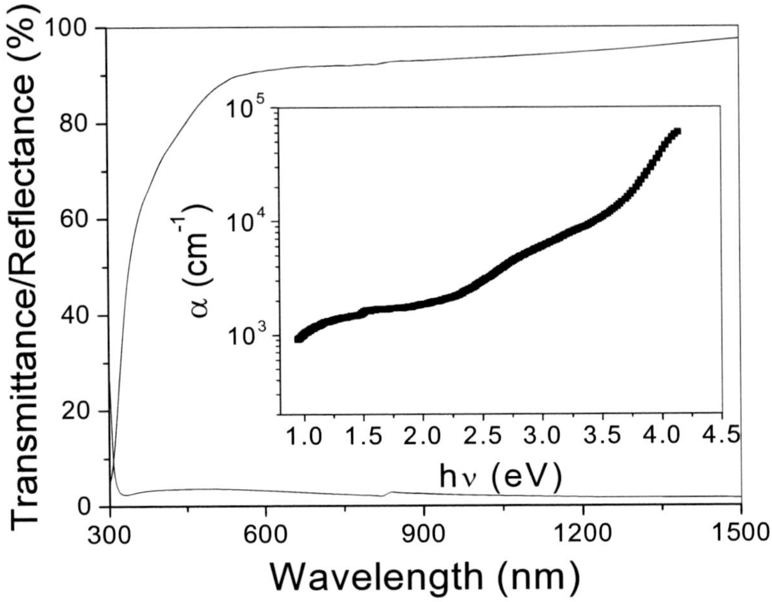

Figure 27. UV-Vis-NIR spectra of reactive D. C. sputtered $CuAlO_2$ thin film. **Inset**: Energy dependence of absorption coefficients.

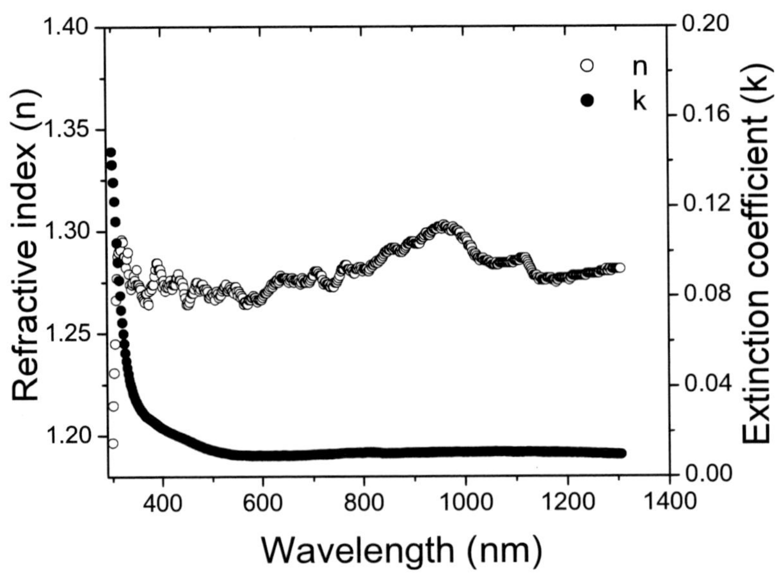

Figure 28. Spectral variation of extinction coefficients and refractive indices of reactive D. C. sputtered $CuAlO_2$ film.

of reactive sputtered $CuAlO_2$ thin film, **Eq. 17** has been used. **Fig. 29** shows the evaluation of direct and indirect bandgap values obtained from extrapolating the linear portion of the graphs to the hν axis. The direct and indirect bandgap values, we have obtained, as 3.90 eV and 1.89 eV respectively, which are comparable to the previously reported values [64, 67] as well as those of D. C. sputtered films obtained by us (cf. **Table 10**), but slightly greater than the previously reported reactive co-sputtered Cu-Al-O thin films prepared by Ong and Gong (2.9-3.3 eV) [146]. This is mainly because of their phase impure films, which contain some amount of CuO within the copper aluminum oxide samples. A comparison between the direct and indirect bandgap values of reactive sputtered thin films with D. C. sputtered films with different post-annealing times are furnished in **Table 15**.

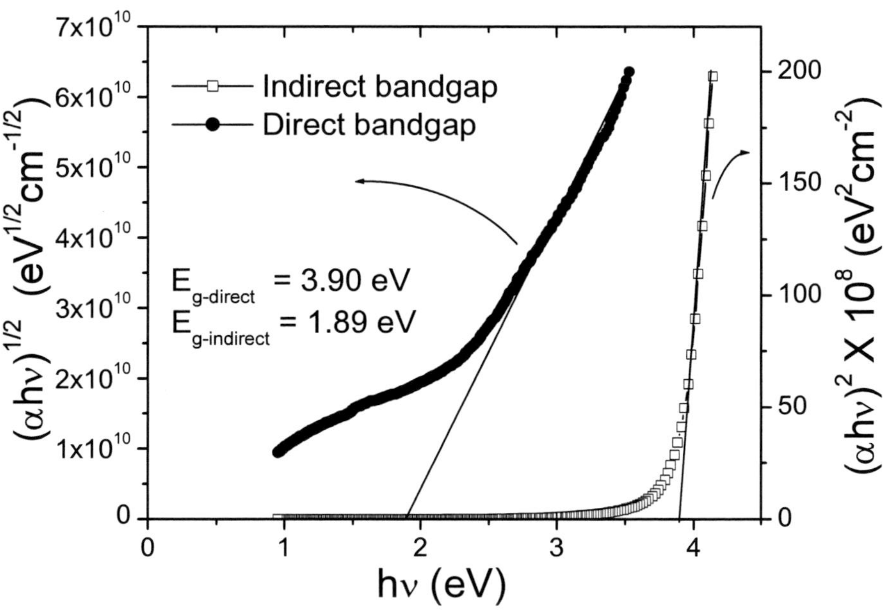

Figure 29. Determination of bandgaps of reactive sputtered $CuAlO_2$ film.

Table 15. Comparison between the bandgap values of $CuAlO_2$ thin films deposited by D. C. and reactive sputtering.

D. C. sputtered films						Reactive D. C. sputtered films	
T_a = 30 min		T_a = 60 min		T_a = 90 min			
$E_{g\text{-direct}}$ (eV)	$E_{g\text{-indirect}}$ (eV)	$E_{g\text{-direct}}$ (eV)	$E_{g\text{-indirect}}$ (eV)	$E_{g\text{-direct}}$ (eV)	$E_{g\text{-indirect}}$ (eV)	$E_{g\text{-direct}}$ (eV)	$E_{g\text{-indirect}}$ (eV)
3.81	2.8	3.7	2.1	3.8	2.32	3.90	1.89

(v) Electrical properties and Hall studies

Temperature variation of the conductivity of reactive D. C. sputtered $CuAlO_2$ thin film has also been studied in the temperature range of 300 K to 550 K according to **Eq. 18**. In this case also the contacts were made by silver paint and the ohmic nature of the contacts were verified accordingly. **Fig. 30** represents ln σ vs. 1000/T plot of the reactive sputtered $CuAlO_2$ thin film on glass substrate. The film thickness was ~ 500 nm obtained from cross-sectional SEM (not shown here). The temperature variation of conductivity of the $CuAlO_2$ thin films were studied below the room temperature by previous authors [64, 67], but no study on high temperature conduction was reported. The straight-line nature of the Arhenius plot indicates the thermally activated conduction as often found in semiconductors. Room temperature conductivity of the film was obtained as 0.22 S cm^{-1}, which is comparable to that obtained by D. C. sputtered films post-annealed for 60 min. From the slope of the graph we get the value of activation energy (E_a) which corresponds to the minimum energy required to transfer carriers from acceptor level to the valence band and the value of E_a comes out as 250 eV, which is comparable to that of D. C. sputtered films post-annealed for 60 min.

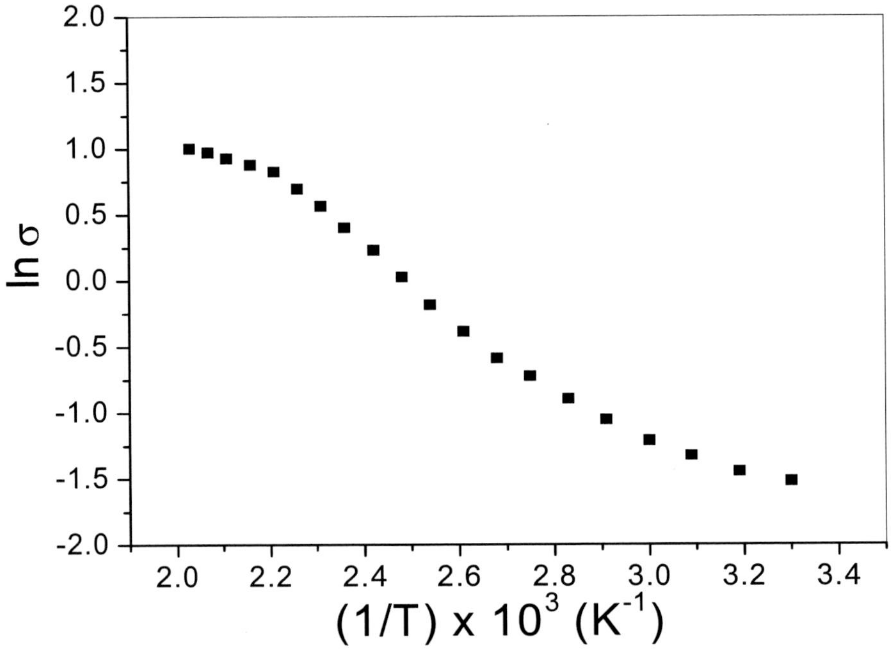

Figure 30. Temperature variation of conductivity of reactive sputtered $CuAlO_2$ thin film.

Hall measurements of reactive D. C. sputtered films were done at room temperature. Hall coefficient of the films was determined to be R_H = 14.1 cm^{-3} C^{-1}, corresponding to carrier density 4.4 x 10^{17} cm^{-3}. Positive value of Hall coefficient confirmed the p-type conductivity of the film. The carrier concentration of this film is comparable to that of D. C. sputtered film post-annealed for 60 min. As far as conductivities of previously reported reactive sputtered copper aluminium oxide films are concerned, Tsuboi et al [147] obtained phase impure copper aluminium oxide films (a mixture of $CuAlO_2$ and Cu_2O) with maximum conductivity around 0.1 S cm^{-1}. A comparison between the electrical parameters of reactive sputtered film and D. C. sputtered films is furnished in **Table 16**.

Table 16. Comparison between the electrical properties of reactive sputtered and D. C. sputtered films.

Synthesis technique	t_a (min)	σ_{RT} (S cm^{-1})	E_a (meV)	R_H ($cm^{-3}C^{-1}$)	n (cm^{-3})
Reactive sputtering	60	0.22	250	+ 14.1	4.4×10^{17}
D. C. Sputtering	30	0.09	270	+ 22.5	2.8×10^{17}
	60	0.16	245	+ 13.9	4.5×10^{17}
	90	0.39	196	+ 4.60	1.2×10^{18}

(vi) Thermoelectric properties

Thermoelectric properties of reactive sputtered $CuAlO_2$ film show almost similar nature as those of D.C. sputtered films (cf. **Fig. 22**). **Fig. 31** shows the temperature dependence of Seebeck coefficient. The room temperature Seebeck coefficient (S_{RT}) is found to be + 115 μV K^{-1}, which lies between that of D. C. sputtered films with t_a = 30 min and 60 min (cf. **Table 12**). Positive values of the Seebeck coefficients also confirm the p-type nature of the films. The Seebeck coefficients range from +115 μV K^{-1} to 520 μV K^{-1}, at 300 K to 470 K respectively. The Fermi energy of reactive sputtered film has been calculated from the slope of the linear portion of the curve in **Fig. 31**, near room temperature, according to the **Eq. 20**. The Fermi energy as obtained is 100 meV, which is slightly less than that of the D. C. sputtered film with t_a = 90 min (cf. **Table 12**). This type of band structure is typical of a non-degenerate semiconducting material with Fermi level lying between acceptor level (which corresponds to the activation energy, E_a = 250 meV) and the top of the valence band.

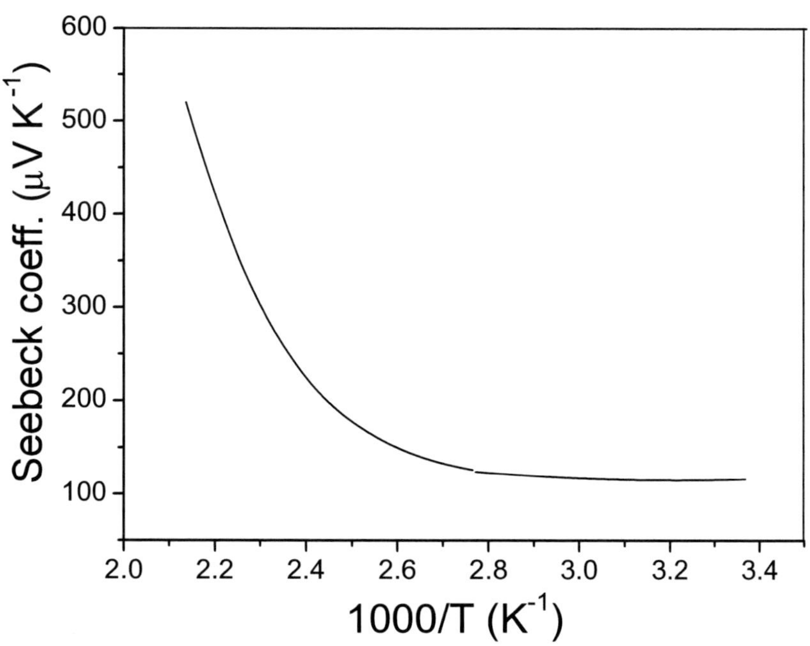

Figure 31. Temperature variation of Seebeck coefficient of reactive sputtered $CuAlO_2$ thin film.

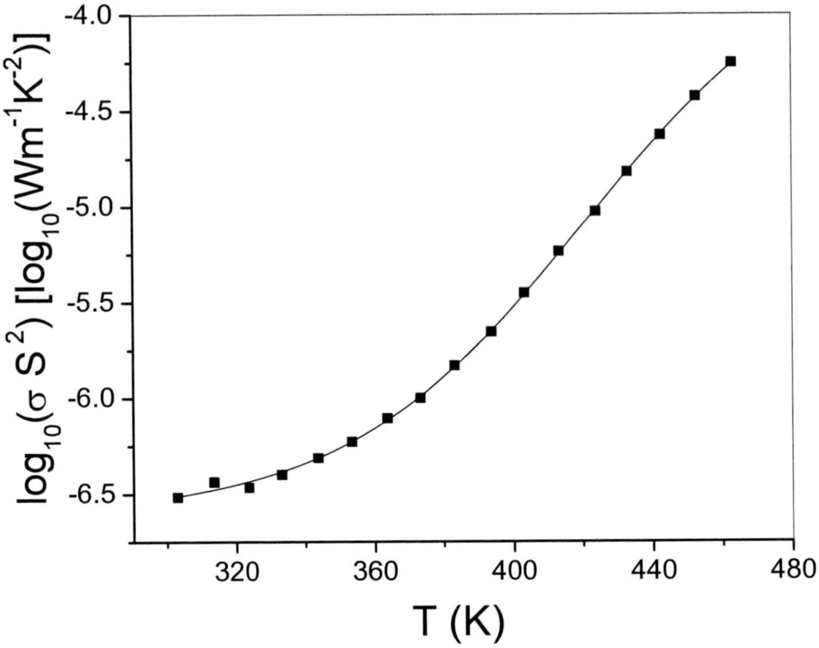

Figure 32. Temperature variation of thermoelectric power factor of reactive sputtered $CuAlO_2$ thin film.

Fig. 32 represents the temperature dependence of thermoelectric power factor (σS^2) of $CuAlO_2$ thin film for the temperature range of 300 K to 460 K. The values range from 3.09 x 10^{-7} $Wm^{-1}K^{-2}$ at temperature around 300 K to 5.5 x 10^{-5} $Wm^{-1}K^{-2}$ around 460 K. These are also comparable to those of D. C. sputtered films. A comparison between different thermoelectric parameters of D. C. sputtered and reactive sputtered films is furnished in **Table 17**.

Table 17. Different thermoelectric properties of D. C. sputtered and reactive sputtered $CuAlO_2$ thin films with different post-deposition oxygen annealing times (t_a).

Synthesis technique	t_a (min)	S_{RT} ($\mu V\ K^{-1}$)	E_f (meV)
Reactive D. C. sputtering	60	+ 115	100
D. C. Sputtering	30	+ 120	200
	60	+ 141	151
	90	+ 230	130

5.3. Properties of Wet-Chemical Dip-Coated Films

(i) Structural properties

Beside physical processes like D. C. sputtering and reactive sputtering, we have also synthesized $CuAlO_2$ thin film by wet-chemical dip-coating process. The experimental procedure is furnished in details in **Section 4.3**. **Fig. 33** shows the XRD pattern of the dip coated copper aluminium oxide thin film on glass substrate. The pattern reveals a strong (006) orientation of $CuAlO_2$ phase. Very small peaks of (012), (107), (00$\overline{12}$), (116) and (119) reflections have also been observed. As the intensities of these peaks are quite small (7 % to 2 %) compared to the (006) peak, so our film shows a preferential (006) orientation. A comparison with vacuum deposited films (PLD-process by Kawazoe et al. [64], Yanagi et al. [67] and sputter-deposited by us [160, 161]) shows similar (006) orientation. As far as the copper aluminium oxide films deposited by solution processes are concerned, Tonooka et al. [153] had not reported any XRD data of the film. But the XRD data of the powdered samples were shown, which depicted the presence of a mixture of $CuAlO_2$, CuO and $CuAl_2O_4$ phases in the sample. But they suggested that the powdered sample prepared by nitrate route (at $1100^{O}C$) would give maximum percentage of $CuAlO_2$ phase with a strong (012) orientation. It is worthwhile to mention that, for XRD pattern of $CuAlO_2$ powdered sample, all the previous reports showed a high (012) orientation, whereas in the thin film, a strong (006) orientation was observed [64, 67, 160]. On the other hand, Ohashi and co-authors [157] reported the XRD pattern of their sol-gel deposited multiphase copper aluminum oxide films which consisted of a mixture of $CuAlO_2$, CuO and Cu_2O. As shown in our XRD pattern (**Fig. 33**), a small peak of $Cu_2Al_2O_4$ phase has been observed [113]. As its intensity is as low as 7 % of the (006) peak of $CuAlO_2$ phase, so it may be concluded that our copper aluminium oxide thin film has very high percentage of $CuAlO_2$ phase with a strong (006) orientation. Also no peaks of starting materials (e.g. CuCl and $AlCl_3$) or any reactant species such as metal oxides have been found in the pattern. It is noteworthy that synthesis of copper aluminium oxide thin films by spray technique [156] at $525^{O}C$ yielded amorphous films. But a transition to crystalline nature was observed at a deposition temperature of $570^{O}C$ with a small (101) reflection of $CuAlO_2$ phase. But this film also not phase pure as it contained small amount of CuO phase as depicted from their XRD pattern [156].

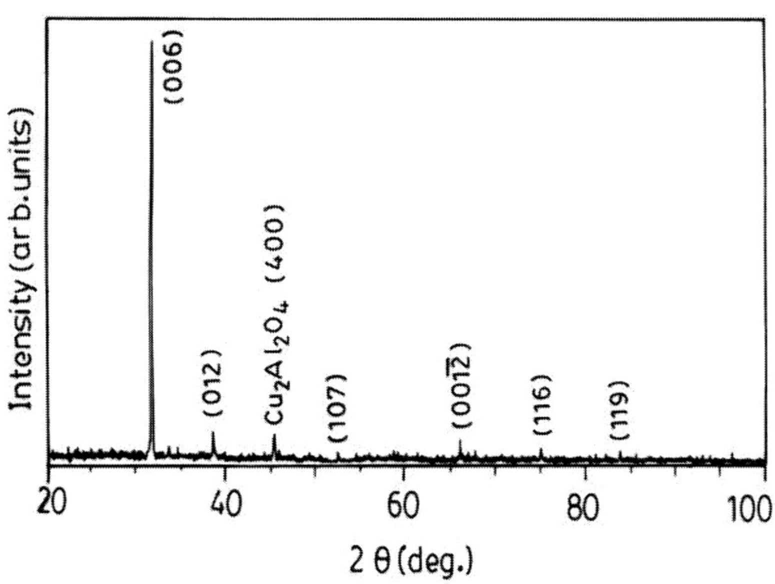

Figure 33. XRD pattern of copper aluminium oxide thin film deposited on glass substrate.

(ii) Surface morphology

Fig. 34 shows the scanning electron micrograph (SEM) of a typical $CuAlO_2$ thin film deposited on glass substrate. Existence of a smooth surface with finer grains and well defined grain boundaries are observed. Also some bigger clusters are also observed to be dispersed on the surface, which resulted due to the agglomeration of finer grains. Cross-sectional SEM reveals the thickness of the film around ~1.5 μm.

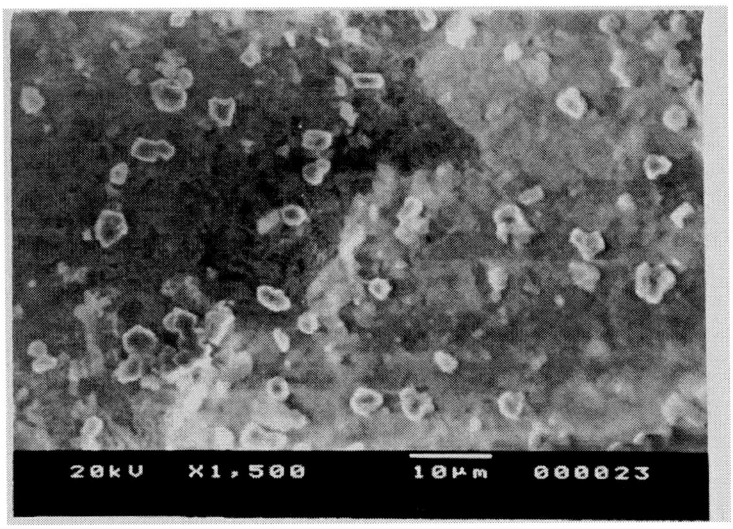

Figure 34. SEM micrograph of copper aluminium oxide thin film on glass substrate.

Previously, Ohashi et al. [157] reported the SEM micrograph of their sol-gel dip-coated copper aluminium oxide film and observed a smooth surface with fine particles but interior of the film was found to be porous with larger grains. On the other hand, Bouzidi et al. [156] reported very smooth surface morphology of their copper aluminium oxide thin film deposited by spray technique at 500 ^{O}C, cross-sectional SEM of which revealed the thickness of their film ~ 1 μm.

(iii) Optical properties

Fig. 35 shows the transmission spectrum of copper aluminium oxide thin film deposited on glass substrate in the wavelength range of 300 nm to 800 nm, taking similar glass as reference. It shows nearly 90 % transmittance in the wavelength range of 500 nm to 800 nm. From the transmittance data, using Manifacier model [254] we have calculated the absorption coefficients (α) at the region of strong absorption by re-writing **Eq. 16** (neglecting reflectance, R) as follows:

$$\alpha = \frac{1}{d}\ln(\frac{1}{T}) \tag{23}$$

where d is the film thickness and T is the transmittance obtained from **Fig. 35**.

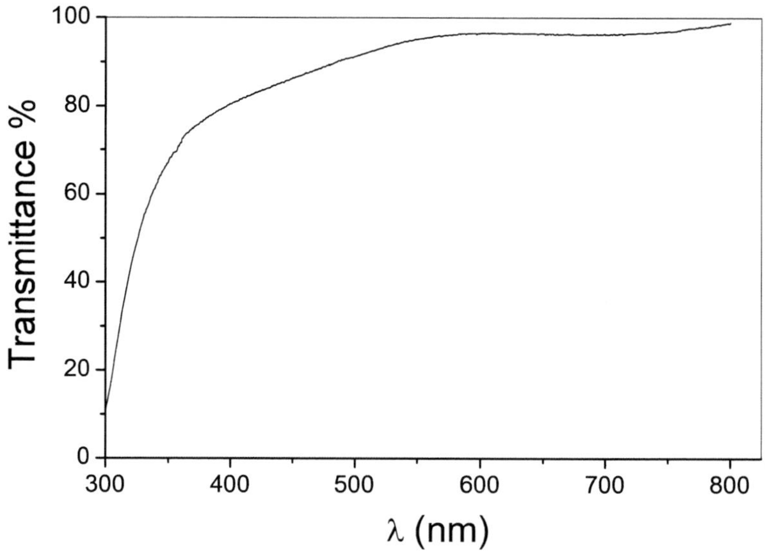

Figure 35. Transmittance vs. wavelength plot of copper aluminium oxide thin film deposited on glass substrate.

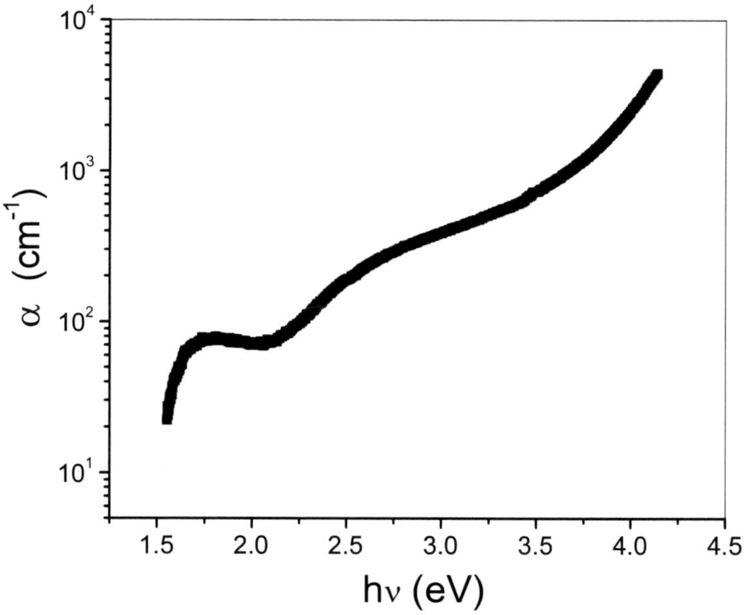

Figure 36. Energy dependence of absorption coefficient of $CuAlO_2$ thin film prepared by dip-coating process.

Fig. 36 shows the spectral variation of absorption coefficient for wet-chemical dip-coated copper aluminium oxide thin film. The value of α varies from 22.0 to 4.5 x 10^2 cm^{-1} in the wavelength range of 300 to 800 nm. **Fig. 37** represents the spectral variation of extinction coefficient (k) according to the **Eq. 14**. The value of k varies from 1.42 x 10^{-3} to 106.23 x 10^{-3}

in the spectral range of 300 to 800 nm. These values are comparable to the D.C. sputtered as well as reactive sputtered films shown in **Fig. 15, 16, 17** and **Fig. 29**. Using these values of α

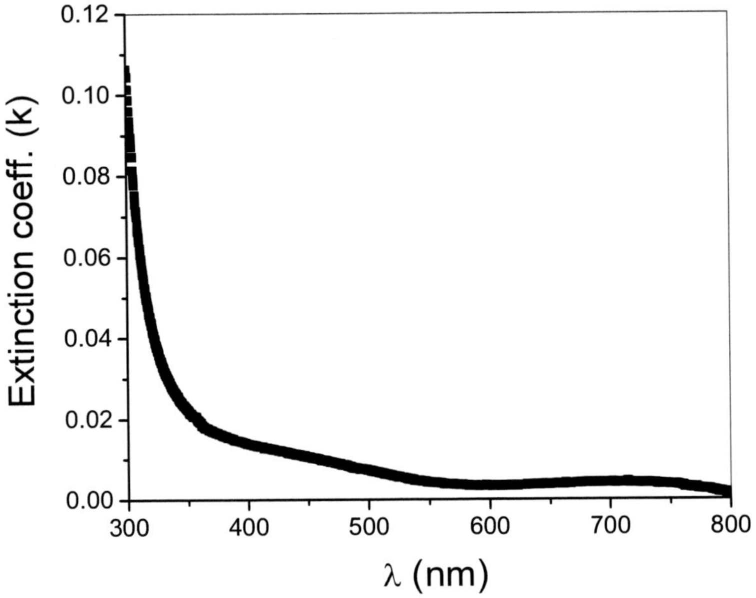

Figure 37. Spectral variation of extinction coefficients (k) of copper aluminium oxide thin film deposited by wet-chemical method.

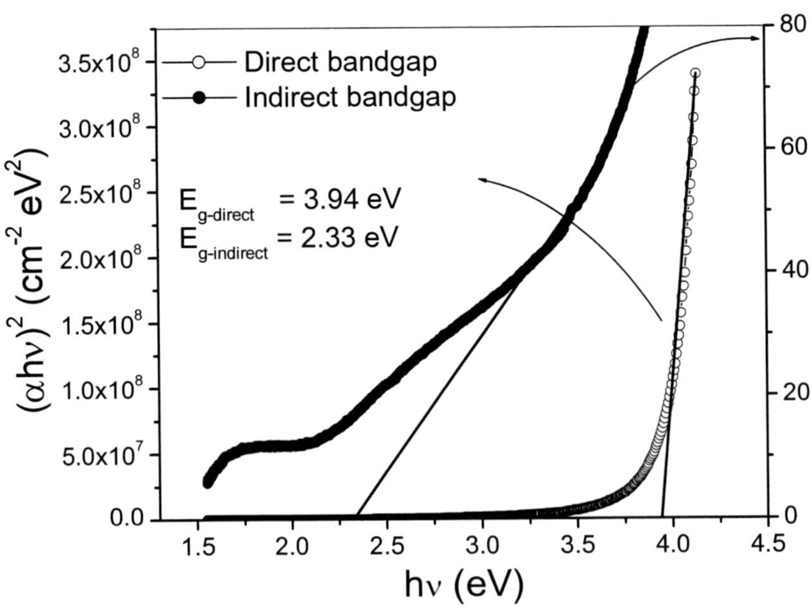

Figure 38. Plot to determine direct bandgap of copper aluminum oxide thin film.

the nature and value of the optical band gap has been determined according to **Eq. 17**. To determine the possible transitions,. $(\alpha h\nu)^{1/n}$ vs. hν were plotted for different values of n. The $(\alpha h\nu)^2$ and $(\alpha h\nu)^{1/2}$ vs. hν plots are shown in the **Fig. 38**. Extrapolating the linear portion of the graphs to the hν axis we have obtained the direct and indirect band gap values as 3.94 eV and 2.33 eV respectively. As far as bandgap value of chemically deposited copper aluminium oxide is concerned, Bouzidi et al [156] reported the direct bandgap for their spray-deposited copper aluminium oxide as 3.87 eV. Also these values nearly agree with the value reported previously by others [64, 67] and us [160, 161]. **Table 18** compares the various optical parameters of chemically deposited copper aluminium oxide films with that of D. C. sputtered and reactive sputtered $CuAlO_2$ films by us.

Table 18. Comparison of different optical parameters of wet-chemical deposited $CuAlO_2$ thin film with that of physically deposited films.

Process	Post-annealing time (min)	Average visible transmittance (%)	Bandgap (E_g)	
			Direct (eV)	Indirect (eV)
Wet-chemical	---	80	3.94	2.33
D. C. sputtering	30	65	3.81	2.80
	60	80	3.70	2.10
	90	75	3.80	2.32
Reactive sputtering	60	85	3.90	1.89

(iv) Electrical properties

Electrical properties of chemically deposited copper aluminum oxide thin films have been studied by standard four-probe methods. All electrical contacts were made by silver paint, which showed linear I-V characteristics over a wide range of voltages and temperatures. **Fig. 39** represents lnσ vs. 1000/T plot of the copper aluminum oxide film on glass substrate from room temperature (300 K) to 570 K. The straight-line nature of the Arhenius plot indicates the thermally activated conduction, as often found in semiconductors. Room temperature conductivity of the film was obtained as 4.0×10^{-3} S cm^{-1}. This value is quite comparable to the previously reported copper aluminum oxide films prepared by chemical routes (5.0×10^{-3} S cm^{-1} by Tonooka et al. [153], 4.0×10^{-3} S cm^{-1} by Bouzidi et al. [156]). As far as $CuAlO_2$ films prepared by physical processes are concerned, this value is one order less than that obtained by Kawazoe et al. [64] (0.095 S cm^{-1}) for their pulsed laser deposited film. Also a comparison with sputter deposited films prepared by us [160, 161], this value comes out to be two orders of magnitude less as shown in **Table 19**. This may be due to the higher number of defect states formed within the film. It is generally observed that the films produced by SGDC route contains higher defect states than that produced by vacuum deposited films. Proper regulation of the deposition parameters as well as intentional substitutional doping of the film are required to increase the conductivity which is the next aim of our work.

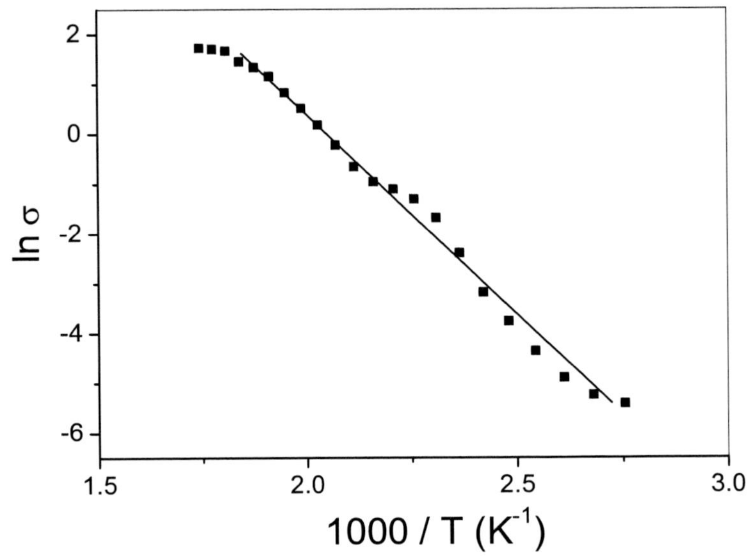

Figure 39. Temperature variation of conductivity of copper aluminum oxide thin films.

Table 19. Comparison between different electrical parameters of $CuAlO_2$ thin films, deposited by various processes.

Process	Post-annealing time (min)	σ_{RT} ($S\ cm^{-1}$)
Wet-chemical	---	0.004
DC sputtering	90	0.39
	60	0.16
	30	0.09
	120	0.055
	150	0.014
Reactive sputtering	60	0.22

Thermoelectric power (TEP) of the $CuAlO_2$ thin films deposited on glass substrates was measured over the temperature range 308 – 488 K. Room temperature Seebeck coefficient was found to be +206 $\mu V\ K^{-1}$. Positive values of Seebeck coefficients confirmed the p-type conductivity of the film. Also Hot-probe measurement confirms the p-type nature of the films.

6. Transparent Junctions

Fabrication of transparent junction is the most important aspect in the field of p-TCO technology. Amongst various junctions, transparent p-n heterojunction diode is the simplest one with rectifying properties. It is also the simplest structure to fabricate. We have synthesized n-ZnO: Al / p-$CuAlO_2$ heterojunction diode on glass substrates with considerable electro-optical properties. The process and results are furnished below.

6.1. Fabrication of All-Transparent Diode

The all-TCO p-n hetero-junction diode having the structure: n-ZnO: Al / p-$CuAlO_2$ were fabricated on glass substrates. The n-type layer was taken as aluminum doped zinc oxide films (ZnO: Al), which was deposited on commercial glass substrates (of size 18 mm X 8 mm) by Sol-Gel-Dip-Coating technique. Thereafter these n-layer coated glasses were used as the substrates for the deposition of p-layer (p-$CuAlO_2$ film) by D. C. sputtering technique. Six independent junctions (1 mm X 1 mm) were fabricated on a single substrate using proper masking. Details of the deposition procedures are furnished below.

Deposition of n-layer: ZnO films were deposited on glass substrates by SGDC route from a starting solution of zinc acetate dihydrate ($Zn(CH_3COO)_2{\cdot}2H_2O$) and isopropyl alcohol (Pr^i-OH). Since zinc acetate has low solubility in isopropyl alcohol, diethanolamine (DEA) was added (with $[DEA]/[Zn^{2+}] = 1.5$) to get transparent solution and to keep the solution stable in dip-coating process. Doping of Al was done by the addition of controlled amount of aluminum nitrate ($Al(NO_3)_3{\cdot}9H_2O$) to the solution. Then the resultant solution was stirred and refluxed, keeping the temperature at 343 K for one hour. The atomic ratio of Al/Zn in the initial solution was varied from 0.32 % to 1.62 % and the concentration of zinc acetate was fixed at 0.85 mol/L. Distilled water (with $[H_2O]/[Zn^{2+}] = 14$) and acetic acid (with $[H^+]/[Zn^{2+}] = 2$) were added for better stability of the solution and to avoid gelation or precipitation. The pH of the solution was kept around 7.0. Lastly, stirred and refluxed solution was aged for half an hour to get the resultant solution. Then the ultrasonically cleaned glass substrates were dipped vertically into the solution and withdrawn at a speed of 8 cm/min to coat them with the required material. The coated substrates were dried at room temperature for 10 minutes and heated at ~ 423 K for 10 minutes in open atmosphere for gelation. This process was repeated for 2-3 times for getting a desired thickness. Finally the films were heated at 573 K for one hour to obtain crystalline ZnO: Al films. The details of the deposition conditions were reported elsewhere [255]. It is to be noted that although the Al concentration in the starting solution was varied from 0.32 % to 1.62 % to get $Zn_{1-x}Al_xO$ films with varied opto-electrical properties, but for the fabrication of the diode, those films were chosen which were having Al concentration of 1.62 % in the starting solution. This is because of the better comparability of the electrical and optical properties of these films with the corresponding p-layer ($CuAlO_2$ films).

Deposition of p-layer: The n-layer coated glass was used as the substrate in the D. C. sputtering process to deposit p-$CuAlO_2$ thin film. Mica masks were used on the n-ZnO: Al coated glass substrates for preferential deposition of p-$CuAlO_2$ layers on desired position. Initially, solid-state reaction between stoichiometric ratios of Cu_2O and Al_2O_3 powder at 1400 K produced $CuAlO_2$ powder. This powder was then pressed into a pellet and was used as a target for D. C. sputtering. The sputtering unit was evacuated by standard rotary-diffusion arrangement upto a base pressure of 10^{-6} mbar. The pellet was arranged properly by aluminum holder to act as upper electrode and the negative terminal of the D. C. power supply unit was connected to it. n-layer coated glass substrates were placed on the lower electrode and connected to the ground of the power supply. The electrode distance was taken as 1.8 cm. Ar and O_2 (3 : 2 vol. ratio) were taken as sputtering gas and the sputtering was done at an elevated substrate temperature (~ 453 K) to achieve high crystallinity in the film. Post-deposition annealing of the film (at 473 K) for 30 min in an O_2 atmosphere (at a pressure of 0.2 mbar) was performed to induce nonstoichiometry in the film for enhancing p-type conductivity. Details of the deposition conditions are furnished in details in **Section 4.1** (cf. **Table 6**). A flow-chart of the diode fabrication process is shown in **Fig. 40** and the corresponding schematic diagram of the diode structure is given in **Fig. 41**.

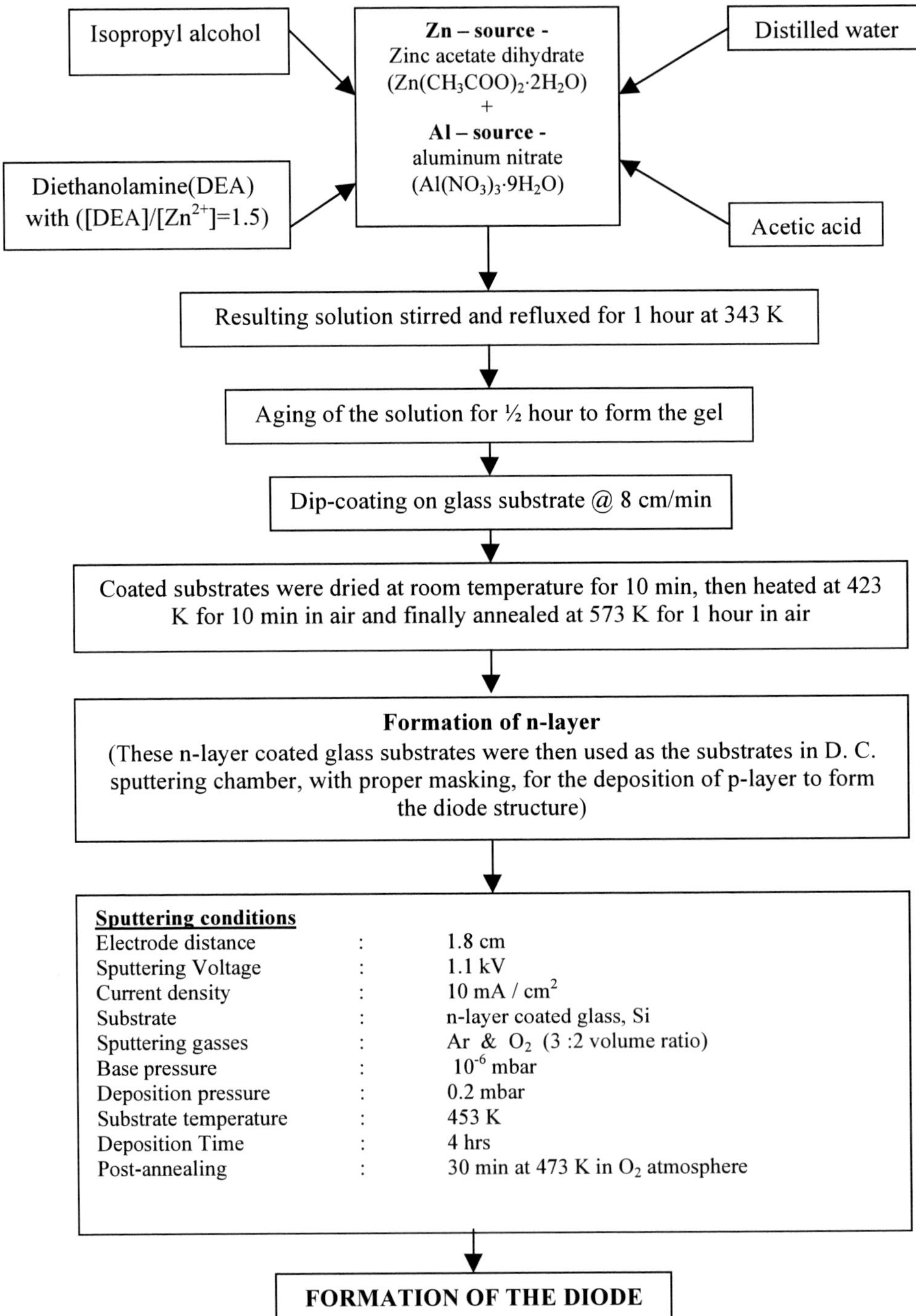

Figure 40. Flow-chart of the diode fabrication.

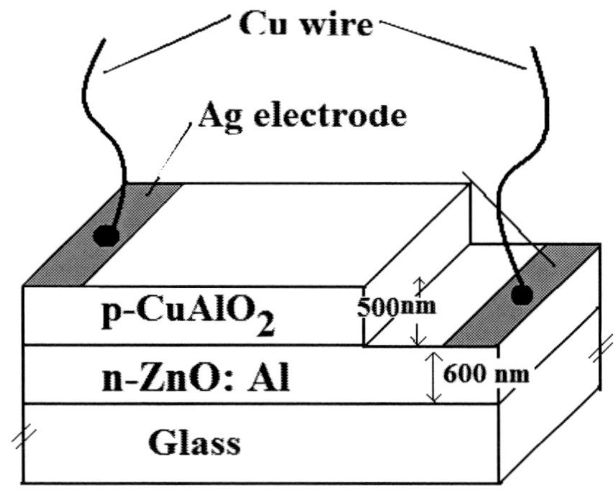

Figure 41. Schematic diagram of n-ZnO: Al / p-$CuAlO_2$ diode structure.

The optical transmittance of the diodes was measured by UV-Vis-NIR spectrophotometer (SHIMADZU-UV-3101-PC). All electrical measurements were done by standard four-probe method using Keithley-6514 electrometer under vacuum condition (~ 10^{-3} mbar). For ohmic contacts, evaporated silver electrodes were used with proper masking in both types of layers, which showed linear I-V characteristics over a wide range of voltages and temperatures. Thereafter the electrical connections were made by Cu leads with silver paints, as shown in **Fig. 41**.

6.2. Characterizations of the Diode

Structural properties of the films were studied by X-ray diffractometer (XRD, BRUKER, D8, ADVANCE) using the Cu K_α (1.5406 Å) radiation. **Fig. 42** shows the XRD patterns of individual layers of $CuAlO_2$ (pattern-a) and ZnO: Al (pattern-b) on glass substrates respectively, deposited under the same conditions as that used for diode fabrication. All the peaks match with the standard JCPDS files (# 35-1401, for CAO [113]) and (# 36-1451, for ZnO) respectively, as shown by the circles and lines in the figure. The XRD pattern for $CuAlO_2$ is similar to that given in **Fig. 8(b)**. Also no peaks of starting materials and any other reactant species have been found which conclusively indicate that the reactants were completely mixed to form the proper phase of the materials. As stated earlier, the XRD pattern of the p-layer is obtained for p-$CuAlO_2$ thin film deposited on bare glass substrate under the same conditions as that used for diode fabrication. But it is worthwhile in mentioning that, here we have not taken into account the change in crystal quality of the p-layer due to the presence of ZnO: Al layer underneath, to fabricate the diode. It must be admitted that the crystal structure of ZnO film might affect the crystal quality of the $CuAlO_2$ in terms of intensity and sharpness of the XRD peaks. And it might not be unreasonable to speculate that the presence of crystalline ZnO film underneath might have improved the crystal quality of $CuAlO_2$ film with respect to bare glass substrate, which, in turn, enhances the formation of rectifying junction.

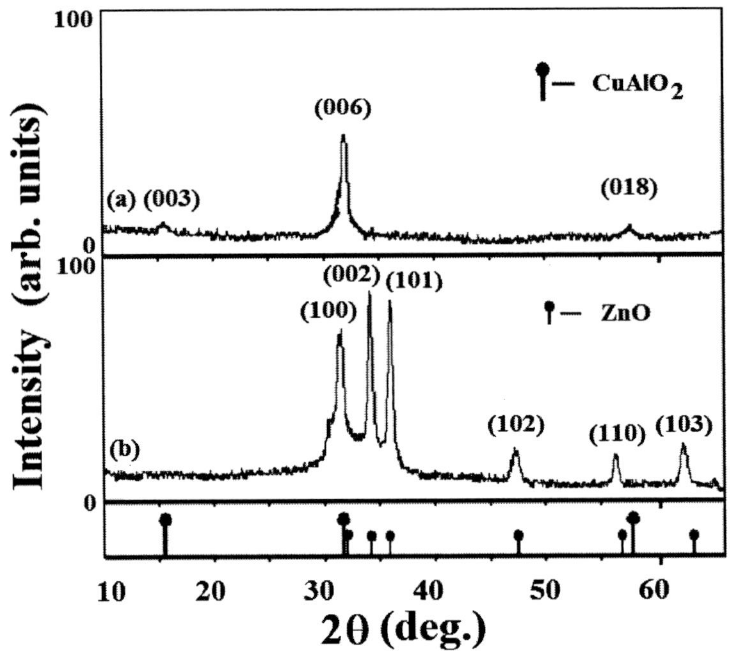

Figure 42. XRD patterns of (a) p-$CuAlO_2$, (b) n-ZnO: Al films. Lines and circles represent the reference patterns of corresponding materials.

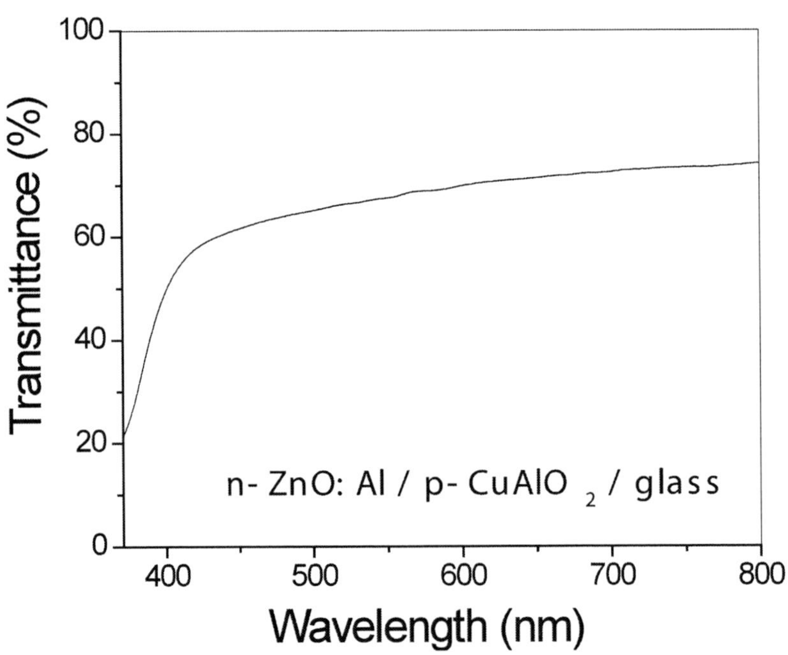

Figure 43. Optical transmision spectra of the n-$Zn_{1-x}Al_xO$ / p-$CuAlO_{2+x}$ diode deposited on glass substrate.

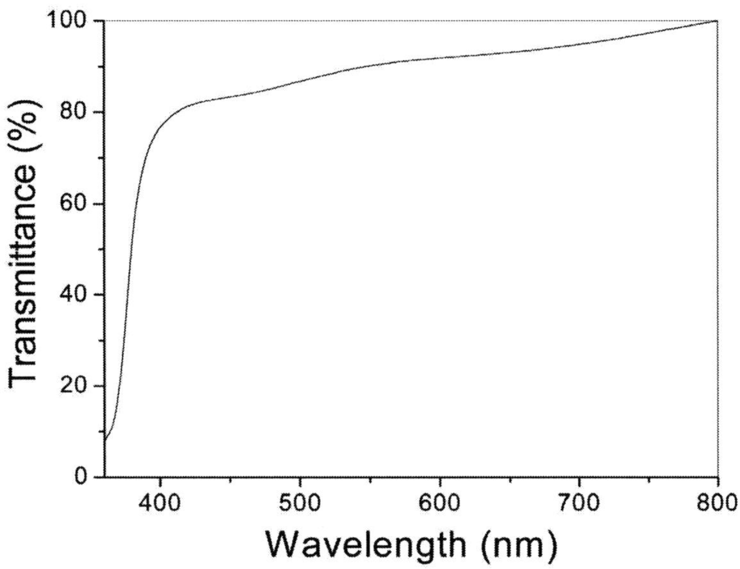

Figure 44. Optical transmission spectra of n-ZnO: Al film.

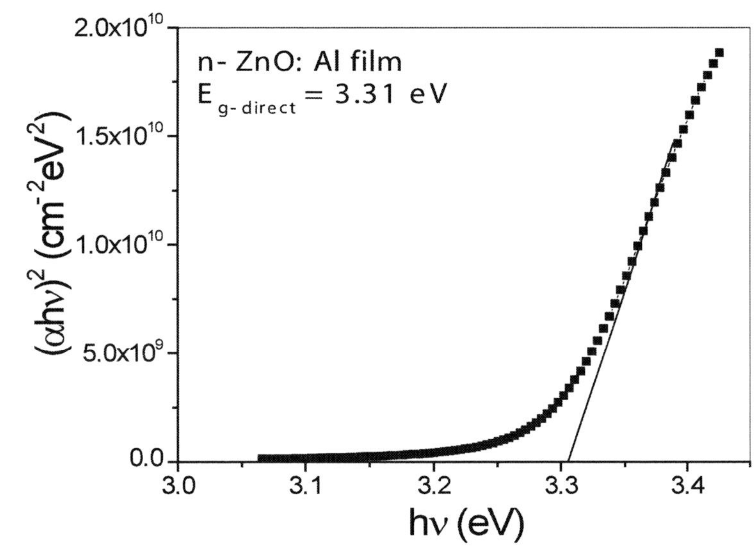

Figure 45. Determination of direct bandgap of n-ZnO: Al film.

The optical transmision spectrum of the n-ZnO: Al / p-$CuAlO_2$ diode is shown in **Fig. 43**. As mentioned earlier, the thickness measurements were done by cross-sectional SEM (not shown here). The thicknesses were found to be 600 nm for ZnO: Al film and 500 nm for $CuAlO_2$ film respectively, making the total device thickness around 1100 nm as shown in the inset of **Fig. 41**. The visible transparency of the diode is around 60 %, which indicates its potential application in 'Invisible Electronics' [65]. It is to be noted that the visible transparency of the individual p-layer with identical deposition condition as that used for diode fabrication is around 75 % as shown in **Fig. 13, Section 5.1**. On the other hand, the visible transparency of the n-layer is around 80% as shown in **Fig. 44**. A comparison of these spectra shows that the starting point of the fundamental absorption region of the diode structure is comparable to that of n-ZnO: Al layer, which is having lower bandgap energy (3.31 eV, as shown in **Fig. 45**). Previously, Tonooka et al. [172] obtained the average visible transmittance of their n^+-ZnO / n-ZnO / p-Cu-Al-O diode around 60 %. As far as other all-oxide transparent diodes are concerned, Sato et al. [66] reported 20 % visible transmittance for their p-NiO / i-NiO / i-ZnO / n-ZnO structure, Kudo et al. [165] obtained 70 % - 80 % visible transmittance for p-$SrCu_2O_2$ / n-ZnO diode, Hoffman and co-authors [170] reported 35 % to 65 % visible transmittance in a p-$CuY_{1-x}Ca_xO_2$ / n-$Zn_{1-x}Al_xO$ / n^+-ITO heterojunction diode, Yanagi et al. [103] obtained 60 % to 80 % transmittance for their p-$CuIn_{1-x}Ca_xO_2$ / n-$CuIn_{1-x}Sn_xO_2$ homojunction diode in the visible region.

Electrical properties of the individual layers have been studied in details and represented in our previous literatures [243, 244, 255]. **Fig. 46** represents the temperature variation of individual n- and p-layers deposited under identical conditions as that during diode fabrication. A comparative study of different electro-optical properties of the individual films is furnished in **Table 20**. For proper fabrication of rectifying junction, a comparable elctro-optical property of the individual p- and n-layers is very important, and in that respects ZnO: Al film is widely used because of its easy controllability of carrier concentration by varying percentage of Al during deposition. This is necessary in order to match the carrier concentrations with those positive holes in p- $CuAlO_2$ which is more difficult to control. Also possibility of low-temperature deposition of crystalline ZnO films on glass as well on as plastic substrates [61] has make ZnO films one of the most promising component for the fabrication of transparent diodes in the field of 'Invisible Electronics'.

Table 20. Different electrical and optical properties of individual p-$CuAlO_2$ layer (cf. Table 11, 12) and n ZnO: Al layers [255].

Film	Direct bandgap ($E_{g\text{-direct}}$) eV)	Room-temperature conductivity (σ_{RT}) (S cm^{-1})	Activation energy (E_a) (meV)	Fermi energy (E_f) (meV)	Carrier concentration (n) (cm^{-3})
n-ZnO: Al	3.31	0.08	550	280	2.6×10^{17}
p-$CuAlO_2$	3.81	0.09	270	200	2.8×10^{17}

The current-voltage characteristics of the transparent diodes have been measured by Keithley-6514 electrometer. For ohmic contacts, evaporated silver electrodes were used with proper masking in both types of layers, which showed linear I-V characteristics over a wide range of voltages and temperatures. Thereafter the electrical connections were made by Cu leads with silver paints. The I-V characteristic of the diode is shown in **Fig. 47**, which shows rectifying properties, indicating proper formation of the junction. The turn-on voltage obtained around 0.8 V. However, it varied from 0.6 V to 1.0 V from junction to junction. This indicates considerable reproducibility of the junctions. The forward-to-reverse current ratio is approximately ~ 30 at ± 4 V. Maximum current obtained at 5 V is around 1 μA and a small leakage current as low as 30 nA was observed at a reverse bias of –4 V. Previously, Tonooka et al. [172] reported the average turn on voltage of their n-ZnO/p-Cu-Al-O diode ~ 0.5 V,

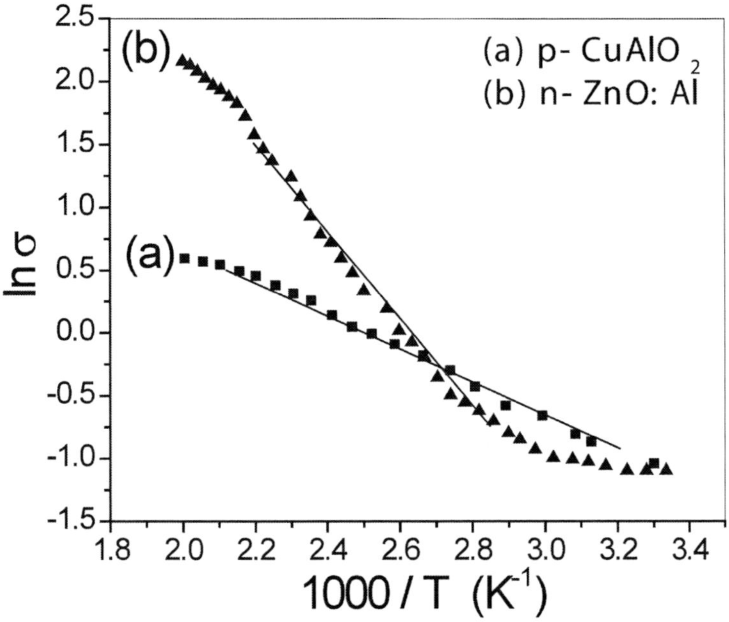

Figure 46. Temperature variation of conductivity of (a) $p-CuAlO_2$ and (b) n-ZnO: Al films.

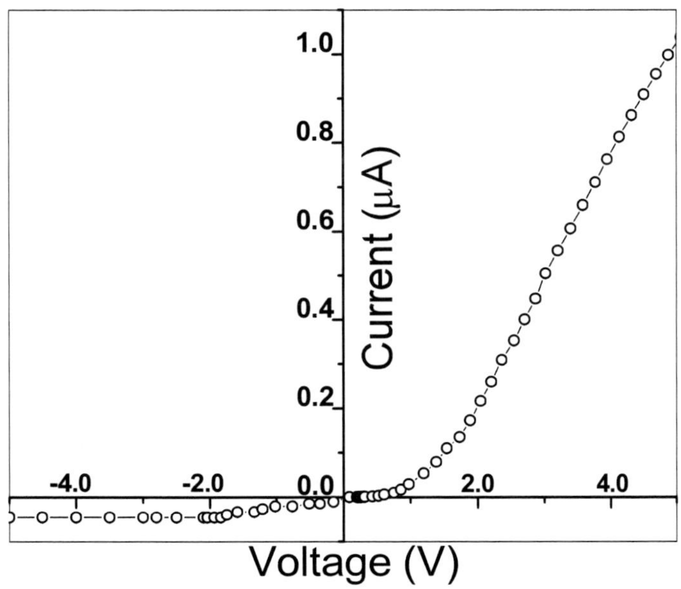

Figure 47. Current–Voltage characteristics of $p-CuAlO_2$ / n-ZnO: Al diode.

which is comparable to ours. Generally, for heterojunction diodes, the structural imperfections at grain boundaries as well as at the interface detoriate the efficiency of the doide [165]. Also the inherent difficulty in manufacturing these all-oxide diodes is that the p and n-layers must be produced under oxidizing and reducing conditions respectively, so that optimal processing for one type is detrimental to the other [110]. All these facts must be addressed with considerable attention for diverse applications of these heterojunction all-oxide transparent diodes in the field of "Invisible Electronics'.

If we probe into the bandstructure of the interface, considering the bandgaps of n-ZnO: Al and p-$CuAlO_2$ as 3.31 eV (cf. **Fig. 45**) and 3.81 eV (direct), 2.8 eV (indirect, cf. **Fig. 18c**) respectively, then the depletion barrier height comes out within the range of 3.3 eV to 2.3 eV (for this calculation, the position of Fermi level of both p and n type materials are obtained from thermo-electric power measurements of the materials. For p-$CuAlO_2$ films this value is around 200 meV (cf. **Table 12**) and for n-ZnO: Al films it is around 280 meV [255]. The activation energy values are obtained from **Fig. 46** and furnished in **Table 20**). But these values are quite larger than that of the observed turn-on voltage, which is around 0.8 V. This inconsistancy between the turn-on voltage and the barrier height may be explained in the following way: Investigation of previous literatures about the band structure calculations of Mattheis [111] and experimental findings of Cava et al. [225] of similar delafossite p-$CuYO_{2+x}$ material, we see the existance of some midgap impurity bands within the material due to the interstitial oxygen doping, which decreases the effective bandgap of the material. In a similar way it can be argued that in our p-$CuAlO_2$ thin films, excess oxygen intercalation and probably some unintentional impurity incorporation may give rise to some new and deep states within the bandgap via self-compensation [256], which further reduces the effective bandgap of the material, so also the barrier height. This might provide an explanation of the low turn-on voltage of the p-$CuAlO_2$ / n-ZnO: Al hetero-junction diode.

7. Nanocrystalline p-$CuAlO_2$ Thin Films

Fabrication of nanostructured p-TCOs, coupled with the already existing and well-known materials of nanostructured n-TCOs, will give an added impetus in the field of "Invisible Electronics" by creating the opportunity for the fabrication of nano-active devices, which can have highly efficient applications in the optoelectronics device technology. We have synthesized nanostructured p-$CuAlO_2$ thin films by D.C. sputtering technique by reducing the deposition time and substrate temperature during deposition. Effect of deposition time on crystallinity, particle size, strain, bandgap etc of the film has been investigated. Also photoluminescence properties of this nanocrystalline material have been reported here.

7.1. Preparation of the Nanostructured Film

(i) $CuAlO_2$ Powder Preparation

Polycrystalline $CuAlO_2$ powder was synthesized in the same procedure as described in **Section 4.1(i)**. At first Cu_2O and Al_2O_3 powder (both 99.99 %) were taken with Cu / Al atomic ratio 1 : 1 and mixed for 1 hr. Then the mixture was heated in alumina boat at 1100°C for 24 hours in air to form the $CuAlO_2$ powder. The sintered body was then reground and pressed into pellets by hydrostatic pressure of about 200 kgf / cm^2. These pellets were placed in aluminum holder by some appropriate arrangement, which was used as the target for sputtering.

(ii) Nanocrystalline $CuAlO_2$ Film Deposition

The sputtering unit was evacuated by standard rotary-diffusion arrangement upto a base pressure of 10^{-6} mbar. The target was then pre-sputtered for 10 min to remove contamination, if any, from the surface and then the shutter was displaced to expose the substrates in the sputtering plasma. Films were deposited on ultrasonically cleaned glass and Si substrates, which were placed on the lower electrode and connected to the ground of the power supply. Before placing into the deposition chamber the glass substrates were cleaned at first by mild soap solution, then washed thoroughly in deionized water and also in boiling water. Finally they were ultrasonically cleaned in acetone for 15 minutes. Si substrates were first immersed in 20 % HF solution for 2 minutes for removing surface oxide layers. Then they were cleaned in deionized water and finally with alcohol in an ultrasonic cleaner. The electrode distance was taken as 1.5 cm. Ar and O_2 (3 : 2 vol. ratio) were taken as sputtering gases. Details of the deposition conditions were described in **Section 4.1**. Only differences from the previous deposition conditions are the deposition time (t_d), which ranges from 3 min to 45 min (instead of 240 min, cf. **Table 6**) and the lower substrate temperature, which was kept at 373 K (instead of 453 K, cf. **Table 6**). This is because, generally at higher substrate temperature, the particles tend to coalesce to form bigger clusters, which is unwanted for the formation of nano-structured films. The variation in the deposition time is done to observe the changes in the nanostructure and optical properties of the films. Also no post-annealing of the films was performed.

7.2. Characterization and Discussion

The target pellets as well as the films were characterized by X-ray diffractometer (XRD, BRUKER, D8 ADVANCE) to observe the proper phase formation of the material. To study the nanotructure of the films, transmission electron microscopy (TEM, HITACHI-H600) analyses were performed. For TEM measurements, the films were directly deposited on carbon coated copper grids. Optical studies have been performed by measuring the transmittance and reflectance of the films, deposited on glass substrates, in the wavelength region 300 nm - 800 nm using a UV-Vis spectrophotometer (HITACHI-U-3410). Photoluminescence studies were performed by HITACHI F-4500 instrument for the films deposited on Si substrates. The thicknesses of the films were measured by optical interferometric process.

Structural properties: The XRD pattern of the synthesized $CuAlO_2$ powder, which was used for target preparation has already been presented in **Fig. 7** and described in **Section 5.1**. The peaks of the powdered material confirm the proper phase formation of the required target material. This target material was then used for the thin film preparation.

Four sets of samples were prepared by D. C. sputtering technique having deposition times (td) as 3 min, 9 min, 15 min, 45 min, to observe any variation in the XRD patterns and its effect on particle size. Another set of sample is also prepared for deposition time 150 min, which is used as the reference bulk material to compare with the nanocrystalline films. This film has almost similar structural and optical properties as that deposited for 240 min and described in Section 5.1. Fig. 48 represents the XRD patterns of sputter-deposited nanocrystalline CuAlO2 thin films on Si substrates with deposition times (td) 15 min (pattern-#1) and 45 min (pattern-#2). For the film with td = 15 min (curve-#1), two broad peaks of (101) and (012) reflections are observed along with two smaller peaks of (107) and (018) reflections. On the other hand for the film deposited in 45 min (curve-#2), a slightly stronger peak of (006) reflections and a small peak of (018) reflections are observed along with the presence of a broad and considerably attenuated hump representing (101) and (012) reflections. It is worthwhile to mention in this connection that in all the previously reported XRD patterns of CuAlO2 thin films by us (cf. Fig. 8) as well as by others [64, 67], a strong (006) orientation were present, whereas the XRD patterns of sintered targets show either a preferred (012) orientation [64] or (101) orientation [67]. Likewise, we have also observed a (101) orientation for the sintered target (cf. Fig. 7). But for the films deposited in 15 min, due to smaller deposition time, the film thickness was quite low (~ 90 nm) and quasi-continuous formation of the film restricted the growth of any preferred orientation. And therefore two broad peaks of (101) and (012) reflections are present and as obvious, resemble close to the sintered target. Also, due to the nanocrystalline nature of the film, the peaks are quite broader and peak intensities are fairly low. On the other hand, for the films with td = 45 min, due to longer deposition time, a slightly stronger (006) orientation in the film is observed, along with two considerably attenuated (101) and (012) peaks, which is similar to that reported previously by Kawazoe, Yanagi, Hosono and co-authors [64, 67], for their pulsed laser deposited CuAlO2 thin films on sapphire substrates. For the films deposited in 150 min, the XRD pattern is almost similar to that shown in Fig. 8. A stronger (006) peak is observed along with two smaller peaks of (003) and (018) reflections. This shows that with increase in the deposition time, the films become more and more (006) oriented. Also it is to be mentioned here, that, in the XRD patterns of the films with td = 3 min and 9 min, due to nanocrystalline nature and low film thicknesses (~ 30 nm for td = 3 min and ~ 60 nm for td =9 min), the intensity of the peaks are very low (i.e almost became indistinguishable from the background noise) and therefore no satisfactory representable results were obtained, and that is why not shown here.

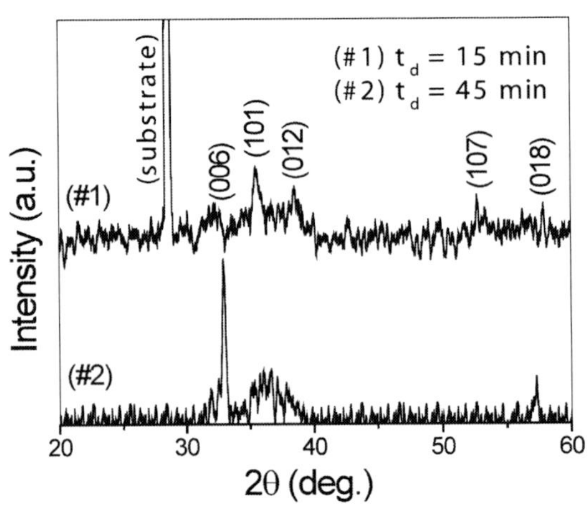

Figure 48. XRD pattern of nano-crystalline $CuAlO_2$ thin film deposited for (#1) 15 min and (#2) 45 min.

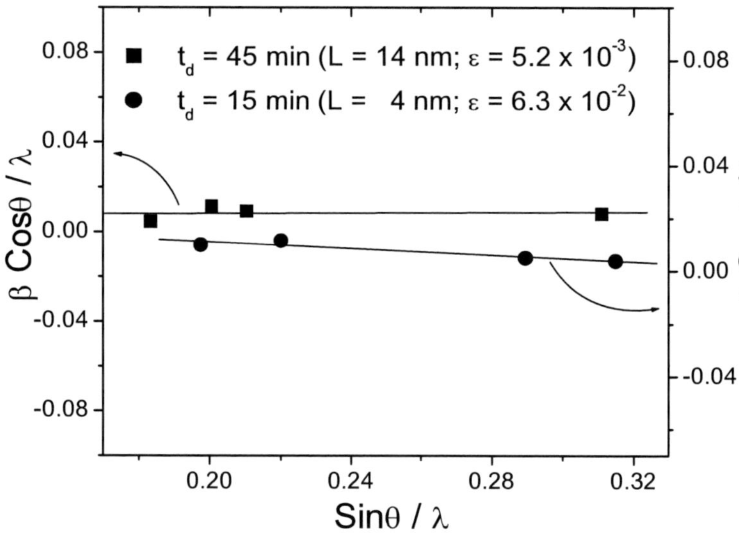

Figure 49. Determination of strain (ε) and particle size (L) of the nano-structured $CuAlO_2$ thin films deposited for 15 min and 45 min.

But the nanocrystalline nature of these films were confirmed from transmission electron microscopic measurements and the structural information were extracted from selected area electron diffraction patterns (SAED), which have been described in the later part of this paper.

The information on strain and the particle size of the deposited films were obtained from the FWHMs of the diffraction peaks, according to the **Eq. 6** given in **Section 5.1**. **Fig. 49** represents the plot of $\frac{\beta Cos\theta}{\lambda}$ vs. $\frac{Sin\theta}{\lambda}$ for the films deposited in 15 min and 45 min. From the slopes and intercepts on y-axes, the strain (ε) and particle size (L) were obtained as 5.2×10^{-3} and 14 nm (for t_d = 45 min) and 6.31×10^{-2} and 4 nm (for the film with t_d = 15 min) respectively. An increase in the particle size with deposition time is observed because of the greater amount of influx of sputtered particles at higher deposition times, leading to the agglomeration of bigger particles. A comparison with these values with bulk film is furnished in **Table 21**.

Table 21. Comparison between the effective particle size and effective strain of D. C. sputtered and reactive D. C. sputtered $CuAlO_2$ thin films.

Deposition time (t_d) (min)	Effective particle size (nm)	Effective strain
15	4.0	6.31×10^{-2}
45	14.0	5.20×10^{-3}
240 (Bulk film. Cf. **Table 14**)	26.0	8.52×10^{-3}

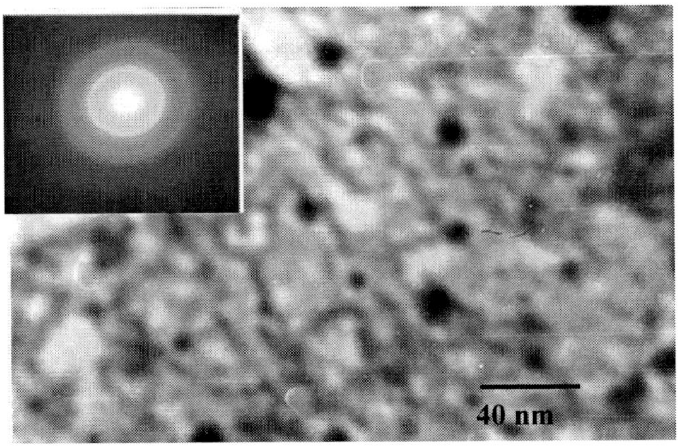

Figure 50(a). TEM micrograph of nano-structured $CuAlO_2$ thin film deposited for 3 min. **Inset**: SAED pattern of the same.

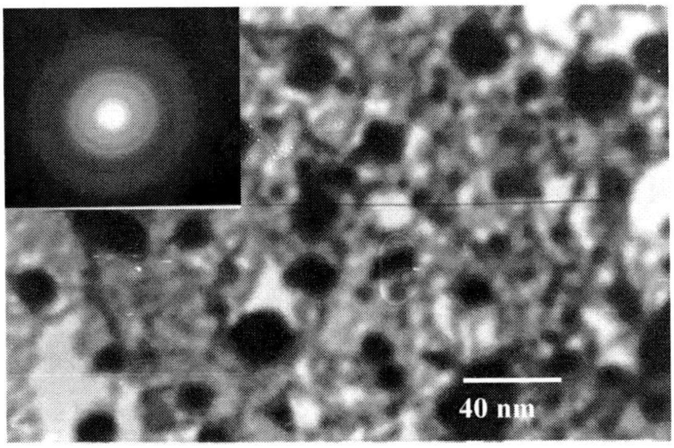

Figure 50(b). TEM micrograph of nano-structured $CuAlO_2$ thin film deposited for 9 min. **Inset**: SAED pattern of the same.

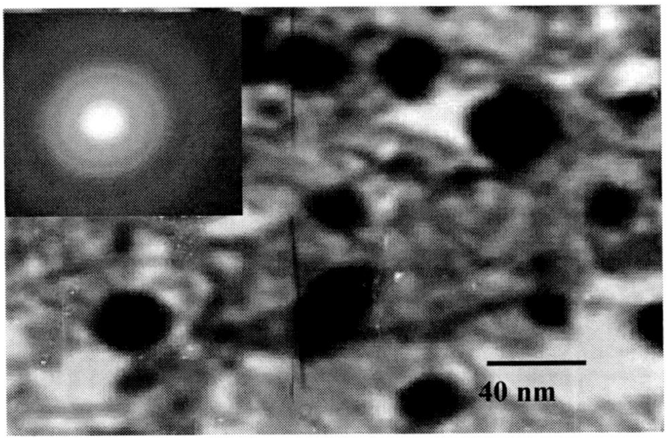

Figure 50(c). TEM micrograph of nano-structured $CuAlO_2$ thin film deposited for 15 min. **Inset**: SAED pattern of the same.

TEM studies: Transmission electron microscopic (TEM) analyses were done for nanocrystalline $CuAlO_2$ thin films with various deposition times (t_d). **Fig. 50(a)**, **(b)** and **(c)** show the TEM micrographs of $CuAlO_2$ films deposited in 3, 9 and 15 min respectively. From the micrographs, the particle sizes (L) are obtained around 8 nm to 12 nm for the films deposited in 3 min (**Fig. 50a**), around 18 to 22 nm for the films deposited in 9 min (**Fig. 50b**) and around 27 to 33 nm for the films deposited in 15 min respectively (**Fig. 50c**). Previously, Gong et al [150] and Gao et al [154] reported the particle size of their nanocrystalline copper aluminium oxide films around 10 nm, which is comparable to our samples deposited in 3 min. Similar to the XRD measurements, here also, from the TEM micrographs we have observed an increase in the average particle size of our nanocrystalline $CuAlO_2$ thin films with increase in the deposition times. And as already mentioned, this is mainly due to the greater amount of influx of sputtered particles, which results into the agglomeration of bigger particles. Thus the average particle size increases with increase in the deposition time as observed in **Fig. 50(a)**, **(b)** and **(c)** and when the deposition time is 45 min and above, the average particle size (L) becomes ~ 60 nm and more (not shown here). It is also note-worthy that there is a difference in the values of the particle size calculated from XRD data (L_{XRD}) and that obtained from TEM micrographs (L_{TEM}) for the films with t_d = 15 min and 45 min. For the films deposited in 15 min, L_{XRD} = 4 nm whereas average $L_{TEM} \approx 30$ nm. On the other hand, for the films deposited in 45 min, these values are 14 nm and 60 nm respectively. This is because the particle size calculated from **Eq. 6** always gives underestimated value as the term 'L', in **Eq. 6** is actually the 'crystallite size' rather than the 'particle size'. When the size of individual crystallites in a polycrystalline sample is less than 100 nm, the term 'crystallite size' is approximately taken to be equal to the 'particle size' [230]. But any individual grain or particle in a sample (whether it is nanocrystalline or else), always contain quite a few number of crystallites and therefore the information extracted from **Eq. 6** about the 'particle size' will always be less than the 'actual' particle size.

As we have observed marked differences of the particle size, measured directly from TEM image and as determined indirectly from X-ray diffraction peak broadening, particularly for films deposited with higher deposition times (t_a = 15 min and 45 min), we suppose that larger particles (as observed by TEM) might consist of a number of smaller crystallites and in that sense, larger particles are not single crystalline.

Selected area electron diffraction pattern (SAED) of the films deposited in 3 min, 9 min and 15 min are shown in the insets of **Fig. 50(a), (b)** and **(c)**. Few diffraction rings are obtained in all the patterns which correspond to the (101) & (202) planes of the films deposited in 3 min, (101) & $(00\overline{12})$ for the films with t_d = 9 min and (101) & (018) for the films deposited in 15 min respectively. The lattice spacings (d) corresponding to these rings in the diffraction patterns were measured with the camera constant of the equipment and the diffraction ring radii were measured from the micrographs [257]. These 'd'-values calculated from all the patterns along with that obtained from XRD measurements were then matched with the theoretical 'd'-values obtained from JCPDS file [113] and compared in **Table 22**. It has been observed that in all the SAED patterns, a (101) orientation is present, which is similar to the target material (c.f. **Fig. 7**) as well as to that of the film deposited for 15 min (as shown in the XRD pattern of **Fig. 48**; curve-**#1**). Therefore, this observation basically indicates the formation of quasi-continuous films consisting of $CuAlO_2$ nano-particles, when the deposition time is 15 min or less, (as has been depicted from TEM micrographs shown in

Fig. 50a, b and **c**), whereas with further increase in the deposition time (i.e. for $t_d \geq 45$ min), the growth mechanism followed a preferred (006) orientation.

Table 22. Comparison between the experimentally obtained d-values from SAED patterns (d_{SAED}) of the nano-crystalline $CuAlO_2$ thin films deposited for 3 min and 9 min and that of XRD patterns (d_{XRD}) for the films deposited for 15 min and 45 min respectively with that given in JCPDS file (d_{JCPDS}).

h k l	d_{SAED} (Å)			d_{XRD} (Å)		d_{JCPDS} (Å)
	t_d = 3 min	t_d = 9 min	t_d = 15 min	t_d = 15 min	t_d = 45 min	
0 0 6	---	---	---	---	2.816	2.820
1 0 1	2.438	2.441	2.450	2.448	2.450	2.437
0 1 2	---	---	---	2.350	2.380	2.376
1 0 7	---	---	---	1.732	---	1.732
0 1 8	---	---	1.618	1.607	1.610	1.612
0 0 $\overline{12}$	---	1.406	---	---	---	1.401
2 0 2	1.220	---	---	---	---	1.225

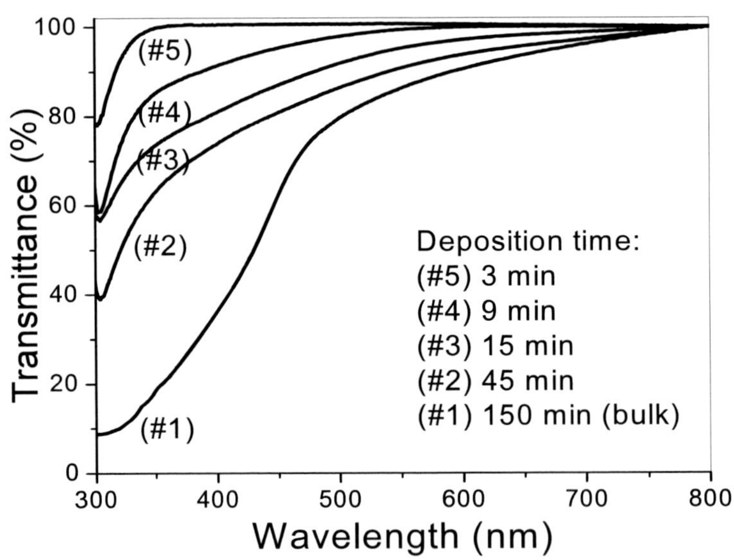

Figure 51(a). Optical transmission spectra of nanocrystalline $CuAlO_2$ thin films.

Optical properties: UV-Vis spectrophotometric measurements of $CuAlO_2$ thin films were done for the samples with deposition times 3 min, 9 min, 15 min, 45 min and 150 min. **Fig. 51(a)** shows the transmittance (T) vs. wavelength graphs of these films deposited on glass substrates taking similar glass as reference. Therefore the spectra are for the films only. Thickness of the films are in the range of 30 nm, 60 nm, 90 nm, 200 nm and 400 nm for the films deposited in 3 min, 9 min, 15 min, 45 min and 150 min respectively. The average visible transmittance of these films increases from 75 % to 98 % with decrease in the deposition time. This is mainly due to the decrease in the film thickness, which leads to lesser scattering and absorption of photons. **Fig. 51(b)** represents the spectral variation of reflectance (R) of the same films deposited on glass substrates. From the transmittance (T) and reflectance (R) data, the absorption coefficients (α) of these films were measured according to the **Eq. 16**. **Fig. 52** represents the spectral variation of α in the visible range. The value of α varies from 8.61×10^2 cm^{-1} for film deposited for 3 min to 2.56×10^4 cm^{-1} for film deposited for 150 min (bulk film)

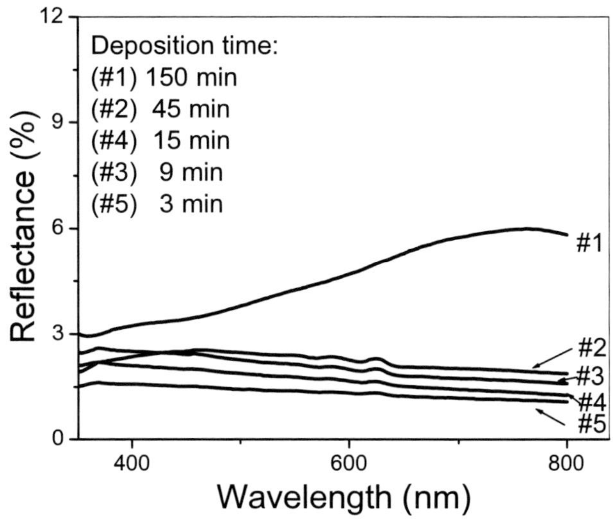

Figure 51(b). Spectral variation of reflectance of the nanocrystalline $CuAlO_2$ films.

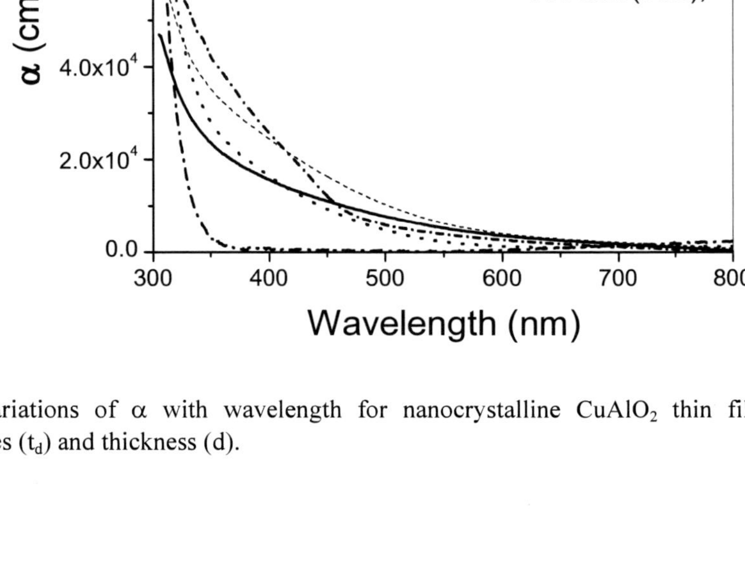

Figure 52. Variations of α with wavelength for nanocrystalline $CuAlO_2$ thin films for various deposition times (t_d) and thickness (d).

at 400 nm wavelength. Also the refractive indices (n) and extinction coefficients (k) of these films were determined according to **Eq. 15** and **14** respectively using the values of α and R. **Fig. 53(a)** and **(b)** show the wavelength vs. n and k plots respectively. Various optical parameters of nanocrystalline $CuAlO_2$ films deposited for different times are compared in **Table 23**.

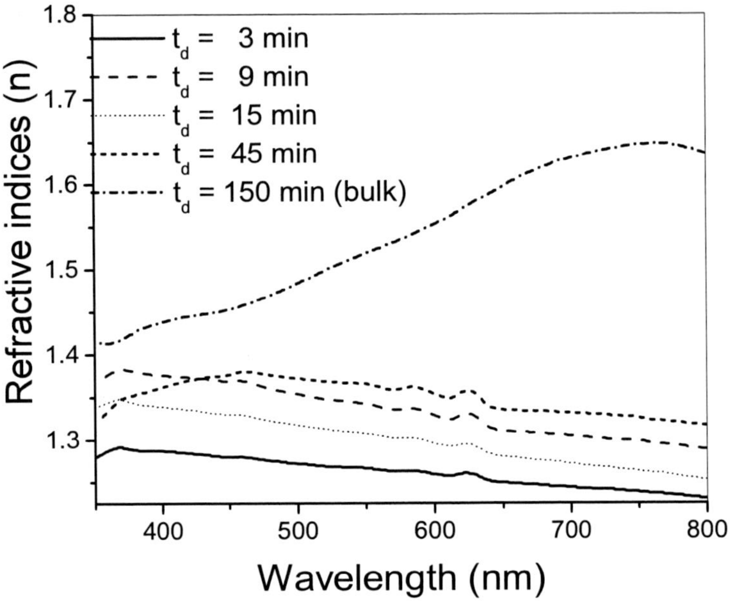

Figure 53(a). Dispersion of refractive indices of nanocrystalline $CuAlO_2$ thin films.

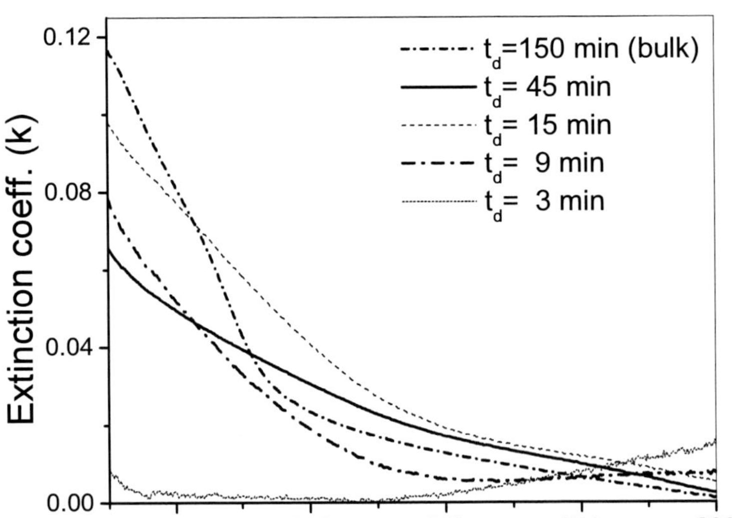

Figure 53(b). Spectral variation of extinction coeff. (k) of nanocrystalline $CuAlO_2$ thin films.

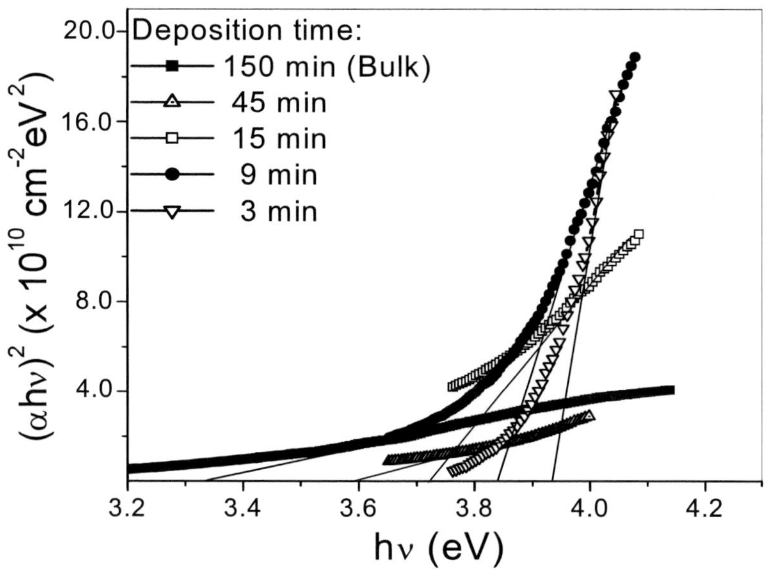

Figure 54. Determination of direct bandgaps of nanocrystalline $CuAlO_2$ thin film with different deposition times.

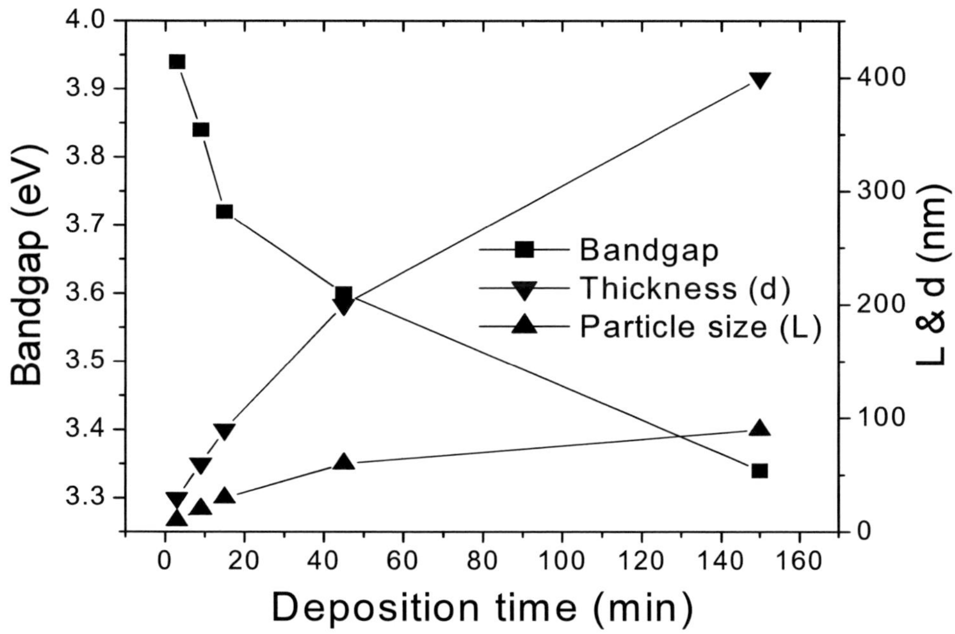

Figure 55. Variation of bandgap, particle size and thickness with deposition time.

In the range of the onset of absorption edge, the absorption coefficients (α) can be described by the relation for parabolic bands according to **Eq. 17**. The $(\alpha h\nu)^2$ vs. $h\nu$ plots for the films with different deposition times (t_d) is shown in **Fig. 54**. The direct allowed bandgap values for the films deposited for 3 min, 9 min, 15 min, 45 min and 150 min are obtained as 3.94 eV, 3.84 eV, 3.72 eV, 3.60 eV and 3.34 eV respectively. The corresponding average particle sizes (L) are 10 nm, 20 nm, 30 nm, 60 nm and greater than 90 nm respectively. The variation of bandgaps, particle size and film thickness with deposition time is shown in **Fig. 55**. Previously, Kawazoe and co-authors [64] reported the direct allowed bandgap of their pulsed laser deposited $CuAlO_2$ thin film around 3.5 eV with an average visible transmittance ~ 60 %. But as far as nanocrystalline $CuAlO_2$ thin films are concerned, Gong and co-authors [150] obtained the direct bandgap of their nanocrystalline Cu-Al-O films as 3.75 eV with an average visible transmittance ~ 35 %. This bandgap value is comparable to our films (3.72 eV) with deposition time 15 min. But the average visible transmittance of our sample is quite higher (~ 80 %) than that of Gong and co-authors (~ 35 %). This is mainly due to the presence of impurity (Cu_2O) in their sample as well as higher thickness (250 nm) of these films than ours (90 nm), leading to the scattering and absorption of photons. On the other hand Gao et al. [154] obtained the direct bandgap and average visible transmittance of their nanocrystalline $CuAlO_2$ thin film as 3.75 eV and 60 % respectively, which is nearly comparable to our values. From **Fig. 55**, we have observed the broadening of the bandgap energy of our nanocrystalline $CuAlO_2$ thin film with the decrease in the deposition time. This may be attributed to the quantum confinement effect put forward by Brus [69] where the size dependency of the bandgap of a semiconductor nanoparticle ($E_{g[nano]}$) is given by the following formula:

$$\Delta E = E_{g[nano]} - E_{g[bulk]} = \frac{h^2}{8\mu^* (\frac{L}{2})^2} - \frac{1.8e^2}{(\frac{L}{2})\varepsilon} \tag{24}$$

where ΔE is the shift of the bandgap with respect to the bulk bandgap $E_{g[bulk]}$, $\frac{L}{2}$ is the radius of the nano-particles (where L is the particle diameter, taken to be equivalent to the particle size, mentioned earlier), μ^* is the reduced mass of electron-hole effective masses and ε is the semiconductor dielectric constant. The first term of the RHS expression in the equation represents the particle-in-a box quantum localization energy and has an $\frac{1}{L^2}$ dependence for both electron and hole. The 2nd term represents the Coulomb energy with an $\frac{1}{L}$ dependence. In the limit of large L, the value of $E_{g[nano]}$ approaches that of $E_{g[bulk]}$. As TEM micrographs (Fig. **50a**, **b** and **c**) reveal that the average particle size of our samples decreases with the decrease in the deposition time (i.e. L ~ 10 nm and ~ 20 nm for t_d = 3 min & 9 min respectively and for t_d = 15 min and 45 min, L ~ 30 nm and ~ 60 nm, c.f. **Table 23**), the observation of bandgap widening in our samples is consistent with the quantum confinement effect explained by the **Eq. 24**. Previously, Gong et al [150] observed similar bandgap widening of their nanocrystalline Cu-Al-O films from bulk material and explained it in terms of the exciton confinement in semiconductor nanocrystals, which produces discrete, excited electronic states having higher oscillator strength and bandgaps as an inverse function of crystallite size [72, 258] The same group also observed a blue-shift of the bandgap of co-sputter-deposited Cu-Al-O films with a variation in the Cu : Al atomic ratio in their sample [146], but whether this was due to the size-dependant bandgap widening of semiconductor nanoparticles is not quite clear for their multiphase (a mixture of CuO and $CuAlO_2$) samples.

Table 23. Variation of average particle size, film thickness and bandgap with the deposition time of nano-crystalline $CuAlO_2$ thin film.

t_d (min)	Average particle size (nm)	Film thickness (nm)	Avg. T (%)	α *(at λ=400 nm)* (cm^{-1})	n *(at λ=400 nm)*	k *(at λ=400 nm)*	Band-gap (eV)
3	~ 10	30	95	8.61×10^2	1.29	0.003	3.94
9	~ 20	60	90	1.63×10^4	1.38	0.05	3.84
15	~ 30	90	80	2.44×10^4	1.34	0.08	3.72
45	~ 60	200	75	1.56×10^4	1.36	0.05	3.60
150 (bulk)	> 90	400	65	2.56×10^4	1.44	0.08	3.34

To examine the size-dependant optical properties of $CuAlO_2$ nanoparticles, the photoluminescence (PL) spectroscopic measurements were also performed at room temperature. The PL spectra shown in **Fig. 56** were obtained with a 210 nm excitation wavelength and the films were deposited on Si substrates. Three spectra shown in the **Fig. 56** are for three samples deposited for 9 min (curve – a), 15 min (curve – b) and 45 min (curve – c) respectively. Three peaks obtained are around 3.56 eV, for curve – c (λ = 348.5 nm), 3.61 eV, for curve – b (λ = 343.6 nm) and 3.66 eV, for curve – a (λ = 339.0 nm) respectively. These peaks may be attributed to the UV near-band edge (NBE) emission [252] of wide bandgap $CuAlO_2$, namely the recombination of free excitons through an exciton-exciton collision process. This observation again indicates the existence of direct transition type bandgap of this material, which is favorable for the optoelectronics applications like light-emitting diodes (LED). It is also note-worthy that, like other widegap semiconductors such as ZnO and LaO(CuS), where excitons can be observed at room temperature, the excitons in $CuAlO_2$ are supposed to have large binding energy (E_b). Although the exact value of the binding energy is not known yet, but the above argument seems reasonable if we get an indirect estimation of the binding energy by the following relation (assuming hydrogen-like model) [123]

$$E_b = \left(\frac{\mu}{m_O \varepsilon_r^2}\right) x\, 13.6\, eV \tag{25}$$

with

$$\frac{1}{\mu} = \frac{1}{m_e^*} + \frac{1}{m_h^*} \tag{26}$$

where μ, m_O, ε_r, m_e^* and m_h^* denote reduced mass, free electron mass, relative dielectric constant, effective masses of electrons and holes respectively. According to **Eq. 25**, large exciton binding energy would result from a small relative dielectric constant (ε_r) and a high-reduced mass (μ) of the excitons. The relative dielectric constants (ε_r) of $CuAlO_2$ thin film is estimated from the reflectance data (**Fig. 51b**), which fall within the range of 1.7 to 3.5, for the films deposited in 9 min, 15 min and 45 min respectively. These values are less than that of LaO(CuS) and ZnO [123, 259]. Therefore the reduced mass in $CuAlO_2$ may be considered to be large enough to generate room-temperature excitons. As has been suggested that the layered-crystal structure is responsible for the stability of excitons in LaO(CuS) [123], following similar argument, it may be considered that the super-lattice structure of $CuAlO_2$ [93-95] is responsible for the large reduced mass of the excitons, which, in turn, produces large binding energy to generate room temperature excitons in this material. Also, a slight blue-shift of the emission peaks are observed with decrease in the deposition time. As already mentioned previously, that a decrease in the average particle size (L) is observed with the decrease in the deposition time, t_d (i.e. for t_d = 9 min, 15 min and 45 min, L ~ 20 nm, ~ 30 nm and ~ 60 nm respectively), therefore this blue-shift may be another indication of experimentally observed bandgap enhancement results from low-dimensional quantum confinement effects.

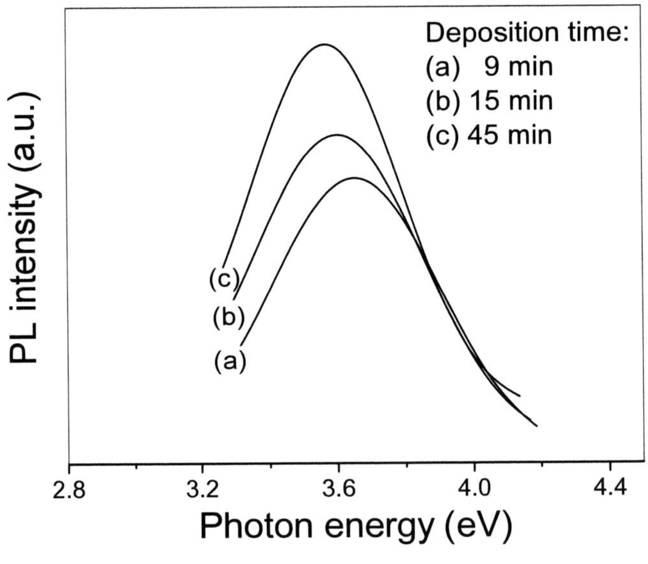

Figure 56. Photoluminescence spectra (PL) spectra of nano-structured $CuAlO_2$ thin films deposited for (a) 9 min, (b) 15 min and (c) 45 min.

To justify further whether quantum confinement is the most likely explanation for the observed blue-shift of the bandgap of our nano-structured $CuAlO_2$ thin films, we have tried to fit **Eq. 24** with reasonable values of μ and ε and the fit was satisfactory (within 5 % of the experimental value). On the other hand, forcing 100 % matching with **Eq. 24** and our data, we have obtained an estimation of effective excitonic mass (μ) around 0.03 m_O (m_O is the free electron mass), which appears reasonable. However, as so far there is no published data of effective masses of carriers of this material, hence, we cannot claim that the agreement is very accurate. But in another way we have indirectly estimated the value of μ from **Eq. 25**: As we have seen from photoluminescence measurements (**Fig. 56**) that our nanostructured $CuAlO_2$ films show exciton absorption at room temperature (300 K), therefore the binding energy (E_b) of excitons must exceed the thermal energy of $k_BT_{(T=300\ K)}$ = 26 meV. So, putting E_b approximately around 30 meV in **Eq. 25** and the value of ε_r as 3.5, obtained from reflectance data (**Fig. 51b**), (this value is comparable to that of other p-type TCO like LaOCuS, which is ~4.0 [123] as stated earlier), the value of μ comes out as 0.03 m_O. This value agrees with the previous one. So we can say that the quantum confinement effect is the most likely explanation for the bandgap widening of our nanostructured $CuAlO_2$ thin films.

Also we have observed slight decrease in the intensities of the peaks with decrease in the particle size. This may be due to the presence of some surface states in our nanostructured material. It is well known that the surface states may seriously influence the PL efficiency in nano-materials due to the high surface-to-volume ratio [260]. Larger the particle size, lesser will be the surface-to-volume ratio and therefore smaller will be the effect of surface states on PL intensity. That is why we have observed an increase in the PL intensity with the increase in the particle size as shown in **Fig. 56**. It must be mentioned here that as there are no reports on the photoluminescence properties of $CuAlO_2$ thin films (as literature survey depicts), therefore the exact mechanism of the different possible transitions is yet to be explored completely and intense research is needed in this direction to explore proper emission mechanism.

Hall measurements could not be performed in all of our samples. But p-type conductivity of the sample deposited for 150 min was established by thermo-power measurement and the positive value of room temperature Seebeck coefficient (S_{RT} = + 93 μV K^{-1}) of this sample confirmed the p-type nature of the film. But for the films deposited for 45 min and less, thermo-power measurement could not be performed and only hot-probe measurements confirmed the p-type conductivity in these films.

8. Field-Emission Properties of $CuAlO_2$ Thin Films

Low-macroscopic field (LMF) emission of electrons from the surface of a thin film to the vacuum in the presence of a macroscopic electric field (mean field between the parallel plates in a capacitor configuration) is currently of much interest due to the potential applications in cold cathode devices.

Here we have discussed the field-emission properties of $CuAlO_2$ thin film synthesized by both D.C. and reactive sputtering, and discussed in details the emission mechanism. As $CuAlO_2$ is a wide bandgap p-type semiconducting material, its field-emission properties may give an additional impetus on the properties of this technologically important material and may open-up a new window in the field-emission technology with a new group of materials other than carbon-based films.

8.1. Description of Apparatus

Field emission measurements were carried out by using a diode configuration consisting of a cathode (the film under test) and a stainless steel tip anode (conical shape with a 1 mm tip diameter) mounted in a liquid nitrogen trapped rotary-diffusion vacuum chamber with appropriate chamber baking arrangement. The measurements were performed at a base pressure of ~ 7 x 10^{-7} mbar. As the substrate glass was non-conducting, the negative terminal of the high voltage D. C. power supply (range is 0 to 5 kV) was connected with the films by silver paint, at least 6 mm away from the position of the anode tip. The Ohmic nature of the contact was checked before field emission experiment. The sheet resistance of our film was few hundred kilo-ohms/□ and the maximum emission current measured, was almost 30 μA. So, during emission process, if there was any voltage drop occurred between the contact and the portion of the sample just under the anode tip, it would be at the most, of the order of few volts, which was quite small compared to the applied voltages (~ kV). Hence we neglected this drop and all the calculations were done with the applied voltages. The tip-sample distance was continuously adjustable to a few hundred μm by spherometric arrangement with a screw-pitch of 10 μm. The tip was first touched to the sample surface and then raised by a controlled amount according to the spherometer-scale attached to the anode. The macroscopic field is calculated from the external voltage applied (V), divided by the anode-sample spacing, d (obtained from spherometric arrangement). The current was measured by Keithley electrometer (model 6514), whose current detection range is 100 aA to 21 mA. A current limit of 1 mA was set to avoid destruction of the films from excessive current flow. The whole surface of the film was visible through the chamber view-port, which enabled us to recognize the electron emission and discharge. It was confirmed that no discharge was taking place between the anode and the sample, so the current detected, was entirely due to cold field-emission process. The schematic diagram of the field-emission apparatus was shown previously [207].

8.2. Field-Emission Properties

Fig. 57 and **Fig. 58** show the emission current (I) vs. macroscopic field (E) curves of D. C. sputtered $CuAlO_2$ thin film (post-annealed for 60 min) and reactive sputtered $CuAlO_2$ thin film (post-annealed for 60 min) respectively Both the films were deposited on glass substrates and the anode-sample separations (d) were 100, 140 and 200 μm respectively for both the films. As obvious, it has been observed that in both cases, the curves are almost identical in nature. The macroscopic field is calculated from the external voltage applied (V), divided by the anode-sample spacing, d (obtained from spherometric arrangement in the field-emission apparatus).

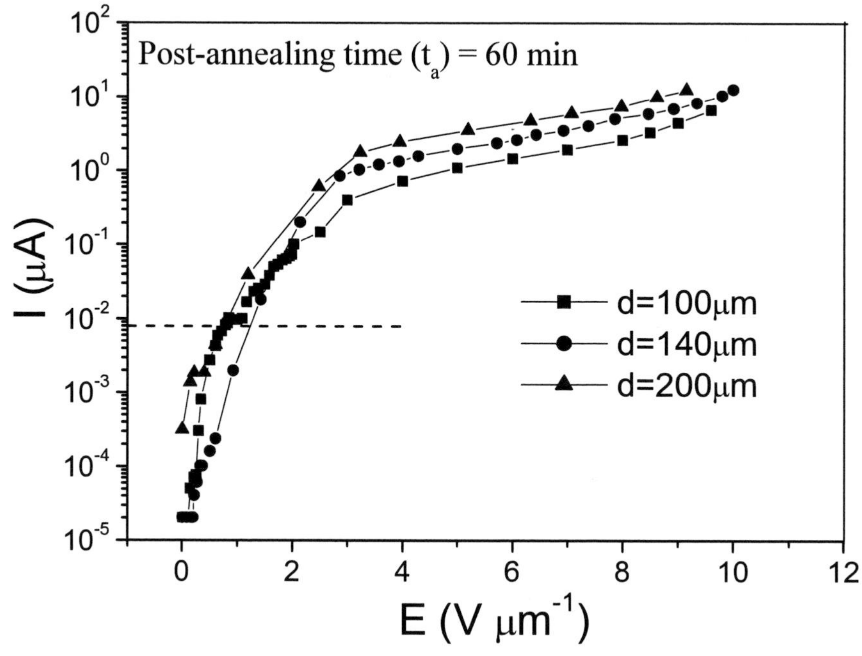

Figure 57. Emission current (I) vs. macroscopic field (E) curves for D. C. sputtered $CuAlO_2$ thin films post-annealed for 60 min, for different anode-sample (d) spacing.

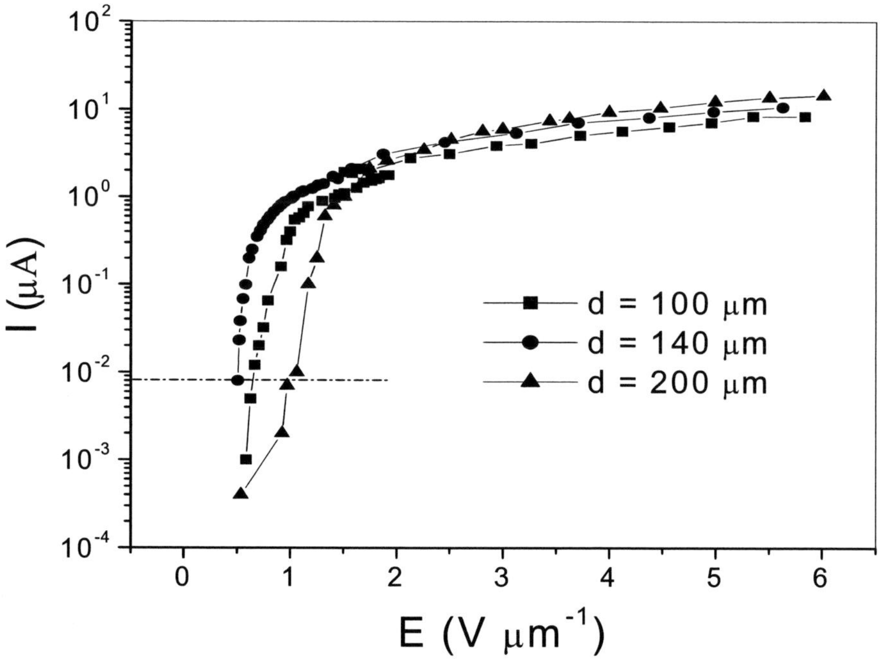

Figure 58. I-E curves for reactive sputtered $CuAlO_2$ thin film.

Theoretically, the emission current I is related to the macroscopic field E by

$$I = A a t_F^{-2} \phi^{-1} (\beta E)^2 \exp\{\frac{-b v_F \phi^{3/2}}{\beta E}\} \tag{27}$$

where, ϕ is the local work-function, β is the field enhancement factor discussed earlier, A is the effective emission area, a is the First F-N Constant (= 1.541434×10^{-6} A eV V^{-2}), b is the Second F-N Constant (= 6.830890×10^{9} $eV^{-3/2}$ V m^{-1}), and v_F and t_F are the values of the special field emission elliptic functions v and t [210], evaluated for a barrier height ϕ. In so-called Fowler-Nordheim coordinates, this equation takes the form:

$$\ln\{\frac{I}{E^2}\} = \ln\{t_F^{-2} A a \phi^{-1} \beta^2\} - \frac{(v_F b \phi^{3/2} \beta^{-1})}{E} \tag{28}$$

An experimental F-N plot is modified by the tangent to this curve, taken in the mid-range of the experimental data. This tangent can be written in the form [261, 262]:

$$\ln\{\frac{I}{E^2}\} = \ln\{r A a \phi^{-1} \beta^2\} - \frac{(s b \phi^{3/2} \beta^{-1})}{E} \tag{29}$$

where r and s are appropriate values of the intercept and slope correction factors, respectively. Typically, s is of the order of unity, but r may be of order 100 or greater. Both r and s are relatively slowly varying functions of 1/E, so an F-N plot (plotted as a function of 1/E) is expected to be a good straight line. The F-N plots of our samples are shown in **Fig. 59** and **Fig. 60**, for the D. C. sputtered and reactive sputtered films respectively. It has been observed that all the I-E curves in the present work are closely fitted with straight lines. This suggests that the electrons are emitted by cold field emission process. The turn-on field, which we define as the macroscopic field needed to get an emission current I = 8.0×10^{-3} μA [cf. **Fig. 57** and **58**], (which corresponds to an estimated macroscopic current density, J_{est} = 1μA/cm^2, where J_{est} = I/A, A = anode-tip area) was found at and around 0.5 to 1.2 V/μm for both the D. C. sputtered and reactive sputtered films. These values are comparable to the conventional low-threshold field emitters like carbon nano-fibres (~1.1V/μm) [86] but quite lower than that of amorphous carbon and DLC films (8 – 20 V/μm), diamond films (5 – 15 V/μm) [82, 85, 263-267], Si-C nanorods (13 – 20 V/μm) [87] etc. It is worthwhile to mention that the definition of the turn-on field is not universal. For carbonaceous emitters like a: C, DLC,

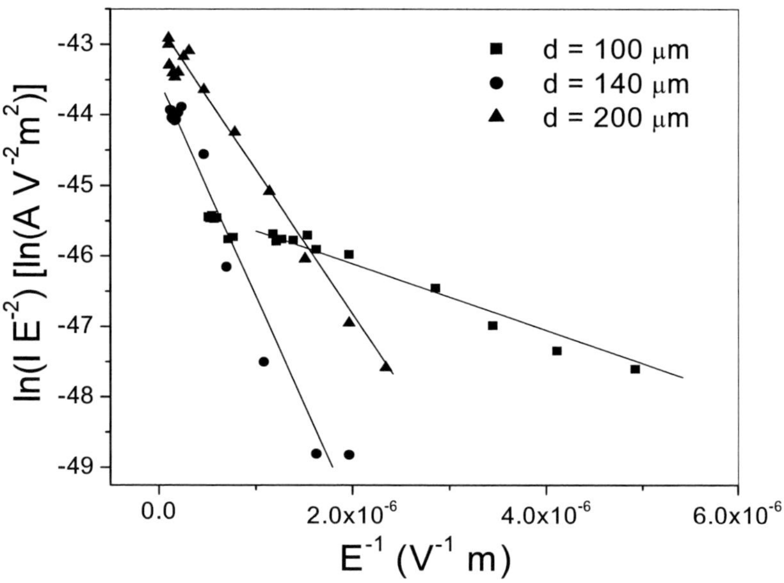

Figure 59. F-N plots of D. C. sputtered $CuAlO_2$ thin film.

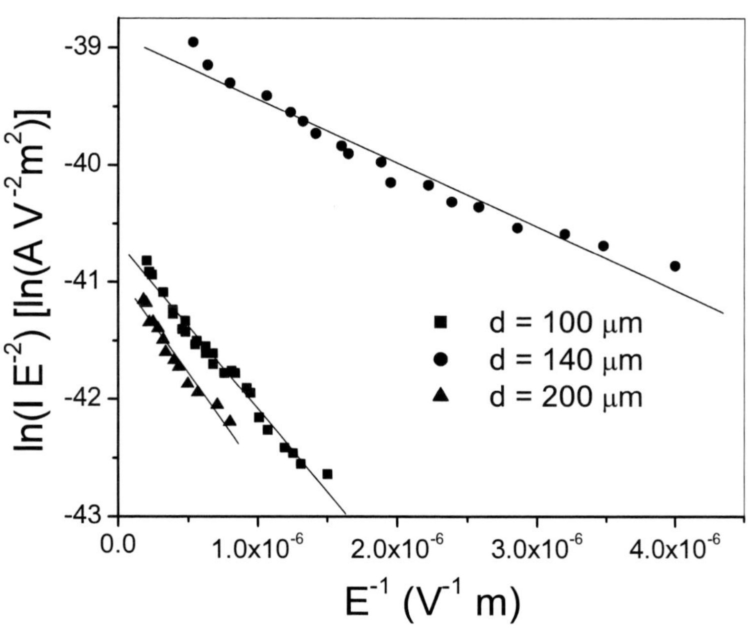

Figure 60. F-N plots of reactive D. C. sputtered $CuAlO_2$ thin film.

diamond etc., some authors [80, 265] had defined it as the field for which the macroscopic current density is $1\mu A/cm^2$. But Hirakuri et al. [82] considered this value as $0.01\mu A/cm^2$, whereas Robertson [263] had chosen this at $0.1\mu A/cm^2$. But for carbon nano-fibres and Si-C nanorods, as the maximum emission current is quite large, the turn-on field is defined at a higher macroscopic current density of the order of 10 $\mu A/cm^2$ [86, 87] to as large as 10 mA/cm^2 [268]. In both the I- E curves (**Fig. 57** and **58**), we have observed a parallel-shift of the curves w. r. t. anode-sample separation (d), but the nature of the curves is identical, i.e. a slight increase in the current with the increase in the anode-sample separation, at a given field, has been observed. For example, in **Fig. 57**, at a field of 4 V/μm, the I-values were found to be 2.5 μA [for d = 200 μm], 1.5 μA [for d = 140 μm] and 0.8 μA [for d = 100 μm] respectively. Similarly, in **Fig. 58**, at a field of 4 V/μm, the I-values were found to be 10.3 μA [for d = 200 μm], 7.6 μA [for d = 140 μm] and5.1 μA [for d = 100 μm] respectively. Similar observation was also reported by Zhou et al. [87], for their β-SiC nanorods. Although they have not given any reason for that, but we suppose that this type of shift observed for our sample, was probably due to the change in the effective emission area of the sample. And this change in the effective emission area w. r. t. 'd', might be related to the geometry of the anode. As mentioned earlier, the anode in our experiment is conical in shape with a tip diameter of 1 mm, therefore the lines of force emanating from the edge of the anode-tip, and terminating to the sample surface, are diverging in nature, whereas the lines of force emanating from the flat surface of the tip are parallel in nature (neglecting the small surface undulations of the highly polished anode-tip). Hence, the effective emission area of the sample becomes an increasing function of the anode-sample separation, 'd', as described schematically in **Fig. 61(a)**. It seems reasonable to consider this argument as a valid one, if we compare this with the experimental findings of Okano et al. [269] and Gröning et al. [270] for their diamond and DLC films respectively. Okano et al. [269] reported that their macroscopic current density was independent of the anode-sample separation. This might be related to the basic constructional differences between their field-emission apparatus and ours. Their field-emission apparatus consisted of a parallel plate arrangement of the anode and the sample, separated by spacers, as shown in **Fig. 61(b)**. So the electric lines of force between the anode and the sample were more or less parallel in nature; hence the effective emission area remained independent of the anode-sample spacing. On the other

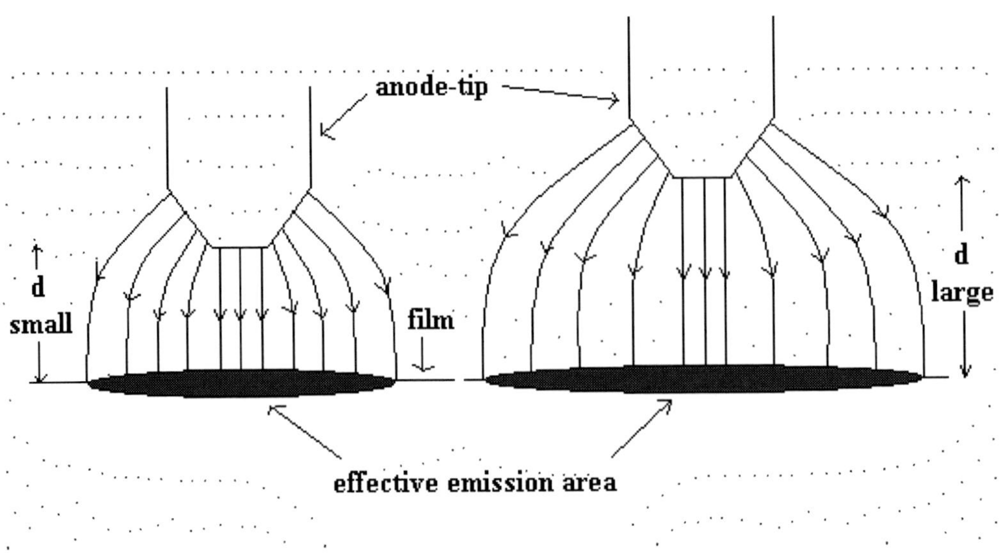

Figure 61(a). Schematic diagram of dependence of effective emission area as an increasing function of anode-sample separation, d. The sketch is not exactly to the scale.

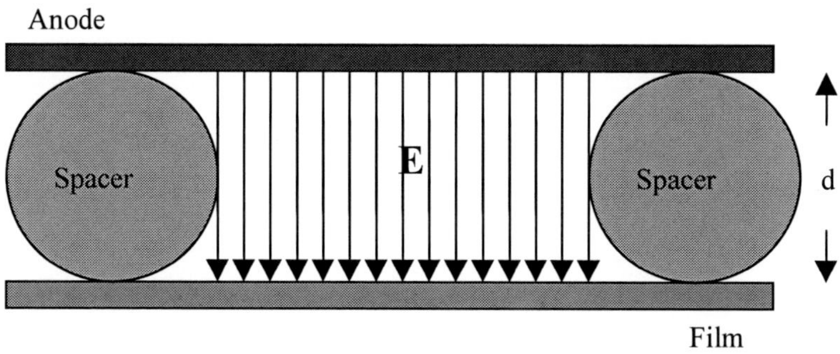

Figure 61(b). Schematic diagram of parallel plate configuration of field-emission apparatus. The sketch is not exactly to the scale.

hand, Gröning et al. [270] used a spherical stainless steel anode-tip in their field-emission apparatus. Hence, lines of force between the anode-tip and the sample in their experiment were diverging in nature. They observed a parallel shift in the I – E curves for their sample before and after raising the field to 50 V/μm for 1 hour of operation, keeping the anode-sample separation fixed. They argued that this parallel shift was due to the increase in the emission area of the sample and this area-enhancement was the result of the morphological changes occurred in the film during operation. Following this point of view, to see whether the area-increment in our samples is due to any morphological changes in the films or not, we have done the experiment in both ways: firstly, with increasing anode-sample spacing and secondly, with decreasing anode-sample spacing. But in both cases similar types of shifts in the I – E curves were observed, indicating no (or almost negligible) morphological changes occurred in the film during operation. So our argument, that in our experimental set-up, the effective emission area of the sample becomes an increasing function of the anode-sample separation, is justified.

According to **Eq. 29**, the slope of the tangent would carry the information of the local work function (ϕ) of the emitter-tip. Assuming an ideal flat emitter with field enhancement factor (β) equal to 1, we have obtained an estimation of the values of ϕ from the F-N plots (**Fig. 59** and **60**) to lie between 1.68×10^{-3} and 5.84×10^{-3} eV for D. C. sputtered films and 1.85×10^{-3} to 4.0×10^{-3} for reactive sputtered films. But the true local work function must be much larger than these values, due to the factor, β, which depends on the shape of the emitter. Forbes et al. [271] determined its value via the 'hemisphere on a post approximation' (**Fig. 62**) as:

$$\beta \approx \frac{0.7L}{R} \tag{30}$$

within the range $30 \leq L/R \leq 2000$. (where, L = height of the post, and R = radius of the hemisphere). Some comments on various models for the determination of field enhancement factor are given in Ref. [271]. In the previous section [cf. **Section 5**, **Table 14**], we have furnished the particle size of both D. C. sputtered and reactive sputtered

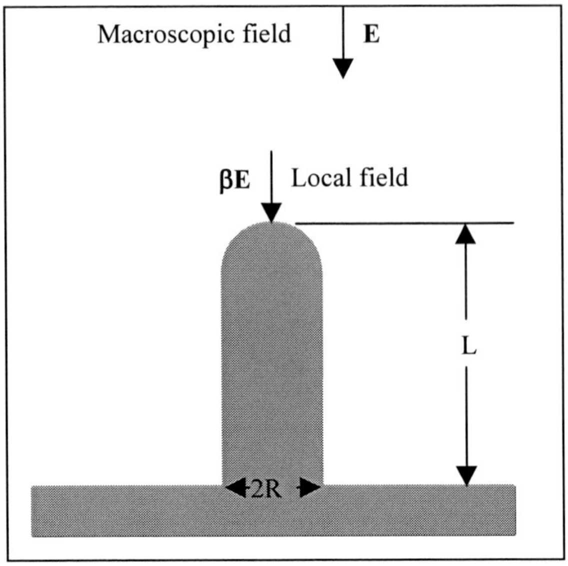

Figure 62. 'Hemisphere on a post' model. Macroscopic field, E, enhanced by a factor, β, determined by the height of the post, L, and radius of the hemisphere, R (Ref. [271]).

$CuAlO_2$ thin films as low as 26 nm and 32 nm respectively, obtained from XRD data, whereas the film thicknesses were around 0.5 μm, obtained from cross-sectional SEM. The nanometric particle sizes of our films are also justified from the SEM and TEM micrographs of the films shown in **Fig. 63(a)** and **63(b)** respectively. SEM image (**Fig 63a**) of reactive sputtered film depicts a smooth surface of the films, indicating very small particle size, beyond the resolution of the SEM used. On the other hand, TEM micrograph (**Fig. 63b**) of the same indicates some cluster formation with particle size roughly around 30 to 40 nm (SEM and TEM images of D. C. sputtered films are almost identical in nature, and hence not shown here). So the emission tip radius (R) of our sample would be of the order of few nanometers (considering the sharp emission tip radii are almost 10 % of the particle size) and assuming the height of the post (L) is equal to the film thickness, the approximate β-value obtained for our samples was around 180. This value of β falls within the range (150 – 300) predicted by Gröning et al. [270], for their N-doped DLC films and they stated that this sort of β-values was not unusual even on mirror like polished copper samples. So the local work functions increased by almost two orders of magnitude, and come out in the order of 0.15 eV to 0.2 eV. But still these are approximate values and quite less than the work function of the bulk material, especially for a wide bandgap, p-type semiconductor like $CuAlO_2$, as shown schematically in **Fig. 64(a)**. So these small values of φ may be treated as some barrier potential (barrier height, H, before Schottky lowering), which would give an estimation of the electron affinity, χ, of the material, depending on the mode of emission [213]. Although, the mechanism of the electron emission from this material is still not quite clear, but the low value of barrier height might be an indication of the dominant non-degenerate conduction-band emission, with an estimated electron affinity of the order of 0.2 eV. This low value of χ might be related to the wide bandgap of our film (E_g = 3.7 eV, for D. C. sputtered films and 3.9 eV for reactive sputtered films (cf. **Table 15**) respectively), as explained previously, by Robertson [81], for diamond films [212, 267], which is a p-type semiconductor having considerable large bandgap (E_g = 5.6 eV). But for a p-type semiconducting material, this type of sustained conduction-band emission is unlikely, unless injection to the conduction band inside the film may take place. This injection may be related to the internal nanostructure of our material.

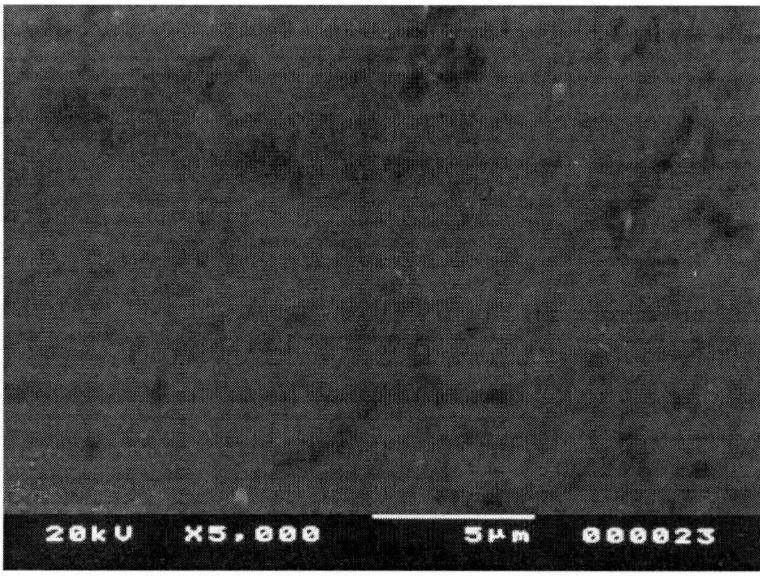

Figure 63(a). SEM micrograph of reactive sputtered $CuAlO_2$ thin film.

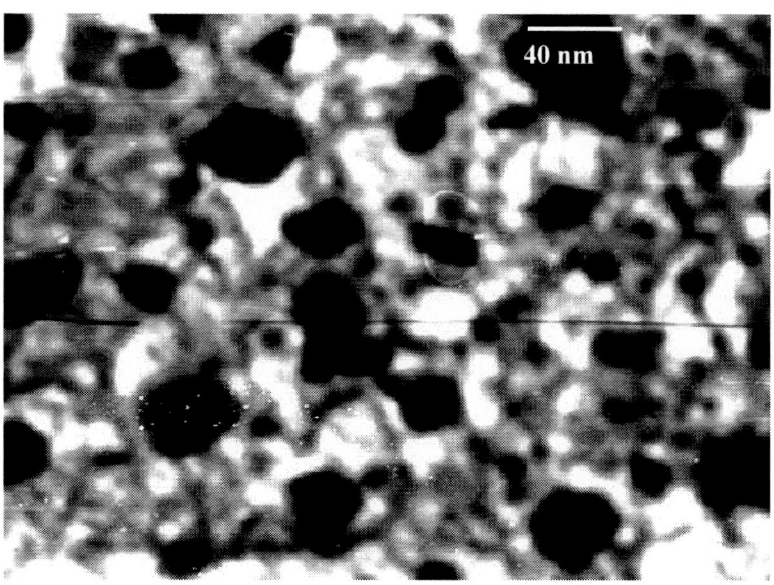

Figure 63(b). TEM micrograph of reactive sputtered $CuAlO_2$ thin film.

So, the 'ENH-material hypothesis', put forward by Forbes [213], that electrically nanostructured heterogeneous (ENH) materials with quasi-filamentary conducting channels inside a less conducting matrix, show low-macroscopic field emission, may be applicable to

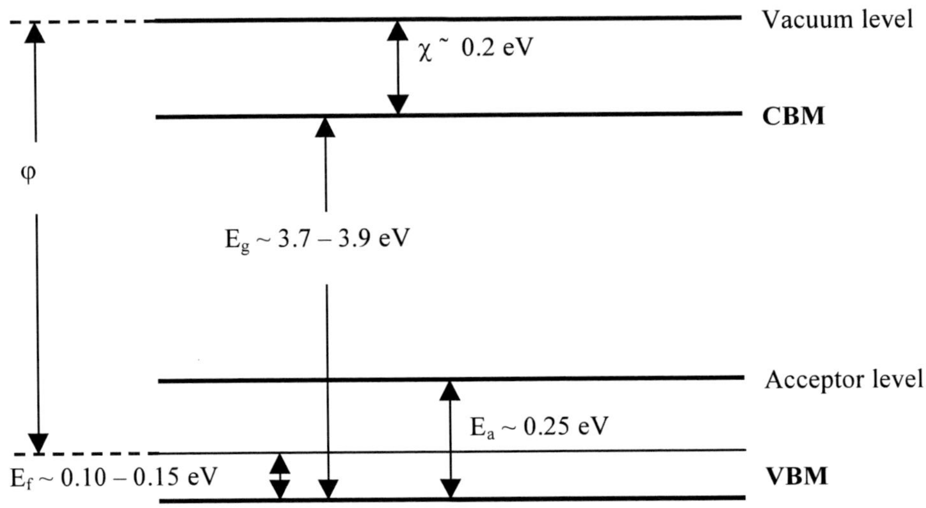

Figure 64(a). Schematic diagram of approximate energy level diagram of $CuAlO_2$ thin film. The energy levels are not exactly to the scale.

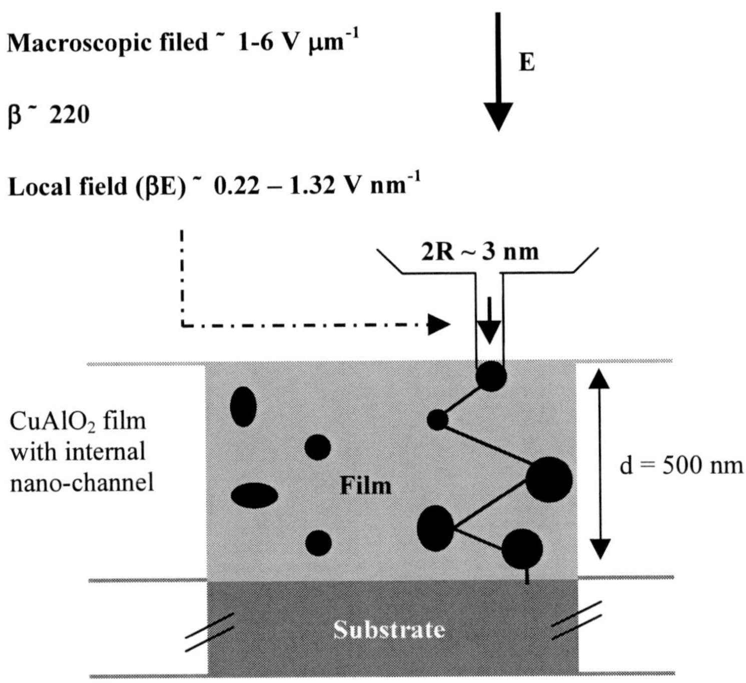

Figure 64 (b). Schematic diagram of ENH-model for $CuAlO_2$ film (After Ref.[213]).

our film also, as the particle size of our film was obtained around 30 to 40 nm with a film thickness of 500 nm (cf. **Fig. 63**). The mechanism is explained schematically in **Fig. 64(b),** where the macroscopic field (E ~ 1 to 6 V/μm) is enhanced due to the field enhancement factor ($\beta \sim 220$) by almost two to three orders of magnitude to produce large local field ($\beta E \sim$ 0.22 to 1.32 V/nm) and thus provides necessary energy for electron tunneling. So, geometrical field enhancement inside as well as at the film / vacuum interface is assumed to be the primary cause of the low-threshold field emission of our films. It is also to be noted that, recently several transparent, wide bandgap oxide-based thin films have been reported to show good field emission properties such as ITO [198], SnO_2 [199, 272], ZnO [200-205, 273] etc. So these results are very important, interesting and encouraging in the sense that these wideband transparent semiconductors can become perfect alternative to the carbon-based films in the area of field-emission displays.

9. Conclusion

Polycrystalline, p-type semiconducting, transparent $CuAlO_2$ thin films were deposited by D. C. sputtering of sintered $CuAlO_2$ powder on Si and glass substrates successfully. The post-deposition oxygen annealing time (t_a) of the films was taken as a variable parameter (from 30 min to 150 min) to observe any change in the optical and electrical characteristics of the films. XRD spectrum confirms the polycrystalline nature of the films with small grain size (~ 26 nm). All the films were highly transparent in the visible region. Both allowed direct and indirect transitions were found to exist in the films. Corresponding direct band gap values were determined to be around 3.7 to 3.8 eV for all the films. P-type conductivity was confirmed from both thermoelectric power and Hall effect measurements. Sputtered-deposited transparent p-type semiconducting $CuAlO_2$ thin film showed fairly high conductivity with a maximum room temperature conductivity in the range of 0.39 S cm^{-1} and a carrier concentration ~ 1.2×10^{18} cm^{-3}, for the films with t_a = 90 min. It appears that, to some extent, post-deposition annealing of the film in oxygen atmosphere controls the p-type conductivity of the film. Compositional analyses reveal an increase in the excess oxygen content within the films with t_a upto 90 min. These values range from 0.5 at % (for t_a = 30 min) to 5.0 at % (for t_a = 90 min) over stoichiometric value, whereas room temperature conductivities (σ_{RT}) increase from 0.09 S cm^{-1} to 0.39 S cm^{-1} (for annealing times 30 to 90 min respectively). This suggests that excess oxygen, within the crystallite sites, may be inducing nonstoichiometry in the film, which, in turn, manifests the improved p-type conductivity of the $CuAlO_2$ thin film. Activation energies for these films range from 270 meV to 196 meV respectively. FT-IR spectra of the films indicated the existence of various bondings among Cu, Al and oxygen.

Thermo-power measurements indicate that $CuAlO_2$ may become a candidate material for thermoelectric conversion. $CuAlO_2$ has a natural superlattice structure, with an effective two-dimensional density of states (along ab-plane). This type of layered-structure material could become a good thermoelectric converter. Room temperature Seebeck coefficients (S_{RT}) are found to be +230, +141 and +120 μV K^{-1} for t_a = 90, 60 and 30 min respectively, with E_f = 130, 151 and 200 meV respectively. An increase in S_{RT} with σ_{RT} is observed, which is expected for superlattice materials. Also, from band picture, it is observed that, higher the conductivity of the film, closer is its Fermi level to the upper edge of the valence band, which is obvious for a p-type material.

We have also successfully deposited polycrystalline p-type semiconducting $CuAlO_2$ thin films on glass and Si (400) substrates by reactive D. C. sputtering of a target, fabricated from a stoichiometric mixture of Cu and Al metal powders. XRD spectrum confirmed the polycrystalline nature of the films with small grain size (~ 32 nm). The films were transparent in the visible region. Both allowed direct and indirect band gaps were found to exist and their corresponding values were 3.75 eV and 1.85 eV respectively. The p-type conductivity was confirmed by positive values of the Seebeck and Hall coefficients. The films showed fairly high room temperature conductivity of the order of 0.22 S cm^{-1}. The carrier concentration in the films was found to be ~ 4.4×10^{17} cm^{-3}. This is due to nonstoichiometric defect attributed to excess oxygen atmosphere introduced into the system during deposition. From EDX analyses, the composition of the film is found to be Cu: Al: O = 1: 1: 2.08, supporting the above argument. FT-IR spectra of the films indicated the existence of various bondings among Cu, Al and oxygen.

Wet-chemical synthesis of transparent copper aluminium oxide thin films has been performed successfully. XRD-pattern confirms the proper phase formation of the film with a strong (006) orientation. SEM micrograph shows existence of a smooth surface with some bigger clusters dispersed on the surface, which resulted due to the agglomeration of finer grains. Cross-sectional SEM reveals the thickness of the film around ~1.5 μm. Optical transmittance spectra depicts almost 90 % transparency of the film in the wavelength range of 500 nm to 800 nm, with a direct allowed bandgap of 3.98 eV. Hot-probe measurement confirms the p-type nature of the film.

The cost-effective fabrication of this technologically important material is extremely important for the large-scale production of device quality films. Low-cost physical routes like D.C. and reactive sputtering as well as chemical synthesis of p-$CuAlO_2$ thin films will enable fabrication of high quality films for diverse device applications.

All-TCO p-n hetero-junction diodes having the structure: n-ZnO: Al / p-$CuAlO_2$ have been successfully fabricated on glass substrates. The current-voltage characteristics of the all-transparent heterojunction diode shows the rectifying properties, indicating proper formation of the junction. Maximum current obtained at 5 V is around 1 μA and the turn-on voltage obtained as ~ 0.8 V. The forward bias current is greater than the reverse bias current by approximately a factor of ~ 30 at ± 4 V. The optical transmision spectra of the n-ZnO: Al / p-$CuAlO_2$ diode showed visible transparency around 60 %, indicating its potential applications in the field of 'Transparent Electronics'.

Nanostructured p-type conducting $CuAlO_2$ thin films have been synthesized by D. C. sputtering with deposition time as the variable parameter. It has been observed from TEM micrographs that for the films deposited with as low as 3 min, particle size is around 10 nm with a film thickness ~ 30 nm. With increase in the deposition time, an increase in the particle size is observed in the films, which is attributed to the agglomeration of smaller particles into bigger ones due to the greater time of exposure of these films into sputtering plasma. Optical transmission spectra of the films show an increase in the average visible transmittance with decrease in deposition time. For the film deposited for 3 min, the average visible transmittance is almost 99 %. This is mainly due to the smaller thickness of the film (~30 nm), which reduces the scattering and absorption of photons within the films. Also a blue-shift or widening of the bandgap of the material is observed with decrease in deposition time. As particle size decreases with decrease in the deposition time, this bandgap broadening is attributed to the quantum confinement effect, where the bandgap of a semiconductor nanoparticle becomes an inverse function of the particle size. Room temperature photoluminescence measurements of this material showed UV bands around 3.56 eV to 3.66 eV for the films deposited for 45 min to 9 min respectively, which arises from the room temperature excitons. The existence of room-temperature excitons in $CuAlO_2$ is supposed to originate from low relative dielectric constant of the material and high reduced mass of the excitons, which produces large exciton binding energy. A blue-shift of the emission peaks is observed with decrease in the particle size, confirming the quantum confinement effect within the $CuAlO_2$ nanoparticles. The p-type conductivity of the films is confirmed by thermo-power as well as hot-probe measurements. This result of the synthesis of nano-crystalline p-$CuAlO_2$ (as well as other nanostructured p-TCO thin films, reported by various groups) will enable to fabricate nano-active devices which may give a new dimension in the field "Invisible Electronics".

Transparent p-$CuAlO_2$ thin films prepared by D. C. and reactive sputtering have been investigated for its field emission properties. The anode-sample distance was varied from 100 to 200 μm. The film showed considerable low turn-on field between 0.5 – 1.2 V/μm. The F-N plots were found to be straight-line in nature which indicates that the electrons are emitted via cold-field emission mechanism. This low macroscopic field emission of the films may be attributed to the internal nanostructure of the films, which creates significant field enhancement inside as well as near the film-vacuum interface. Also secondary effect, such as, the presence of surface states creating field enhancement, may not be ruled out. This finding might open up a new direction in the field emission technology, and a new type of materials (such as, different TCOs) might become a promising candidate for low-threshold field emitter.

10. Future Directions

First and foremost future course of our research will be to increase the conductivity of these p-TCO materials. The maximum conductivity of our p-$CuAlO_2$ film is almost two orders of magnitude less than that of commercially available n-TCO films. So this may put hindrance in the formation of effective active devices for large-scale production. It is found that nonstoichiometric oxygen intercalation within the material has its limitation to increase the conductivity of the film. Excess oxygen intercalation, beyond an optimum value, is found to deteriorate the film quality. So intentional doping of the material is the obvious step to increase the conductivity of the film. Identification of proper dopant and doping procedure will be the focus of our future work. Several theoretical articles have been published so far [163, 164, 227, 228, 274], indicating various doping materials and procedures to enhance the electrical characteristics of this material but no experimental work has yet been reported, as far as literature survey goes. Therefore, doping of p-$CuAlO_2$ thin film for superior device quality films is an important area of research for the development of "Transparent Electronics".

Another interesting area of research is the cost-effective fabrication of transparent junctions, without compromising its electro-optical properties. This is important for the large-scale production of various junctional devices with diverse applications. Here we have used a combination of cost-effective physical and chemical routes but future work will be aimed to improve the opto-electrical properties of these transparent junctions.

Another area of research, which is not yet explored completely, but has tremendous potential, is the thermoelectric properties of $CuAlO_2$ films. Being a layered-structured material, this material as well as other delafossite materials can become very good candidate for thermoelectric converters. Recently, Park et al [275] have reported a significant increase in the thermoelectric properties of this material for Ca substitution in Al sites at high temperature. They have observed the highest value of power factor around 7.82×10^{-5} Wm^{-1} K^{-2} for $CuAl_{0.9}Ca_{0.1}O_2$ at 1140 K. If proper studies can be done on the thermoelectric properties of these types of superlattice materials, new horizon may open up in the field of thermoelectric converters.

Also keeping an eye in the tremendous progress in nanotechnology, fabrication and characterization of nano-structured p-$CuAlO_2$ as well as other p-TCO thin films may become an important field of work, because of new and interesting properties exhibited by these nano-materials. Proper fabrication procedure to get reproducible nano-materials is the most important future work. Also in-depth studies of the photoluminescence properties of p-$CuAlO_2$ nano-particles will be another area of research, which is needed to be explored properly. Fabrication of nano-structured p-TCOs, coupled with the already existing and well-known materials of nano-structured n-TCOs, will give an added impetus in the field of "Invisible Electronics" for the fabrication of nano-active devices, which can have high-efficient applications in the optoelectronics device technology.

Field-emission property of $CuAlO_2$ thin films is a completely new area of research, which has tremendous opportunities. This material showed very low turn-on field comparable to most of the carbonaceous low-threshold field-emitters like CNT, DLC, diamond, a:C, SiC-nanorods etc. So these types of TCO materials may become promising alternative to the existing materials in the field of FED technology. But, proper emission mechanism in these materials is not very clear till date and very good scopes are there to properly investigate the emission mechanism so that the material properties can be tuned accordingly to get better field-emission properties of these films.

Also recent study showed that p-$CuAlO_2$ can become a good candidate for ozone sensors. Zheng and co-authors [276] reported that $CuAlO_2$ has a selective and reversible response to ozone gas at room temperature. All existing commercial semiconductor ozone sensors are of n-type [277-280]. This study demonstrated the feasibility of developing an inexpensive p-type transparent ozone sensor. Hence, transparent p–n junction ozone sensors may be fabricated using the p-$CuAlO_2$ and existing n-TCO such as In_2O_3.

Photocatalytic hydrogen evolution over delafossite $CuAlO_2$ is another interesting report published recently by Koriche et al [281]. This group proposed a new photochemical system for water reduction based on p-$CuAlO_2$ and S^{2-} as hole scavenger. They have used co-precipitation method, a new synthetic route, to synthesis $CuAlO_2$, which increased the surface to volume ratio and delivered a highest H_2 production. This report is very interesting and shows newer applications of delafossite p-$CuAlO_2$ material.

Also, recently Kizaki and co-authors [282] proposed a materials designing procedure to get $CuAlO_2$-based dilute magnetic semiconductors. Ab-initio calculations showed that Fe- and Mn-doped $CuAlO_2$-based dilute magnetic semiconductors possess high-Curie-temperature ferromagnetic characteristics. Being a natural p-type transparent semiconductor without intentional doping, $CuAlO_2$ can easily be used for the host of dilute magnetic semiconductors. Also, most importantly, the delafossite structure of $CuAlO_2$ has the advantage of possessing two cation-sites, Cu^{+1} and Al^{+3} sites, for possible magnetic ion substitution. O–Cu–O dumbbell-sites in delafossite $CuAlO_2$ can be partially replaced with magnetic ions. Due to this coordination one can realize new ferromagnetic dilute magnetic semiconductors from the standpoint of the hybridization of orbitals between 3d orbitals with the impurities and 2p orbitals with the oxygen in $CuAlO_2$.

Therefore, it will not be an exaggeration to say that next decade will see the renaissance of delafossite materials and various new, interesting and novel technological applications with these materials are on the verge of exploration.

Acknowledgement

The authors wish to gratefully acknowledge the financial assistance of the Department of Science and Technology, Govt. of India, during the execution of the work.

References

[1] Badekar, K. *Ann. Phys. (Leipzig)* 1907, 22, 749.
[2] Lampert, C. M. *Sol. Energy Mater.* 1981, 6, 1-41.
[3] Chopra, K. L.; Major, S.; Pandya, D. K. *Thin Solid Films* 1983, 102, 1-46.
[4] Hamberg, I.; Granqvist, C. G. *J. Appl. Phys.* 1986, 60, R123-R160.
[5] Holland, L. Vacuum Deposition of Thin Films; Wiley: New York, NY, 1958; pp. 492.
[6] Cachet, H.; Gamard, A.; Campet, G.; Jousseaume, B.; Toupance, T. *Thin Solid Films* 2001, 388, 41-49.
[7] Wendt, R.; Ellmer, K. *Surf. Coat. Technol.* 1997, 93, 27-31.
[8] Hartnagel, H. L.; Dawar, A. L.; Jain, A. K.; Jagadish, C. *Semiconducting Transparent Thin Films*; IOP Publishing Ltd.: Bristol and Philadelphia, PA, 1995.
[9] Ginley, D. S.; Bright, C. *MRS Bull.* August 2000, pp.15.
[10] Cava, R. J.; Philips, J. M.; Kwo, J.; Thomas, G. A.; van Dover, R. B.; Carter, S. A.; Krajewski, J. J.; Peck Jr., W. F.; Marshall, J. H.; Rapkin, D. H. *Appl. Phys. Lett.* 1994, 64, 2071-2073.
[11] Philips, J. M.; Cava, R. J.; Thomas, G. A.; Carter, S. A.; Kwo, J.; Siegrist, T.; Krajewski, J. J.; Marshall, J. H.; Peck Jr., W. F.; Rapkine, D. H. *Appl. Phys. Lett.* 1995, 67, 2246-2248.
[12] Freeman, A. J.; Poeppelmeier, K. R.; Mason, T. O.; Chang, R. P. H.; Marks, T. J. *MRS Bull.* August 2000, pp.45.
[13] Gordon, R. G. *MRS Bull.* August 2000, pp.52.
[14] Mochel, J. M. *U.S. Patent* No. 2,564,706, 1947.
[15] Sinclair, W. R.; Peters, F. G.; Stillinger, D.W.; Koonce, S. E. *J. Electrochem. Soc.* 1965, 112, 1096.
[16] Lytle, W. O.; Junge, A. E. *U. S. Patent* No. 2,566,346 (1951).
[17] Gordon, R. G. *U. S. Patent* No. 4,146,657, 1979.
[18] Banerjee, A. N.; Kundoo, S.; Saha, P.; Chattopadhyay, K. K. *J. Sol-Gel Sci. Technol.* 2003, 28, 105-110.
[19] Banerjee, A. N.; Maity, R.; Kundoo, S.; Chattopadhyay, K. K. *Phys. Stat. Solidi A* 2004, 201, 983-989.

[20] Major, S.; Banerjee, A.; Chopra, K. L. *Thin Solid Films* 1984, 122, 31-43.

[21] Minami, T.; Nanto, H.; Takata, S. *Jpn. J. Appl. Phys*., Part 2: Lett. 1984, 23, L280-L282.

[22] Hu, J.; Gordon, R. G.; *Sol. Cells* 1991, 30, 437-450.

[23] Vijayakumar, P. S.; Blaker, K. A.; Welting, R. D.; Wong, B.; Halani, A. T.; Park, C. *U. S. Patent* No. 4,751,149, 1988.

[24] Choi, B. H.; Im, H. B.; Song, J. S.; Yoon, K. H. *Thin Solid Films* 1990, 193, 712-720.

[25] Hu, J.; Gordon, R. G. *J. Appl. Phys*. 1992, 72, 5381-5392.

[26] Mochel, J. M.; *U. S. Patent* No. 2, 564,707, 1951.

[27] Holland, L.; Siddall, G. *Vacuum III* 1955.

[28] Avaritsiotis, J. N.; Howson, R. P. *Thin Solid Films* 1981, 77, 351-357.

[29] Groth, R. *Phys. Stat. Solidi* 1966, 14, 69.

[30] Nozik, A. J.; *U. S. Patent* No. 3,811,953, 1974.

[31] Haacke, G.; Mealmaker, W. E.; Siegel, L. A. *Thin Solid Films* 1978, 55, 67-81.

[32] Vossen, J. L. *RCA Rev*. 1971, 32, 269.

[33] Enoki, H.; Nakayama, T.; Echigoya, *J. Phys. Stat. Solidi A* 1992, 129, 181.

[34] Minami, T.; Sonohara, H.; Takata, S.; Sato, H. *Jpn. J. Appl. Phys*., Part 2: Lett. 1994, 33, L1693-L1696.

[35] Phillips, J. M.; Kwo, J.; Thomas, G. A.; Carter, S. A.; Cava, R. J.; Hou, S. Y.; Krajewski, J. J.; Marshall, J. H.; Pack, W. F.; Rapkin, D. H.; van Dover, R. B. *Appl. Phys. Lett*. 1994, 65, 115-117.

[36] Otabe, T.; Ueda, K.; Kudoh, A.; Hosono, H.; Kawazoe, H. *Appl. Phys. Lett*. 1998, 72, 1036-1038.

[37] Dali, S. E.; Jayachandran, M.; Chockalingam, M. J. J. Mater. Sci. Lett. 1999, 18, 915-917.

[38] Edwards, D. D.; Mason, T. O. Goutenoire, F.; Poeppelmeier, K. R.; *Appl. Phys. Lett*. 1997, 70, 1706-1708.

[39] Minami, T.; Takata, S.; Kakumu, T.; Sonohara, H. *Thin Solid Films* 1995, 270, 22-26.

[40] Minami, T.; Kakumu, T.; Shimokawa, K.; Takata, S. *Thin Solid Films* 1998, 317, 318-321.

[41] Omata, T.; Ueda, N.; Ueda, K.; Kawazoe, H. *Appl. Phys. Lett*. 1994, 64, 1077-1078.

[42] Kammler, D. R.; Mason, T. O.; Young, D. L.; Coutts, T. J. *J. Appl. Phys*. 2001, 90, 3263-3268.

[43] Minami, T. *MRS Bull*. August 2000, pp. 38.

[44] Lewis, B. G.; Paine, D. C. *MRS Bull*. August 2000, pp. 22.

[45] Granqvist, C. G.; Azens, A.; Hjelm, A.; Kullman, L.; Niklasson, G. A.; Ronnow, D.; Mattson, M. S.; Veszelei, M.; Vaivars, G. *Sol. Energy* 1998, 63, 199-216.

[46] Kammler, D. R.; Edwards, D. D.; Ingram, B. J.; Mason, T. O.; Palmer, G. B.; Ambrosini, A.; Poeppelmeier, K. R. In Photovoltaics for the 21st Century; Kapur, V. K.; McConnel, R. D.; Carlson, D.; Ceasan, G. P.; Rohatgi A.; Ed.; *The Electrochem. Soc. Proc*. **99-11**; Pennington, NJ, 1999; pp. 68.

[47] Lee, S. H.; Hwang, K. H.; Joo, S. K. In Electrochromic Materials; Ho, K. C.; McArthur D. A.; Ed.; *The Electrochem. Soc. Proc*. **94-2**; Pennington, NJ, 1994; pp. 290.

[48] Lugg, P. S.; Bommarito, S.; Bailey, J.; Budd, K.; Cullen, P.; Chen, K.; Hardy, L. C.; Nachbor, M. In Solid State Ionic Devices; Wachsman, E.D.; Akridge, J. R.; Liu, M.; Yamazoe, N.; Ed.; *The Electrochem. Soc. Proc*. **99-13**; Pennington, NJ, 1999; pp. 284.

[49] Nakato, Y.; Kai, K. I.; Kawabe, K. *Sol. Energy* Mater. *Sol. Cells* 1995, 37, 323-335.
[50] Emons, T. T.; Li, J.; Nazar, L. F. J. *Am. Chem. Soc.* 2002, 124, 8517.
[51] Hutchins, M. G.; McMeeking, G. D. *Biosensor Patent* No. 90,27,607.2, 1990.
[52] He, Y.; Kanicki, *J. Appl. Phys. Lett.* 2000, 76, 661.
[53] Wager, J. F. *Science* 2003, 300, 1245.
[54] Lampert, C. M. *Solar Energy Materials* 1990, 21, 191.
[55] Grivas, C.; Mallis, S.; Boutsikaris, L.; Gill, D. S.; Vainos, N. A.; Chandler, P. J. *Laser Phys.* 1998, 8, 326.
[56] Moschovis, K.; Gagaoudakis, E.; Chatzitheodoridis, E.; Kiriakidis, G.; Mailis, S.; Tzamali, E.; Vainos, N. A.; Fritzsche, H. *Appl. Phys. A* 1998, 66, 651.
[57] Moller, S.; Perlov, C.; Jackson, W.; Taussig, C.; Forest, S. R. *Nature* 2003, 426, 166.
[58] Hayes, R. A.; Feenstra, B. J. *Nature* 2003, 425, 1038.
[59] Cornia, R. L.; Fenn, J. B.; Memarian, H.; Ringer, R. In *Proc. 41st Annual Technical Conference;* Soc. of Vacuum Coaters; Boston, MA, 1998; pp. 452.
[60] Cairns, D. R.; Witte, R. P.; Sparacin, D. K.; Sachsman, S. M.; Paine, D. C.; Crawford, G. P.; Newton, R. *Appl. Phys. Lett.* 2000, 76, 1425.
[61] Banerjee, A. N.; Ghosh, C. K.; Chattopadhyay, K. K.; Minoura, H.; Sarkar, A. K.; Akiba, A.; Kamiya, A.; Endo, T. *Thin Solid Films* 2006, 496, 112-116.
[62] Seager, C. H.; McIntyre, D. C.; Warren, W. L.; Tuttle, B. A. *Appl. Phys. Lett.* 1996, 68, 2660.
[63] Prince, M. W. J.; Gross-Holtz, K. O.; Muller, G.; Cillessen, J. B.; Giesbers, J. B.; Weening, R. P.; Wolf, R. M. *Appl. Phys. Lett.* 1996, 68, 3650.
[64] Kawazoe, H.; Yasukawa, M.; Hyodo, H.; Kurita, M.; Yanagi, H.; Hosono, H. *Nature* 1997, 389, 939.
[65] Thomas, G. *Nature* 1997, 389, 907.
[66] Sato, H.; Minami, T.; Takata, S.; Yamada, T. *Thin Solid Films* 1993, 236, 27.
[67] Yanagi, H.; Inoue, S.; Ueda, K.; Kawazoe, H.; Hosono, H.; Hamada, N. *J. Appl. Phys.* 2000, 88, 4159.
[68] Efros, A. L.; Efros, A. L. *Sov. Phys. Semicond.* 1982, 16, 772.
[69] Brus, L. E. *J. Chem. Phys.* 1984, 80, 4403.
[70] Cox, A. J.; Louderback, J. G.; Bloomfield, L. A. *Phys Rev. Lett.* 1993, 71, 923.
[71] Alivisatos, A. P. *Science* 1996, 271, 933.
[72] Heath, J. R. *Science* 1995, 270, 1315.
[73] Demel, T.; Heitmann, D.; Grambow, P.; Ploog, K. *Phys. Rev. Lett.* 1990, 64, 2559.
[74] Li, Q.; Wang, C. *Appl. Phys. Lett.* 2003, 83, 359.
[75] Demel, T.; Heitmann, D.; Grambow, P.; Ploog, K. *Phys. Rev. Lett.* 1991, 66, 2657.
[76] Klein, D. L.; Roth, R.; Lim, A. K. L.; Alivisatos, A. P.; McEuen, P. L. *Nature* 1997, 389, 699.
[77] Feldheim, D. L.; Keating, C. D. *Chem. Soc. Rev.* 1998, 28, 1.
[78] Fink, R. L.; Li Tolt, Z.; Yaniv, Z. *Surf. Coat. Technol.* 1998, 108-109, 570.
[79] May, P. W.; Hohn, S.; Wang, W. N.; Fox, N. A. *Appl. Phys. Lett.* 1998, 72, 2182.
[80] Baek, Y. G.; Ikuno, T.; Ryu, J. T.; Honda, S. I.; Katayama, M.; Oura, K.; Hirao, T. *Appl. Surf. Sci.* 2002, 185, 243.
[81] Robertson, J. Diamond Relat. Mater. 1996, 5, 797.
[82] Hirakuri, K.; Yokoyama, T.; Enomoto, H.; Mutsukura, N.; Friedbacher, G. *J. Appl. Phys.* 2001, 89, 8253.

[83] Yanagi, H.; Kawazoe, H.; Kudo, A.; Yasukawa, M.; Hosono, H. *J. Electrocerem.* 2000, 4, 427.
[84] May, P. W.; Hohn, S.; Wang, W. N.; Fox, N. A. *Appl. Phys. Lett.* 1998, 72, 2182.
[85] Baek, Y. G.; Ikuno, T.; Ryu, J. T.; Honda, S. I.; Katayama, M.; Oura, K.; Hirao, T. *Appl. Surf. Sci.* 2002, 185, 243.
[86] Chen, C. F.; Lin, C. L.; Wang, C. M. *Appl. Phys. Lett.* 2003, 82, 2515.
[87] Zhou, X. T.; Lai, H. L.; Peng, H. Y.; Au, F. C. K.; Liao, L. S.; Wang, N.; Bello, I.; Lee, C. S.; Lee, S. T. *Chem. Phys. Lett.* 2000, 318, 58.
[88] Tate, J.; Jayaraj, M. K.; Draeseke, A. D.; Ulbrich, T.; Sleight, A. W.; Vanaja, K. A.; Nagrajan, R.; Wager, J. F.; Hoffman, R. L. *Thin Solid Films* 2002, 411, 119.
[89] Nagrajan, R.; Duan, N.; Jayaraj, M. K.; Li, J.; Vanaja, K. A.; Yokochi, A.; Draeseke, A.; Tate, J.; Sleight, A. W. *Int. J. Inorg. Mater.* 2001, 3, 265.
[90] Nagrajan, R.; Uma, S.; Jayaraj, M. K.; Tate, J.; Sleight, A. W. *Solid State Science* 2002, 4, 787.
[91] Norton, D. P. *Mat. Sci. Engg.* 2004, R 43, 139.
[92] Banerjee, A. N.; Chattopadhyay, K. K. *Prog. Cryst, Growth and Charac. Mater.* 2005, 50, 52-105.
[93] Shannon, R. D.; Rogers, D. B.; Prewitt, C. T. *Inorg. Chem.* 1971, 10, 713.
[94] Prewitt, C. T.; Shannon, R. D.; Rogers, D. B. *Inorg. Chem.* 1971, 10, 719.
[95] Rogers, D. B.; Shannon, R. D.; Prewitt, C. T.; Gilson, J. L. *Inorg. Chem.* 1971, 10, 723.
[96] Hahn, H.; Lorent, C. Z. *Anorg. Allg. Chem.* 1955, 279, 281.
[97] Benko, F. A.; Koffyberg, F. P. *J. Phys. Chem. Solids* 1984, 45, 57.
[98] Ishiguro, T.; Kitazawa, A.; Mizutani, N.; Kato, M. *J. Solid State Chem.* 1981, 40, 170.
[99] Ishiguro, T.; Ishizawa, N.; Mizutani, N.; Kato, M. *J. Solid State Chem.* 1982, 41, 132.
[100] Ishiguro, T.; Ishizawa, N.; Mizutani, N.; Kato, M. *Acta. Cryst.* 1983, B39, 564.
[101] Ueda, K.; Hase, T.; Yanagi, H.; Kawazoe, H.; Hosono, H.; Ohta, H.; Orita, M.; Hirano, M. *J. Appl. Phys.* 2001, 89, 1790.
[102] Yanagi, H.; Hase, T.; Ibuki, S.; Ueda, K.; Hosono, H. *Appl. Phys. Lett.* 2001, 78, 1583.
[103] Yanagi, H.; Ueda, K.; Ohta, H.; Orita, M.; Hirano, M.; Hosono, H. *Solid State Comm.* 2001, 121, 15.
[104] Kohler, B. V.; Jansen, M. Z. *Anorg. Allge. Chem.* 1986, 543, 73.
[105] Shimode, M.; Sasaki, M.; Mukaida, K. *J. Solid State Chem.* 2000, 151, 16.
[106] Robertson, J.; Peacock, P. W.; Towler, M. D.; Needs, R. *Thin Solid Films* 2002, 411, 96.
[107] Ingram, B. J.; Mason, T. O.; Asahi, R.; Park, K. T.; Freeman, A. J. *Phys. Rev. B* 2001, 64, 155114.
[108] Duan, N.; Sleight, A. W.; Jayaraj, M. K. Tate, *J. Appl. Phys. Lett.* 2000, 77, 1325.
[109] Nagrajan, R.; Draeseke, A. D.; Sleight A. W.; Tate, J. *J. Appl. Phys.* 2001, 89, 8022.
[110] Jayaraj, M. K.; Draeseke, A. D.; Tate, J.; Sleight, A.W. *Thin Solid Films* 2001, 397, 244.
[111] Mattheiss, L. F.; *Phys. Rev. B 1993*, 48, 18300.
[112] Crottaz, O.; Kubel, F.; Schmid, H. J. *Solid State Chem.* 1996, 122, 247.
[113] Powder Diffraction File; Joint Committee on Powder Diffraction Standards, *ASTM*; Philadelphia, PA, 1967; Card # 35-1401.
[114] Yanagi, H.; Park, S.; Draeseke, A. D.; Keszler, D. A.; Tate, J. *J. Solid State Chem.* 2003, 175, 34.

[115] Ueda, K.; Inoue, S.; Hirose, S.; Kawazoe, H.; Hosono, H. *Appl. Phys. Lett.* 2000, 77, 2701.
[116] Kudo, A.; Yanagi, H.; Hosono, H.; Kawazoe, H. *Appl. Phys. Lett.* 1998, 73, 220.
[117] Teske, C. L.; Müller-Buschbaum, H. Z. *Anorg. Allg. Chem.* 1970, 379, 113.
[118] Boudin, S.; Felser, C.; Studer, F. *Solid State Science*s 2003, 5, 741.
[119] Hiramatsu, H.; Ueda, K.; Ohta, H.; Orita, M.; Hirano, M.; Hosono, H. *Thin Solid Films* 2002, 411, 125.
[120] Palazzi, M. *Acad. Sci.,* Paris, C. R. 1981, 292, 789.
[121] Ishikawa, K.; Kinosita, S.; Suzuki, Y.; Matsuura, S.; Nakanishi, T.; Aizawa, M.; Suzuki, Y. *J. Electrochem. Soc.* 1991, 138, 1166.
[122] Takano, Y.; Yahagi, K. –I.; Sekizawa, K. *Physica B* 1995, 206 & 207, 764.
[123] Ueda, K.; Inoue, S.; Hosono, H.; Sarukura, N.; Hirano, M. *Appl. Phys. Lett.* 2001, 78, 2333.
[124] Ueda, K.; Hosono, H. *J. Appl. Phys*. 2002, 91, 4768.
[125] Takase, K.; Koyano, M.; Shimizu, T.; Makihara, K.; Takahashi, Y.; Takano, Y.; Sekizawa, K. *Solid State Comm*. 2002, 123, 531.
[126] Hiramatsu, H.; Orita, M.; Hirano, M.; Ueda, K.; Hosono, H. *J. Appl. Phys*. 2002, 91, 9177.
[127] Hiramatsu, H.; Ueda, K.; Takafuji, K.; Ohta, H.; Orita, M.; Hirano, M.; Kamiya, T.; Hosono, H. *J. Appl. Phys*. 2003, 94, 5805.
[128] Hiramatsu, H.; Ueda, K.; Ohta, H.; Hirano, M.; Kamiya, T.; Hosono, H. *Appl. Phys. Lett.* 2003, 82, 1048.
[129] Ueda, K.; Hiramatsu, H.; Ohta, H.; Hirano, M.; Kamiya, T.; Hosono, H. *Phys. Rev. B* 2004, 69, 155305.
[130] Ueda, K., Hosono, H. *Thin Solid Films* 2002, 411, 115.
[131] Zhu, W. J.; Huang, Y. Z.; Dong, C.; Zhao, Z. X. *Mater. Res. Bull.* 1994, 29, 143.
[132] Inoue, S. –I.; Ueda, K.; Hosono, H.; Hamada, N. *Phys. Rev. B* 2001, 64, 245211.
[133] Furukawa, Y.; Ikeda, S.; Kumagai, K.; Mori, K.; Takano, Y.; Sekizawa, K. *Phys. Rev. B* 2000, 62, 15598.
[134] Takano, Y.; Mori, K.; Koizumi, K.; Ozaki, H.; Sekizawa, K. *J. Alloys Compd.* 1998, 275-277, 447.
[135] Sekizawa, K.; Takano, Y.; Mori, K.; Yahagi, T. *Czech. J. Phys*. 1996, 46, 1943.
[136] Park, S.; Keszlor, D. A.; Valencia, M. M.; Hoffman, R. L.; Bender, J. P.; Wager, J. F. *Appl. Phys. Lett.* 2002, 80, 4393.
[137] Wu, W. J.; Huang, Y. Z.; Wu, F.; Dong, C.; Chen, H.; Zhao, Z. X. *Mater. Res. Bull.* 1994, 29, 505.
[138] Yanagi, H.; Park, S.; Draeseke, A. D.; Keszler, D. A. J. Tate, J. *Solid State Chem.* 2003, 175, 34.
[139] Iglesias, J. E.; Pachali, K. E.; Stoinfink, H. J. *Solid State Chem*. 1974, 9, 6.
[140] Maissel, L. I.; Glang, R. Handbook of Thin Films Technology; McGraw-Hill: New York, NY, 1970.
[141] Chrisey, D. B.; Hubler, G. K. *Pulsed Laser Deposition of Thin Films*; Wiley-Interscience: New York, NY, 1994.
[142] Chopra, K. L. Thin Film Phenomena; McGraw-Hill: New York, NY, 1969.
[143] Anderson, J. C. *The Use of Thin Films in Physical Investigations*; Academic: New York, NY, 1966.

[144] Kawazoe, H.; Yanagi, H.; Ueda, K.; Hosono, H. *MRS Bull.* August 2000, pp. 28.

[145] Stauber, R. E.; Perkins, J. D.; Parilla, P. A.; Ginley, D. S. *Electrochem. Solid State Lett.* 1999, 2, 654.

[146] Ong, C. H.; Gong, H. *Thin Solid Films* 2003, 445, 299.

[147] Tsuboi, N.; Takahashi, Y.; Kobayashi, S.; Shimizu, H.; Kato, K.; Kaneko, F. *J. Phys. Chem. Solids* 2003, 64, 1671.

[148] Alkoy, E.M.; Kelly, P. J. *Vacuum* 2005, 75, 221.

[149] Wang, Y.; Gong, H. *Advanced Materials CVD* 2000, 6, 285.

[150] Gong, H.; Wang, Y.; Luo, Y. *Appl. Phys. Lett.* 2000, 76, 3959.

[151] Wang, Y.; Gong, H.; Zhu, F.; Liu, L.; Huang, L.; Huan, A. C. H. *Mater. Sci. Engg. B* 2001, 85, 131.

[152] Kim, D. –S.; Choi, S. –Y. *Phys. Stat. Solidi (A)* 2005, 202, R167.

[153] Tonooka, K.; Shimokawa, K.; Nishimura, O. *Thin Solid Films* 2002, 411, 129.

[154] Gao, S.; Zhao, Y.; Gou, P.; Chen, N.; Xie, Y. *Nanotechnology* 2003, 14, 538.

[155] Shy, J. H.; Tseng, B. H.; *J. Phys. Chem. Solids* 2005, 66, 2123.

[156] Bouzidi, C.; Bouzouita, H.; Timoumi, A.; Rezig, B. *Mater. Sci. Engg. B* 2005,118, 259.

[157] Ohashi, M.; Iida, Y.; Morikawa, H. *J. Am. Cerem. Soc.* 2002, 85, 270.

[158] Dloczik, L.; Tomm, Y.; Könenkamp, R. *Thin Solid Films* 2004, 451–452, 116-119.

[159] Shahriari, D. Y.; Barnabè, A.; Mason, T. O.; Poeppelmeier, K. R. *Inorg. Chem.* 2001, 40, 5734.

[160] Banerjee, A. N.; Kundoo, S.; Chattopadhyay, K.K. *Thin Solid Films* 2003, 440, 5 – 10.

[161] Banerjee, A. N.; Maity, R.; Chattopadhyay, K. K. *Materials Letters* 2003, 58, 10 - 13.

[162] Ghosh, P. K.; Banerjee, A. N.; Maity, R.; Chattopadhyay K. K. (unpublished data).

[163] Yoshida, H. K.; Koyanagi, T.; Funashima, H.; Harima, H.; Yanase, A. *Solid State Comm.* 2003, 126, 135.

[164] Koyanagi, T.; Harima, H.; Yanase, A.; Yoshida, H. K. J. *Phys. Chem. Solids* 2003, 64, 1443.

[165] Kudo, A.; Yanagi, H.; Ueda, K.; Hosono, H.; Kawazoe, H.; Yano, Y. *Appl. Phys. Lett.* 1999, 75, 2851.

[166] Ohta, H.; Kawamura, K.; Orita, M.; Sarukura, N.; Hirano, M.; Hosono, H. *Electron. Lett.* 2000, 36, 984.

[167] Ohta, H.; Kawamura, K.; Orita, M.; Hirano, M.; Sarukura, N.; Hosono, H. *Appl. Phys. Lett.* 2000, 77, 475.

[168] Ohta, H.; Orita, M.; Hirano, M. *J. Appl. Phys.* 2001, 89, 5720.

[169] Hosono, H.; Ohta, H.; Hayashi, K.; Orita, M.; Hirano, M. *J. Cryst. Growth* 2002, 237–239, 496.

[170] Hoffman, R. L.; Wager, J. F.; Jayaraj, M. K.; Tate, J. *J. Appl. Phys.* 2001, 90, 5763.

[171] Jayaraj, M. K.; Draeseke, A. D.; Tate, J.; Hoffman, R. L.; Wager, J. F. *Mat. Res. Soc. Proc.* 2001, 666, F4.1.

[172] Tonooka, K.; Bando, H.; Aiura, Y. *Thin Solid Films*, **445** (2003) 327.

[173] Hwang, D. –K.; Bang, K. –H.; Jeong, M. –C.; Myoung, J. –M. *J. Cryst. Growth* 2003, 254, 449.

[174] Tüzemen, S.; Xiong, G.; Wilkinson, J.; Mischuck, B.; Ucer, K. B.; Williams, R. T. *Physica B* 2001, 308-310, 1197.

[175] Aoki, T.; Hatanaka, Y.; Look, D. C. *Appl. Phys. Lett.* 2000, 76, 3257.

[176] Butkhuzi, T. V.; Bureyev, A. V.; Georgobiani, A. N.; Kekelidze, N. P.; Khulordava, T. G. *J. Cryst. Growth* 1992, 117, 366.
[177] Wager, J. F. *Science* 2003, 300, 1245.
[178] Prins, M. W. J.; Grosse-Holz, K.-O.; Müller, G.; Cillessen, J. F. M.; Giesbers, J. B.; Weening, R. P.; Wolf, R. M. *Appl. Phys. Lett.* 1996, 68, 3650.
[179] Prins, M. W. J.; Zinnemers, S. E.; Cillessen, J. F. M.; Giesbers, J. B. *Appl. Phys. Lett.* 1997, 70, 458.
[180] Hoffman, R. L.; Norris, B. J.; Wager, J. F. *Appl. Phys. Lett.* 2003, 82, 733.
[181] Masuda, S.; Kitamura, K.; Okumura, Y.; Miyatake, S.; Tabata, H.; Kawai, T. *J. Appl. Phys.* 2003, 93, 1624.
[182] Carcia, P. F.; McLean, R. S.; Reilly, M. H.; Nunes, Jr. G.; *Appl. Phys. Lett.* 2003, 82, 1117.
[183] Nomura, K.; Ohta, H.; Ueda, K.; Kamiya, T.; Hirano, M.; Hosono, H. *Science* 2003, 300, 1269.
[184] Ohta, H.; Nomura, K.; Hiramatsu, H.; Ueda, K.; Kamiya, T.; Hirano, M.; Hosono, H. *Solid-State Electronics* 2003, 47, 2261.
[185] Banerjee, A. N.; Chattopadhyay, K. K. *J. Appl. Phys* 2005, 97, 084308.
[186] Jacob, A.; Parent, C.; Boutinaud, P.; Le Flem, G.; Doumerc, J. P.; Ammar, A.; Elazhari M.; Elaatmani, M. *Solid State Comm.* 1997, 103, 529.
[187] Hasegawa, S.; Nishida, S.; Yamashita, T.; Asahi, H. *Thin Solid Films* 2005, 487, 260.
[188] Hasegawa, S.; Nishida, S.; Yamashita, T.; Asahi, H. J. *Cerem. Processing Res.* 2005, 6, 245.
[189] Yu, K.; Zhang, Y.; Luo, L.; Zhu, Z. *Proc. SPIE-The Int. Soc. Optical Engg.* 2005, 5774, 353.
[190] Liu, B.; Bando, Y.; Tang, C.; Xu, F.; Hu, J.; Golberg, D. *J. Phys. Chem. B* 2005, 109, 17082.
[191] Liu, B. D.; Bando, Y.; Tang, C. C.; Xu, F. F.; Golberg, D. *J. Phys. Chem. B* 2005, 109, 21521.
[192] Yue, S. L.; Gu, C. Z.; Shi, C. Y.; Zhi, C. Y. *Appl. Surf. Sci.* 2005, 251, 215.
[193] Wang, Y. X.; Li, Y. A.; Feng, W.; Li, W. Q.; Zhao, C. H.; Liu, L. H.; Feng, K. C.; Zhao, Y. N. *Appl. Surf. Sci.* 2005, 243, 394.
[194] Tong, Y.; Cong, H.; Chen, Z.; Cheng, H. *Appl. Phys. Lett.* 2005, 86, 233104.
[195] Liu, C.; Hu, Z.; Wu, Q.; Wang, X.; Chen, Y.; Lin, W.; Sang, H.; Deng, S.; Xu, N. *Appl. Surf. Sci.* 2005, 251, 220.
[196] Tang, Y. B.; Cong, H. T.; Zhao, Z. G.; Chen, H. M. *Appl. Phys. Lett.* 2005, 86, 153104.
[197] Chen, S. –C.; Chen, C. –F.; Chattopadhyay, S.; Chen, K. –H.; Chen, L. –C *Appl. Phys. Lett.* 2005, 87, 073109.
[198] Olesik, J.; Malachowski, M. *Thin Solid Films* 2002, 422, 193.
[199] Baranauskas, V.; Fontana, M.; Guo, Z. J.; Ceragioli, H. J.; Peterlevitz, A. C. *Sensors and Actuators B: Chemical* 2005, 107, 474.
[200] Jo, S. H.; Lao, J. Y.; Ren, Z. F.; Farrer, R. A.; Baldacchini, T.; Fourkas, J. T. *Appl. Phys. Lett.* 2003, 83, 4821.
[201] Dong, L.; Jiao, J.; Tuggle, D. W.; Petty, J. M.; Elliff, S. A.; Coulter, M. *Appl. Phys. Lett.* 2003, 82, 1096.
[202] Xu, C.; Sun, X. Int. J. *Nanotechnology* 2004, 1, 452.

[203] Yang, H. Y.; Lau, S. P.; Yu, S. F.; Huang, L.; Tanemura, M.; Tanaka, J.; Okita, T.; Hng, H. H. *Nanotechnology* 2005, 16, 1300.
[204] Maity, R.; Banerjee, A. N.; Chattopadhyay, K. K. *Appl. Surf. Sci.* 2004, 236, 231.
[205] Maity, R.; Das, S.; Mitra, M. K.; Chattopadhyay, K. K. *Physica E* 2005, 25, 605.
[206] Yang, Y. H.; Wang, C. X.; Wang, B.; Xu, N. S.; Yang, G. W. *Chem. Phys. Lett.* 2005, 403, 248.
[207] Banerjee, A. N.; Chattopadhyay, K. K. *Appl. Surf. Sci.* 2004, 225, 10.
[208] Banerjee, A. N.; Ghosh, C. K.; Das, S.; Chattopadhyay, K. K. *Physica B* 2005, 370, 264.
[209] Fowler, R. H.; Nordheim, L. *Proc. Roy. Soc. Lond.* 1928, A119, 173.
[210] Murphy, E. L.; Good Jr., R. H. *Phys. Rev.* 1956, 102, 1464.
[211] Schwettman, H. A.; Turneaure, J. P.; Waites, R. F. *J. of Appl. Phys.* 1974, 45, 914.
[212] Latham, R. V. *IEEE Trans. Elec. Insul.* 1983, EI-18, 194.
[213] Forbes, R. G. *Solid-State Electronics* 2001, 45, 779.
[214] Forbes, R. G. J. *Vac. Sci. Technol. B* 1999, 17, 526.
[215] DeVault, D. *J. Chem. Educ.* 1944, 21, 526.
[216] Bailar Jr., J. C.; Emeléus, H. J.; Sir Ronald, N.; Trotman-Dickenson, A. F. *Comprehensive Inorganic Chemistry*; Pergamon: New York, NY, 1973.
[217] Leung, Y. K.; Khan, A. K.; Kos, J. F.; Koffyberg, F. P. *Proc. Nat. Conf. Solar Energy*; Solar Energy Society of Canada, Montreal, 1981; pp. 124.
[218] Dahl, J. P.; Switendick, A. C. *J. Phys. Chem. Solids* 1966, 27, 931.
[219] Hayashi, M.; Katsuki, K. *J. Phys. Soc. Jpn.* 1950, 5, 380.
[220] Kleinman, L.; Mednick, K. *Phys. Rev. B* 1980, 21, 1549.
[221] Otabe, T.; Ueda, K.; Kudoh, A.; Hosono, H.; Kawazoe, H. *Appl. Phys. Lett.* 1998, 72, 1036.
[222] Kofstad, P. *Nonstoichiometry, Diffusion, and Electrical Conductivity in Binary Metal Oxides;* Wiley – Interscience, 1972, pp.19.
[223] Koumoto, K.; Koduka, H.; Seo, W. S. *J. Mater. Chem.* 2001, 11, 251.
[224] Cava, R. J.; Jr. Peck, W. F.; Krajewski, J. J.; Cheong, S. W.; Hwang, H. Y. *J. Mater. Res.* 1994, 9, 314.
[225] Cava, R. J.; Zandbergen, H. W.; Ramirez, A. P.; Tagaki, H.; Chien, C. T.; Krajewski, J. J.; Jr. Peck, W. F.; Waszczak, J. V.; Meigs, G.; Roth, R. S.; Schneemeyer, L. F. J. *Solid State Chem.* 1993, 104, 437.
[226] Jayaraj, M. K.; Draeseke, A.; Tate, J.; Duan, N.; Sleight, A. W. *Proceedings of the MRS Workshop on Transparent Conductive Oxide*, Denever, CO; 19 – 20 June, 2000.
[227] Lalić, M. V.; Filho, J. M.; Carbonari, A. W.; Saxena, R. N.; Moralles, M. *J. Phys.: Condens. Matter* 2002, 14, 5517.
[228] Lalić, M. V.; Filho, J. M.; Carbonari, A. W.; Saxena, R. N. *Solid State Communication* 2003, 125, 175.
[229] Benko, F. A.; Koffyberg, F. P. *J. Phys. Chem. Solids* 1987, 48, 431.
[230] Cullity, B. D. *Elements of X-ray Diffraction*, Addison-Wesley: CA, 1978.
[231] Quadri, S. B.; Skelton, E. F.; Hsu, D.; Dinsmore, A. D.; Yang, J.; Gray, H. F.; Ratna, B. R. *Physical Review B* 1999, 60, 9191.
[232] Nakamoto, Infrared Spectra of Inorganic and Coordinated Compounds, John Wiley & Sons, Inc.: New York, NY, 1963.
[233] Tarte, P. *Spectrochimica Acta* 1967, 23A, 2127.

[234] Tarte, P. *Mém. Acad. Roy. Belg.* 1965, 35.
[235] Tarte, P.; Home, J. P. *Acta Cryst.* 1963, 16, 227.
[236] Tarte, P. *Spectrochimica Acta* 1962, 18, 467.
[237] Pankove, J. I. *Optical Processes in Semiconductors*, Prentice-Hall. Inc. 1971.
[238] Moss, T. S. *Optical Properties of Semiconductors,* Butterworths: London, 1959.
[239] Demichelis, F.; Kaniadakis, G.; Tagliaferro, A.; Tresso, E. *Applied Optics* 1987, 26, 1737.
[240] Szczyrbowski, J.; Schmalzbauer, K.; Hoffmann, H. *Thin Solid Film* 1986, 137, 169.
[241] Gurumurugan, K.; Mangalaraj, D.; Narayandass, Sa. K. *Thin Solid Films* 1994, 251, 7.
[242] Bardeen, J.; Blatt, F. J.; Hall, L. H. *Proc. Atlantic City Photoconductivity Conference*, J. Wiely, Chapman and Hall, 1956.
[243] Banerjee, A. N.; Ghosh, C. K.; Chattopadhyay, K. K. *Sol. Energy* Mat. *Solar Cells* 2005, 89, 75.
[244] Banerjee, A. N.; Maity, R.; Ghosh, P. K.; Chattopadhyay, K. K. *Thin Solid Films* 2005, 474, 261.
[245] Hicks, L. D.; Dresselhaus, M. S. *Phys. Rev. B* 1993, 47, 12727.
[246] Hicks, L. D.; Herman, T. C.; Dresselhaus, M. S. *Appl. Phys. Lett.* 1993, 63, 3230.
[247] Lin-Chung, P. J.; Reinecke, T. L. *Phys. Rev. B* 1995, 51, 13244.
[248] Broido, D. A.; Reinecke, T. L. *Phys. Rev. B* 1995, 51, 13797.
[249] Terasaki, I.; Sasago, Y.; Uchinokura, K. *Phys. Rev. B* 1997, 56, 12685.
[250] Masuda, Y.; Ohta, M.; Seo, W. S.; Pitschke, W.; Koumoto, K. J. *Solid State Chem.* 2000, 150, 221.
[251] Wang, Y.; Rogado, N. S.; Cava, R. J.; Ong, N. P. *Nature* 2003, 423, 425.
[252] Willardson, R. K.; Beer, A. C. Semiconductors and Semimetals; Vol. 8, Academic Press: New York and London, 1971.
[253] Park, K.; Ko, K. Y.; Seo, W. –S. J. *Euro. Cerem. Soc.* 2005, 25, 2219.
[254] Manifacier, J. C.; Gasiot, J.; Fillard, J. P. *J. Phys. E* 1976, 9, 1002.
[255] Maity, R.; Kundoo, S.; Chattopadhyay, K. K. *Solar Energy Materials & Solar Cells* 2005, 86, 217.
[256] Kröger, F. A. *The Chemistry of Imperfect Crystal*; North Holland, Amsterdam, 1974.
[257] Hirsch, P.; Howie, A.; Nicholson, R. B.; Pashley, D. W.; Whelan, M. J. Electron Microscopy in Thin Crystals; Robert E. Krieger Publishing Co.: Huntington, NY, 1977.
[258] Lu, Z. H.; Lockwood, D. J.; Baribeau, J. M. *Nature* 1995, 378, 258.
[259] Reynolds, D. C.; Collins, T. C. *Excitons: Their Properties and Uses*; Academic: New York; 1981.
[260] Yao, B. D.; Chan, Y. F.; Wang, N. *Appl. Phys. Lett.* 2002, 81, 757.
[261] Forbes, R. G. *Ultramicroscopy* 1999, 79, 11.
[262] Forbes, R. G. In *Proceedings of 4th Moscow International ITEP School of Physics*; Suvorov, A. L.; Abov, Y. G.; Firsov, V. G.; Ed.; Academprint: Moscow, 2001; pp 62.
[263] Robertson, J. *Thin Solid Films* 1997, 296, 61.
[264] Kotoh, S.; Kikuchi, T.; Muruyama, T.; Kobayashi, T.; Kanaya, M. *Jpn. J. Appl. Phys.* 2002, 41, 3924.
[265] Satyanarayana, B. S.; Hart, A.; Milne, W. I.; Robertson, *J. Appl. Phys. Lett.* 1997, 71,1.
[266] Gupta, S.; Weiss, B. L.; Weiner, B. R.; Pilione, L.; Badzian, A.; Morell, G. *J. Appl. Phys.* 2002, 92, 3311.

[267] Hirakuri, K. K.; Kurata, T.; Mustukura, N.; Friedbacher, G.; Ohuchi, M. *J. Appl. Phys.* 2000, 87, 1.
[268] Oh, S. J.; Cheng, Y.; Zhang, J.; Shimoda, H.; Zhou, O. *Appl. Phys. Lett.* 2003, 82, 2521.
[269] Okano, K.; Koizumi, S.; Silva, S. R. P.; Amartunga, G. A. J. *Nature* 1996, 381, 140.
[270] Gröning, O.; Küttel, O. M.; Gröning, P.; Schlapbach, L. *Appl. Surf. Sci.* 1997, 111, 135.
[271] Forbes, R. G.; Edgcombe, C. J.; Valdrè, U. *Ultramicroscopy* 2003, 95 57.
[272] Luo, S.; Chu, P. K.; Zhang, Z. Di, M.; Liu, W.; Lin, C.; Fan, J.; Wu, X. *Appl. Phys. Lett.* 2006, 88, art.# 013109.
[273] Marathe, S. K.; Koinkar, P. M.; Ashtaputre, S. S.; More, M. A.; Gosavi, S. W.; Joag, D. S.; Kulkarni, S. K. *Nanotechnology* 2006, 17, 1932.
[274] Lalic, M. V.; Mestnik-Filho, J. *Hyperfine Interactions* 2004, 158, 89.
[275] Park, K.; Ko, K. Y.; Seo, W. –S. *Mater. Sci. Engg. B* 2006, 129, 1.
[276] Zheng, X. G.; Taniguchi, K.; Takahashi, A.; Liu, Y.; Xu, C. N. *Appl. Phys. Lett.* 2004, 85, 1728.
[277] Sauter, D.; Weimar, U.; Noetzel, G.; Mitrovics, J.; Gopel, W. *Sens. Actuators B* 2000, 69, 1.
[278] Kim, S.-R.; Hong, H.-K.; Kwon, C. H.; Yun, D. H.; Lee, K.; Sung, Y. K. *Sens. Actuators B* 2000, 66, 59.
[279] Gagaoudakis, E.; Bender, M.; Douloufakis, E.; Katsuarakis, N.; Natsakou, E.; Cimalla, V.; Kiriakidis, G. *Sens. Actuators B* 2001, 80, 155.
[280] Bender, M.; Gagaoudakis, E.; Douloufakis, E.; Katsuarakis, N.; Natsakou, E.; Cimalla, V.; Kiriakidis, G.; Fortunato, E.; Nunes, P.; Marques, A.; Martins, R. *Thin Solid Films* 2002, 418, 45.
[281] Koriche, N.; Bouguelia, A.; Aider, A.; Trari, M. *International Journal of Hydrogen Energy* 2005, 30, 693.
[282] Kizaki, H.; Sato, K.; Yanase, A.; Katayama-Yoshida, H. *Physica B* 2006, 376-377, 812.

In: Materials Science Research Trends
Editor: Lawrence V. Olivante, pp. 133-149
ISBN: 978-1-60021-654-1

Chapter 2

ATOMISTIC ANALYSIS OF CRYSTAL PLASTICITY IN A COPPER NANOWIRE DURING TENSILE LOADING

***R. S. McEntire*[1,2] *and Y. L. Shen*[1]**
[1]Department of Mechanical Engineering, University of New Mexico, Albuquerque, NM 87131
[2]Sandia National Laboratories, Albuquerque, NM 87185

Abstract

Plastic deformation in a copper crystal is modeled using three dimensional atomistic simulations. The primary objective is to gain fundamental insight into the deformation features in face-centered-cubic materials in the form of a nanowire under tensile loading. An initial defect is utilized in the molecular statics model to trigger plasticity in a controlled manner. A parametric study is then performed by varying the atomic interaction range for the Morse interatomic potential used in the model. The simulation parameters are employed such that dislocation slip behavior and/or phase transformation can be observed without the influence of an unstable surface state of the specimen. We focus on tensile loading along a low-symmetry orientation where single slip prevails upon yielding. When the interaction distance is small, slip is seen to be the dominant deformation mechanism. A slight increase in the interaction range results in phase transition from the FCC structure to a BCC structure. Re-orientation of the BCC lattice also occurs at later stages of the deformation via a twinning operation. The phase transition mechanism is further enhanced if the nanowire is attached to a flat substrate parallel to the initial close-packed plane. The mechanisms of dislocation evolution, phase transformation, and crystal re-orientation features are discussed.

Introduction

Atomistic simulations are being increasingly utilized for gaining fundamental understandings of microscopic features in crystalline solids. There have been extensive efforts for employing molecular dynamics simulations to study the deformation mechanisms in nano-scale metallic crystals. For stand-alone crystals in the form of a nanowire with a large surface-to-volume ratio, the surface energy is found to play a very significant role in affecting the mechanical response, leading to possible phase transformation, yield asymmetry, lattice reorientation, and the shape memory effect [1-6]. Previous simulation studies have mainly focused on tensile loading along high-symmetry crystallographic directions including <100>, <110> and <111>. Low-symmetry orientations remain largely unexplored. The present work concerns plastic deformation induced by tensile loading along a low-symmetry orientation of a copper crystal. Within the framework of conventional dislocation plasticity, this type of loading typically leads to "single slip" during the initial stages of yielding.

In addition to "testing" the new loading direction of the nanowire, the present study also aims at implementing a technique used previously in two-dimensional simulations: embedding an initial point defect in the crystal to activate the dislocation event [7,8]. This approach, together with the employment of a pairwise interatomic potential, have been shown to render a ductile behavior with dislocation emission facilitated at a prescribed location. Aside from its physical implication (i.e., point defect leading to dislocation nucleation during loading), this approach serves as a convenient way in atomistic simulation for triggering plasticity in a controlled manner.

Furthermore, in this work we conduct *parametric* molecular statics simulations by varying the cutoff distance (maximum range of atomic interaction) of the pairwise potential. It is recognized that in nanoscale samples, the surface phenomenon and the "intrinsic" atomic mechanism mutually affect one another, which gives rise to an apparent material response. Results obtained from typical simulations are inevitably a *combined* effect with those various contributions. This hinders our understanding of the *individual* effects at the fundamental level. It is with this appreciation that we limit our attention to nearest neighbor interaction in this work, for the purpose of suppressing the surface effect and for gaining fundamental insight into the influence the interaction range may have on the deformation behavior.

Computational Model

Figure 1 shows a schematic of the model setup. Atoms are packed into an FCC crystal, and the nanowire takes the form of a rectangular bar to be subjected to tensile stretching. (The actual atomic arrangements can be seen in the figures below). We consider tensile loading along the direction $[7\,\overline{10}\,\bar{3}]$. This low-symmetry orientation is arbitrarily chosen to avoid multiple slips when plastic yielding is first activated. The model dimensions are characterized by the outer edges of the atomic spheres in the three directions: l = 128.0 Å, w = 24.73 Å and t = 17.19 Å. Note the thickness (t) direction is $[\bar{1}\,\bar{1}\,1]$. As schematically shown in Fig. 1, an initial point defect in the form of a self-interstitial is placed at an octahedral site at the geometrical center of the model and allowed to equilibrate with its surrounding atoms before the loading steps commence. This local disturbance forces plastic deformation to initiate at the prescribed location (thus avoiding the deformation initiation point normally caused by the artificial "gripping" constraint at the ends of the wire).

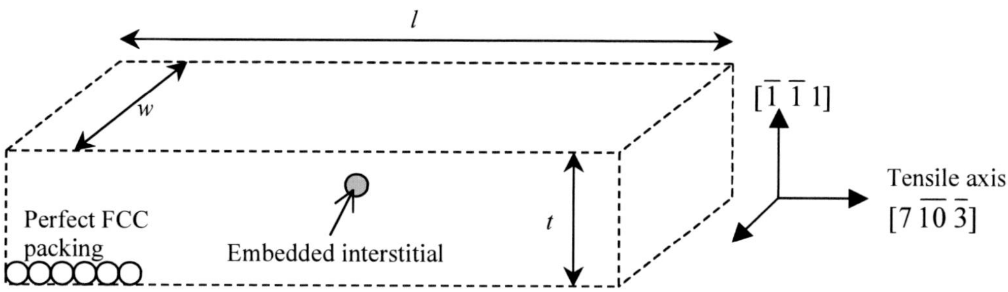

Figure 1. The rectangular model geometry for nanowire simulations.

The interatomic potential employed is the Morse potential. This is a potential of the form:

$$V = -V_0\left[e^{-2a(r-r_0)} - 2e^{-a(r-r_0)}\right], \tag{1}$$

where r is the interatomic spacing and the parameters r_0, a and V_0 are determined by fitting the equation to experimental data of the equilibrium lattice parameter, cohesive energy and bulk modulus of copper [9-11]. The parameters thus obtained are: $r_0 = 2.56$ Å, $a = 1.399$ Å^{-1} and $V_0 = -0.581$ eV. In this analysis, we consider potentials with the same parameters except the cutoff distance, r_i, beyond which the atoms were not allowed to interact. In particular, we examine two cutoff distances: 1.325 r_0 and 1.335 r_0. Note that both of these cutoff distances are less than the distance between an internal atom and its second-nearest neighbors in the regular FCC structure. However, it will be shown below that they give rise to qualitatively different deformation behaviors, due mainly to their different influences on the disturbance caused by the initial defect. It is also noted that this treatment does not generate an unstable surface (because all surface atoms interact with only the nearest neighbors initially and are thus at equilibrium). Therefore, the specifically oriented FCC crystal model is in a mechanically stable state at the beginning of the loading process. Although the present setup is not as realistic as simulations performed using many-body potentials, it offers an opportunity for a parametric analysis on how the detailed crystallographic features can be affected by the modeling parameters.

The molecular statics simulation was carried out by prescribing a small displacement in the tensile direction on all end face atoms at each loading step. Lateral displacement of these end atoms was not allowed. Atoms on the face opposite to the prescribed displacement atoms were fixed. The side boundary atoms in the rest of the specimen were not constrained. In response to each prescribed loading step all atomic points were allowed to iteratively reach their new equilibrium positions. The overall load was calculated by adding the force components along the tensile direction pertaining to the atoms where the displacement is prescribed.

In addition to the simple tensile stretching described above, we also performed simulations with the same crystal model attached to a substrate material. The objective is to study interface mediated plasticity in the crystal under nominal uniaxial loading. (Note in many actual nanostructures, the material is bonded to other components on at least one side.) A previous two-dimensional atomistic simulation study has shown that the elimination of a free surface facilitated by the substrate delays plastic yielding of the film by restricting dislocation activities [12]. Here we extend the previous two-dimensional work to three dimensions using the model crystal shown in Fig. 1. The substrate material is not explicitly included in the model. Instead, a boundary condition was applied to all atoms in the bottom layer, a $(\bar{1}\,\bar{1}\,1)$ plane, so these atoms were allowed only in-plane motion with any out-of-plane displacement prohibited. This simulates bonding to an ideal flat substrate, but all interface atoms still have a free-sliding capability along the interface.

Results and Discussion

Overall Load-Displacement Response

Figure 2 shows the simulated overall load-displacement curves for all cases considered. The symbols labeled along the curves correspond to the associated figure numbers of specific atomic snapshots. It is seen that a small modification of the modeling parameter can lead to a large difference in overall material behavior. Generally all four cases show ductile behavior after the initial elastic response. A sharp drop in load occurs upon initial plastic yielding. The atomic interaction range r_i = 1.325 r_0 results in relatively earlier fracture (vanishing load-bearing capability) than the case of r_i = 1.335 r_0 which does not experience final fracture over the strain range presented. The incorporation of a flat substrate has a larger influence on the result for the case with a greater interaction range. It is worth mentioning that, if there is no initial point defect included in the model (not shown), the specimen either showed brittle fracture after the elastic regime or yielding was forced to occur at the artificial gripping site at the end of the specimen.

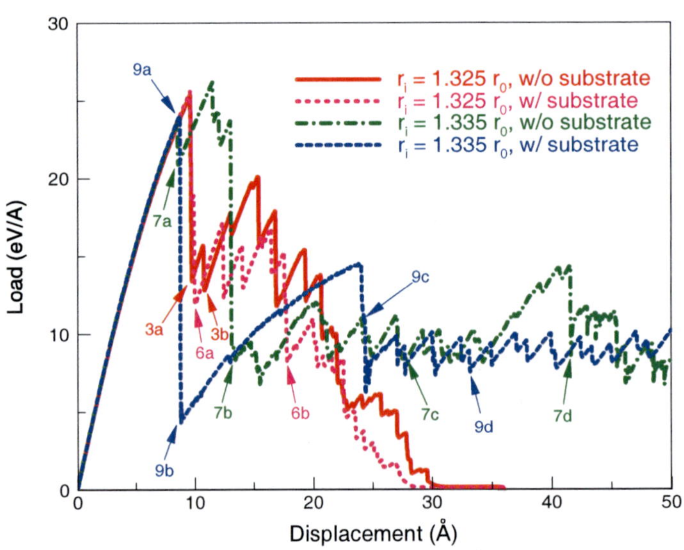

Figure 2. Simulated overall load-displacement curves under tensile stretching along the $[7\,\overline{10}\,\bar{3}]$ orientation. Labels are associated with Figures of corresponding atomic snapshots.

The Case of $r_i = 1.325\ r_0$

The atomic snapshots in the figures below are color-coded to display their relative energy state in the deformed configuration. The energy of each atom is normalized with the number of its nearest neighbors. This treatment enables a "clean" presentation of atoms and any structural change can be more clearly viewed on the surface. Attention is first focused on the cases where the cutoff distance is 1.325 r_0. Without the influence of the "substrate" (i.e., the curve shown as "w/o substrate," meaning no interfacial constraint on the bottom atoms), the crystal shows an elastic response up to about 25 eV/Å, after which a sudden drop in load occurs. It is observed from the atomic snapshot in Fig. 3(a) that this load reduction is associated with the slip operation along the $(1\bar{1}1)[011]$ primary system (with the greatest Schmid factor of 0.4702). Furthermore, this process was initiated from the initial point defect at the center of the specimen, which illustrates the capability of this approach in prompting the slip to be activated at a specified location. Upon a brief elastic response beyond point 3a in Fig. 2, another load reduction is observed, and the atomic snapshot in Fig. 3(b) demonstrates the continued slip along the primary system. Further deformation will lead to a mixture of lattice rotation, intermittent short elastic response, conjugate slip and local atomic bond breaking, which eventually causes fracture of the specimen when the overall displacement is at about 30 Å.

The detailed mechanism of how the self-interstitial evolves into the slip operation deserves attention. Figure 4(a) shows the top view (viewing direction along $[11\bar{1}]$) of an internal section of the $(\bar{1}\bar{1}1)$ plane containing the initial interstitial atom. The figure shows a snapshot when the specimen is deformed to shortly before the first overall load reduction shown in Fig. 2. The interstitial is highlighted by an arrow in the figure. Due to the elastic deformation at this moment, the interstitial is being accommodated by its surrounding atoms.

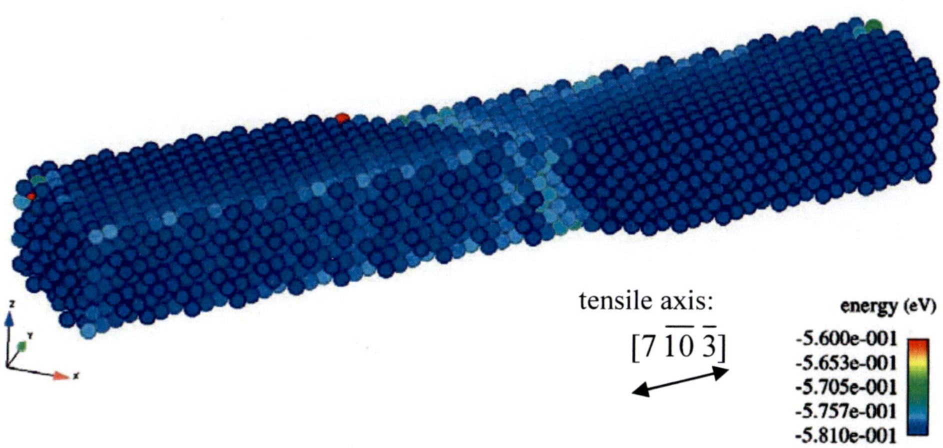

(a)

Figure 3. Continued on next page.

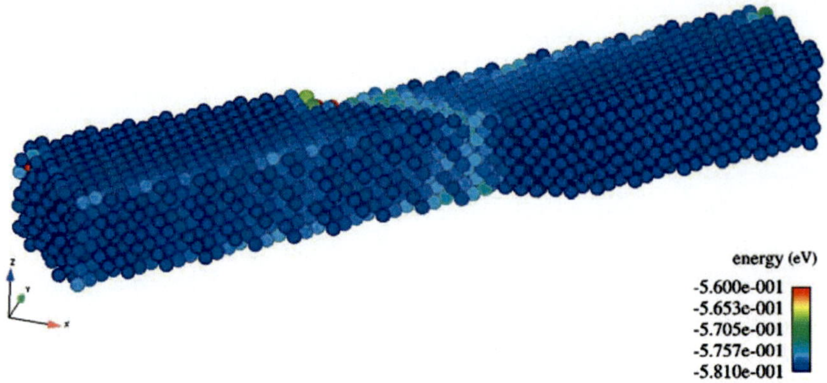

(b)

Figure 3. Atomic snapshots corresponding to points (a) 3a and (b) 3b labeled along the load-displacement curve in Fig. 2, for the case of 1.325 r_0 cutoff distance with no substrate.

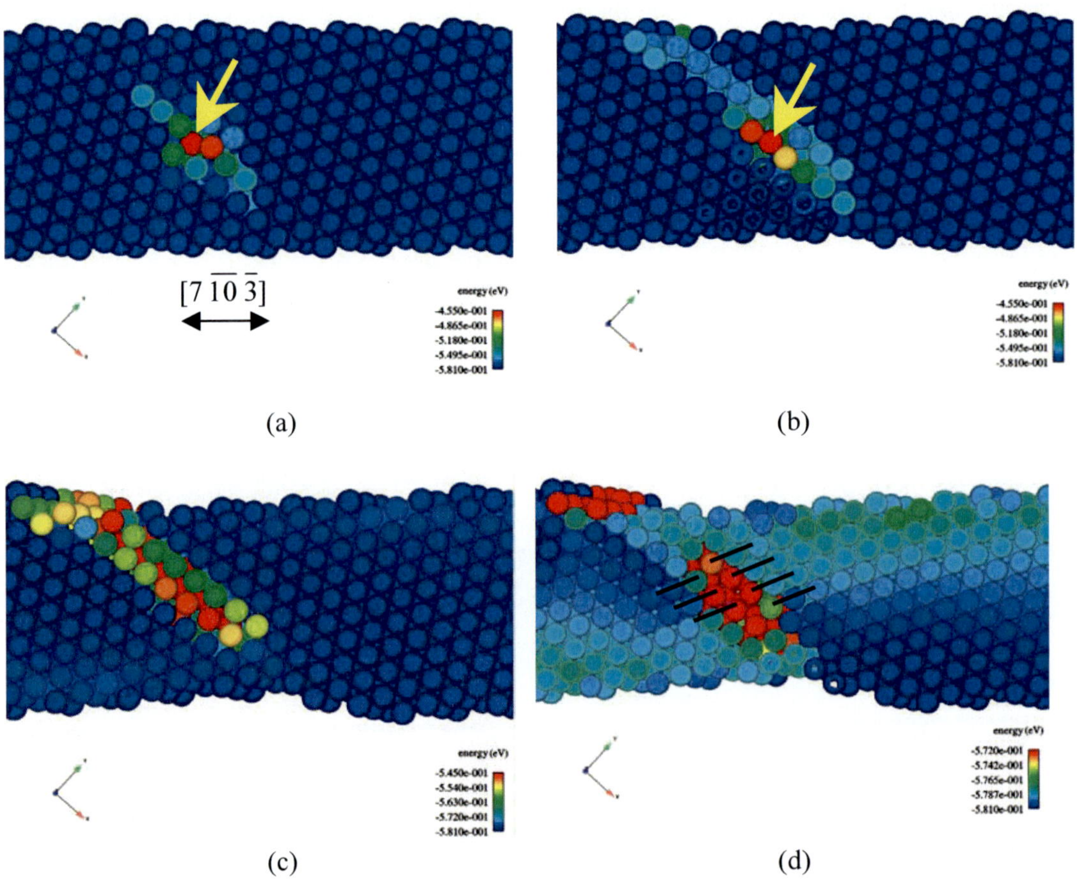

Figure 4. (a) and (b) Snapshots of atomic arrangement along an internal section of $(\bar{1}\,\bar{1}\,1)$ plane passing through the initial interstitial atom (highlighted by arrow) shortly before the first reduction of overall load during tensile stretching. (c) and (d) Snapshots of atomic arrangement along an internal section of $(\bar{1}\,\bar{1}\,1)$ plane, one atomic layer above those in (a) and (b), during the first reduction of overall load during tensile stretching.

Subsequently, Fig. 4(b), the atom is seen aligned with its neighbors along the close-packed [011] direction. A linear array of atoms having high energy states can be observed. Figures 4(c) and 4(d) show another internal section, parallel to, but one atomic layer above, those in 4(a) and 4(b), during the first overall load reduction. Several parallel lines along $[1\bar{1}0]$ were drawn in Fig. 4(d), illustrating staggered arrays across the "discontinuity" which is a screw dislocation along the slip direction [011]. The formation of screw dislocation can also be demonstrated by the Burgers circuit operation on the side surface of the specimen.

The incipient plastic yielding process can now be depicted by the schematics in Fig. 5, where the dislocation slip mechanism and the resulting crystal shape change are shown. Upon the incorporation of the interstitial into a line of atoms along the maximum shear direction, a small bulge at the end of the atomic line on the side surface becomes the starting point for forming the surface step, from which a pair of screw dislocations evolves. The slip of these dislocations leads to plastic yielding of the crystal.

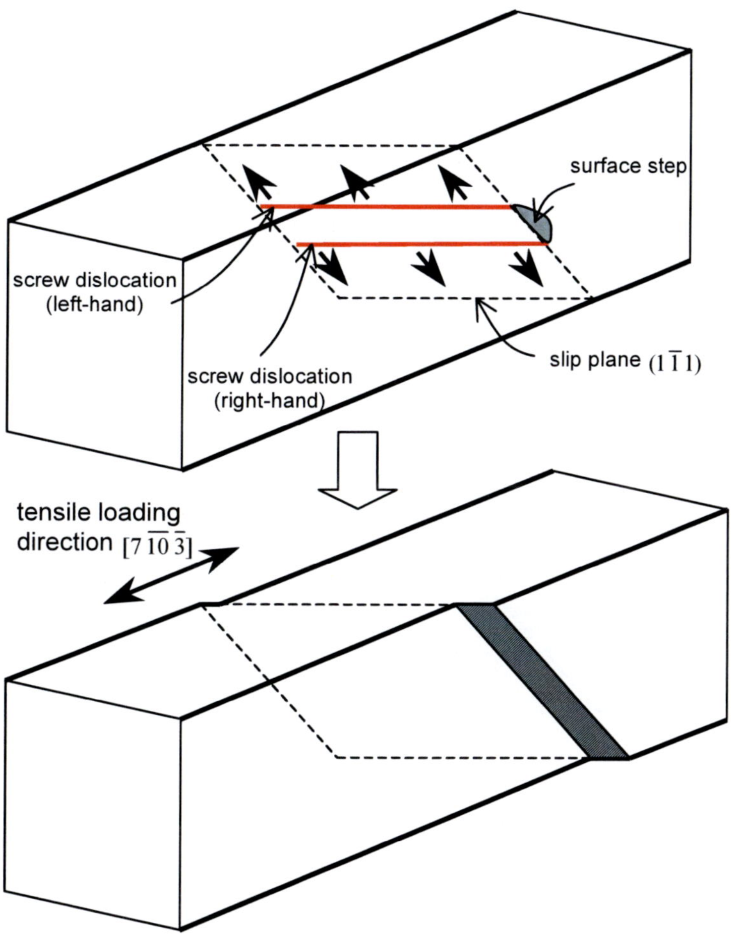

Figure 5. Schematics illustrating the dislocation slip mechanism and crystal shape change at the beginning stages of plastic deformation during loading.

In the case of 1.325 r_0 with the interfacial constraint (shown as "w/ substrate" in Fig. 2), the first load reduction after initial elastic response occurs at a similar overall displacement as in the previous case. Figure 6(a) shows the slip, originated from the initial point defect, along the same primary system $(1\bar{1}1)[011]$. In general, the entire load-displacement response is quite similar to the case with no substrate. This similarity arises from the fact that surface steps created as a result of the slip are on the side surfaces and not the bottom plane (Figs. 3 and 6). Therefore the flat interface treated in the substrate model does not significantly influence the deformation features. Figure 6(b) is another snapshot at a later stage of the deformation in the "w/ substrate" model.

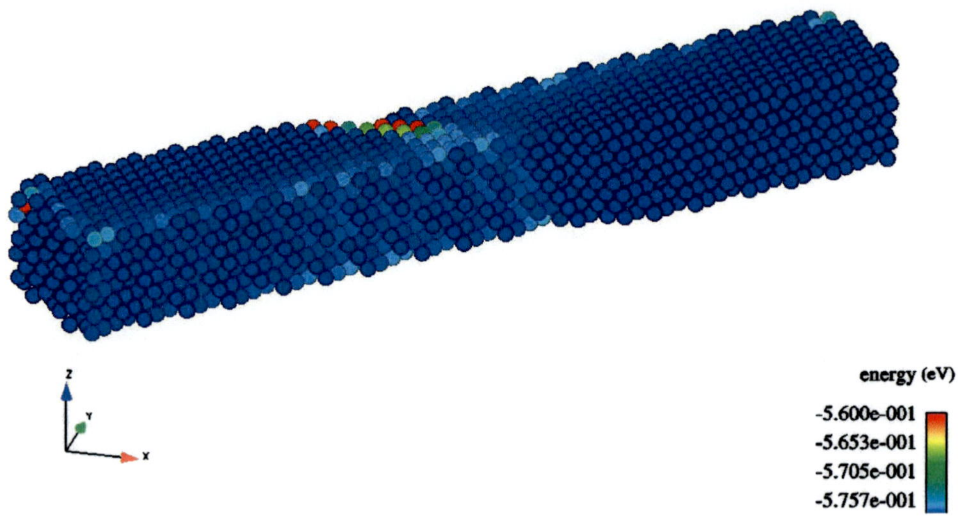

(a)

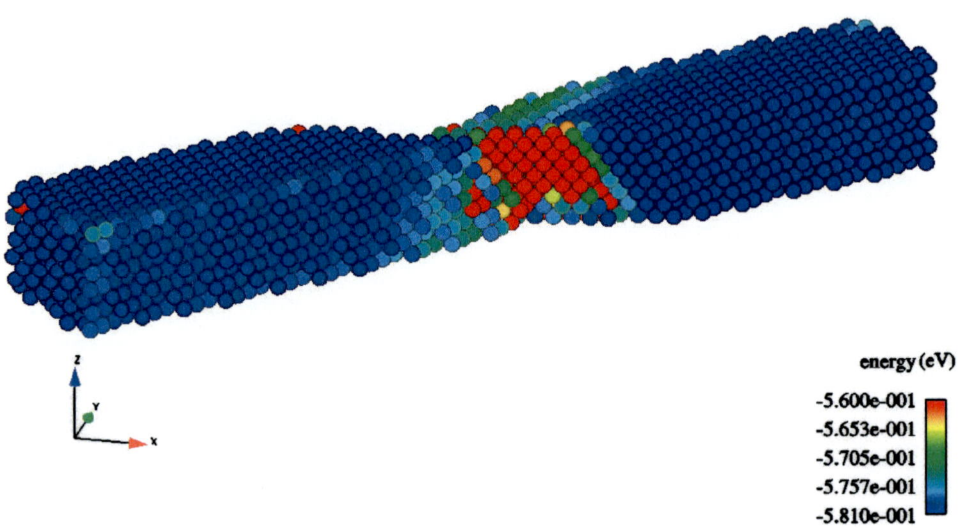

(b)

Figure 6. Atomic snapshots corresponding to points (a) 6a and (b) 6b labeled along the overall-load displacement curve in Fig. 2 for the case of 1.325 r_0 cutoff distance with substrate.

The results presented thus far showed that dislocation plasticity in the nanowire structure can be conveniently studied with the present approach. It has been argued, on the basis of comparison studies using the Lennard-Jones or Morse pair potentials and the many-body embedded atom potential [13,14], that while simulations involving only pair potentials generally yield brittle behavior, ductile materials *must* be described with a many-body potential [15,16]. In the current work we have illustrated that, with the incorporation of an initial point defect, the employment of a pairwise interatomic potential *is* able to render a ductile behavior. In previous 2D simulations applying the same technique [7,8,12], the initial point defect was observed to evolve into a pair of edge dislocations with opposite senses. In the present 3D study, the formation of two opposite screw dislocations is seen. This is believed to be a useful simulation methodology to induce local plasticity in a controlled manner, which is of interest for studying the interaction between dislocations, and between a dislocation and other microstructural features such as grain boundaries, interfaces, and second-phase particles. Another implication of the present results, which is of fundamental importance, is that an existing point defect in the crystal can serve as a source for dislocation nucleation during deformation. Although our study follows previous 2D simulations by making use of a self-interstitial atom, in actual crystalline materials vacancies are a more common form of point defects. The possible effect of vacancies in this type of atomistic simulation is worthy of further investigation.

The Case of $r_i = 1.335\ r_0$

Attention is now turned to models using the 1.335 r_0 cutoff distance. For the case without substrate, the overall load-displacement curve in Fig. 2 shows that the first load reduction occurs at a slightly smaller displacement, compared with the previous case of 1.325 r_0. The extent of the drop is also smaller. Figure 7(a) shows the atomic snapshot right after this drop. There is seemingly a tendency for slip along the $(1\bar{1}1)$ plane. Upon further deformation, a very large scale load reduction occurs, the corresponding snapshot of which is shown in Fig. 7(b). Apparently some fundamentally different deformation mechanisms have been involved. Detailed analyses reveal that regions A and B highlighted in Fig. 7(b) now have a BCC crystal structure, while other parts remain to be FCC. Regions A and B, however, have different orientations: the tensile axis is close to a <411> type direction in A and in B it is roughly along <110>. Regions A and B have {110} and {100} planes, respectively, roughly parallel to the bottom surface of the specimen. The load-displacement curve in Fig. 2 also shows that, in the present case, the specimen can be stretched to a very large strain without failure. This is associated with the varying crystallographic features in the specimen, as illustrated in Fig. 7(c). (Note in Figs. 7(c) and 7(d) a different viewing angle is used, where the line of sight is along the original $[11\bar{1}]$ of the FCC crystal, Fig. 1.) In Fig. 7(c) the initial region A has grown to the edge of the specimen. However, a region AA has now formed near the middle section and is propagating toward the specimen edge to replace A. Region AA is found to have the BCC structure with a {110} plane roughly parallel to the bottom surface and a {110} direction along the tensile axis. At a later stage, the other half of the specimen has started to undergo phase transition with a new BCC region being formed, Fig. 7(d). Therefore, we have observed that, with a slight increase in range of the atomic interaction (from 1.325 r_0 to 1.335 r_0), plastic deformation through regular dislocation slip tends to be suppressed at the cost of increased structural transformation.

The atomic configuration at the onset of plastic yielding (Fig. 7(a)) can be analyzed in greater detail. Figures 8(a) shows an internal section, approximately parallel to the plane containing the tensile axis and the vertical z-direction, at the vicinity of the initial interstitial atom shortly before the first load reduction leading to point 7a in Fig. 2. The initial interstitial atom is highlighted in the figure. A similar internal section corresponding to point 7a in Fig. 2 (and thus Fig. 7(a)) is shown in Fig. 8(b). In Fig. 8(a) the interstitial is still discernible as highlighted, but it is in the process of being accommodated into a line of close-packed atoms. The process has been completed in Fig. 8(b). This line of overcrowded atoms, bulging out on the top and bottom surfaces of the specimen, results in the formation of a screw dislocation.

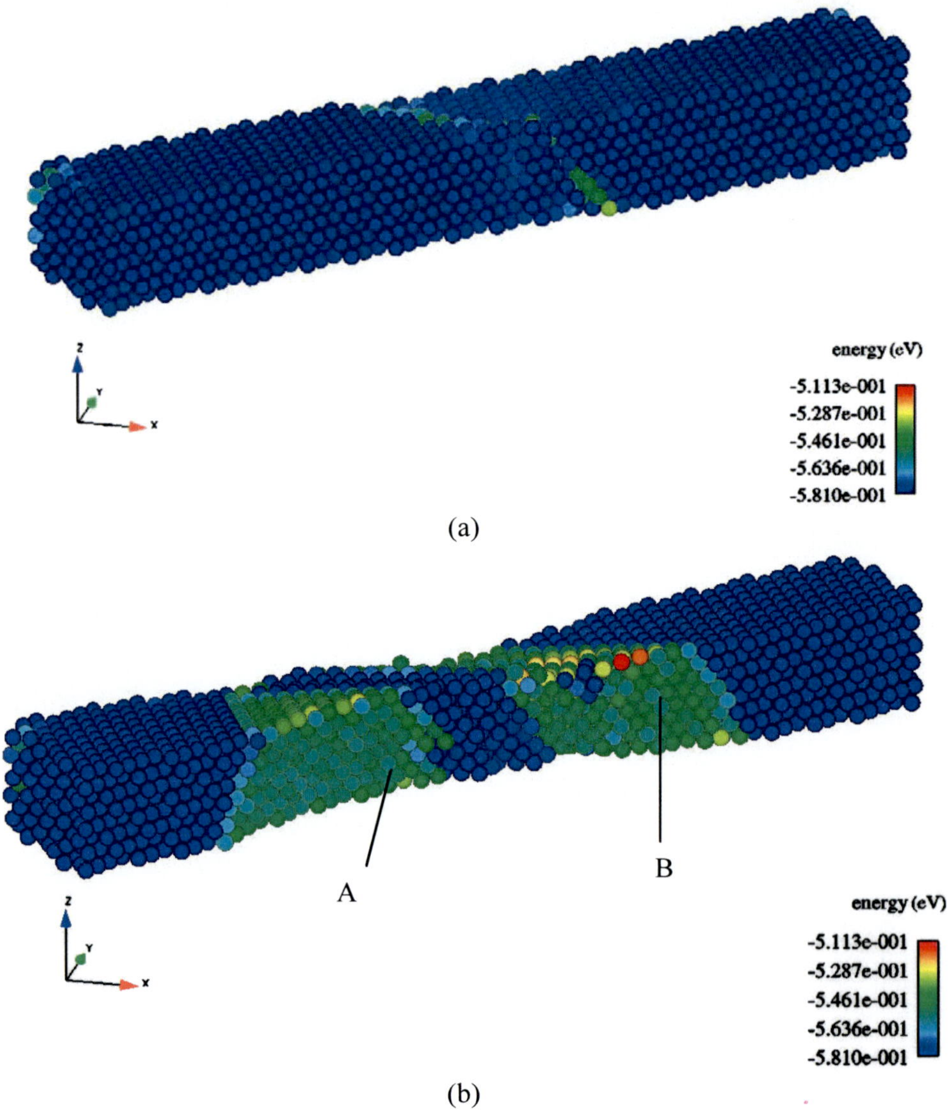

Figure 7. Continued on next page.

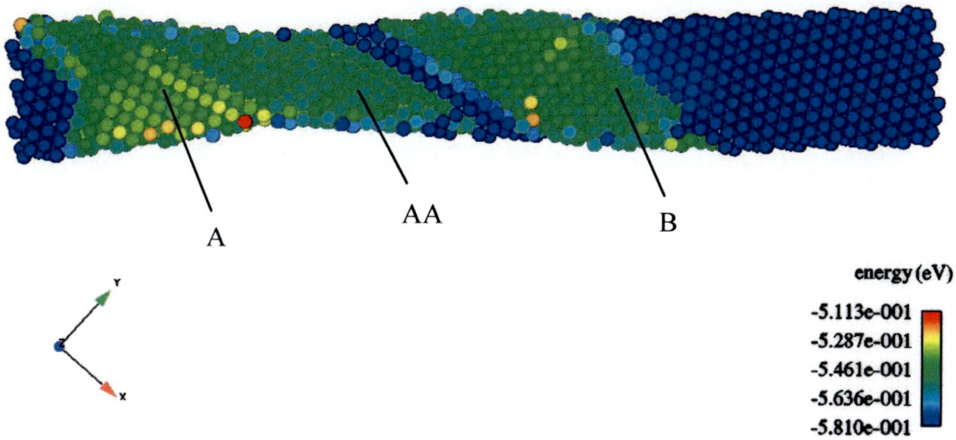

(c)

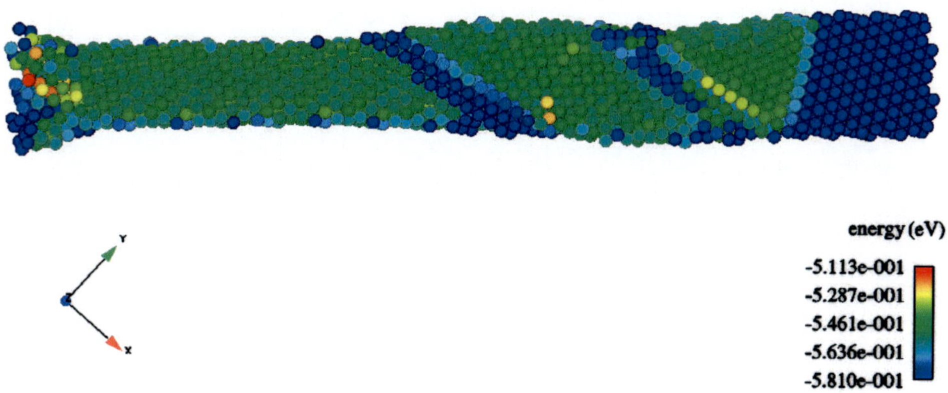

(d)

Figure 7. Atomic snapshots corresponding to points (a) 7a, (b) 7b, (c) 7c and (d) 7d labeled along the overall-load displacement curve in Fig. 2 for the case of 1.335 r_0 cutoff distance with no substrate. Note that (c) and (d) have a different viewing angle than (a) and (b).

Figure 8(c) shows the top end of the screw dislocation. Note that Fig. 8(c) shows the same snapshot as Fig. 7(a), but closer and with a different view angle. The screw dislocation line is identified to be along $[\bar{1}01]$, and its potential slip plane is $(1\bar{1}1)$. However, the Schmid factor of this slip system is 0.3617 which is significantly smaller than that of the primary slip system $(1\bar{1}1)[011]$. Even though this slip system has a dislocation available, because it has a relatively low Schmid factor, the dislocation in Fig. 8(c) stays immobile for the time being and an elastic response ensues (see Fig. 2, after point 7a). At a later stage (the peak load in the green curve in Fig. 2), the screw dislocation starts to glide and quickly moves out of the nanowire. A surface step is thus created as seen in Fig. 8(d).

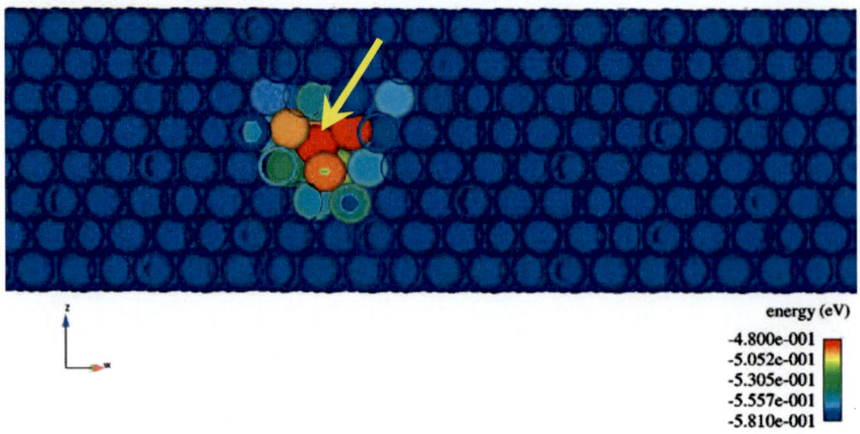

(a)

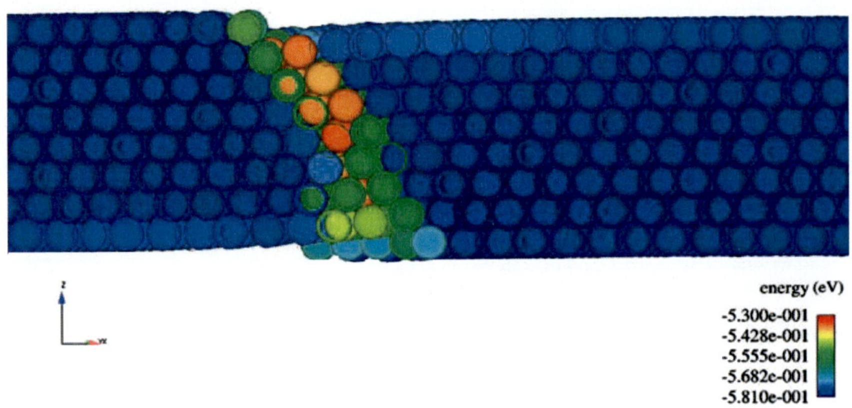

(b)

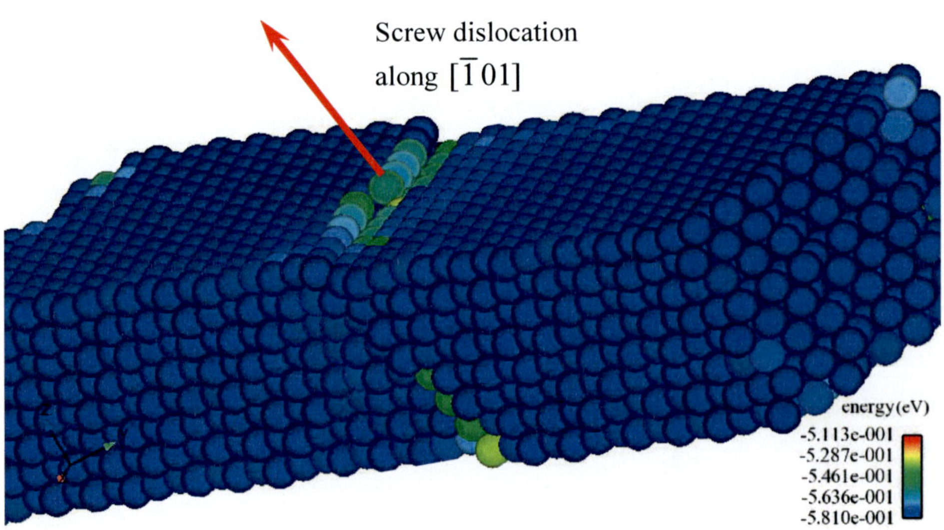

(c)

Figure 8. Continued on next page.

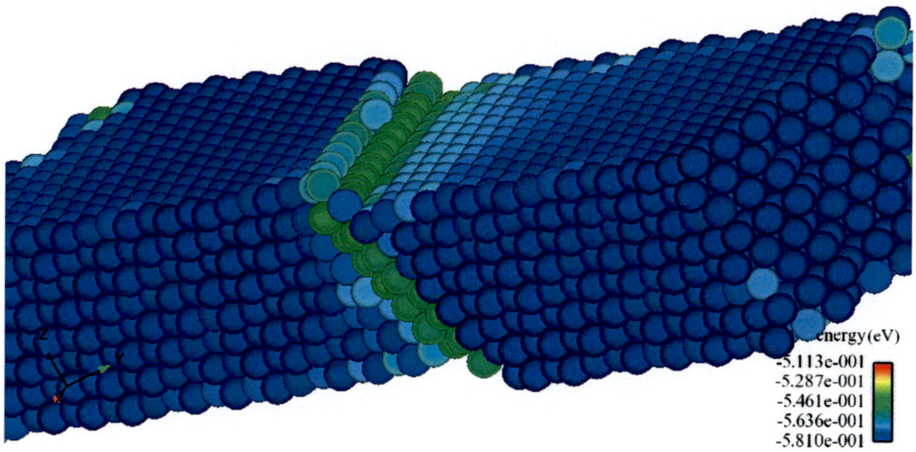

(d)

Figure 8. (a) and (b) Snapshots of atomic arrangement along an internal section roughly parallel to the plane containing the tensile axis and the vertical z-direction shortly before (part (a)) and right after (part (b)) the first reduction of the overall load. The initial interstitial atom is highlighted by the arrow. (c) and (d) Close-up views of the top surface right after the first load reduction (part (c)) and right after the second load reduction (part (d)).

The analysis above illustrates the interesting change in dislocation evolution and slip behavior due solely to a slight alteration of the atomic interaction range. While slip has occurred, the dominant mechanism of plastic deformation in the case of $r_i = 1.335\ r_0$ is phase transformation as seen in Fig. 7. The FCC to BCC transformation occurs after the screw dislocation slips out of the specimen, at which time there is no longer any apparent defect inside the crystal. The formation of the BCC phase in regions A and B observed in Fig. 7 (b) appears to be a coordinated movement of atoms at locations having significant elastic distortion resulting from the previous deformation. It is worth pointing out that, from theoretical calculations [17,18], the cohesive energy of BCC copper is only slightly below (within about 1%) that of FCC copper with an increasing atomic volume gradually favoring an FCC to BCC transition. It has also been theoretically calculated that uniaxial tensile loading can lead to the FCC to BCC transition [19]. The FCC to BCC transition in region A (Fig. 7(b)) may also be viewed as a variant of the Bain distortion [20], where the interchangeability of the FCC and BCT (body-centered tetragonal) lattices serves as the basis of the transition. Note that in this form of transition, upon being transformed from the FCC structure to BCC, the original {111} planes in the FCC lattice become {110} type planes in the BCC lattice. This type of geometric relationship was also observed in our simulations (Fig. 7(b), region A). In our model with the substrate effect included, the original $(\bar{1}\,\bar{1}\,1)$ plane at the bottom of the FCC crystal is treated as the interface plane. Therefore, it offers a unique opportunity for further examination of this FCC to BCC transition, as illustrated in the atomic snapshots shown in Fig. 9 below.

Figure 9(a) corresponds to the point along the load-displacement curve right before the first load reduction (Fig. 2). A structural change is just about to commence, which also starts from the region where the initial interstitial is located. The phase transition quickly spreads through almost the entire specimen (Fig. 9(b)), leading to a very dramatic load drop seen in

Fig. 2. The crystal structure in the transformed region is confirmed to be BCC with the same orientation as in region A in Figs. 7(b) and 7(c). Upon further deformation to point 9c, a significant change in deformation pattern has started from one end of the specimen (Fig. 9(c)). In Fig. 9(d), two regions, C and D, can be clearly identified. Both regions have the BCC structure, with C showing the same orientation as in region A in Fig. 7 and D the same orientation as in region AA in Fig. 7. The boundary between regions C and D advances from the right end of the copper wire toward the left (Fig. 9(d)), with region D eventually covering the entire specimen.

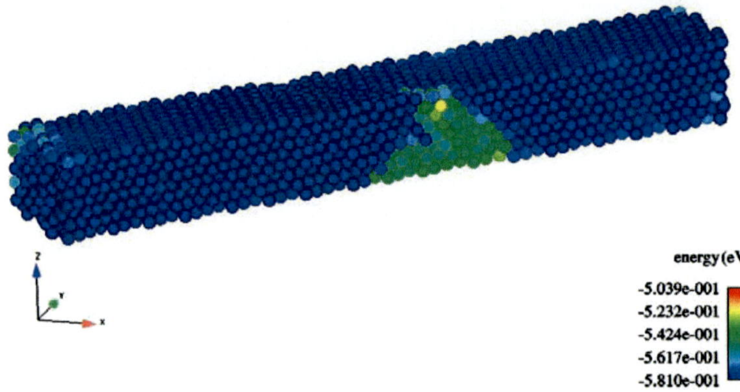

(a)

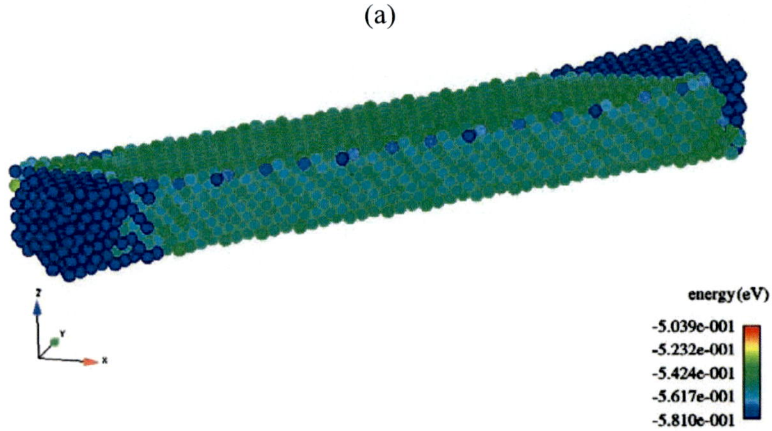

(b)

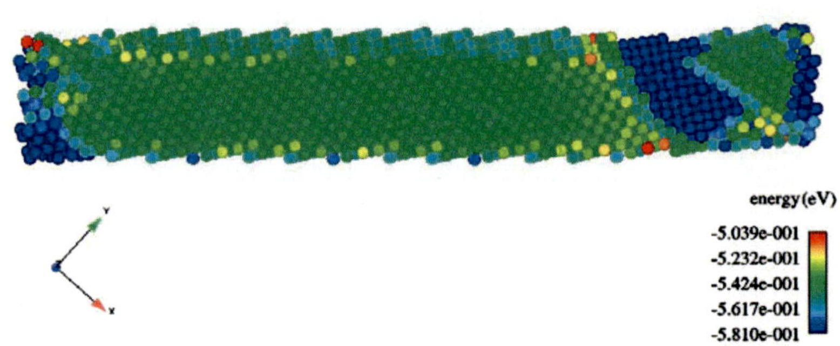

(c)

Figure 9. Continued on next page.

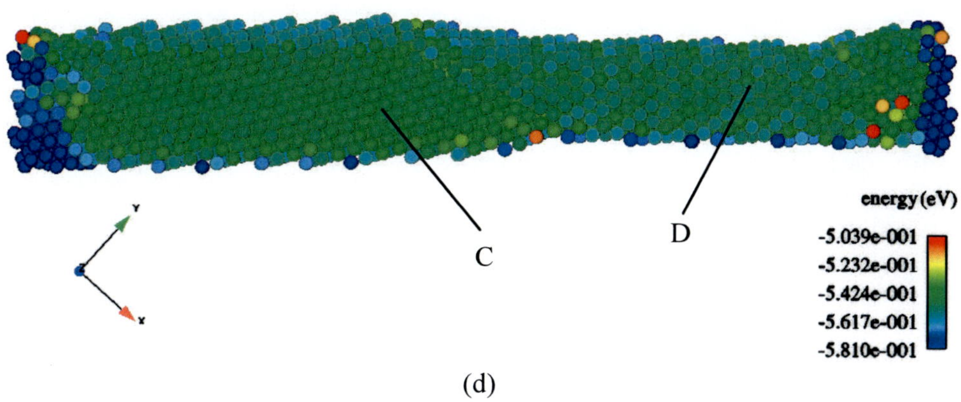

Figure 9. Atomic snapshots corresponding to points (a) 9a, (b) 9b, (c) 9c and (d) 9d labeled along the overall-load displacement curve in Fig. 2 for the case of 1.335 r_0 cutoff distance with no substrate. Note that (c) and (d) have a different viewing angle than (a) and (b).

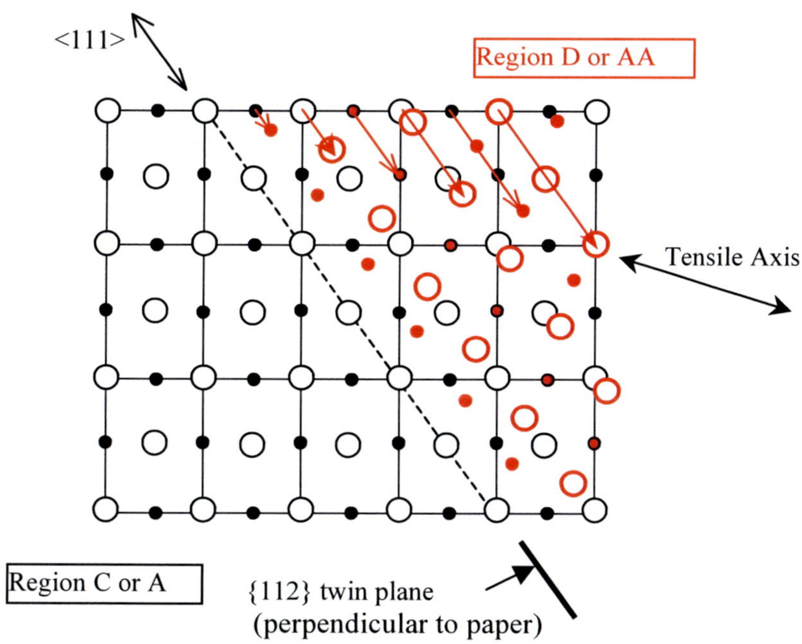

Figure 10. Schematic showing the deformation twinning mechanism in the transformed BCC structure [21]. The figure lies in a {110} plane. The twinning plane is a {112} perpendicular to the paper and the twinning direction is <111>. The lower-left and upper-right portions correspond to regions C and D, respectively, in Fig. 9(d). These portions also apply to regions A and AA, respectively, in Fig. 7(c).

The reorientation of the BCC crystal at the boundary between regions C to D observed in Fig. 9 is through a {112}<111> mechanical twinning mechanism, as depicted in Fig. 10. The figure shows an exposed {110} plane, with open circles representing atoms lying in the plane and filled circles representing atoms in an adjacent parallel plane immediately above (or below). The tensile loading direction of the nanowire, as well as the crystallographic twin boundary and twinning direction, are indicated in the figure. In our simulated specimen, coordinated atomic movement results in the twinning boundary advancing from right to left. As a consequence, the re-orientation from regions C to D (in Fig. 9(d)) and from regions A to AA (in Fig. 7(c)) takes place. Note that during the process, the plane of the schematic in Fig. 10 remains to be {110}, which is parallel to the interface with the simulated substrate in the model. It is clear that the bottom surface of the specimen (a {111} plane in the original FCC and a {110} plane in the transformed BCC) serves as an anchor plane for the initial FCC to BCC transition and subsequently aids in the orientation adjustment in a more structured way as compared to the case of Fig. 7.

Conclusion

In this work we have carried out atomistic simulations of uniaxial tensile loading of a copper crystal in the form of a nanowire, focusing on crystal defect mechanisms and their correlation with the overall mechanical response. The technique of embedding an initial self-interstitial in the model was utilized. With this treatment, the use of a pair potential is seen to be able to yield a ductile behavior, avoiding brittle fracture or the boundary effect. This is believed to be a useful strategy for computationally studying the interaction between dislocations and other crystalline features. We have illustrated, with a parametric analysis employing different atomic interaction ranges, that the deformation behavior in the specimen can be altered in a dramatic fashion. Without the surface energy effect, the embedded point defect in the model prompted the initiation of plastic deformation at a prescribed location. Detailed analyses of atomic configurations revealed that, as the deformation progresses, the initial interstitial atom was gradually accommodated by a line of atoms along a close-packed direction, leading to a geometric condition favoring the formation of screw dislocation. When the interatomic potential is confined to a small range, Schmid-type slip prevails. If a slightly longer-range interaction is allowed, however, the overall deformation is accommodated predominantly by the phase transition from FCC to BCC followed by re-orientation of the BCC lattice through progressive mechanical twinning. This latter process (phase change and then re-orientation) is much enhanced in the substrate-attached specimen. The present study demonstrated the high degree of sensitivity of atomistic movement in responding to the varying range of the interaction force in a metallic nanostructure. It is not possible to obtain this type of information utilizing more realistic many-body interatomic potentials. The present findings also aid in fundamental understanding of how an existing point defect can evolve into one or more dislocations during deformation.

References

[1] Kang, J.-W.; Hwang, H.-J. 2001, 12, 295-300.
[2] Diao, J.; Gall, K.; Dunn, M. L.; Zimmerman, J. A. *Acta Mate*. 2006,54, 643-653.
[3] Liang, W.; Zhou, M.; Ke, F. *Nano Lett*. 2005, 5, 2039-2043.
[4] Wang, J.; Huang, H. *Appl. Phys. Lett*. 2006, 88, 203112.
[5] Diao, J.; Gall, K.; Dunn, M. L. *Nano Lett*. 2004, 4, 1863-1867.
[6] Park, H. S.; Zimmerman, J. A.; *Phys. Rev. B* 2005, 72, 054106.

[7] Popova, M.; Shen, Y.-L.; Khraishi, T. A. *Mol. Simul.* 2005, 31, 1043-1049.
[8] Popova, M.; Shen, Y.-L.; *J. Comput. Theo. Nanosci.* 2006, 3, 448-452.
[9] Phillips, R. Crystals, *Defects and Microstructures* – Modeling Across Scales; Cambridge University Press: Cambridge, 2001; p. 206.
[10] McEntire, R. S.; Shen, Y.-L. *Molecular Simulation* 2006, 32, 857-867.
[11] McEntire, R. S.; Shen, Y.-L. in Mechanics of Nanoscale Materials and Devices, *Mater. Res. Soc. Symp. Proceedings*; 2006; vol. 924, paper number 0924-Z02-04.
[12] Shen, Y.-L. *J. Mater. Res.* 2003, 18, 2281-2284.
[13] Holian, B. L.; Voter, A. F.; Wagner, N. J.; Ravelo, R. J.; Chen, S. P.; Hoover, W. G.; Hoover, C. G.; Hammerberg, J. E.; Dontje, T. D. *Phys. Rev. A* 1991, 43, 2655-2661.
[14] Wagner, N. J.; Holian, B. L.; Voter, A. F. *Phys. Rev. A* 1992, 45, 8457-8470.
[15] Heino, P.; Hakkinen, H.; Kaski, K. *Europhys. Lett.* 1998, 41, 273-278.
[16] Heino, P.; Hakkinen, H.; Kaski, K. *Phys. Rev. B* 1998, 58, 641-652.
[17] Lu, Z. W.; Wei, S.-H.; Zunger, A. *Phys. Rev. B* 1990, 41, 2699-2703.
[18] Chelikowsky, J. R.; Chou, M. Y. *Phys. Rev. B* 1988, 38, 7966-7971.
[19] Milstein, F.; Farber, B. *Phys. Rev. Lett.* 1980, 44, 277-280.
[20] Bain, E. C. *Trans. AIME* 1924, 70, 25-46.
[21] Honeycombe, R. W. K. *The Plastic Deformation of Metals*; Edward-Arnold: London, 1984, 2nd Ed., p. 211.

In: Materials Science Research Trends
Editor: Lawrence V. Olivante, pp. 151-195
ISBN: 978-1-60021-654-1

Chapter 3

ADVANCES IN MATERIALS ENGINEERING USING STATE-OF-THE-ART MICROSTRUCTURAL CHARACTERIZATION TOOLS

Jian Li
CANMET-Materials Technology Laboratory, 568 Booth Street, Ottawa, Ontario, Canada

Abstract

Progress in materials science and engineering is closely related to material characterization. Materials performance is highly dependent on its microstructure. Microstructural characterization has long surpassed the optical microscopy era. Advanced techniques including scanning electron microscopy (SEM) and transmission electron microscopy (TEM) have been well integrated into routine characterization excises. Other microscopy techniques like electron probe microanalyzer, Auger, X-ray photon spectroscopy (XPS) and secondary ion mass spectroscopy (SIMS) are also well recognized in the past years. In recent years, the focused ion beam (FIB) microscope has gradually evolved into an important microstructure characterization instrument. The combination of high-resolution imaging and stress-free site-specific cross sectioning provides valuable microstructure information both at the specimen surface and beneath. In addition, FIB techniques are often the preferred method to prepare TEM specimens, which, in many circumstances, are impossible to make by any other conventional methods. In this chapter, various FIB microscopy applications in microstructural characterizations will be discussed using practical examples in our recent research.

1. Introduction

Focused ion beam microscope was invented in the mid 1980s exclusively for use in the semiconductor industry. In 1987, the total number of FIB systems was estimated to be about 35 and they were only used for mask repair, lithography (to replace electron-beam lithography), implantation doping of semiconductors, ion-induced deposition for circuit repair or rewiring [1]. The number of FIB systems has drastically increased in recent years, and their applications have extended into materials and biological sciences.

The development of enhanced imaging resolution of single ion-beam FIB systems and the introduction of dual-beam FIB systems has extended their applications in various fields of materials research [2-5]. A typical FIB microscope contains a liquid metal ion source that produces a fine beam of Ga ions. The primary Ga ion beam is accelerated by 30-50 keV. The finely focused ion beam is directed towards the features of interest on the targeted sample. The incident ion beam will sputter atoms off the sample surface either in ionic form, accompanying secondary electron emission, or in neutral form. Depending on the application, the beam current can be adjusted to as high as 40 nA for rapid ion beam milling or as low as 1 pA for high-resolution ion beam imaging (up to 5 nm ion-beam imaging resolution can be achieved in some FIB systems). Site-specific micro-depositions (e.g., either metallic or glass) and micro-etching can also be achieved by the interaction of the primary ion beam with the deposition (and etching) gas introduced into the system. Figure 1 shows a schematic diagram of a typical single-beam FIB microscope.

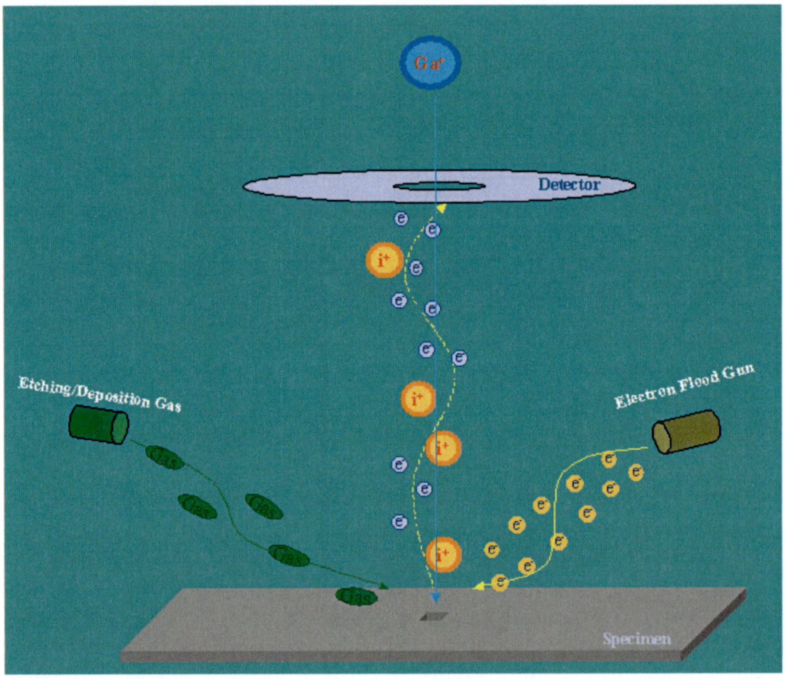

Figure 1. Schematic diagram of a typical FIB system.

Similar to a typical SEM, FIB microscopes can be used to produce high-resolution images directly from either as-received samples or mechanically polished surfaces. The primary gallium ion beam can produce enhanced crystallographic contrast using the secondary electron (SE) generated from the specimen surface. The FIB secondary electron yield is strongly dependant on the crystallographic orientation of individual crystals on the surface due to the very limited penetration depth of the primary gallium ion beam (only a few nanometers under typical operating conditions). Most of the sputtered atoms from the sample surface are ionized. The secondary ion (SI) particles are also collected by the detector to produce secondary ion images. Since the secondary particle ionization yield is strongly dependent on the local chemistry of the specimen, the FIB secondary ion images can provide valuable information related to the local chemistry. In addition to high-resolution imaging on sample surfaces, small features on the sample surfaces can be cross-sectioned in-situ using the primary gallium ion beam. The stress-free FIB cross-sections can be imaged by tilting the specimen.

The FIB microscope can also be used as a powerful tool to prepare TEM specimens, and this has been recognized as one of its most important applications. Various FIB-TEM specimen preparation techniques have already been reported [6-8]. In this paper, we will briefly review the currently available techniques, and demonstrate a method of using the FIB to prepare TEM specimens from very small samples.

2. An Overview of Focused-Ion Beam Techniques and Inovative Applications

The application of focused-ion beam microscopes in materials science can be categorized as: high-resolution imaging, TEM specimen preparation, micro-machining and micro-deposition. The vast majority of FIB microscopes acquired in the past years have been mainly used to prepare TEM specimens while the unique imaging capability using the primary Ga ion beam has frequently been overlooked. This becomes especially true with the availability of the more capable dual-beam systems. The high-resolution SEM (usually FEG SEM) column tends to take over the "imaging job" while the Ga ion beam in the FIB columns are often regarded as the dedicated "milling machine". Although the modern FEG SEM columns could achieve higher ultimate imaging resolution, the unique FIB images are still beneficial in many aspects. The heavy Ga ions, although accelerated to 50 keV, can only penetrate a few nanometers into the specimen (depending on the material's properties). This makes FIB imaging extremely surface sensitive. Strong crystallographic contrast can be obtained directly from the metallographically polished surface.

2.1. High-Resolution Imaging

Similar to a conventional SEM, FIB microscopes can produce high-resolution secondary-electron and secondary-ion images directly from an as-received sample surface. In many metallurgical applications, samples are mounted and mechanically polished for microstructure investigation. A typical SEM study would require metallographic etching as shown in Figure 2, however using FIB, such etching may not be necessary. Careful metallographic polishing is needed to obtain high-quality FIB images. Apart from potential surface/subsurface mechanical damage introduced during sample polishing, any surface oxidation or contamination will also have a significant effect on FIB imaging. However, under certain circumstances, the gallium ions can also be used to sputter off the surface oxide and contaminants prior to imaging.

The interaction between the gallium ion beam and the metal surface depends on many factors, such as acceleration voltage, beam raster parameters and material properties (including crystallographic orientation). Figure 3 shows a typical FIB image of annealed low-carbon steel prepared by metallographic polishing only (without etching). Similar contrast (although weaker) may also be obtained using SEM back-scattered electron (BSE) imaging mode if the steel surface is properly electro-polished. However the contrast and image quality of FIB images are far superior. In addition, electropolishing tends to be problematic when dealing with multi-phase materials, and in some cases (e.g. corrosion), it should be avoided since it could disturb or even dissolve the corrosion products.

Figure 2. SEM secondary electron images of polished and etched surface.

Figure 3. FIB SE image showing strong crystallographic contrast on an as-polished low-carbon steel sample.

Figure 4. FIB SE images of an area on a metallurgically polished low-carbon steel surface with various sample tilt angles.

When imaging crystalline specimens, the contrast of FIB images is very sensitive to crystallographic orientation [9]. Certain grains with less dense atomic planes parallel to the imaging surface could result in the Ga ion channeling deeper into the substrate and reducing both secondary-ion and secondary-electron emissions. These grains appear "darker". For example, in face-centered cubic (FCC) aluminum, grains with {100} and {110} plane parallel to the sample surface appear to be "dark" [9]. However, partly due to the size of the Ga ion, the incident angle for channeling to occur is very small. As soon as the incident angle changes slightly (e.g. 2^0), the contrast of some grains starts to show noticeable change. Such sensitivity is demonstrated using the following simple experiment. An identical area on a polished steel specimen is imaged using various specimen tilt angles. Suppose that, at no specimen tilt, the ion beam strikes a grain with crystallographic plane ($h_1k_1l_1$) parallel to the surface. By tilting the specimen, the incident angle undergoes a change equivalent to imaging a different grain with crystallographic plane ($h_2k_2l_2$) parallel to the surface. Figure 4 shows a series of images with different tilt angles. As small as 2^0 tilt has resulted in some changes in contrast of certain grains. This high sensitivity is very useful to detect small amount of plastic deformation. Each individual grain in a fully annealed crystalline material has its designated crystallographic orientation. The orientation should not vary across each grain. When a specific amount of plastic deformation is applied, the dislocations sweep across grains and forms cell walls or subgrains depending on the deformation condition. This will result in small changes in crystallographic orientation within the grain. Usually, the orientation changes across each grain are cumulative due to the local stress tensor (at each location within the grain) can be assumed identical, and the mechanical properties at each location within the grain should be the same. Thus, with a given degree of deformation, the orientation change within each grain can be significant enough to be detected by FIB secondary-electron imaging. Grains with a certain degree of plastic strain appear to be of non-uniform in contrast. Examples of FIB images of plastically deformed microstructures are shown in Figure 5. Two samples were cut from a fully annealed interstitial-free (IF) steel sheet. Plastic deformation was applied to one of the samples by means of a cup drawing test. The test imposed 53% major strain (along the drawing direction) and 25% minor strain (in the cup tangential direction) to the sheet. Both samples were mounted and polished using a metallographic polishing routine. Figure 5(a), shows the typical equi-axed ferrite grain in the annealed IF steel, but the microstructure is completely different in the deformed sample (Figure 5b). Within each grain, the contrast varies indicating that changes in crystallographic orientation resulted in the formation of some kind of dislocation structure or subgrains.

We have used FIB imaging to evaluate the effect of hydrostatic testing on existing cracks in pipelines. Concerns about stress-corrosion cracking (SCC) in the pipeline industry have increased in recent years due to an increase in the frequency of pipeline failures. The SCC-related failures have occurred not only in natural gas pipelines but also in pipelines transporting oil. In 2003, the US Department of Transportation issued an advisory notice to all US pipeline owners and operators to assess their pipeline SCC risk in both high-pH and low-pH environments [10]. Pipeline SCC inspections/assessments are often carried out by hydrostatic testing over decades. In a typical hydrostatic test, sections of pipelines are pressurized using water up to 110% of the materials' specified minimum yield strength (SMYS) and held for a designated length of time. The hydrostatic test is very effective in detecting near-critical cracks [11], however there are concerns that sub-critical sized cracks may grow larger and some blunt dormant cracks may be re-activated during hydrostatic tests.

Figure 5. FIB images of annealed (left) and deformed (right) IF steel.

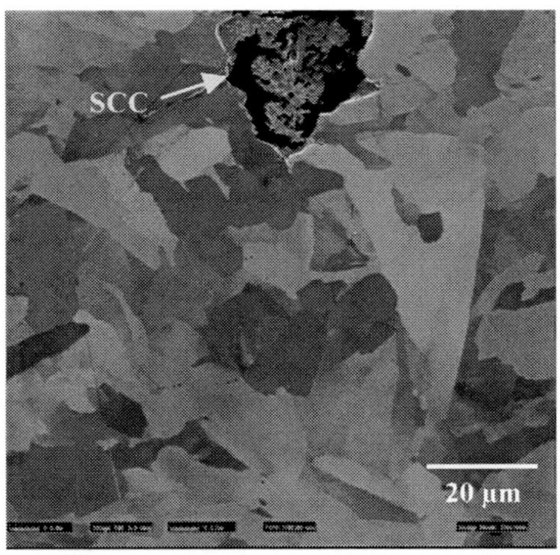

Figure 6. SCC tip from a sample with no hydrostatic test.

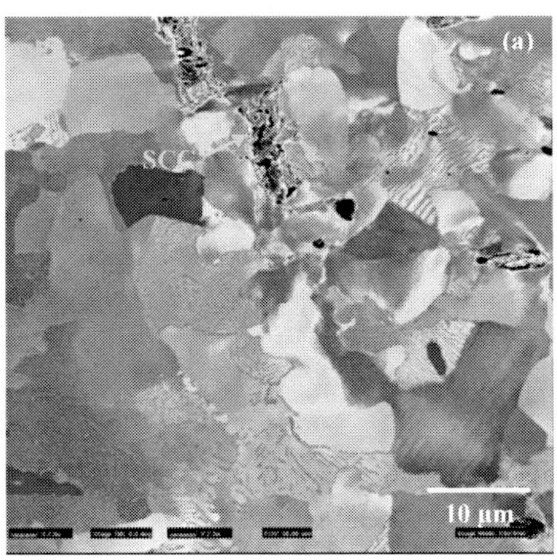

Figure 7. Continued on next page.

Figure 7. Plastic zones at large SCC tips found in hydrostatic tested pipes.

Figure 8. Montage of a large SCC showing the SCC re-activation and propagation from the earlier hydrostatic test. A: Plastic zone at the SCC tip due to 1994 hydrotest, B: SCC tip and plastic zone resulting from the hydrotest in 2004.

To study the effect of hydrostatic testing on the substrate material near existing crack tips, sections of in-service pipeline containing colonies of SCC cracks were investigated. Two hydrostatic tests were performed in 1994 and 2004. A small sample containing SCC cracks was cut out, mounted in low shrinkage Epoxy resin, polished and finished with 0.05 μm colloidal silica using a special sample-preparation routine suitable for imaging using a FIB microscope. Figure 6 shows a typical FIB image taken at an SCC crack-tip from a sample that had not been subjected to hydrostatic testing. The crystallographic orientation contrast indicated no apparent plastic zone ahead of the crack tip. Figure 7 shows a crack tip from a pipe subjected to hydrostatic tests; this sample shows a plastic zone near the crack tip. The existence of a deformed zone near the crack tips could promote further SCC propagation. Figure 8 shows a montage of a section of a long crack. The top portion of this crack is much wider than the distinctively thinner bottom portion. Apparent plastic zones exist at both the transition zone and the thin crack tip. It is reasonable to suggest that the thicker SCC existed prior to the hydrostatic test in 1994 and became inactive for a period of time (dormant crack before 1994 hydrostatic testing). Signs of the hydrotest in 1994 are marked by the plastic deformation in the transition zone. This SCC became re-activated and started to propagate again after the 1994 hydrostatic test, and the 2004 hydrostatic test created the second plastic zone at the crack tip. During this study, the same crack tip was also imaged using a conventional optical microscope and SEM. These plastic zones are not visible using either the optical microscope or SEM. Details of the study are published elsewhere [12].

2.2. FIB Cross-Sectioning

Focused-ion beam microscopes are also used as precision ion-milling machines in microscopic scale. Microscopic features appearing on the sample surfaces can be cross-sectioned using the primary gallium ion beam. High-resolution images of the cross sections can be obtained by tilting the sample in single beam FIB systems, or more conveniently using the electron beam in modern dual-beam systems. Prior to FIB sectioning, a strip of metallic coating is deposited in the FIB to protect the surface features from ion beam erosion during milling. Figure 9(a) shows a FIB cross-section of an aluminum metal-matrix composite (MMC). The thin reaction layer between the SiC particle and aluminum matrix is shown clearly in Figure 9(b). The typical sizes of FIB cross-sections range from 10-100 μm. Very small cross-sections tend suffer from re-deposition. Large FIB sections are limited not only by extensive FIB milling time (especially on materials of higher atomic numbers and most ceramic materials), but also by the difficulty in creating high-quality cross-sectioned surfaces. The primary reason for this difficulty is that the relatively coarse ion beams that are frequently used to minimize milling time usually result in cross-sections with a "curtain" effect. Lower beam current should be used to obtain high-quality cross-sections, but when milling in large scale, during the extensive milling time, the beam instability and stage drift could diminish the advantage. The largest cross-section accomplished by the author was 500 μm in width on a galvanealed steel sheet to assess the quality of the Zn coating. In this case, the major milling job was performed using high current (up to 40 nA) and small sections of ~100 μm in width were "polished" using a lower beam current (1.5 nA). If larger cross-sections are desired, a combination of mechanical polishing and FIB imaging could be used.

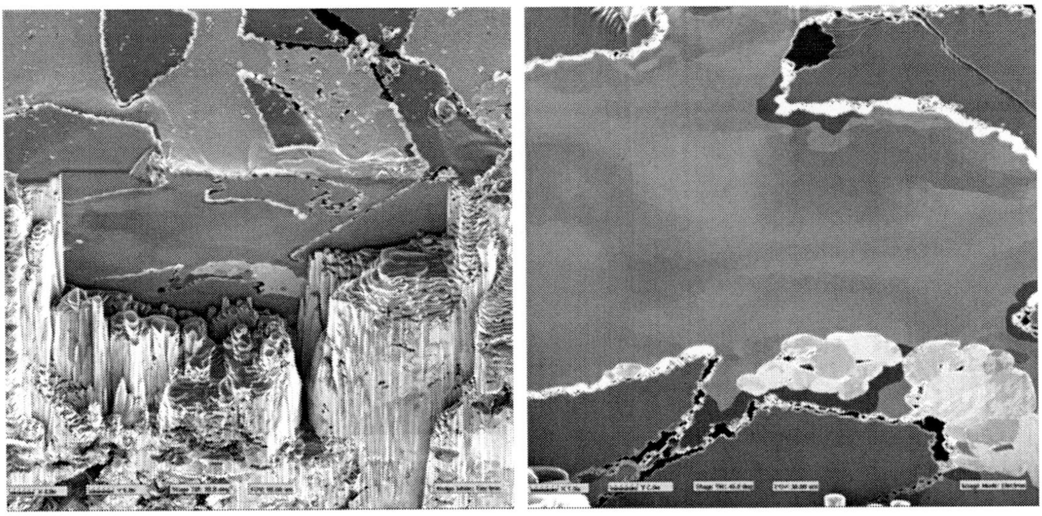

Figure 9. (a) A FIB trench of a specific feature on an aluminum MMC. (b) High-resolution FIB image showing thick aluminum oxide beneath the surface at this location.

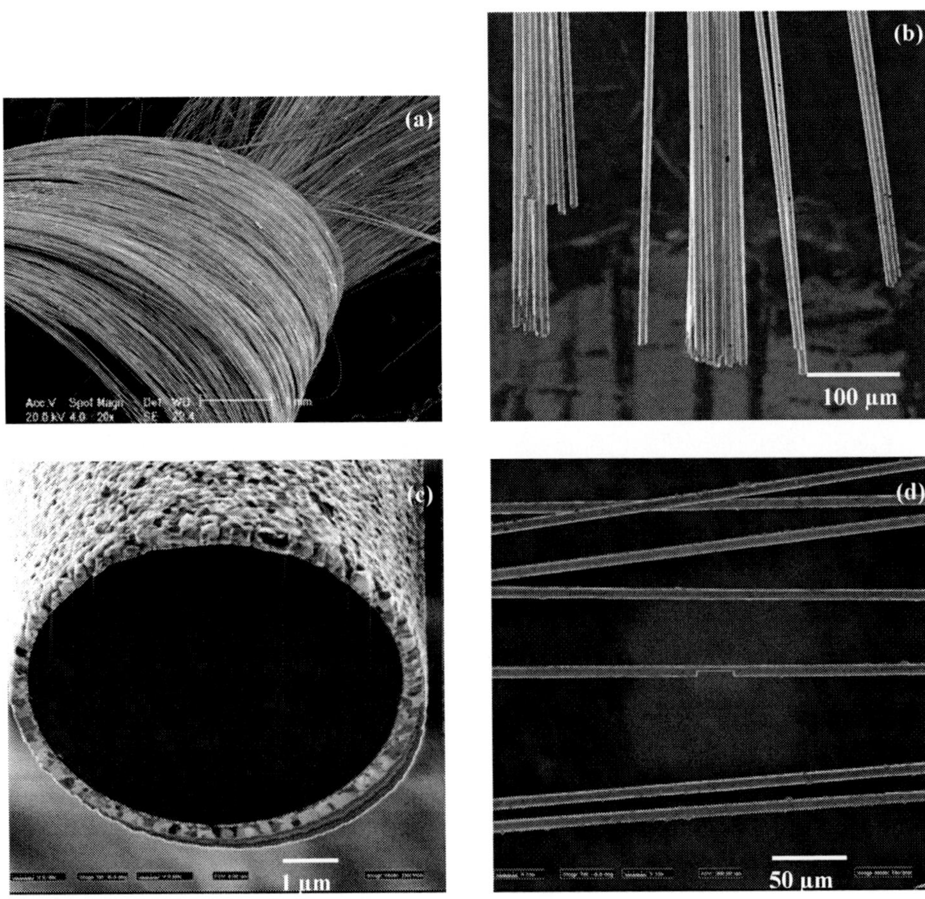

Figure 10. Continued on next page.

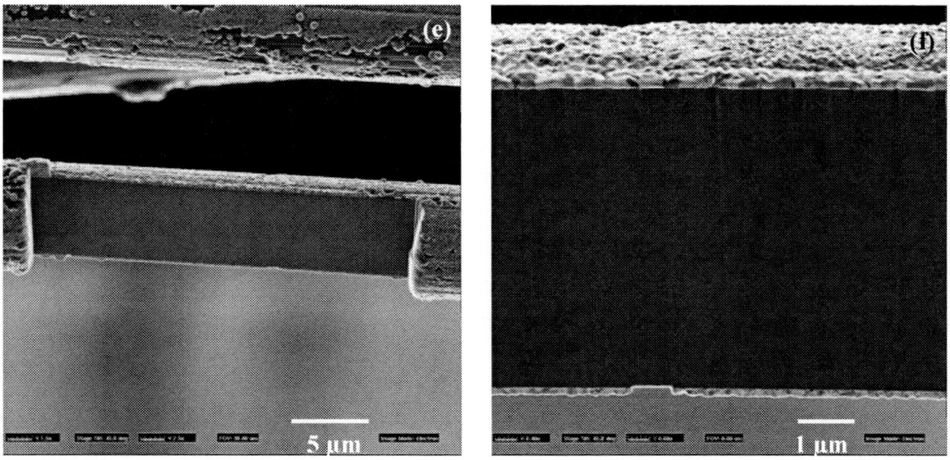

Figure 10. (a) Low-magnification SEM image of fiber bundle. (b) FIB image of the fibers. (c) FIB cross-section and image showing fine-grained coating microstructure. (d-e) FIB cross-section along the fiber longitudinal direction to evaluate coating uniformity.

In some cases, preparing a metallurgical polished surface is not an option because the samples are either too small or too fragile. Under these circumstances, stress-free FIB sectioning would be the only appropriate technique to reveal the microstructure. Figure 10 shows the examination of the microstructure of a thin metal coating on fine carbon fibers (about 10 µm in diameter). A cross-section of a selected fiber was made in the FIB, and the microstructure was investigated in-situ. The high-resolution FIB secondary electron image in Figure 10(c) shows the fine columnar grain structure with no detectable voids or delamination. FIB cross-sections were made along the fiber's longitudinal direction to examine coating uniformity along the length of the fiber [Figures 10(d-e)].

2.3. TEM Sample Preparation

Preparing high-quality TEM specimens is of paramount importance in TEM studies. FIB microscopes have become powerful tools in TEM specimen preparation [2,6-8], and the techniques used have been evolving rapidly. In recent years, a more advanced "lift-out" technique has demonstrated unique advantages in cases where mechanical preparation is difficult or impossible [6]. A major advantage of the lift-out technique is that TEM specimens can be made directly from bulk specimens. One problem, however is that the TEM sample cannot be re-thinned. The lift-out technique has more recently been combined with the "H-bar" technique [13]. Instead of lifting out an electron-transparent specimen, a much thicker specimen (typically 3-4 µm in thickness) is lifted out from the bulk. This specimen, containing the feature of interest, is then mounted onto a carrier using a biological micromanipulator and thinned in a FIB microscope. Although the currently available techniques can be applied to almost all cases of TEM specimen preparation, they are ineffective in some special cases. In this section, after these currently available techniques are reviewed, and a new technique suitable for making small and fragile fiber and powder TEM specimens is demonstrated.

2.3.1. Conventional H-bar Technique

The conventional H-bar technique has been in use as a major FIB-TEM specimen preparation routine since the introduction of FIB microscopes [14]. As illustrated in Figure 11, small samples (about 2.5x1.0x0.5 mm) are cut from the bulk specimen using a precision diamond cut-off wheel. Both sides of this small specimen are then carefully polished with a "tripod polisher" [15] in order to produce a flat surface and minimize the mechanical damage induced by cutting. The specimen is then mounted onto a modified TEM grid using an epoxy adhesive and allow sufficient time to cure. Mounted samples are usually polished to less than 100 μm in thickness to reduce FIB milling time. Some thought must be given as to the final thickness before FIB thinning. The thinner the sample, the less the FIB milling time will be needed, but if the sample is too thin, it may have insufficient mechanical strength, and residual mechanical damage from the mechanical polishing could lead to problems. The mounted specimen is then loaded into a FIB microscope for precise ion-beam thinning.

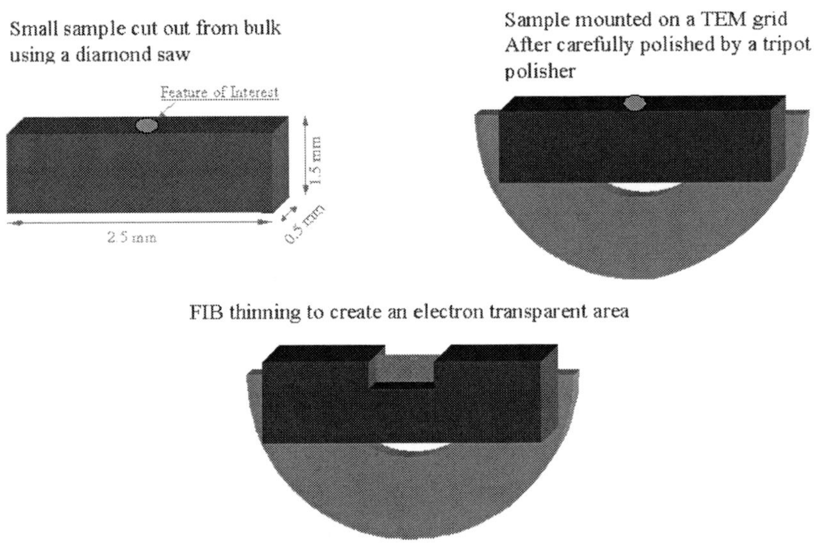

Figure 11. Schematic diagram of conventional FIB-TEM specimen preparation technique.

This technique has found many applications, especially for non-site-specific TEM specimens, and is useful for making most TEM specimens. Once the sample is mounted onto the copper TEM grid, FIB sectioning and imaging is used to identify the area of interest, as shown in Figure 12. An electron-transparent TEM sample can subsequently be made by thinning the backside of the specimen. However, care must be taken during diamond saw cutting and various stages of mechanical polishing to ensure the integrity of the area of interest. Currently, with significantly increased FIB applications, the H-bar FIB-TEM specimen preparation technique has become inadequate. For example, in stress-corrosion cracking (SCC) studies, diamond saw cutting and mechanical polishing could introduce excessive stress that may lead to crack propagation or alteration of the chemical composition of corrosion products in SCC.

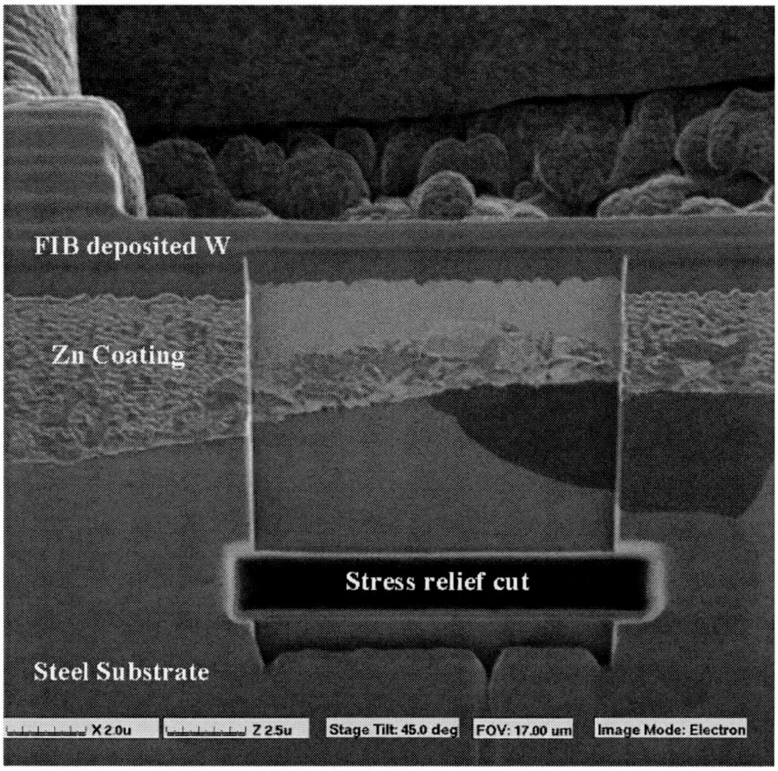

Figure 12. An example of the H-bar technique to make a TEM specimen from galvannealed steel.

2.3.2. The "Lift-out" Technique

The application of the lift-out technique to FIB-TEM specimen preparation was revolutionary [6,8,16-18]. The main advantage of this technique is that TEM specimens can be made directly from bulk samples; cutting and mechanical polishing, although beneficial in some occasions, are not required. This is particularly useful for fragile samples that cannot be prepared by any mechanical means. The identified feature of interest on the surface is protected by FIB-deposited metal (tungsten or platinum); FIB milling around the targeted area creates a membrane with a typical thickness of less than 100 nm depending on the material. This thin-membrane (TEM foil) containing the feature of interest is then cut free from the bulk. Subsequently, either an in-situ or an ex-situ lift-out process is used to transfer the thin membrane to a carbon-coated TEM grid using a micromanipulator. A schematic diagram of this technique is shown in Figure 13. The preparation of a lift-out TEM specimen affects only a relatively small local region typically on the order of 50 x 50 x 50 μm in size. The bulk sample is virtually unaffected, so this technique is considered to be non-destructive.

The lift-out technique enables rapid TEM specimen preparation with minimal mechanical damage. The TEM specimens have reduced spurious bulk X-ray signal due to significantly reduced bulk material around the electron-transparent membrane. This is particularly beneficial for TEM investigations of magnetic materials as there is less mass of the specimen to be attracted to the TEM pole pieces. In addition, samples that are fragile or sensitive to contamination can be prepared using this technique.

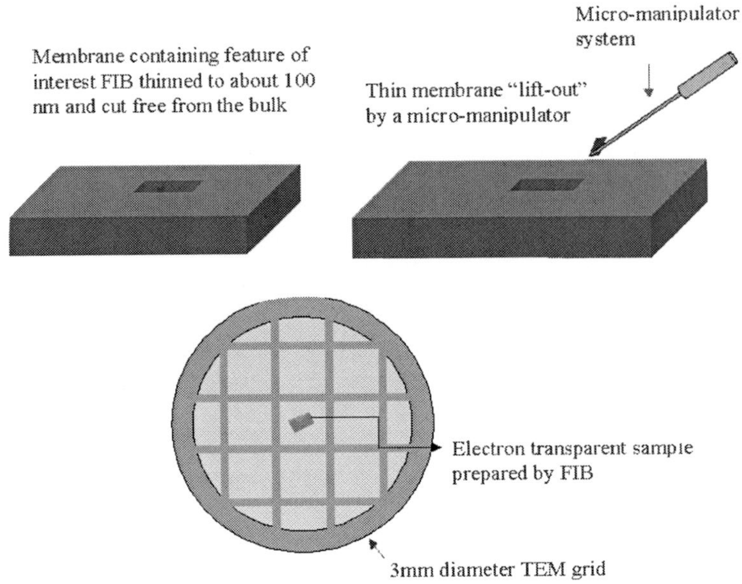

Figure 13. Schematic diagram of lift-out TEM specimen preparation technique.

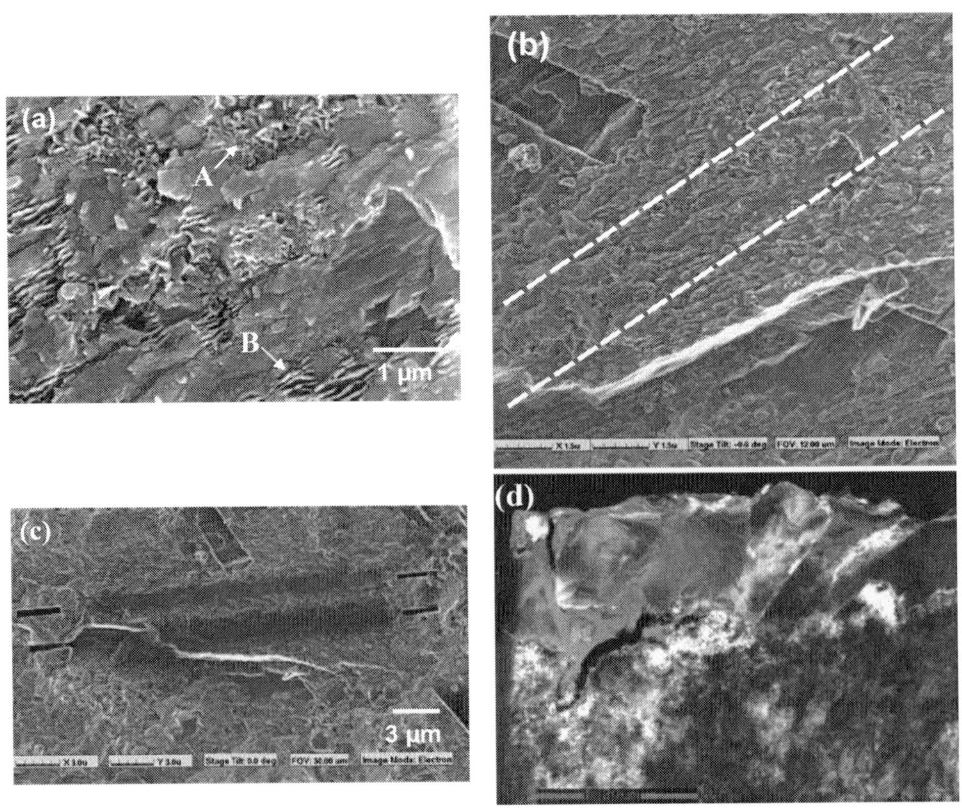

Figure 14. Preparation of TEM specimen from a Mars meteorite, (a) high-resolution SEM image showing feature of interest found on the surface (arrowed), (b) identified cluster of interest in FIB, (c) Features of interest protected by FIB-deposited tungsten, (d) Lift-out TEM specimen.

To demonstrate the viability of the lift-out technique, we undertook a challenging case of precision specimen preparation - preparing TEM sections through the controversial features in ALH84001, the NASA Mars meteorite that reportedly contains evidence of former Martian life. Considering the nature of the specimen, the FIB had the best chance of producing such sections using a lift-out procedure. Since the features were only 20-30 nm in the smallest dimension and lay on the specimen surface [Figure 14(a)], a practice meteorite of asteroid origin was first examined to determine whether imaging could result in significant surface material removal, or if sputtering effects during FIB trenching could create unforeseen problems. Neither effect proved insignificant. These features of interest were located in the FIB as shown in Figure 14(b). After careful FIB tungsten deposition along these lines [Figure 14(c)], the primary section was thinned and cut free. A micromanipulator specifically designed for FIB section lift-out was used to transfer the thin TEM foil over a formvar-coated grid and carefully pressed down so that the film held the section firmly due to the large contact area. The lift-out TEM section is shown in Figure 14(d).

The lift-out technique has proven to be an efficient and versatile TEM specimen preparation method especially in cases when the conventional H-bar technique is not applicable. However, TEM foil thickness cannot be measured accurately in FIB systems. The degree of final FIB thinning is very dependant on the operator's experience. Once the TEM specimen is made and properly transferred to the TEM grid, it is not possible to perform any further FIB re-thinning even though imaging or analysis requirements may suggest it.

2.3.3. The "H-bar Lift-out" and "Plan-view Lift-out" Techniques

The potential risk of losing a TEM specimen during the ex-situ lift-out process can be quite high. The electrostatic charge at the tip of the glass needle used to liftout thin foils could repel the tiny TEM foil and cause it to "fly-off" the needle tip. Also, too much charge at the needle tip could make it very difficult to "unload" the specimen onto the TEM grid. In contrast, modern dual-beam FIB systems are usually equipped with an in-situ lift-out tool has a relatively high success rate. Recent work by Patterson *et al.* [13] combined the "H-bar" and "lift-out" techniques. Using this combined method, a much thicker slab on the order of ~5 μm in thickness (usually 20x10x4 μm) containing the feature of interest is cut free from the bulk specimen and transferred to a TEM grid using the same micromanipulator as shown in Figure 13. Figure 15 shows a small steel specimen that has been cut free and is ready to be lifted out. The small sample is then carefully positioned onto a carrier, which is glued on the edge of a TEM grid, then thinned to electron transparency in a FIB microscope. TEM samples produced using this technique can be further thinned in the FIB if required. In addition, there is less chance of losing a TEM specimen during the lift-out process. In the authors' opinion, this is by far the most versatile and convenient ex-situ lift-out technique under almost all circumstances. The in-situ lift-out technique is very similar to the H-bar lift-out technique. The Omniprobe probe tip is positioned to touch the FIB milled sample; FIB metal deposition is used to attach the probe to the sample before the sample is lifted out. The lifted out sample is attached to the TEM grid by FIB metal deposition before final FIB thinning.

The H-bar lift-out technique is particularly powerful when used to prepare plan-view TEM specimens. In many TEM investigations, plan-view samples are required to characterize areas of interest. For example, a TEM foil containing the crack tip of interest would be extremely valuable to gain understanding of the mechanism of SCC. However, a site-specific plan-view TEM sample is nearly impossible to prepare with conventional methods. In a recent study, a site-specific plan-view TEM specimen was successfully prepared using the H-bar lift-out technique [19,20]. The implications of the plan-view lift-out technique on the study of SCC will be illustrated in section 3.1.

Figure 15. FIB secondary electron image of a 5 μm thick steel sample cut loose and lifted out.

2.3.4. The "Direct Lift-out" Technique for Ultra-fine Specimens

Although proven to be extremely powerful and versatile, the commonly used FIB-TEM specimen preparation techniques present difficulties when dealing with small and delicate samples such as fine powders and fine and/or fragile fibers. Once the sample dimensions approach the size of the usual lift-out specimen, the application of conventional lift-out techniques becomes difficult.

Early work by Cairney and Munroe [10] demonstrated a method for preparing TEM specimens from fine FeAl and WC powders. In their experiment, powders were first embedded in a low-viscosity epoxy resin. A TEM specimen was successfully prepared using the conventional H-bar technique by treating the hardened resin (embedded with the powder particles) as a bulk specimen. However, a TEM sample prepared using this method contained significant amounts of epoxy, which can be problematic in many ways during TEM examination. In addition, a significant amount of residual stress could be introduced to the particles during the resin-curing process.

In our recent study [21], we evaluated the coating integrity of fine nickel-coated carbon fibers (about 10 μm in diameter). The entire cross-section of the fiber was to be made electron transparent for TEM analysis. Some of the coatings to be studied were extremely fragile or even flaky. The "resin embedding" technique [22] could have caused unacceptable mechanical damage due to shrinkage during the resin solidification and curing process. None of the currently available techniques, which were summarized by T. Malis et. al. [23], were deemed likely to provide artifact-free TEM samples of these coated fibers. Even the very versatile technique of diamond-knife ultramicrotomy, normally excellent for cross-sectional TEM specimens, would have produced mechanically damaged or "shattered" cross-sections, and coating delamination would have been highly likely. We found that the fine glass needle of the micromanipulator could pick up much larger pieces than previously reported [21]. The fibers were first carefully cut to 4-5 mm in length and transferred onto the inner edge of a TEM copper grid using a micromanipulator under an optical microscope. A minimal amount of low-shrinkage epoxy was used to mount the bottom of the fiber to the edge of the grid. The fiber fixed onto the grid was then directly cut and thinned in the FIB microscope as shown in the schematic diagram in Figure 16. Figure 17 shows the FIB and TEM images of the fiber made using the direct lift-out technique. There are several advantages of this new technique:

1). The entire cross-section of the fiber can be made electron transparent for TEM observation if required.
2). No mechanical damage is introduced during the TEM specimen preparation process.
3). Compared with the conventional lift-out technique, this technique reduces the risk of potential mechanical damage by the micromanipulator when the specimen is thin and fragile.

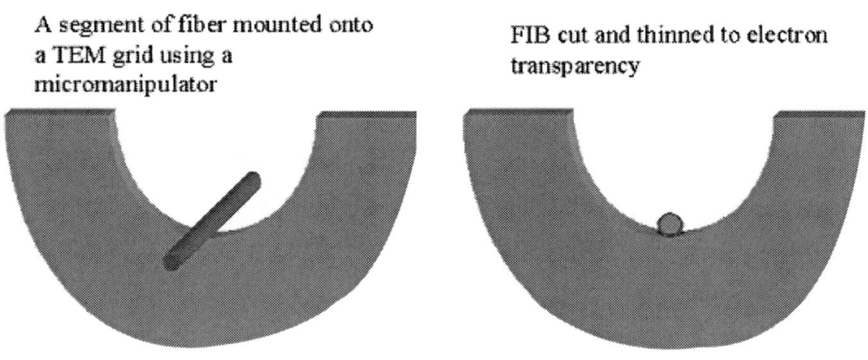

Figure 16. Schematic diagram showing the direct lift-out TEM sample preparation.

This lift-out technique has been recently used to make TEM samples from fine powders. Using this method, no resin embedding (as used by Cairney and Munroe [21]) is required. The powder is simply spread onto a clean surface, and small particles are lifted-out and mounted directly onto the edge of the copper grid. Sometimes, powder can be gently crushed in order to obtain very small pieces. Small particles, of the order of ~5 μm, can be made electron transparent in the FIB with minimal milling effort (time). Figure 18 shows a small particle mounted onto a copper grid and ready to be thinned.

In summary, the conventional FIB-TEM specimen preparation technique is a simple and straightforward technique that is suitable under many circumstances. The lift-out technique not only provides the capability to prepare TEM samples with minimum mechanical damage and minimal contamination, but is also capable of producing site-specific TEM specimens. TEM specimens can be prepared either perpendicular or parallel to the sample surface (plan-view lift-out). The "direct lift-out" technique further facilitates TEM sample preparation of small and/or fragile specimens such as fine fibers and powders. The FIB is the only technique that can produce site-specific, parallel-sided TEM samples with nearly no contamination. However, the selection of TEM specimen techniques is not only material dependent but also based on the type of TEM analysis to be performed [6].

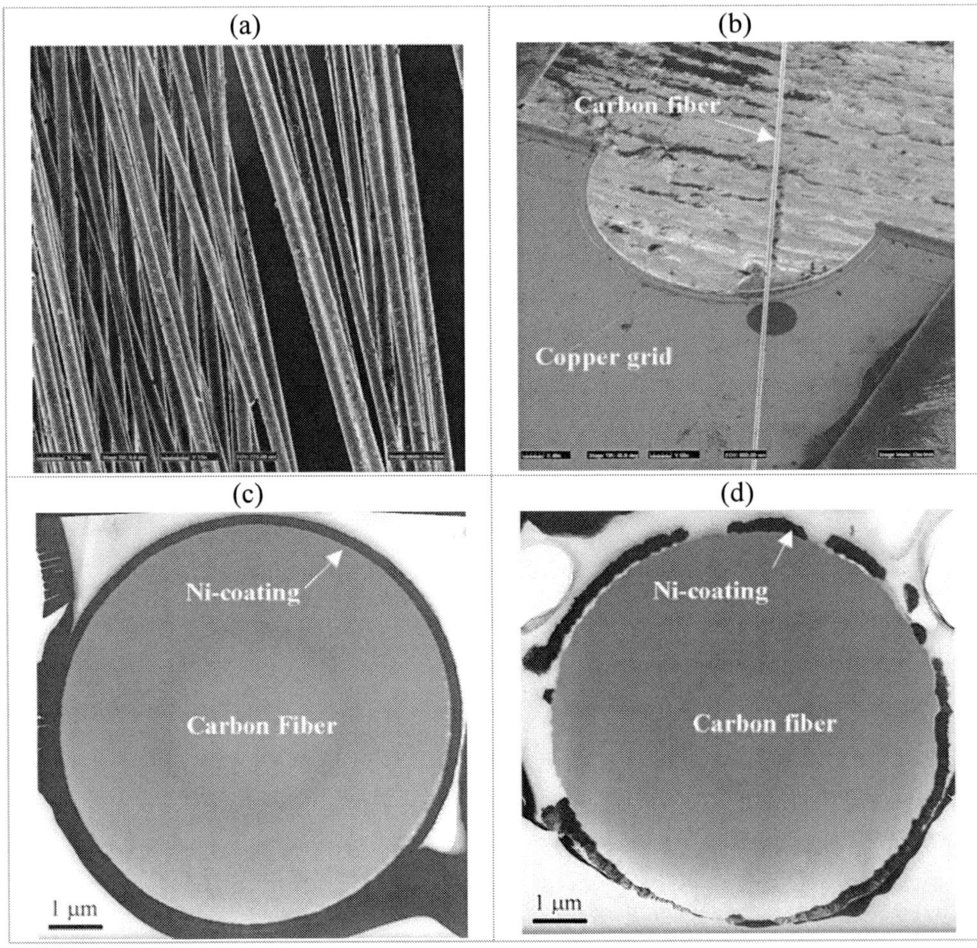

Figure 17. FIB and TEM images of successfully prepared TEM specimens using the direct lift-out technique. (a) FIB low magnification image of a bundle of fine fibers. (b) FIB image of a mounted fiber on a copper grid. (c) A low-magnification TEM image of a fiber with good coating quality. (d) Low-magnification TEM image showing a fiber with poor coating quality.

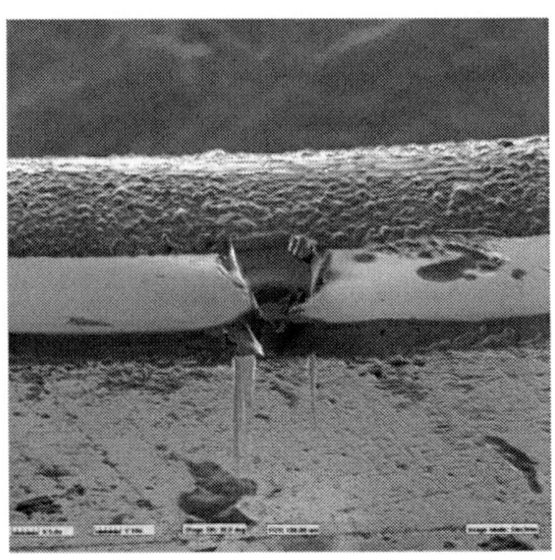

Figure 18. Small mineral samples directly lifted out and mounted onto a copper grid.

2.4. Beware of Artifacts

There have been numerous reports and discussions about artifacts induced by the gallium beam [e.g. 6]. As the number of FIB systems increases and their applications extend into various field of materials engineering, this is of no surprise. Throughout the years in the author's materials engineering research, in both the private sector and the current Canadian national materials research lab, we have encountered a wide range of engineering materials and processes. Our FIB experience started about 10 years ago in the early days when it was just beginning to be introduced into materials research. We have gradually experienced, learned, understood and tried hard to cope with most of the commonly occurring FIB-induced artifacts. In many cases, we have learned hard lessons.

Among the FIB induced artifacts we have encountered are:

1. the "curtain effect",
2. gallium phase formation on the sample surface,
3. beam-induced damage on FIB prepared TEM specimens,
4. beam induced grain growth in nano-crystalline materials,
5. beam damage to most of the HCP materials, and
6. redeposition of materials around the target.

Figure 19. FIB drills on silicon using a 670 pA beam. (a) a well-aligned beam, (b) a poorly aligned beam.

The root causes of FIB-generated artifacts are often complicated. aside from the nature of the materials and the type of FIB work to be performed, the control of the gallium-ion beam is of paramount importance; however, the control of the ion beam has been overlooked by many FIB users. Different FIB systems are designed with different routines to control the milling parameters. It is impossible to generate a universal recipe for efficient milling and minimizing artifacts. In general, controlling the ion beam includes: beam current, beam dwell time (per pixel), pixel spacing (beam overlap), re-trace and refresh time; in addition, one of the most of important factors is the beam shape. When performing fine milling, the beam shape should be checked frequently (with each aperture and beam condition change). One should definitely eliminate "beam tail" prior to fine milling. A tight, stable and well-aligned beam is essential to reduce most of the FIB-induced artifacts. Figure 19 shows examples of FIB hole-drills on silicon using (a) a nearly perfectly aligned and (b) a poorly aligned ion beam. The beam current used in the drilling experiment is 670 pA which is typical when performing fine ion milling or final polishing prior to imaging. In both cases, the ion beams were able to provide well-focused images prior to drilling actions. In reality, the well-aligned beam with minimum beam spread is able to provide high-quality ion beam milling and polishing, while the beam tail of the poorly aligned beam will result in a noticeable "curtain effect" and frequently beam damage to beam-sensitive specimens. In some materials with hexagonal crystal symmetry (e.g. Zn and Zr), artifacts resulting from ion-beam milling are almost inevitable. Figure 20 shows a FIB cross-section of a galvanealed Zn coating on an interstitial-free (IF) steel. Ion-beam damage appears as dark speckles in the coating microstructure. It seems that fine grains at the Fe-Zn interface (Γ phase) and the Zn-rich phase (ξ phase) near the coating surface are less prone to ion-beam damage and can be resolved as shown in Figure 20(a). The majority of the coating suffers noticeable damage. A few more imaging passes using a 32 pA ion beam worsens the damage, as shown in Figure 20(b).

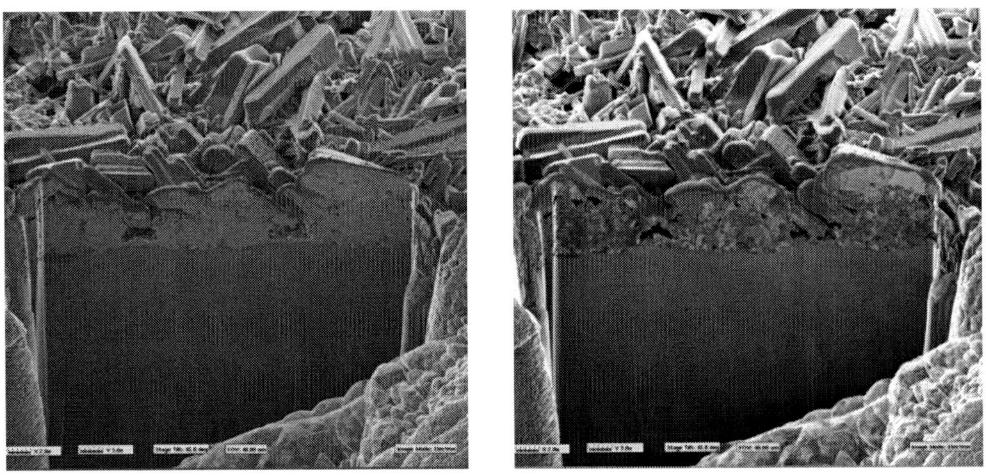

Figure 20. FIB cross-section of a galvanealed Zn coating on IF steel. (a) polished with a well-aligned beam of 210 pA and imaged with a 32 pA beam, (b) a few more imaging passes using 32 pA causes more damage to the coating.

It was noticed that using FIB to perform semiconductor IC chip circuit modification of Cu-based interconnect is generally problematic [24,25] due to the significantly different sputter rate of Cu metal interconnects. The sputter rate variation is related to Cu grain orientation. Measurements of FIB sputter rates on single crystal Cu specimens [26] show a sputter rate variation of about four times between fast milling orientations, such as (111), and slow milling orientations, such as (110). This difference in sputter rate is not limited only to Cu, but has also been observed in certain Au and Ni based systems. The slow sputtering of the (110) orientation is not only attributed to the likelihood of Ga ion beam channelling, but also associated with the formation of an anomalous metal–gallium (M_xGa_y) phase during FIB milling under conditions in which the incident FIB beam hits the specimen at angles far from glancing and closer to normal incidence. During FIB milling of Cu, some grains become dark in color; they grow and spread as the ion dose increases. Later TEM analysis indicated that a thin layer of Cu_3Ga phase formed on the surface of slow milling grains [with (110) plane parallel to the specimen surface]. The Cu_3Ga phase does not form on some other orientations such as (111) [27, 28]. The appearance of such Ga-rich phase is also found to appear on low-carbon steel as shown in Figure 21. These Ga-rich phases tend to initiate and grow from certain grain boundaries. Thus, when performing FIB imaging, attention to beam parameters is necessary to avoid the formation of this artefact.

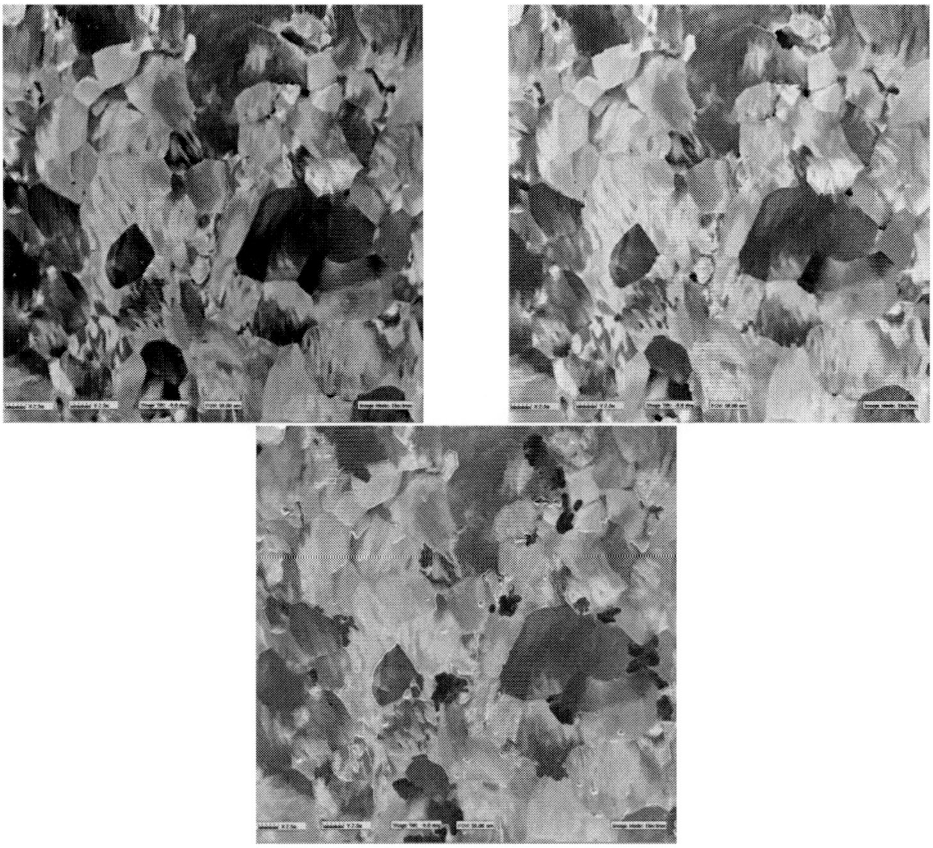

Figure 21. Fe_xGa_y phase formation on low carbon steel during FIB imaging. (a) first image pass, (b) multiple image passes under 1.5 nA beam current, (c) multiple passes under 6 nA beam.

3. Practical Examples of Materials Engineering Using Inovative Microstructural Characterization

3.1. The Study of Stress Corrosion Cracking

3.1.1. Background of SCC in Pipelines

In the pipeline industry, stress-corrosion cracking (SCC) from the external (soil side) surface has been one of the prime concerns since it was first observed in 1965 [29]. Most natural gas pipelines are buried underground. Although they are protected by special wrapping tape and tar, the combination of ground movement and internal gas pressure fluctuation exerts unpredictable stress on the pipe lines. The stress states are also functions of position and time. As the protective materials outside the pipes age, they tend to crack leading to direct contact between the sometimes highly stressed steel pipelines and ground moisture and soil. Corrosion usually starts in the form of pitting. Larger pits result in local stress concentration that could lead to SCC initiation. Most SCCs propagate very slowly. However, they are very difficult to identify prior to major failures that can range from leaks to explosions. In the past years, pipeline failures have increased. This is due to both the aging of existing pipelines and the drive to push for higher transmission rates (higher pressure needed).

There have been significant efforts to understand the mechanisms and factors that control SCC initiation and propagation [e.g. 30-32]. It is commonly agreed that the crack propagation mode is mainly influenced by the local environment. Under high-pH conditions, the SCC cracks tend to propagate along steel grain boundaries ("classical" or "intergranular" SCC). Under near-neutral pH, SCC propagates trangranularly ("low pH", "non-classical" or "transgranular" SCC).

There is still much debate on the mechanisms controlling SCC initiation and propagation. Parkins [33] suggested that the evidence of intergranular attack (IGA) in the absence of stress, together with dissolution kinetics, indicate that intergranular stress-corrosion cracking (IGSCC) of ferritic steels in various environments with passivation behavior was caused by a dissolution mechanism. Film rupture could play an important role in crack growth rate; plastic strain of appropriate magnitude in the metal beyond the penetrating tip could prevent filming, thereby enhancing IGA. In addition, the localization of dissolution in grain-boundary regions can be enhanced by the presence of segregates or precipitates. He also suggested that carbon appears to be one of the most significant elements contributing to the SCC propagation in low-carbon steel. As the carbon content was reduced by decarburizing in wet hydrogen, SCC resistance was greatly increased [34,35]. Wang and Atrens in their TEM study of SCC propagation in X-52 and X-65 steels [36,37], suggested that, under high-pH conditions, SCC was mostly intergranular along ferrite-ferrite grain boundaries. A high concentration of Mn was found to exist between the primary ferrite and the cementite lamellar. However, sulfur and phosphorus were not detectable in their TEM EDS analysis, suggesting that the commonly expected species, S and P, may not be responsible for preferential dissolution of the grain boundaries. In contrast, the transgranular cracking is generally believed to be related to dilute, near-neutral pH environments that do not produce passivating films and allow the dissolution of crack tips and sides (walls along the crack) accompanied by the permeation of hydrogen into the steel [38].

In our recent work, we have performed a series of careful investigations of SCCs found on the surface of an existing X52 steel pipeline. Microstructural features of an X-52 linepipe taken out of service were characterized using the optical microscope, the SEM, the FIB and the TEM in order to understand the SCC propagation mechanism leading to this failure.

3.1.2. Expermental

Using the ultrasonic inspection tool, a colony of SCC was detected on an existing gas pipeline built in 1960. This section of pipe was cut out and shipped to our laboratory for further investigation (Figure 22). The pipeline has a diameter of 406.4 mm (16 in.) and was made of

X-52 steel with a wall thickness of 6.35 mm (0.25 in.). The chemical composition is given in Table 1. A small sample cut along the longitudinal direction of the pipe was mounted in low-shrinkage epoxy resin under vacuum. The mounted specimen was ground and polished using an alcohol-based polishing medium in order to minimize contamination. It was then etched using a 2% Nital solution prior to optical and SEM examinations. The sample was re-polished for FIB imaging and subsequent site-specific TEM sample preparation. The selected crack tip was extracted from the polished sample surface using a FIB microscope for further TEM examination.

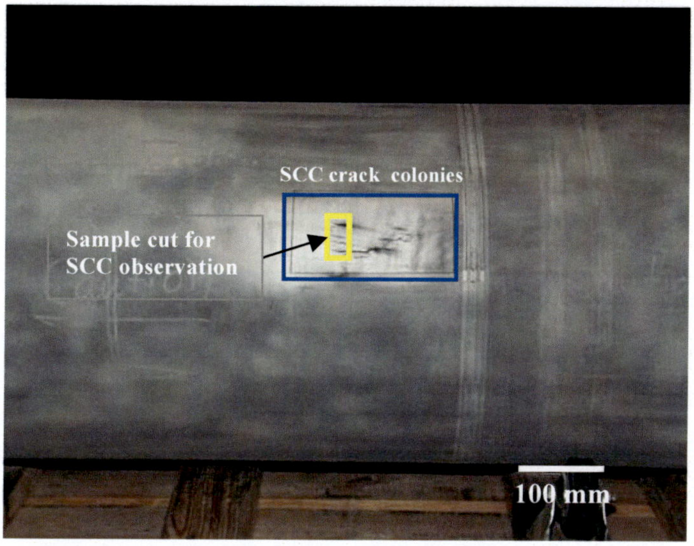

Figure 22. A section of the pipe showing SCC crack colony.

Table 1. Chemical Composition of X-52 Steel (wt%)

C		Mn		P		S		Si
Min	**Max**	**min**	**max**	**Min**	**max**	**Min**	**max**	
0.29	0.34	1.10	1.45	0.04	0.05	0.04	0.06	0.04

3.1.3. Observations and Interpretations

Figure 23 shows typical optical micrographs of an SCC found on a metallurgical polished cross-section. The environment where this section of pipe was exposed to is determined to be in the "high-pH" range. Intergranular cracking seems to be the dominant cracking mode. The higher magnification optical micrograph [Figure 23(b)] shows that the crack propagated along the ferrite grain boundaries when there were no neighboring pearlite grains. Cracks have also been found to propagate across some pearlite clusters.

Higher resolution images were obtained using an SEM. As shown in Figure 24(a), the SCC propagates either around or through the pearlitic grains. Figure 24(b-d) suggests that whether the crack propagates along the boundaries between ferrite and pearlite, or fractures, through the pearlite structure, depends on the geometric orientation of the pearlitic lamellae and the crack propagation direction. When the pearlite lamellae are aligned near parallel to the main crack propagation direction, cracks tend to advance through the pearlite structure. Otherwise, the crack travels along the ferrite-pearlite grain boundaries. Earlier reports by Fessler et.al. [39] and Eiber [40] suggested that the SCC cracks are predominantly at proeutectoid ferrite-proeutectoid ferrite grain boundaries. Cracks also propagate through proeutectoid ferrite/pearlite grain boundaries in some rare instances. Danielson [41], in his study of X-52 steel, reported that the preferred crack path is the ferrite/ferrite boundary. Much less frequently, the path is ferrite/pearlite, and the rarest of all is the fracture across pearlite, which is a hard and brittle phase. However, he gave no further explanation of this important cracking mechanism.

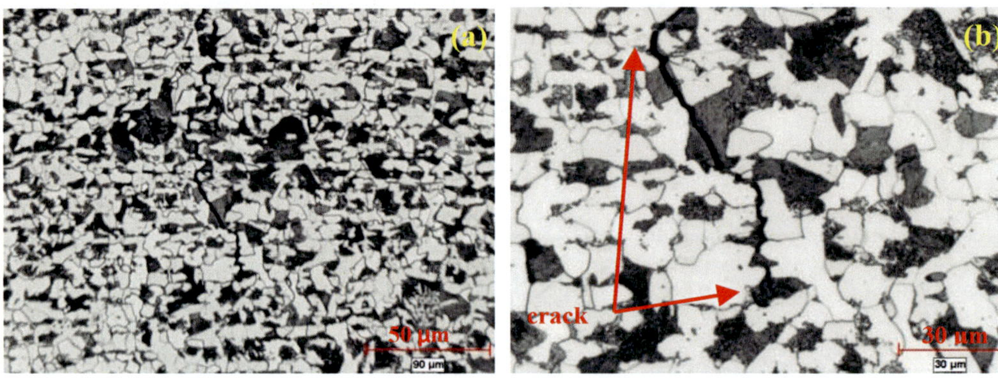

Figure 23. Optical micrographs showing a SCC crack in X-52 pipeline.

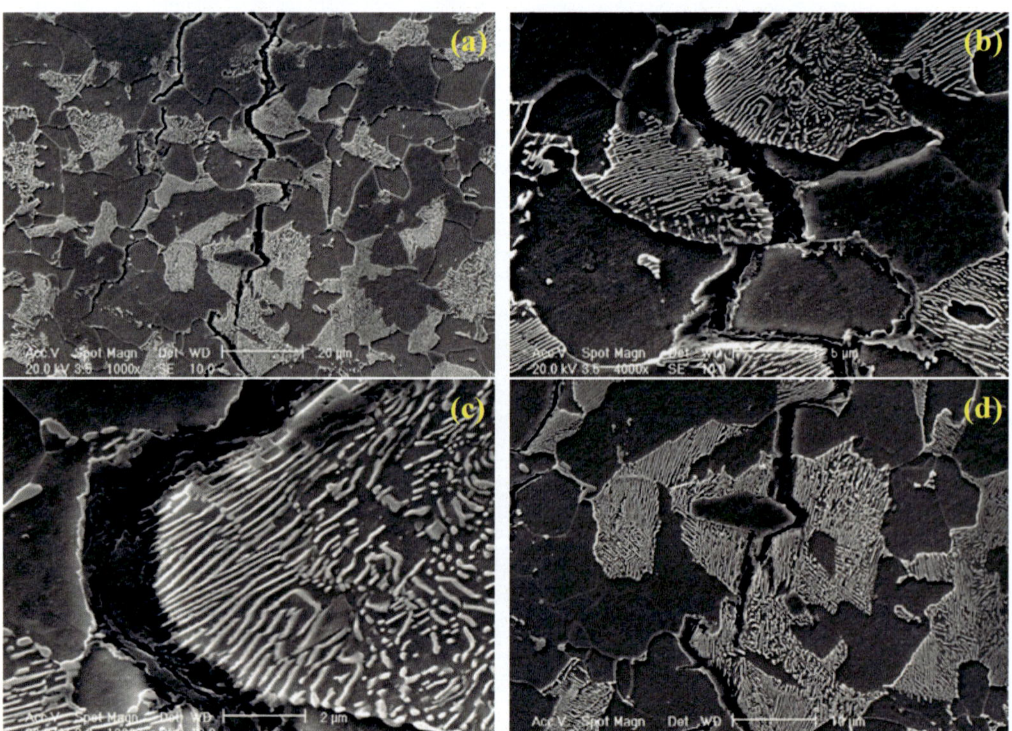

Figure 24. SEM images showing SCC penetration through pearlite structure.

Cracking through the relatively strong pearlitic structure could also be related to local chemistry (environment) and physical properties (mainly the electrochemical potential difference between cementite [Fe_3C] and ferrite). Local stress tensor is also a factor. Green and Parkins [42] reported that Fe_3C could act as efficient points for cathodic discharge, facilitating the dissolution of adjacent ferrite because cementite has low hydrogen evolution over-potential. Local galvanic reaction in the pearlitic structure, if it occurs, weakens the local microstructure and facilitates the SCC propagation. The smaller pearlite grains could be consumed (1corroded), and the crack could continue to propagate intergranularly along ferrite grain boundaries until it encounters the next pearlite grain. If the pearlite grain is large, and the pearlite lamellae align nearly parallel to the crack propagation direction, dissolution could occur along the boundaries between cementite (Fe_3C) and pearlitic ferrite. The crack could "cut through" the pearlite structure (transgranular fracture).

Meanwhile, if hydrogen was produced, it could be transported along the dominant path i.e., through the ferrite grains as well as along the ferrite grain boundaries and the pearlitic ferrite/cementite interface, of which the latter may be the most important since cementite can provide hydrogen-trapping sites. Thus, the corroded pearlite grain could be weaker than the neighboring grain boundaries between pearlite and ferrite. This could lead to transgranular SCC through some of the large pearlite grains. However, the local stress state would also play an important role. Hence, under certain conditions, the fracture mode could depend on the pearlite volume fraction, the grain size and the orientation of pearlite grains. In some extreme circumstances, the entire pearlite grain can be corroded leaving only the cementite skeleton, as shown in Figure 25.

Figure 25. Corroded pearlite grain in a large SCC crack.

Studying the microstructure using optical microscopy and SEM usually requires metallographic etching which can disturb or even dissolve some of the corrosion products. As demonstrated in earlier sections, the FIB microscope can be used as a high-resolution imaging tool, and has demonstrated advantages in imaging metallurgical specimens [2, 43-45]. The FIB secondary-electron images provide unique crystallographic contrast even with a carefully mechanically polished surface, similar to the electron channeling contrast in SEM that frequently requires electropolishing [5,46]. In our study, high-resolution FIB images are assembled in some local regions in order to assess local microstructure and crack propagation route. Figure 26 shows a mosaic of FIB images detailing the intergranular nature of the SCC in this region. The FIB secondary-ion images show enhanced contrast with the presence of oxide, which is particularly useful in SCC studies. Figure 27 shows FIB secondary-ion images in the SCC crack-tip zone in a metallurgically polished cross-section. The very fine crack appears to propagate through a pearlite grain, as shown in Figure 27(a). It appears that only one lamella of pearlitic ferrite has been attacked between the adjacent cementite lamellae that act as effective corrosion barriers to the neighboring lamellae.

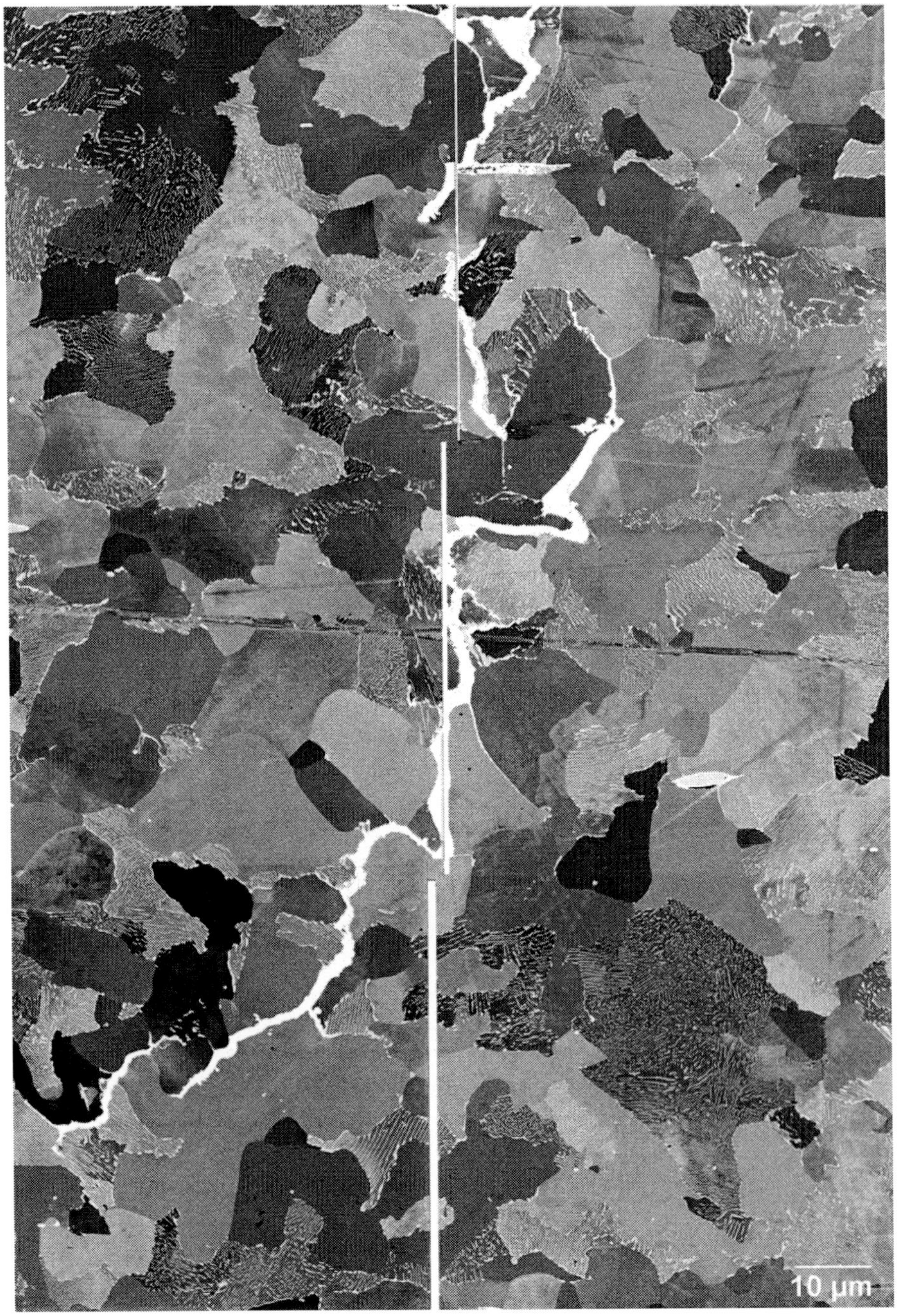

Figure 26. High-resolution mosaic FIB images of SCC.

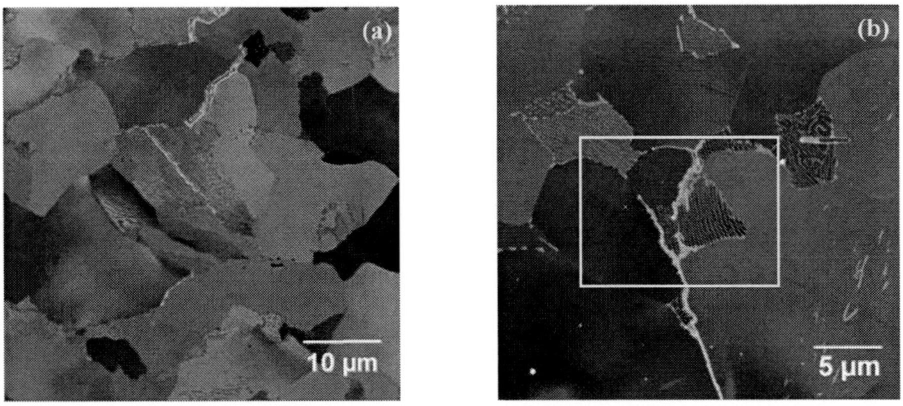

Figure 27. FIB secondary-ion images showing details of region around crack tips.

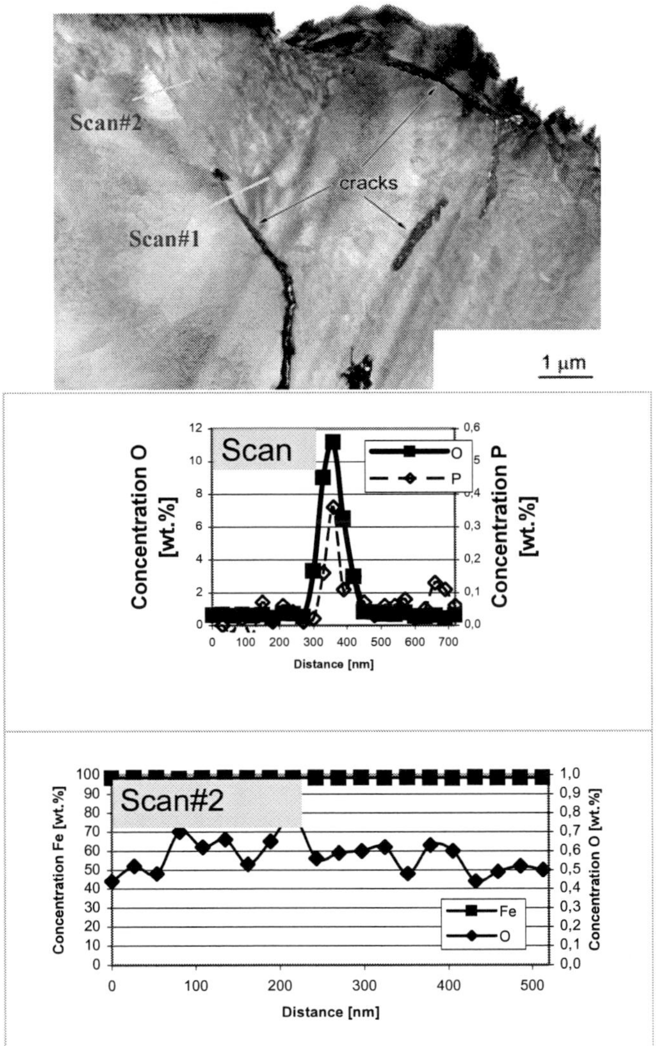

Figure 28. STEM image of crack-tip region and EDS line scans.

Detailed chemical composition analysis using TEM, especially around the crack tip, is of great importance in understanding the micro-cracking mechanism. The widely used electropolishing method could lead to the dissolution and contamination of corrosion product and alter the local chemistry near the crack tip. Traditional ion milling presents great difficulty in providing site-specific TEM specimens. The FIB is probably the most powerful tool for making stress-free and site-specific TEM specimens with minimal contamination and surface distortion. As shown in previous sections, FIB-TEM specimen preparation techniques and their advantages and disadvantages are well documented [7,22,47]. In this study, a TEM specimen with one-to-one correspondence to the crack tip shown in Figure 6b was prepared with the ex-situ "plan-view lift-out" technique. Details of the TEM specimen-preparation technique can be found in one of our earlier publications [19].

Figure 28 shows a general view of the region examined in a Philips CM20 FEG TEM. The low-magnification STEM image shows a very close correspondence to the targeted area shown in Figure 27(b). It is important to note that, during the FIB final thinning, there is always some, although very limited, material removed from the surface in order to clean up the surface re-deposition resulting from FIB milling. Thus, the minor discrepancy between the FIB plan-view image and TEM images is to be expected. Energy-dispersive x-ray (EDS) line scan #1 across the SCC crack, confirms the presence of oxygen and a trace of P in the crack [Figure 28(b)]. No other elements are detectable in the crack. However, EDS line scan #2, across the primary ferrite grain boundary that was not corroded (ahead the SCC tip), showed no P [Figure 28(c)]. The weak oxygen signal in line scan #2 should come from the surface oxide on the TEM foil.

As indicated by Parkins [35], as the carbon content increases, carbon steels are more susceptible to SCC attack. In intergranular SCC, carbide precipitates at grain boundaries could act as effective cathodic discharge points to facilitate the dissolution of adjacent ferrite. However, EDS line scan #1 across the stress-corrosion crack showed only oxygen and phosphorus (and Fe) inside the crack. No carbides were detected at the corroded ferrite-ferrite grain boundary. Further, EDS line scan #2 along the un-attacked grain boundaries shows neither phosphorus nor carbon. Although Parkins [35] suggested that phosphorus alone could not promote intergranular SCC in ferritic steel, the implication of phosphorus in the SCC grain boundary in the vicinity of the stress-corrosion crack tip needs more investigation. In the absence of high carbon concentration at grain boundaries as proposed by Green and Parkins [42] the phosphorus could have played an important role in this intergranular SCC phenomenon, such as by causing grain boundary embrittlement ahead of the crack tip. Hydrogen could also be formed close to the crack tip due to electrochemical and chemical reactions, based on the assumption that the crack-tip environment under certain conditions may consist of hydrogen in atomic form that could be produced by electrochemical and chemical reactions generated by an occluded cell effect inside the crack. According to Li's model [48], the anodic dissolution and hydrolysis reaction inside the crack produces H^+. When the potential inside the crack is low enough, at a certain pH level, a reduction reaction is possible (i.e., $H^+ + e \rightarrow H_2$ or $H_{ad} + H^+ + e \rightarrow H_2$). The potential between the metal and the solution close to the crack tip could be low enough for the hydrogen production process to occur. Even if the electrochemical potential was not low enough, hydrogen could also be produced as a consequence of the so-called dissociative chemisorptions of water molecules within the crack enclave solution. The reaction could proceed in the following two steps [49]: (I) $H_2O + 2M = MH + MOH$; (II) $H_2O + MH = MOH + H_2$. Step II might be the source of the penetrating hydrogen. The reactions could occur only at active sites, i.e., free of adsorbed oxygen and passive films. Such conditions may exist at the bottom of growing pits and at crack tips, as shown by Smialowska [49]. The existence of phosphorus around the crack tip in this study could act as permeation promoters [50], affecting the hydrogen diffusion and permeation rate. Hydrogen diffuses into the metal along the dominant transportation paths, through the ferrite grains as well as along the ferrite grain boundaries and the interface of ferrite and pearlite [51]; this may weaken the atomic bond and facilitate intergranular cracking. In addition, the high carbon content yields a high volume fraction (36% based on image analysis) of pearlite in this X-52 linepipe steel. The galvanic reaction between the ferrite-pearlite grain boundaries may contribute significantly to the intergranular SCC.

3.1.3. Summary

The present investigation has shown that:

1. Investigation of SCC propagation mechanism in this study not only involves extensive knowledge, but also requires a complete set of advanced characterization instruments and innovative characterization techniques.
2. Microstructural examination of SCC in an X52 linepipe steel taken out of field service suggested that the galvanic reaction between cementite (Fe_3C) and adjacent ferrite played an important role in crack propagation.
3. Although intergranular fracture is the dominant SCC propagation mode in this X-52 linepipe, transgranular cracks were found in some pearlite grains when the lamellae were favorably oriented relative to the crack path.
4. The phosphorus detected close to the tip of the stress corrosion crack could have contributed to the SCC propagation.

3.2. Investigation of Wear Resistance of an Aluminum A390 Alloy

3.2.1. Background

The hypereutectic aluminum alloy A390 is a conventional alloy used to make pistons in automobile engines and engine blocks for high-performance race cars and some luxury cars. The engine pistons work under fairly aggressive conditions, experiencing high-speed sliding wear at relatively high temperatures. Wear characteristics under various conditions need to be evaluated. One important aspect to study is the effect of oxygen and moisture on the wear characteristics of this alloy. In this study, the material layers underneath the worn surfaces of hypereutectic Al-Si-Cu alloy (A390) subjected to dry-sliding wear in air and argon atmospheres were characterized. The samples were tested at a constant load of 10 N and a sliding velocity of 1 m/s using a block-on-ring tribometer. The counterface material was an SAE 52100 bearing steel. The wear rate of the alloy tested in an argon atmosphere (3.05×10^{-5} mm^3/m) was 10 times lower than that of the sample tested in air (2.96×10^{-4} mm^3/m). The subsurface microstructures generated under the two different test environments were characterized using an SEM, an EPMA, a FIB microscope and a TEM. Cross-sectional TEM specimens were prepared using a FIB "lift-out" technique. TEM analysis indicated that the tribolayers formed on the sample tested in air contained significant amounts of iron, aluminum and oxygen. In addition, the tribolayers formed in air were hard and appeared to be severely fractured indicatiing their brittleness due to the large amount of oxide present. On the contrary, much smaller amounts of iron and oxygen were found in the tribolayers formed in argon, which were a mechanical mixture of mainly ultra-fine-grained aluminum (~100 nm) and silicon; these tribolayers were more stable on the contact surfaces, which reduced the wear rates of A390.

In past years, extensive studies on the wear characteristics of Al-Si alloys have been carried out [e.g. 52-59]. In these studies, wear resistances of different alloys with various silicon contents were tested against various wear counterparts under different loads and sliding speeds. Jasim and Dwarakadasa [60] studied the wear resistance of Al-Si alloys containing 3-22% silicon using dry-sliding wear. They reported that the wear rates were a function of silicon content and did not depend on the initial microstructure or the distribution of the silicon phase. They also claimed that, in the alloys with lower silicon contents, a subsurface deformation layer was formed in which the initial silicon particles were severely fragmented into small spherical particles. However, in high-silicon alloys, the subsurface region did not show significant plastic deformation in the direction of sliding. Wear characteristics of aluminum alloys with hard reinforced particles were thoroughly reviewed by Hutchings et al. [61]. Various wear maps illustrating the correlation between the test conditions (i.e., load and speed) and the wear mechanisms have been reported (e.g. [62-65]). It has been widely accepted that the wear resistance under dry-sliding wear conditions is closely related to the formation and the stability of the tribolayers on the contact surfaces. Under certain circumstances, the formation and removal of the tribolayers during sliding wear depend not only on the sliding speed and load but also on the atmospheric conditions [66].

Wear tests to this alloy were performed in an air and argon atmosphere to determine if the presence of air (oxygen) increases wear rate (by a change in wear mechanism). Detailed microstructural analyses of the tribolayers and the subsurface material layers underneath the tribolayers were performed using an SEM, EPMA, FIB microscope and TEM. The site-specific TEM specimens were prepared using a FIB.

As demonstrated in previous sections, FIB techniques have already found many applications in microstructural characterizations [3-5,67-69]. When used as high-resolution imaging tool, FIB secondary electron (SE) images provide enhanced crystallographic contrast similar to the electron channeling contrast in SEM. High-resolution FIB imaging on the cross-sections polished perpendicular to the surfaces of the samples subjected to sliding wear tests provide valuable information on the subsurface microstructure. The TEM specimens were prepared directly from the worn surfaces for in-depth microanalyses.

3.2.2. Experiments

The block-on-ring type of dry-sliding wear tests were used to evaluate the wear resistance of this alloy. The chemical composition of the A390 alloy is shown in Table 2. During the test in air, the relative humidity (RH) in the environmental chamber built around the block-on-ring tribometer was kept constant at RH = 5 ± 2 %. The tests in argon atmosphere were performed in order to study the wear resistance in an inert environment. During these tests, the chamber was first flushed with compressed argon gas, and then the tests were carried out under a continuous flow of argon gas that was directly blown (at an exit pressure of 2 psi) onto the samples. The samples were machined in the form of rectangular blocks with dimensions of 5 x 10 x 10 mm. The counterface ring material was an SAE 52100 bearing steel (HRC 61) and the composition is also listed in Table 2. The test load, sliding speed, and sliding distance were 10 N, 1 m/s and 4000 m, respectively. Table 3 summarizes the wear test results. The wear rate in argon (3.05 x 10^{-5} mm^3/m) was approximately an order of magnitude lower than that in air (2.96 x 10^{-4} mm^3/m). The lower wear rate in argon was accompanied by a lower coefficient of friction (COF) value (0.29 ± 0.02), than that measured in the dry air atmosphere (0.57 ± 0.08). In addition, the COF curve in argon was smoother than the COF curve in air that fluctuated between 0.5 and 0.7.

Table 2. Chemical composition of the A390 alloy and counterface SAE5120 steel in wt%

Composition of A390 aluminum alloy										
Si	**Cu**	**Fe**	**Mg**	**Mn**	**Ni**	**Pb**	**Sn**	**Sr**	**Ti**	**Zn**
18.4	4.0	0.23	0.57	0.07	0.02	<0.01	<0.01	<0.002	0.05	0.1
Composition of SAE 5120 steel										
C	**Mn**	**Si**	**Cr**	**Fe**						
0.98-1.1	0.25-0.45	0.15-0.3	1.3-1.6	Balance						

Table 3. Wear rates and coefficient of friction of A390 in air and Argon atmosphere

	In air	**In Argon**
Wear rate (mm^3/m)	2.96 x 10^{-4}	3.05 x 10^{-5}
Coefficient of friction	0.57 ± 0.08	0.29 ± 0.02

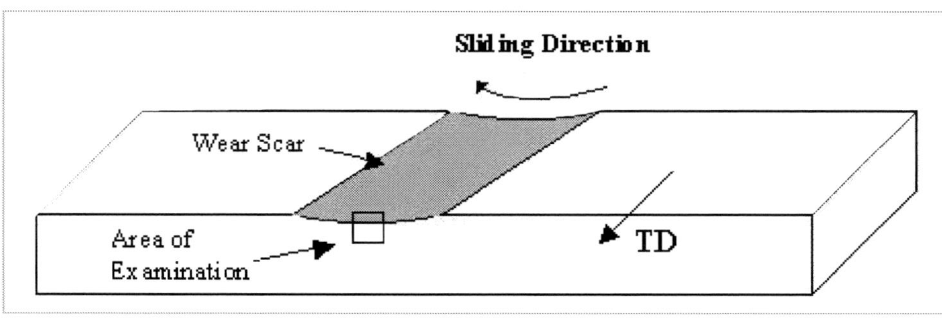

Figure 29. Schematic diagram showing the sample mounting plane for microstructural examination.

Initial metallographic analyses of the tribolayers were carried out using SEM, EPMA and FIB plan-view imaging on the cross-sections prepared as shown in Figure 29. The worn surfaces were electroplated with a thin layer of Ni to protect the tribolayers from damage during subsequent metallurgical sectioning and polishing. The samples were carefully polished and finished with 0.05 μm colloidal silica solution. Figure 30 shows the typical microstructure of the A390 alloy. The back-scattered electron (BSE) image and EDS spectra indicated several phases [A: $CuAl_2$, B: $Al_{15}(Fe,Mn)_3Si_2$, C: $Al_5Mg_8Cu_2Si_6$, and D: Si]. The tribolayers and the underlying fine-grained region were further analyzed by a TEM.

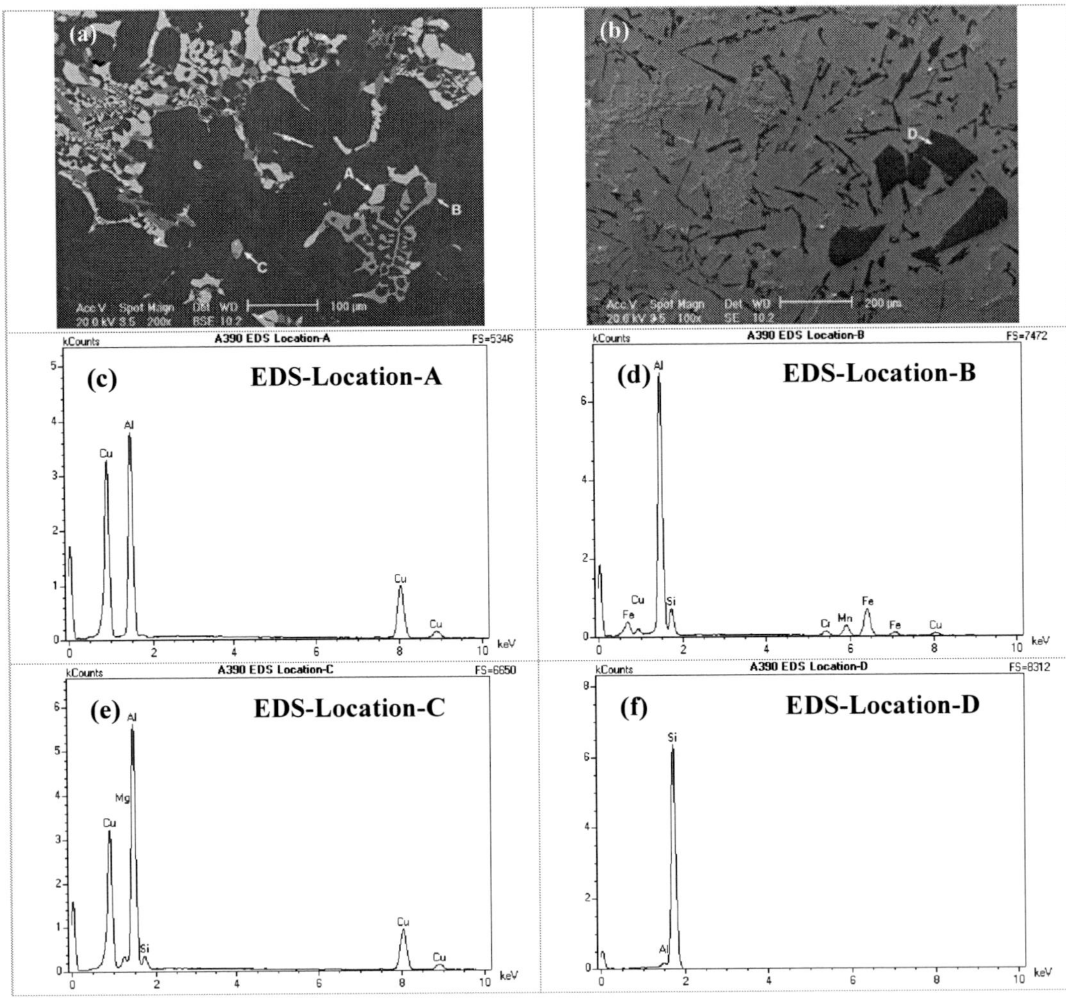

Figure 30. Microstructure and constituent phases in A390 alloy: (a) BSE image, (b) SE image, (c)-(f) EDS spectra.

3.2.3. Microstructural Analyses

SEM examination of the cross-section suggested more or less uniform tribolayers formed on the contact surfaces during the wear tests in argon and air (shown in Figure 31). These tribolayers were of similar thickness but showed some morphological differences. Figure 31(a) and (b) show low- and high-magnification cross-sectional views of tribolayers formed in an argon atmosphere. The thickness of the tribolayer in this section was approximately 10 μm. The tribolayer was almost featureless. The hardness of the tribolayers formed in argon atmosphere was 400 ± 80 kg/mm^2 (HV, 25 g). Figure 31(c) and (d) show cross sections of the tribolayers formed in air (5% RH) under the same wear conditions. The thickness of the tribolayer in this section was about 15 μm. The tribolayers formed in air were harder [800 ± 100 kg/mm^2 (HV, 25g)] and more brittle, as is evident from the network of cracks [Figures 31(c) and (d)].

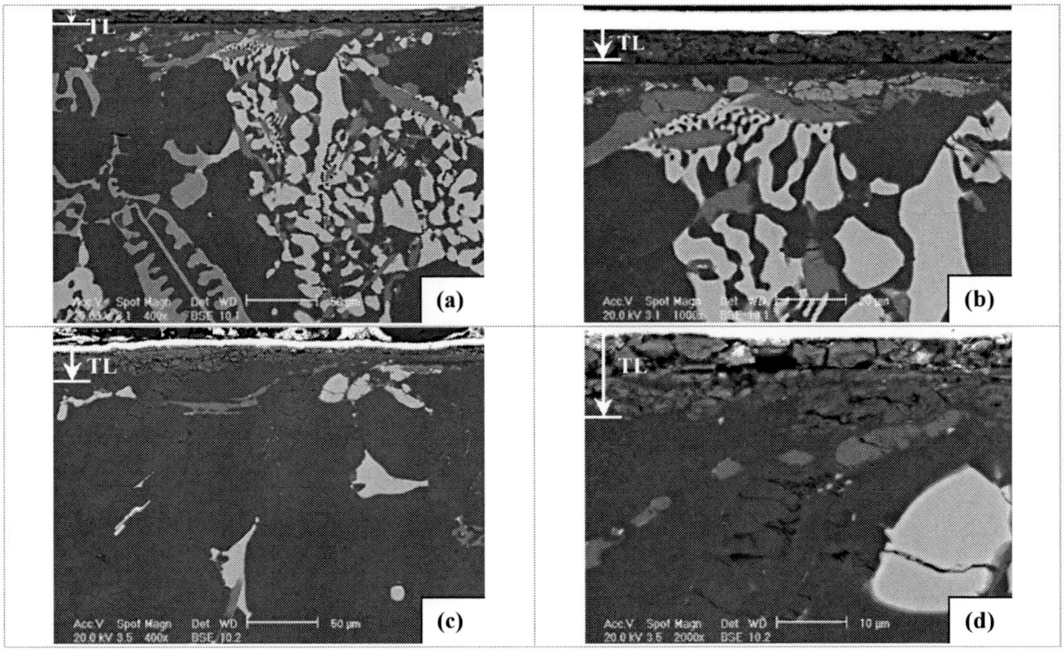

Figure 31. SEM-BSE images of the tribolayer (TL) morphology. Wear in argon (a) and (b), wear in air (c) and (d).

EPMA analyses were conducted in order to obtain the overall chemical composition of the tribolayers. These compositions are the averages of measurements taken at eight randomly selected points within the tribolayers (Table 4). The tribolayers formed on the sample tested in air were rich in iron (24.8%) and oxygen (32.5%), and contained about 35.1% aluminum. The tribolayers formed in argon contained much less iron (1.2%) and oxygen (13.3%) with 64.8% aluminum.

Table 4. Tribolayer chemical composition measured by EPMA (wt%)

	O	Si	Fe	Al
In Air	32.5 +/- 1.6	4.9 +/- 0.1	24.8 +/- 4.4	35.1 +/- 6.2
In Argon	13.3 +/- 2.0	8.6 +/- 1.5	1.2 +/- 0.2	64.8 +/- 2.6

The measured hardness and quantitative EPMA result indicated that the iron-rich tribolayers formed in air were harder than those formed in argon, but they were more brittle as they exhibited evidence of fragmentation. Consequently, they were easily removed from the contact surfaces. On the other hand, the tribolayers formed in argon were less susceptible to fragmentation and removal from the contact surfaces, as supported by the lower wear rate of the sample. It can be suggested that the tribolayers formed in argon provided better protection against wear during the sliding contact. Furthermore, the average COF value measured in argon (0.29 ± 0.02) was lower than that measured in air (0.57 ± 0.08). Higher COF value combined with large fluctuations from the average during the test in air can be attributed to the entrapment of the hard debris particles detached from the tribolayers at the sliding interface.

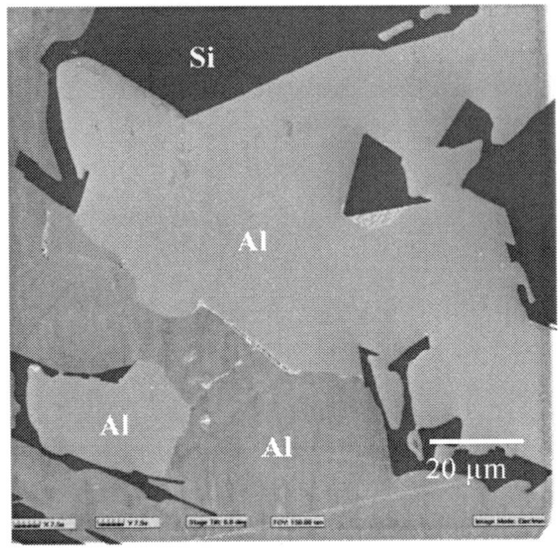

Figure 32. FIB secondary-electron images taken at the center of the A390 specimen.

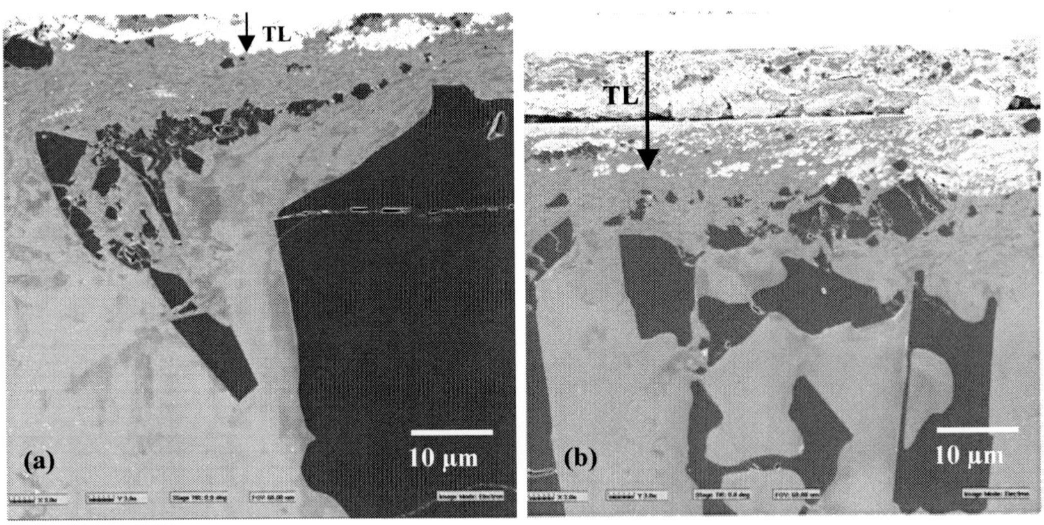

Figure 33. FIB SE images showing subsurface damage zone after wear tests. (a) wear in air, (b) wear in argon.

The polished cross-sections of the worn surfaces were imaged using a FIB microscope. The FIB secondary-electron image taken from the center of the specimen away from the tribolayer (Figure 32) depicted a microstructure consisting of equi-axed and undeformed Al grains. Severe mechanical damage was found in regions beneath the contact surfaces as shown in Figures 33(a) and (b) in both air and argon atmospheres. The thicknesses of the severely damaged aluminum substrates were similar in both samples and ranged between 10 and 15 μm. Plastic flow along the sliding direction has resulted in apparent fragmentation of silicon particles to less than 2 μm in some cases, as shown in Figure 34. The aluminum matrix within the damaged subsurface zone had an ultra-fine-grained (UFG) structure as shown in Figure 35. The average grain size was estimated to be 100 nm as was later confirmed by TEM.

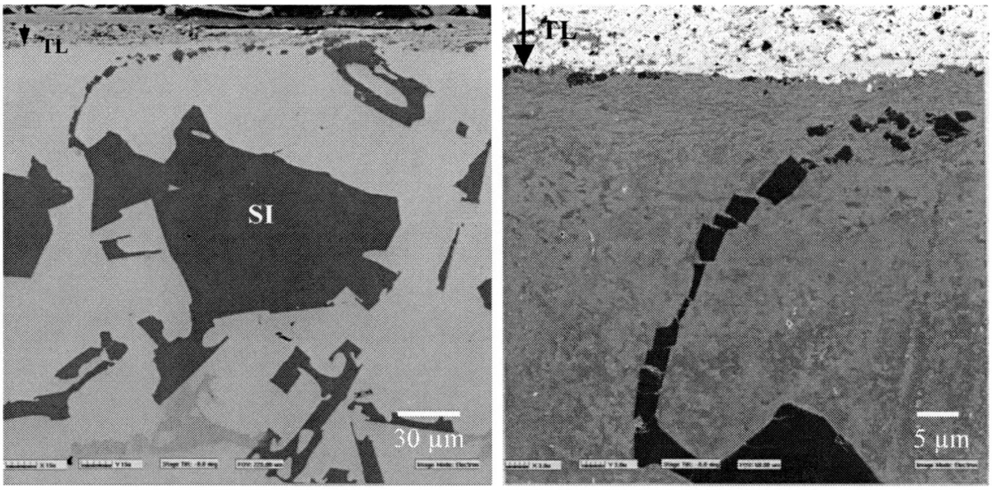

Figure 34. FIB SE images showing fragmentation of Si particle in the subsurface zone (in air). Micrograph (b) represents a higher magnification image of part of micrograph (a).

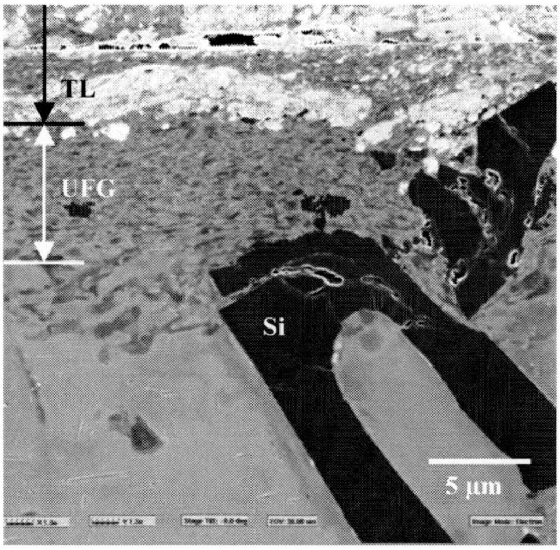

Figure 35. Ultra-fine-grain microstructure formed underneath the tribolayer on sample worn in argon.

TEM specimens were extracted from the cross-sections normal to the worn surfaces using the FIB "lift-out" technique. To prevent contamination, the TEM specimen was mounted onto the TEM copper grid using the FIB-deposited tungsten prior to the final FIB thinning. No glue was used as normally required in an ex-situ lift-out process. Figure 36 shows the FIB lift-out TEM sample preparation process of the sample worn in air.

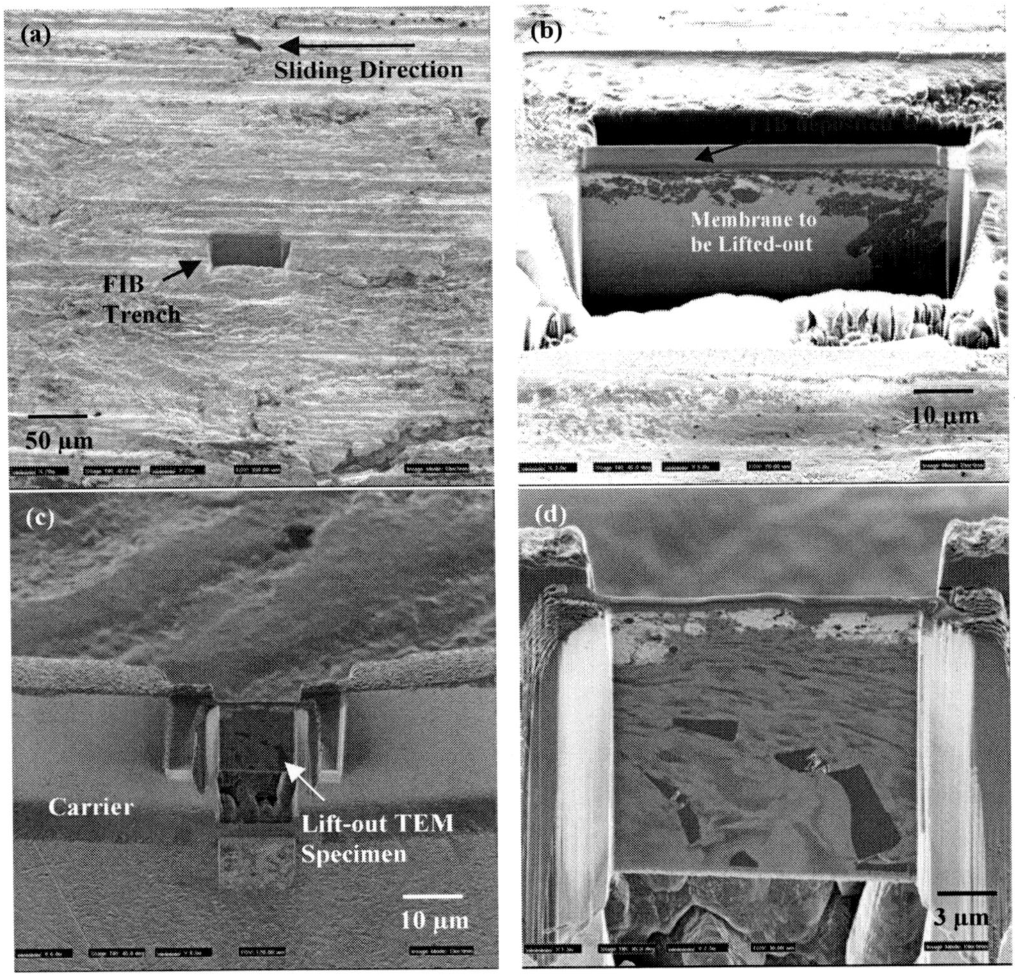

Figure 36. FIB "lift-out" TEM specimen preparation. (a) FIB-cut small trench on the worn surface, (b) Second trench cut to isolate the thin membrane to be lifted-out, (c) The lifted-out membrane mounted onto a TEM grid (carrier), (d) High-resolution FIB image of the TEM foil prior to the final FIB thinning.

Figure 37 shows bright-field TEM images and a typical diffraction pattern taken from the tribolayer generated on the surface of the sample tested in air. The oxidized regions in the tribolayer are indicated with arrows in Figures 37(a) and (c). Four electron diffraction patterns were collected within the oxidized regions in the tribolayer. According to these patterns, the oxides appear to have an amorphous structure [Figure 37(d)]. The amorphous oxide particles contained large amounts of iron, oxygen and silicon, as shown in the EDS maps presented in Figure 38, but further work is needed to determine the exact stoichiometry of the oxides.

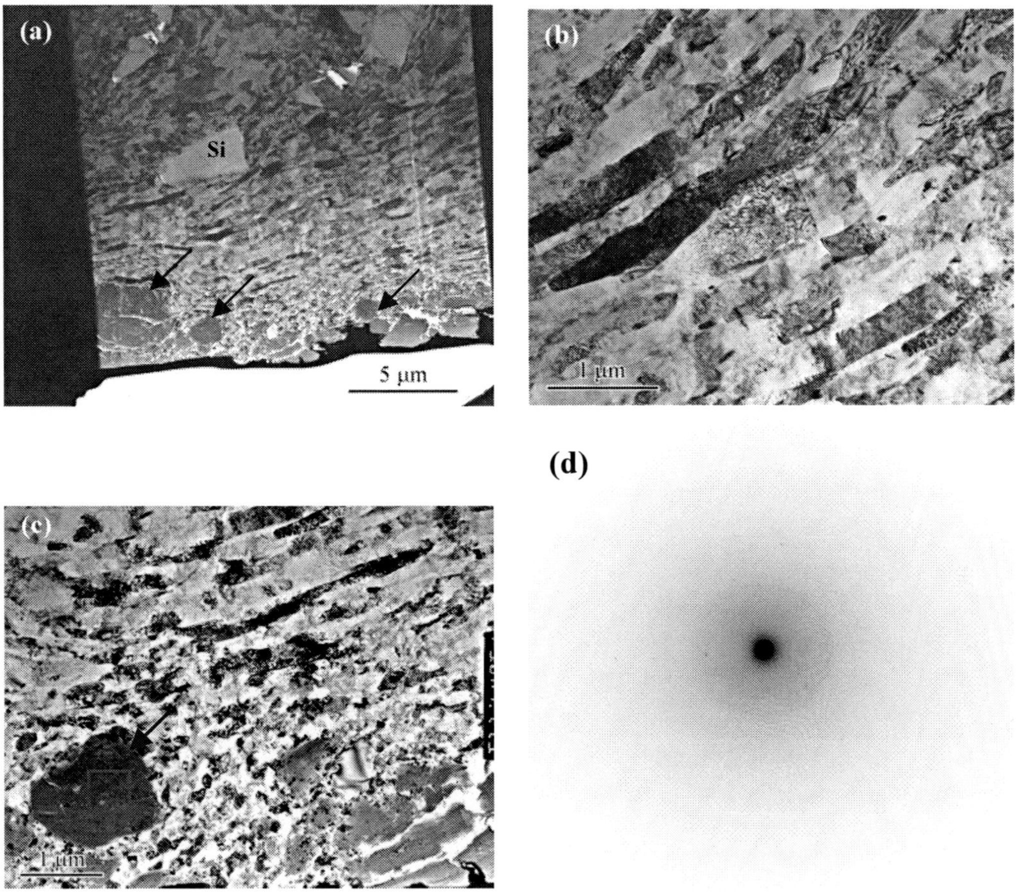

Figure 37. Bright-field TEM images and selected area electron diffraction pattern of the sample worn in air, (a) low-magnification image, large oxide particles (regions) are labeled with arrows, (b) highly deformed fine-grained structure, (c) oxide in tribolayer where the diffraction pattern were taken, (d) electron diffraction pattern from the oxide labeled in (c).

Table 5. Comparison of Tribolayers formed Under the Two Wear Conditions

In air	In argon
Cracked tribolayer	Compact tribolayer
Amorphous oxide of about 10 μm in thickness	Amorphous oxide oxide of about 10 μm in thickness
High iron and high oxygen content	Low Fe and low oxygen content
Mechanically mixed with Al substrate	Mechanically mixed with Al substrate

The morphological features of the tribolayer formed on the sample tested in argon are shown in Figures 39(a)-(c). A typical diffraction pattern of the tribolayer formed in argon, presented in Figure 39(d), shows spot reflections arranged into a ring pattern that belong to pure aluminum and silicon. An oxygen signal was also detected from the particles in the tribolayer, but again no crystalline oxide could be identified. Evidence of oxides in the tribolayer formed in argon can be seen in the EDS maps in Figure 40. Thus, the small amount of amorphous oxide was uniformly mixed with highly deformed aluminum grains and fractured silicon particles in the tribolayer. Small metallic iron particles were also observed in the tribolayer as can be seen in Figure 40(c). However, the overall iron content in the tribolayer was very low which is in agreement with the EPMA analysis. Table 5 summarizes the characteristic features of the tribolayers formed in both air and argon atmospheres.

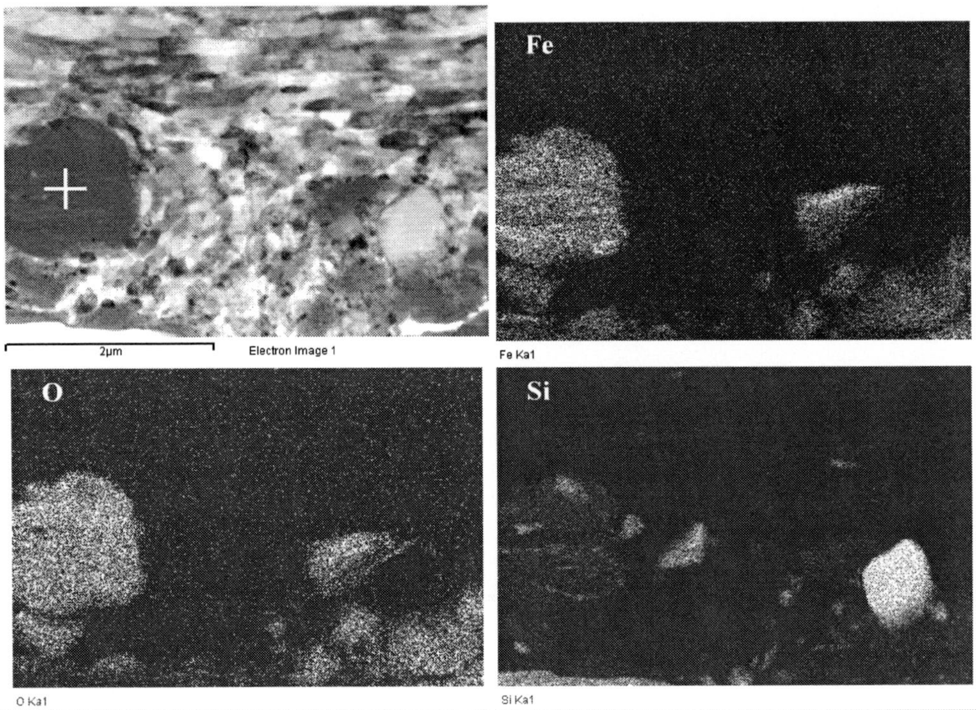

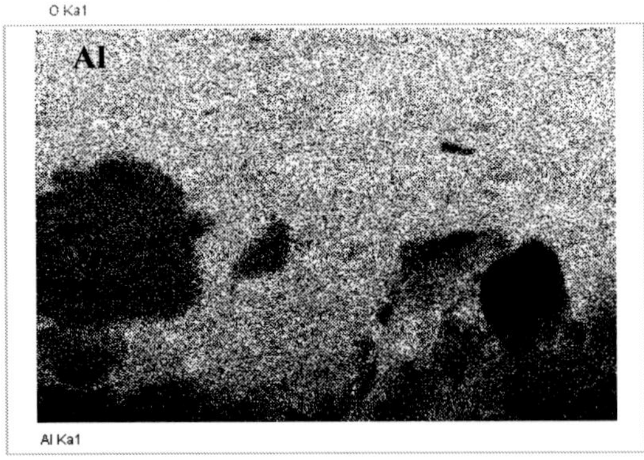

EDS Point Analysis of Oxide Particle

	O	Al	Si	Fe
wt.%	32.4	42.0	9.7	15.9
at.%	48.1	37.0	8.2	6.8

Figure 38. STEM image, EDS maps and semi-quantitative analysis of tribolayer formed on sample worn in air.

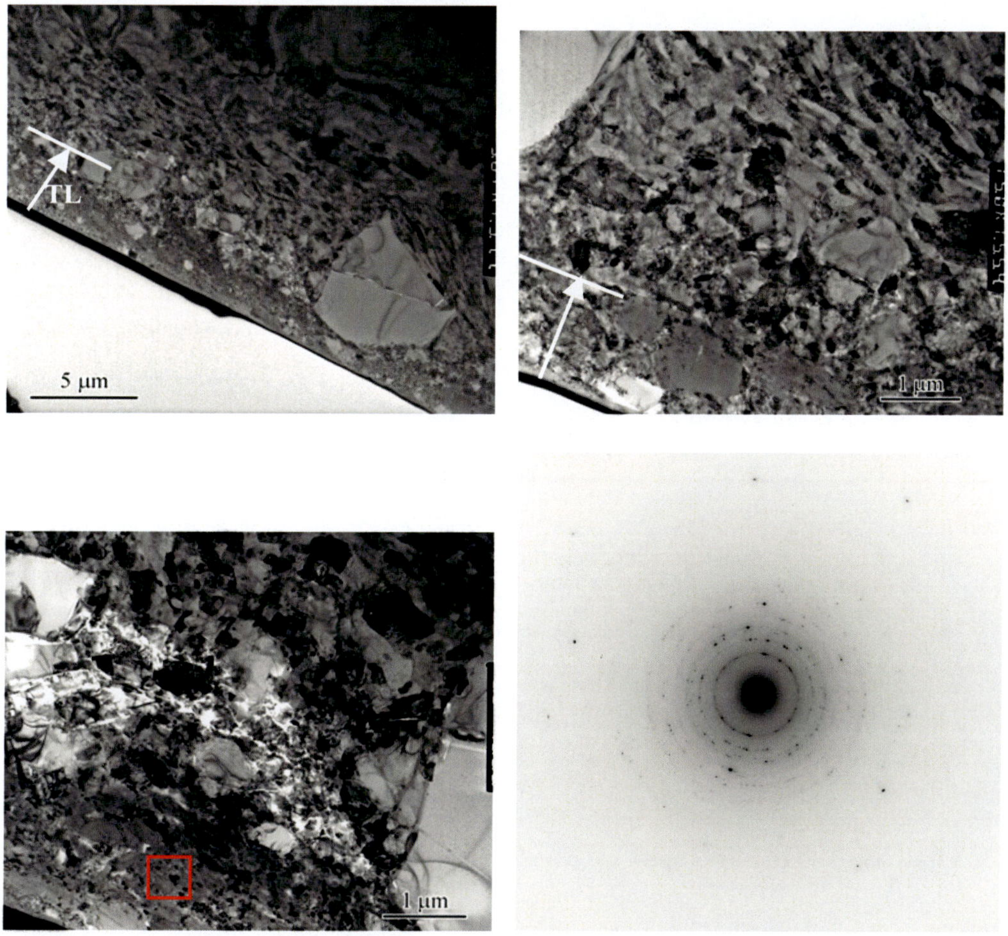

Figure 39. TEM images and SADP from the sample worn in argon, (a) and (b), bright field TEM images, tribolayer is labeled with arrows, (c) location of SADP taken from the tribolayer, (d) SADP from location labeled on (c).

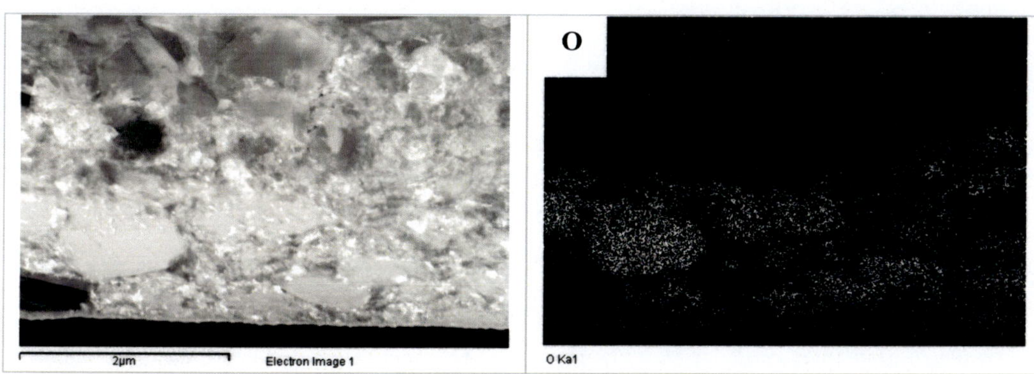

Figure 40. Continued on next page.

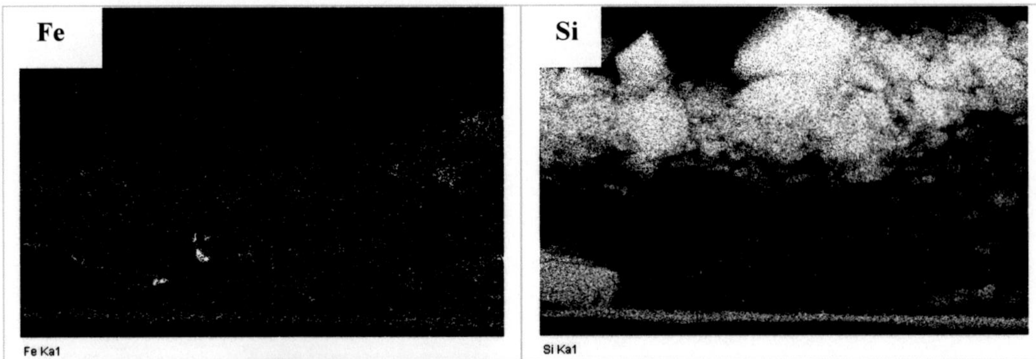

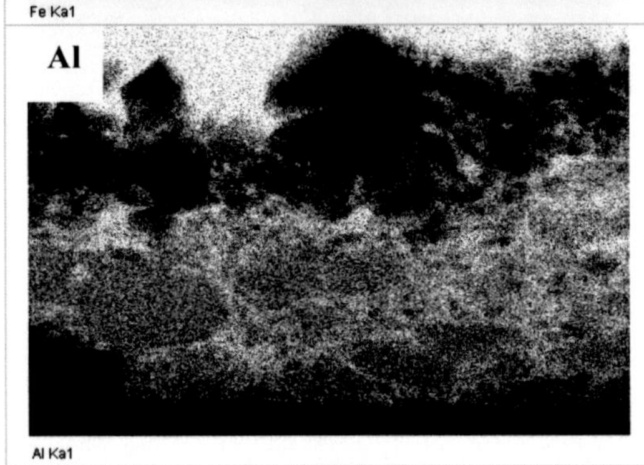

EDS Point Analysis of Oxide Particle

	O	Al	Si	Fe
wt.%	6.3	55.3	37.7	0.7
at.%	10.3	54.0	35.3	0.3

Figure 40. STEM image, EDS maps and semi-quantitative analysis of the tribolayer formed on sample worn in argon.

3.2.3. Analysis of Wear Mechanisms

As shown in previous sections, the wear rate of A390 in air is higher than that in argon. This difference is closely related to the difference in chemical composition of the tribolayers formed under the two test atmospheres. The tribolayers formed in air contained much more oxygen (32.5 wt%) and iron (24.8 wt%) than the tribolayer formed on the sample worn in argon (13.3 wt% of oxygen and 1.2 wt% iron). Ultra-fine aluminum grains were developed in the subsurface layers just beneath the tribolayers due to severe plastic deformation under both atmospheric conditions. The silicon particles that are supposed to strengthen the matrix were severely fractured in the deformed layer in both samples.

EPMA and TEM analyses indicated that aluminum was the main constituent of the tribolayers formed in argon. Although the tests were carried out under a continuous argon gas flow, some oxidation is inevitable. The TEM electron diffraction suggested that the oxide was an amorphous aluminum oxide. The amorphous oxide particles were mixed with ultra-fine-grained (100 nm) aluminum and silicon particles (<2 μm) in the tribolayer. The tribolayers formed in argon also contained a very small amount of iron (in the order of 1 wt%) as detected by both EPMA and TEM EDS analyses. Occasionally, fine metallic iron particles were also found. On the other hand, the tribolayers formed in air were a mixture of oxides of substrate (Al and Si) and the counterface (Fe) materials. Large amount of iron was detected by both EPMA and TEM analyses. This high iron concentration was associated with the higher concentration of oxygen (Table 4).

Li *et al.* [70,71] indicated that mechanical mixing between the surface oxides and the substrate aluminum may occur in the tribolayer. It is interesting to find that the oxide particles in the tribolayer of the argon sample are mixed with ultra-fine grains of aluminum and silicon particles, while the sample worn in air only contains amorphous Al-Fe-Si oxide, which was hard and brittle. This indicates that the mechanically mixed tribolayer has been oxidized during wear in air due to the exposure to a higher-oxygen-containing environment.

The formation of a fine-grained aluminum zone underneath the tribolayer is rather interesting. The relatively soft subsurface aluminum matrix has experienced severe plastic deformation within this zone. The severe damage can be represented, for example, by the fragmentation and flow of a thin silicon arm as shown in Figure 34. Based on both FIB and TEM images, the severe deformation of the subsurface aluminum has resulted in a very fine-grained microstructure, which extends about 10 μm beneath the tribolayer under both conditions. Early reports by Li *et al.* [72-74] suggested that dynamic recrystallization (or recovery) could occur at room temperature even for relatively impure aluminum alloys if certain thermo-mechanical processing was applied to reduce matrix solute iron content. However, the relatively high dislocation density in the fine-grained zone (shown in Figure 37), indicated that recrystallization might not have occurred in these areas due to insufficient stored work at this temperature [75]. On the other hand, one can argue that the severe plastic deformation could have resulted in dynamic recrystallization at a certain stage of this dynamic wear process. The recrystallized microstructure is further deformed after plastic deformation. The wear process is a dynamic process that, in these cases, involves the formation and removal of tribolayers. Obviously, the counterface steel managed to transfer onto A390 in the wear process undergone in air. The transfer mechanism itself is rather important but remains unknown so far. However, at the end, such Fe (or Fe_xO_y) transfer resulted in a hard, brittle and unstable tribolayer which was removed much faster than the aluminum oxide later (formed under Argon). The effect and contribution of the fine-grained aluminum under-layer is now known at this stage. The main contribution of this study can be summarized as follows:

High-resolution FIB images provided valuable microstructural information on the tribolayer and subsurface zones formed under the contact surfaces. High-quality site-specific TEM specimens were also successfully prepared using the FIB lift-out technique.

The thickness of the tribolayer varies from place to place. However, on both samples, the thickness is on the order of 10 μm. Their chemical compositions are significantly different: The air sample contains a large amount of iron (24.8 wt%) and oxygen (32.5 wt%) while the argon sample contains very small amount of iron (1.2 wt%) and less oxygen (13.3 wt%) than that of the air sample. The oxides formed in the tribolayers in both air and argon atmospheres appear to be amorphous. Small metallic iron particles are detected in the tribolayers of the argon sample.

The microstructures of the tribolayers formed under the two test conditions are different. The tribolayer formed on the sample worn in air appeared to be severely fractured as an indication of its brittleness due to a high amount of oxide presence. On the other hand, the tribolayers on the argon sample contained a much smaller amount of oxide and are mixed with the substrate aluminum.

Ultra-fine grained aluminum layers, which extended up to 10 μm, have been formed beneath the tribolayers under both wear conditions. These fine-grained structures were presumably formed due to severe subsurface plastic deformation because the microstructural investigations did not show signs of recrystallization under the current wear conditions.

4. Conclusion

Advances in materials engineering depend heavily on high-level microstructural characterization using a combination of state-of-the-art instruments. New characterization techniques and methodologies are critical in understanding foundamentals of materials properties. Recent developments in focused ion beam microscopy have become increasingly important in microstructural characterization.

Acknowledgements

The author gratefully acknowledges the efforts of the following individuals for their contributions of research and consultations leading to the completion of this book chapter.

Valery Guertsman, Pei Liu, Jason Lo, Mimoun Elboujdaini, Winston Revie, Wenyue Zheng, Tom Malis and Jennifer Jackman (Materials Technology Laboratory, Natural Resources Canada), Michael Phaneuf, David Mayer (Fibics Incorporated), Louise Weaver (University of New Brunswick), Shigeo Saimoto (Queen's University at Kingston), Ming Gao (Blade Energy) and Alan Alpas (Univeristy of Windsor).

References

[1] Melngailis, J. J. *Vac. Sci. Technol. B,* 1987, vol. 5(2), 469-495.
[2] Phaneuf, M. W. *Micron*, 1999, vol. 30, 277-288.
[3] Phaneuf, M. W.; Li, Jian and Malis, T. *Microscopy and Microanalysis*, 1998, vol. 4, 492-493.
[4] Li, Jian, *Journal of Metal*, 2006, vol.58 no.3, 27-31.
[5] Li, Jian, McMahon, G.S. and Phaneuf, M.W. *Proceedings of the Microscopically Society of Canada*, 2001, vol. XXVIII, 26-27.
[6] Giannuzzi, L.A. and Stevie, F.A. *Micron*, 1999, vol. 30, 197-204.
[7] Anderson, R. *Proceeding of Microscopy and Microanalysis*, 2002, vol.8, 44-45.
[8] Shankar, S. Riddle, Y.W. and Makhlouf, M.M., *Metallurgical and Materials Transactions A,* 2003, vol. 34A, 705-707.
[9] Li, Jian, Phaneuf, M.W. and Saimoto, S., *Conference of Microscopy Society of Canada*, 1998, Ottawa, Ontario.
[10] Office of Pipeline Safety, "SCC Threat Advisory Notice", October 2003.
[11] Leis, B.N., Galliher, R.D., Sutherby, R.L., and Sahney, R., *Proceedings of the IPC* 2004, Calgary, Alberta, Canada, October 4-8, 2004, Paper No: IPC04-0665.
[12] Li, Jian, Elboujdaini, M. and Liu, P., *Proceeding of Microscopy and Microanalysis* 2006, Chicago, August, 2006, vol.12, 1306-1307.
[13] Patterson, R.J., Mayer, D., Weaver, L. and Phaneuf, M.W., *Proceeding of Microscopy and Microanalysis* 2002, 2002, vol.8, 566-567.

[14] Longo, D.M., Howe and Johnson, W. C., Ultramicroscopy, 1999, vol. 80, (2), 69-84.
[15] Anderson, R. and Klepeis, S.J., *Materials Research Society Symposium Proceedings, Pittsburgh*, 1997, 480, 187.
[16] Prenitzer, B.I., Giannuzzi, L.A., Newman, K., Brown, S.R., Irwin, R.B., Shofner, T.L. and Stevie, F.A, *Metallurgical and Materials Transactions A,* 1998, vol. 29A, 9, 2399-2406.
[17] Langford, R.M., Petford-Long, A.K. and Gnauck, P., *Microscopy and Microanalysis*, 2002, vol. 8, (Suppl. 2), 46-47.
[18] Gasser, P., Klotz, U.E., Khalid, F.A. and Beffort, O., *Microscopy and Microanalysis*, 2004, vol.10, 311-316.
[19] Li, Jian, Elboujdaini, M. and Gertsman, V.Y., Gao, M. and Katz, D.C., Aerospace Materials and Manufacturing: Development, *Testing and Life Cycle Issues,* ed. P.C. patnaik, 2004, 515-523.
[20] Elboujdaini, M., Li, Jian, Gertsman, V., Gu, G., Revie, W., Gao, M. and Katz, D., *NACE International Conference*, CORROSION/2004, Paper#04554; New Orleans, March 2004.
[21] Li, Jian, Gertsman, V.Y. and Lo, J., *Microscopy and Microanalysis*, 2003, vol.9 (2),888-889.
[22] Cairney, J.M. and Munroe, P.R., *Materials Characterization*, , 2001, 46, 297-304.
[23] Malis,T., Carpenter, G.J.C., Dionne, S., Botton, G.A. and Phaneuf, M.W., Industrial *Applications of Electron Microscopy*, ed. Li, Zhigang R., Marcel Dekker Inc. New York, 2002.
[24] Herschbein, S., Fisher, L.S., Kane, T.L., Tenney, M.P. and Shore, A.D., *Proc. ISTFA*, 1998, 127.
[25] Casey, J.D., Noll, K.E., Shuman, R., Chandler, C., Megordan, M., Phaneuf, M. and Li, Jian, *LSI Testing Symposium*, 2000, Osaka, Japan.
[26] Phaneuf, M.W., Li, Jian, Shuman, R.F. Noll, K. and Casey, J.D. Jr., US patent #6,641,705, issued November 04, 2003.
[27] Shuman, R.F., Noll, K., Casey, J.D. Jr., U.S. Patent 6,322,672 B1, issued November 2001.
[28] Phaneuf, M.W., Li, Jian and Casey, J.D. Jr., *Microscopy and Microanalysis*. 2002, vol. 8 (2), 52-53.
[29] Townsend, H. E. Jr., *Materials Performance*, 1972, vol. 11, 33-37.
[30] Gu, B., Luo, J.L. and Mao, X., *Corrosion*, 1999, 55(1), 96-106.
[31] Qiao, L.J., Luo, J.L. and Mao, X., *J. Mater. Sci. Lett.*, 1997, vol.16, 516-520.
[32] Wang, S.H., Chen, W., Jack, T., King, F. and Fessler, R. R., *Proc. 2000 Intern. Pipeline Conference*, 2000, 1005-1012.
[33] Parkins, R.N., Corrosion, 1996, 52 (5), 363-374.
[34] Long, L.M. and Uhlig, H.H., J. *Electrochem.* Soc., 1965, vol.112, 964.
[35] Parkins, R.N., J. *Iron St. Inst.* 1952, 172 (9), 149.
[36] Wang, J.Q. et al., *Acta Mater.*, 1998, 46 (16), 5677-5678.
[37] Wang, J.Q. and Atrens, A., *Corrosion Science*, 2003, 45, 2199-2217.
[38] Parkins, R.N., NACE International, *Corrosion*/2000, Paper No.00363.
[39] Fressler, R.R., Groeneveld, T.P. and Elsea, A.R., *NACE-5, SCC and Hydrogen Embrittlement of Iron Base Alloys*, ed. Hochman and Staehle, 1977, 135-146.
[40] Eiber, R.J., MH-2-95, *National Energy Board*, Calgary, Alberta, November 1996, 19.

[41] Danielson, M.J. and Jones, R.H., Corrosion 2001, NACE International Houston, TX, Paper 01211.

[42] Green, J.A.S. and Parkins, R.N., *Corrosion*, 1968, vol. 24, 66.

[43] Phaneuf, M.W., and Li, Jian, *Microscroscopy and Microanalysis*, 2000, vol. 6 (2), 524-525.

[44] Wang, Y.Z., Revie, W, Phaneuf, M.W. and Li. Jian, *Fatig. Fract. Eng. Mater. Struct.*, 1999, vol. 22, 251-256.

[45] Phaneuf, M.W., *FIB For Materials Science Applications – A Review*, in Introduction to Focused Ion Beams: Theory, Instrumentation, Applications and Practice; eds. Giannuzzi, L.A. and Stevie, F.A., Academic Press, NY. 2004, ISBN: 0-387-23116-1

[46] Phaneuf, W. and Botton, G.A., *Proc. Fontevraud 5 – Contribution of Materials Investigation to Problems Encountered in Pressurized Water Reactors*, SFEN Paris, France 2002, 117 – 128.

[47] Katz, D.C., Gao, M., Sherstan, R., Elboujdaini, M. and Li, Jian, *Proceeding of 42nd Annual Conference of Metallurgists*, 2003, 293-297.

[48] Li, R. and Ferreira, M.G.S., *Corrosion Science*, 1996, vol. 38 (2), 317-327.

[49] Szklarska-Smialowska, Z., *NACE-5, SCC and Hydrogen Embrittlement of Iron Base Alloys*, ed. Hochman and Staehle, 1977, 30-36.

[50] Westbrook, J.H., *NACE-5, SCC and Hydrogen Embrittlement of Iron Base Alloys*, ed. Hochman and Staehle, 1977, 26-29.

[51] Tau, L. and Chan, S.L.I., *Materials Letters*, 1996, vol. 29,143-147.

[52] Deuis, R.L., Subramanian, C. and Yellup, J.M., *Composites Science and Technology*, 1997, vol. 57 (4), 415-435.

[53] Yasmin, T., Khalid, A.A. and Haque, M.M., *Journal of Materials Processing Technology*, 2004, vol. 153-154 (10), 833-838.

[54] Wilson, S. and Alpas, A.T., *Wear*, 1999, vol. 225-229, 440-449.

[55] Dwivedi, D.K., Arjun, T.S., Thakur, P., Vaidya, H. and Singh, K., *Journal of Materials Processing Technology*, 2004, vol. 155 (3), 323-328.

[56] Dwivedi, D.K., *Materials Science and Engineering A*, 2004, vol. 382 (1-2), 328-334.

[57] Jiang, J.Q. and Tan, R.S., *Wear*, 1996, vol. 195 (1-2), 106-111.

[58] Wang, F., Liu, H., Ma, Y. and Jin, Y., *Materials and Design*, 2004, vol. 25 (2), 163-166.

[59] Torabian, H., Pathak, J.P. and Tiwari, S.N., *Wear*, 1994, vol. 172 (1), 49-58.

[60] Jasim, K.M. and Dwarakadasa, E.S., *Wear*, 1987, vol. 119 (1), 119-130.

[61] Hutchings, I.M., Wilson, S. and Alpas, A.T., *Comprehensive Composite Materials*, Elsevier Science Ltd, ISBN: 0-080437214, vol. 3, 501-519.

[62] Wilson, S. and Alpas, A.T., *Wear*, 1997, vol. 212, 41-49.

[63] Andrews, J.B., Seneviratne, M.V., Zier, K.P. and Jett, T.R., *Proc. Conf. Wear of Materials* 1985, ed. K.C. Ludema, ASME, New york, 1985, 180-185.

[64] Riahi, A.R. and Alpas, A.T., *Wear*, 2001, vol. 251, 396-1407.

[65] Elmadagli, M. and Alpas, A.T., *Proceeding of Light Metals and Metal Matrix Composites Symposium, 43rd Conference of Metallurgist*, Hamilton, Ontario, Canada, August, 2004, 127-139.

[66] Sarvanan, R.A., Lee, J.M. and Kang, S.B., *Metallurgical and Materials Transactions A*, 1999, vol. 30, 2523-2538.

[67] Li, Jian, Elboujdaini, M., Gertsman, V. Y., Gao, M. and Katz, D. C., *Canadian Metallurgical Quarterly*, 2005, vol.44 (3), 331-338.

[68] Li, Jian, Elboujdaini, M., Fang, B., Revie, R.W. and Phaneuf, M.W., *Corrosion*, 2006, vol. 62 (4), 316-322.

[69] Li, Jian, Dionne, S. and Malis T., *Materials Characterization*, 2006, vol. 57, 64-70.

[70] Li, X.Y. and Tandon, K.N., *Wear*, 1999, vol. 225-229 (1), 640-648.

[71] Li, X.Y. and Tandon, K.N.,*Wear*, 2000, vol. 245 (1-2), 148-16.

[72] Jian Li and S. Saimoto, *Materials Science and Engineering A.*, vol. A234-236, 1997, 1019-1022.

[73] Li, Jian and Saimoto, S., *Materials Science Forum*, 1998, vol.273-275, 459-464.

[74] Saimoto, S., Li, Jian, Langelaan, G., Diak, B.J. and Shimizu, J., *Textures and Microstructures*, 1996, vol. 26-27, 245-262.

[75] Humphery, F.J. and Hatherly, M., *Recrystallization and Related Annealing Phenomena*, Pergamon Press, 1995.

In: Materials Science Research Trends
Editor: Lawrence V. Olivante, pp. 197-250
ISBN: 978-1-60021-654-1

Chapter 4

High-Rate and Low-Temperature Film Growth Technology Using Stable Glow Plasma at Atmospheric Pressure

Hiroaki Kakiuchi, Hiromasa Ohmi and Kiyoshi Yasutake
Department of Precision Science and Technology, Graduate School of Engineering, Osaka University, 2-1 Yamada-oka, Suita, Osaka 565-0871, Japan

Abstract

To fabricate high-quality functional thin films at very high deposition rates on large-sized substrates, we have proposed an atmospheric-pressure plasma chemical vapor deposition (AP-PCVD) technique. In the AP-PCVD process, stable glow plasma of gas mixtures containing carrier gases and source gases is generated at atmospheric pressure, and is effectively used to deposit thin films. Since the partial pressure of source gases can be high, the deposition rate is significantly increased. In the AP-PCVD system, combination of the rotary electrode and 150-MHz very high frequency (VHF) power supply makes it possible not only to stably generate high-density atmospheric-pressure plasma but also to suppress ion impingement upon the film surface. The AP-PCVD system equips a gas circulation system connected with the reaction chamber for efficiently collecting and removing particles that float around the plasma region. By virtue of these noble characteristics of the system, it has become possible to fabricate high quality films at extremely high deposition rates.

In this article, the basic concept and principle of the AP-PCVD technique are described first. Then, some of the fundamental research results on the property of atmospheric-pressure plasma and the elemental technologies for the AP-PCVD system are given. To evaluate the performance of the AP-PCVD system, we have deposited silicon (Si) films using silane (SiH_4) diluted with hydrogen (H_2) and helium. The deposition rate, morphology, and structural and electrical properties of the deposited Si films are discussed as functions of the deposition parameters, such as VHF power, SiH_4 and H_2 concentrations, and substrate temperature. The results show that homogeneous amorphous Si films having smooth surface and cross-sectional morphology can be successfully formed at unprecedented high rates. When the ratio of H_2 to SiH_4 and/or the substrate temperature is increased, polycrystalline and single crystalline films grow on a variety of substrate materials, such as Si and SiO_2, even at temperatures lower than in conventional deposition techniques. It is shown that the VHF power is a very important deposition parameter, which dominates the dissociation of SiH_4 molecules and the structural relaxation of a growing film. Note that the plasma gas temperature, including rotational and vibrational temperatures of molecules, and high-density atomic hydrogen in the atmospheric-pressure plasma can supply considerable physical and chemical energies to the film-growing surface, enhancing the film-forming reactions even at low temperatures.

1. Introduction

The thin film formation by plasma-enhanced chemical vapor deposition (PECVD) technique has been generally performed using high-frequency or microwave discharges under vacuum conditions (< 10^2 Pa). However, the deposition rate is low because of the low partial pressure of source gases in the plasma atmosphere. It is also understood that damages and impurities are incorporated in the resulting films due to the collision of highly-accelerated ions against the film surface and inner walls of the vacuum chamber. Moreover, due to the requirement of a vacuum system and the space limitations defined by the dimensions of a vacuum chamber, the deposition processes become expensive, particularly when large-sized substrates have to be coated.

On the other hand, during the last decades, there has been a steady increase in the utilization of physical and chemical properties of plasmas generated at atmospheric pressure for various applications such as film deposition, ultraprecision machining, surface processing and material treatment. Since atmospheric-pressure plasma can serve as an efficient source of atomic and molecular active components, it is useful in both increasing deposition rate and lowering substrate temperature for the fabrication of functional thin films. At atmospheric pressure, the mean free path of atoms and molecules is of the order of 0.1 μm, far smaller than that at low pressures. Thus, plasma generated at atmospheric pressure is confined within a narrow gap region between an electrode and a substrate. This is advantageous for the preparation of high-quality functional thin films, because the impurity incorporation into a deposited film from the inner wall of the chamber can be prevented. Additionally, operating plasma at atmospheric pressure requires practically no vacuum devices, so that the integration of the plasma process into production lines is simplified and batch processing can be avoided. Thus, in the case of surface processing or material treatment using nontoxic and/or non-pyrophoric gases, atmospheric-pressure plasma processes are considered to be profitable from the viewpoint of suppressing cost of manufacturing equipment and constructing a large-scale system. However, a discharge under atmospheric pressure is easy to shift to a thermal plasma, and hence, it tends to be locally concentrated when an excessive electric power is supplied. Therefore, to apply atmospheric-pressure plasma to industrial uses, the development of a new technology, which enables the generation of stable atmospheric-pressure plasma in a wide area, is indispensable.

We have developed an atmospheric-pressure plasma chemical vapor deposition (AP-PCVD) technique [1], which is a novel high-rate and low-temperature film growth method for functional materials. In the AP-PCVD process, combination of a cylindrical rotary electrode and a 150 MHz very high-frequency (VHF) power supply makes it possible to generate stable glow plasma at atmospheric pressure. Since the beginning of development of the AP-PCVD technique, we have been studying high-rate and low-temperature fabrication technologies for hydrogenated amorphous silicon (a-Si:H), polycrystalline silicon (poly-Si) and epitaxial silicon (epi-Si) films [2–7], which have been receiving increasing attention for the production of high-performance electronic devices, such as solar cells, thin film transistors and ultra large scale integrated circuits. Using the AP-PCVD technique, we have also successfully achieved high-rate depositions of silicon carbide and silicon nitride films in both amorphous and crystalline state at relatively low temperatures [8–10].

The content of this article is organized as follows. In section 2, we first outline the basic concept and principle of the AP-PCVD technique. Then, we show some research results on the property of atmospheric-pressure plasma (section 3) using a simple setup for fundamental investigations. On the basis of the results, we describe in section 4 the details of the AP-PCVD system that we have developed for the preparation of high quality functional thin films, together with elemental technologies. To demonstrate the effectiveness of the AP-PCVD system, we have deposited Si films in both amorphous and crystalline states using silane (SiH_4) diluted with hydrogen (H_2) and helium (He). In section 5, we discuss the deposition rate, morphology, and structural and electrical properties of the deposited films as functions of the deposition parameters, such as VHF power, gas composition and substrate temperature. Section 6 is the conclusion.

2. Atmospheric-Pressure Plasma CVD

2.1. Thin Film Deposition Using Atmospheric-Pressure Plasma

The structure and property of a thin film are determined by the internal parameters of plasma, such as form of radicals, flux of radicals onto the film-growing surface, and the surface reactions activated by substrate heating. Thus, it is difficult to achieve high-rate and low-temperature growths of high-quality functional thin films only by empirically improving the conventional plasma-enhanced techniques. In other words, to increase the deposition rate and reduce the growth temperature, it is indispensable to change the plasma kinetics essentially by introducing a new concept or a new principle. The utilization of different energy source from substrate heating is particularly important, because the surface reactions must be accelerated when the deposition rate is increased and/or the substrate temperature is reduced. In the light of such considerations, we have developed AP-PCVD technique.

The film deposition by AP-PCVD is performed at a pressure of 1×10^5 Pa using high partial pressure of source gases diluted with base gases, such as He, Ar and H_2. The electrode surface is coated with insulating materials to prevent the emission of secondary electrons from the surface, which enables the generation of stable nonequilibrium plasma in a wide area. Since the partial pressure of source gases is high, the deposition rate is inevitably increased. The kinetic processes in gas phase and at the film-growing surface should also change when atmospheric-pressure plasma is used instead of the conventional low-pressure plasmas, as described in the following.

First, it is considered that molecule-molecule interactions take a significant part in the decomposition process of source gas molecules in atmospheric-pressure plasma, although electron-molecule collisions should initiate their decomposition. The frequent interactions of gas phase species with the surface also affect the whole plasma kinetics, because the surface kinetics contributes to the creation and destruction of a number of species. Thus, the form of precursors in the AP-PCVD process, which contribute to the film growth, may be different from that in the conventional PECVD process.

Secondly, the high collision frequency of accelerated ions versus other neutral species results in the higher gas temperature, including rotational and vibrational temperatures of molecules, and the smaller kinetic energy of ions than those of low-pressure plasma. The translational temperature of atmospheric-pressure plasma can supply considerable thermal energy to the film-growing surface. It is also expected that the rotational and vibrational energies of heated precursor molecules enhance the chemical reactions at the film-growing surface. In addition, the smaller kinetic energy of ions leads to the reduction of ion damage in the films. Moreover, it is well known that atomic hydrogen plays important roles in the deposition and surface treatment processes of Si and related materials, such as termination of dangling bonds, crystallization of amorphous components and etching of the surfaces. Thus, high-density atomic hydrogen generated in atmospheric-pressure plasma also acts effectively as an energy source through the interactions with the film-growing surface. Therefore, the utilization of atmospheric-pressure plasma is considered to be useful to reduce the substrate temperature for the formation of good-quality films.

2.2. Utilization of VHF Electric Field

In the conventional PECVD processes, 13.56-MHz radio frequency (RF) is mostly employed as an excitation frequency. It is well known, however, that employing higher frequencies is effective not only for the formation of better quality films [11] but also for the improvement of deposition rates [12]. At high frequencies, ionic species cannot follow the polarity change of the electric field and are trapped between an electrode and a substrate. This leads to the suppression of ion impingement upon the film-growing surface and the reduction of damage in the films. The trapping of ions and electrons also leads to the increase in the plasma density. In the case of AP-PCVD, although the collision of electrons and ions with neutral species must be considered, employing higher frequency than 13.56 MHz is still effective to obtain a higher power density in the plasma-generating gap.

To quantitatively evaluate the effects of excitation frequency and pressure on both kinetic energy and oscillation amplitude of electrons and ions, let us assume binary collisions between charged particles A and neutral ones B in a uniform plasma without electrodes. Using the classical hard-sphere model, the equation of motion for particles A can be expressed as

$$m_A(dv/dt) = qE - m_A \nu v, \tag{1}$$

where m_A is the mass of particle A, v is the particle velocity, q is the charge of particle, E is the strength of external electric field, and ν is the elastic collision frequency for momentum transfer between particles A and B. Considering steady-state solutions, and taking into account that v and E vary respectively as $v = v_0 \exp(j\omega t)$ and $E = E_0 \exp(j\omega t)$, where ω is the angular frequency of the external electric field and j is the imaginary unit, $j = (-1)^{1/2}$, equation (1) is rewritten as

$$j\omega\, m_A v_0 = qE_0 - m_A \nu v_0. \tag{2}$$

From equation (2), v_0 is given by

$$v_0 = qE_0(\nu - j\omega) / m_A(\nu^2 + \omega^2), \quad (3)$$

and thus, the absolute value of v_0 is

$$|v_0| = qE_0 / m_A(\nu^2 + \omega^2)^{1/2}. \quad (4)$$

Furthermore, since the position of particle, x, is given by $dx/dt = v$, the oscillation amplitude of charged particles A, $|x_0|$, can be derived from equation (3) as

$$|x_0| = qE_0 / m_A\omega(\nu^2 + \omega^2)^{1/2}. \quad (5)$$

On the other hand, ν can be expressed as

$$\nu = N_B Q_{AB} v. \quad (6)$$

In equation (6), the density of particles B, N_B, is

$$N_B = P / kT, \quad (7)$$

where P is the pressure, k is the Boltzmann's constant, and T is the gas temperature. In addition, the hard-sphere collision cross-section, Q_{AB}, can be written as

$$Q_{AB} = \pi(r_1 + r_2)^2, \quad (8)$$

where r_1 and r_2 are the radiuses of particles A and B, respectively. From equations (4), (6), (7) and (8), the value of ν can be estimated in order of magnitude. As a result, we can obtain the values of $|v_0|$ and $|x_0|$ using equations (4) and (5), respectively.

Calculations of $|v_0|$ and $|x_0|$ in pure He have been carried out for two frequencies: 13.56 and 150 MHz. Since the energy transferred to electrons is, as is well known, a function of E/P, the value of breakdown field varies with pressure. Besides, the field strength for maintaining plasma is considered to be smaller than the breakdown field. Thus, realistic values of field in steady-state plasma should be used for E_0 in the calculations. However, to simplify the calculations, the value $E_0 \approx 1 \times 10^6$ V/m, which can be roughly estimated as the breakdown field for P = 1×10^5 Pa from the Paschen's curve [13], has been used. The gas temperature is assumed as $T \approx 800$ K.

Figures 1 and 2 show the kinetic energies, $(1/2)m_A|v_0|^2$, and oscillation amplitudes, $|x_0|$, of an electron and a He ion as a function of pressure. In general vacuum conditions ($P \leq 1 \times 10^2$ Pa), increasing frequency is effective on the reduction of both kinetic energies and oscillation amplitudes of charged particles. By increasing pressure, since the field-induced motions of charged particles are interrupted by collisions, both the kinetic energies and the oscillation amplitudes tend to decrease. However, the kinetic energies become independent of excitation frequency under the constant electric field strength, while increasing frequency still affects the oscillation amplitudes. From Fig. 1, the kinetic energy of a He ion at $P = 1 \times 10^5$ Pa is approximately 0.29 eV, which has no remarkable difference from the thermal energy at 800 K (~ 0.1 eV). Furthermore, from Fig. 2, the oscillation amplitude of electron at 150 MHz is as small as 0.7 mm. To obtain the field strength of the order of 1×10^6 V/m and generate a stable plasma at atmospheric pressure, we consider that adopting a gap size of 1 mm or less is appropriate. When an excitation frequency of 150 MHz is used, electrons can be trapped in the narrow gap region, and thus, the generation of high-density atmospheric-pressure plasma is expected.

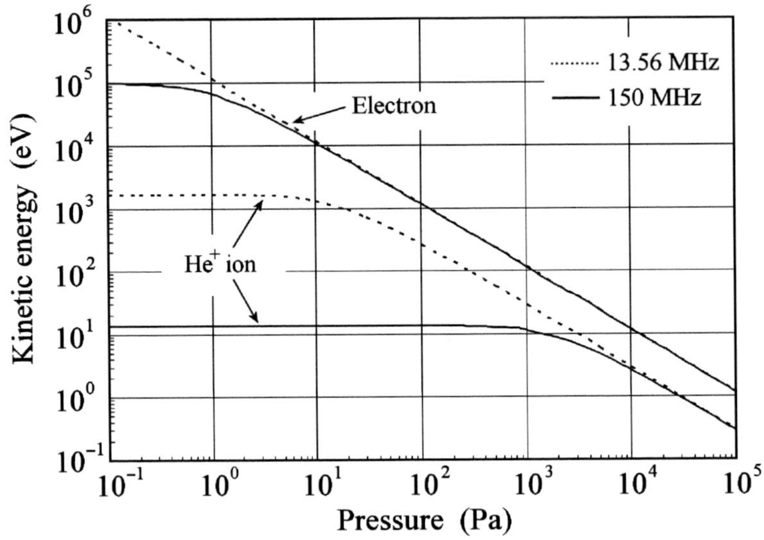

Figure 1. Kinetic energies of an electron and a He ion for the excitation frequencies of 13.56 and 150 MHz as a function of pressure. E_0 is 1×10^6 V/m.

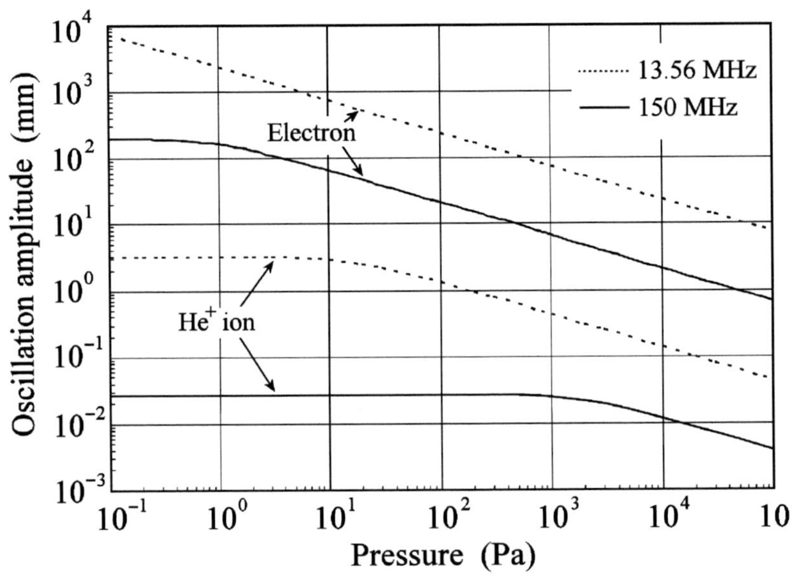

Figure 2. Oscillation amplitudes of an electron and a He ion for the excitation frequencies of 13.56 and 150 MHz as a function of pressure. E_0 is 1×10^6 V/m.

2.3. Utilization of rotary electrode

We have considered that it is difficult to adapt the electrode constructions in conventional low-pressure plasma reactors as it is for the generation of stable glow plasma at atmospheric pressure. Thus, we have proposed the utilization of a cylindrical rotary electrode as one of the methods that enable the generation of stable atmospheric-pressure plasma. In a schematic illustration in Fig. 3, the features of utilizing a cylindrical rotary electrode are summarized as follows.

(a) By rotating electrode, source gas molecules are carried by high-speed viscous flow and homogeneously introduced into the plasma region between the electrode and a substrate.
(b) The electrode surface can be sufficiently cooled, so that a large electric power can be supplied without thermal damage of the electrode.
(c) Particles generated from the chemical reactions in the plasma can be easily removed out of the plasma region.

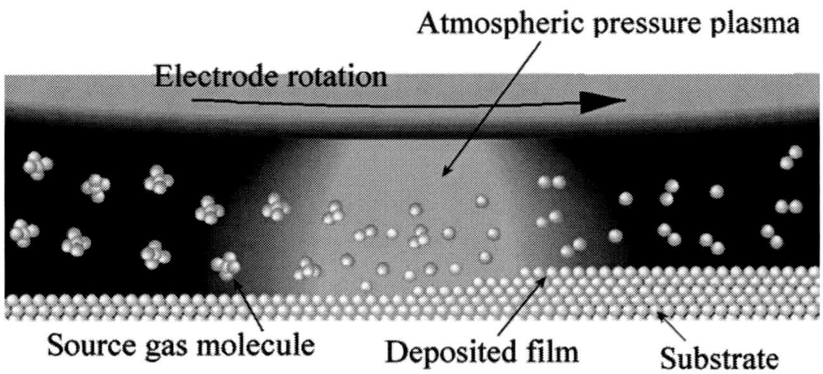

Figure 3. Schematic illustration of atmospheric-pressure plasma CVD with a rotary electrode.

Item (a) is described in further detail in the light of hydrodynamics as follows. When a gas is fed to the small gap between a conventional fixed electrode and a substrate at atmospheric pressure, both of them serve as obstacles for flow of gas due to the viscosity. On the contrary, in the case that the electrode itself rotates, the electrode serves not as an obstacle but as a motive force for accelerating the flow of gas. Therefore, it is possible to supply the gas efficiently into the gap. Additionally, a stable and homogeneous flow of gas can readily be obtained as far as the gap is uniformly kept.

Item (b) is concerned with the supply of sufficient electric power which is suitable for the feed of a large amount of source gases. To obtain a high deposition rate, it is necessary to sufficiently feed source gas molecules while supplying enough electric power to the electrode. In general, the surface of a fixed electrode may be excessively heated and damaged when a large electric power is applied because the gas temperature of atmospheric-pressure plasma may become much higher than that of conventional low-pressure, nonequilibrium plasma. In the case of using a rotary electrode, however, only part of the total surface area of the electrode is in contact with the plasma, and most part of the electrode surface is cooled by the process gas in the atmosphere. Therefore, the electrode surface is free from local heating and the accuracy of the gap is maintained even if a large electric power is applied.

Item (c) can be explained from the viewpoint of hydrodynamics as well as item (a). Particles formed by plasma-induced gas phase reactions during deposition are a source of contamination to be eliminated. In the conventional PECVD process, since the mean free path of atoms and molecules is almost equivalent to the gap size, particulate contaminations easily reach the substrate surface without colliding with other atoms and/or molecules. In the AP-PCVD process, however, particles diffuse in gas phase, frequently colliding with other atoms and/or molecules. Therefore, under the existence of a uniform viscous flow, particles can readily be carried away from the plasma area. To avoid the adhesion of particles onto the substrate surface, however, it is indispensable to remove the particles that float around the plasma edge on the downstream side.

As mentioned above, the utilization of a rotary electrode is an effective means to apply AP-PCVD to the practical use. Since the rotary electrode used in this study has a cylindrical shape, atmospheric-pressure plasma is confined in the gap region as shown in Fig. 4. By scanning a substrate, it is possible to make a film on a large-sized substrate.

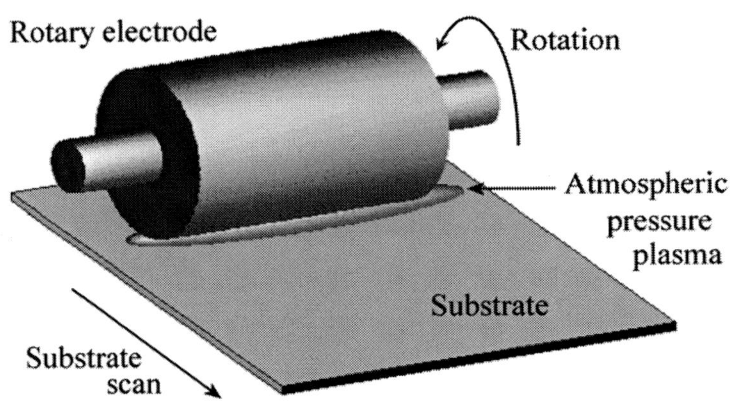

Figure 4. Schematic illustration of substrate scanning.

3. Fundamental Characterization of Atmospheric-Pressure Plasma

3.1. Experimental Methods

In an attempt to confirm the role of excitation frequency and insulating coating of the electrode surface, we have studied VHF and RF plasmas generated at atmospheric pressure by a couple of plasma diagnostic measurements: optical emission spectroscopy (OES) and plasma impedance. The experiments are concerned with plasma generated in pure He which is widely used as a common constituent of atmospheric-pressure plasma. Although He is a chemically unreactive gas, it is possible to utilize the results for the design of the AP-PCVD chamber, which uses chemically more complex plasmas for thin film depositions. In a series of experiments, an apparatus designed for fundamental characterizations of atmospheric-pressure plasma was used. The results and some discussions are described here.

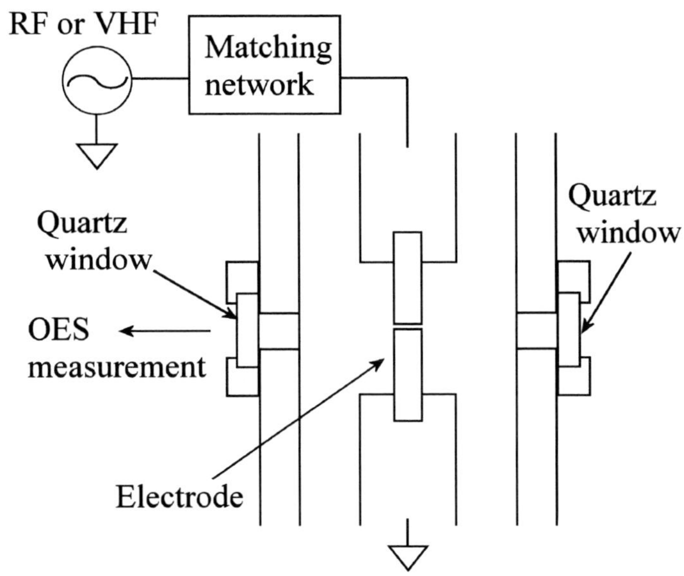

Figure 5. Schematic illustration of the experimental setup for the fundamental characterization of atmospheric-pressure plasma.

Figure 5 shows a schematic illustration of the experimental setup. The construction of the apparatus was determined from the viewpoint of the convenient modeling of parasitic capacitance, inductance and resistive losses due to skin effect conduction for the analysis of plasma impedances. The apparatus was equipped with impedance matching networks for optimizing the electric power transfer from the VHF (145 MHz) or RF (13.56 MHz) power supply to the plasma. Either by VHF and RF excitation, atmospheric-pressure plasma was generated in the gap region between diode-type electrodes with 10 mm diameter and 1 mm spacing. As shown in the figure, the upper electrode was powered and the bottom one was grounded. Using the same apparatus, both VHF and RF plasmas could be generated. This enabled us to investigate the influence of excitation frequency solely while keeping other parameters constant. Both the powered and grounded electrodes were made of aluminum. To investigate how the surface material of the electrode affects the plasma characteristics, a different set of electrodes, which surfaces were coated with alumina (Al_2O_3) of approximately 0.15 mm thickness, were also prepared. The apparatus could be evacuated by a rotary pump equipped with a variable throttle valve. During experiments, He was supplied to the apparatus through a mass flow controller, and the pressure was kept at atmospheric pressure by adjusting the variable throttle valve.

As for plasma diagnostic study, we performed OES measurements, which can readily yield qualitative information on the chemical nature of the species that are present in the plasma. A fused-silica lens with a focal length of 150 mm was used to collect the total emission from the plasma through a quartz window. An optical fiber was used to guide the light from the plasma to a monochromator (NIKON, G-250) with a focal length of 250 mm and a resolution of 0.5 nm at 50 μm of slit width. Intensities of the spectra were detected by a photo-multiplier, and the signals were sent to a personal computer after A/D conversion. We measured the emission spectra in the wavelength range of 200–800 nm.

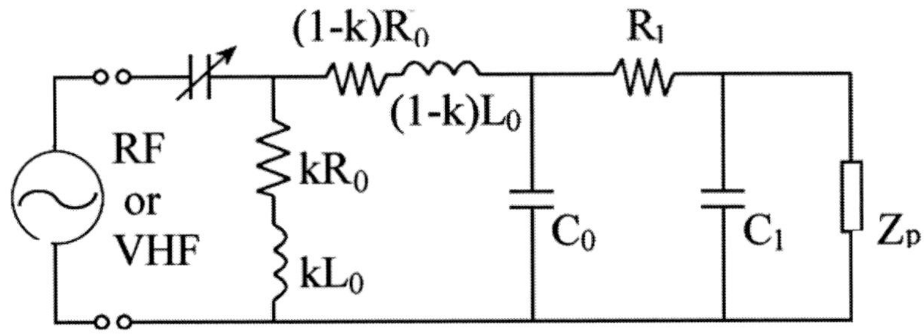

Figure 6. A simplified electrical circuit diagram of the matching network and the chamber, including resistive losses due to skin effect conduction and the parasitic capacitance in the chamber.

To obtain additional diagnostic information, plasma impedance was measured to compare the VHF and RF plasmas. Figure 6 shows a simplified electrical circuit diagram of the matching network and the apparatus. A variable capacitor C_0 and a fixed inductor L_0 make up the matching network. Resistive losses due to skin effect conduction in the matching network and the apparatus are accounted for by the series resistor R_0 and R_1, and the parasitic capacitance of the upper electrode with respect to the grounded apparatus wall is included in the parallel capacitor C_1. The plasma load is indicated by the impedance Z_p, having a real and an imaginary component, $Z_p = R + jX$. The matching networks for VHF and RF are designed to be able to normally match the power supply to the plasma load when plasma is present in the apparatus. The current and voltage waveforms were measured using current and voltage probes on the center conductor of the power line, which was connected with the upper electrode in the apparatus. The signals were fed to a Tektronix TDS644B oscilloscope. When VHF or RF power is applied, a part of the power is dissipated by the skin resistance. The values of skin resistance, R_0 and R_1 in Fig. 6, were estimated by comparing the calculation and measurement of the frequency dependences of amplitude reflectance of input power signals around 145 and 13.56 MHz without plasma operation. The amplitude reflectance of input power signals was measured by a network analyzer. Using the obtained values R_0 and R_1, and based on the current and voltage measurements and the reflectance of electric power during plasma operation, the plasma impedance could be calculated using the electrical equivalent circuit shown in Fig. 6.

3.2. OES Spectra

When plasma is generated in a mixture of an inert gas with trace quantities of a second gas, certain portions of the spectrum of the minority gas are strongly emitted in addition to the spectrum of the inert gas. The enhanced emission from the minority gas is due to electronic transitions originating from excited levels lying high above the ground state of the atom or molecule. This enhancement is firstly attributed to the high electron temperature in the plasma of the inert gas. It is generally recognized, however, that selective excitation occurs by a

Penning collision between a molecule of the minority gas and an inert gas atom in an energetic, long-lived metastable state. He has two such metastable states, 2^1S (20.6 eV) and 2^3S (19.8 eV), and additional long-lived highly energetic species such as the metastable helium molecule He_2 ($a^3\Sigma_u^+$, 14.6–17.4 eV), the molecular ion He_2^+ (18.8–21.6 eV), and the atomic ion He^+ (24.5 eV) [14]. In the Penning collision process, the energy transfer takes place so efficiently from the electrons in the active He species that even an impurity molecule with the amount of less than 1 ppm can be detected.

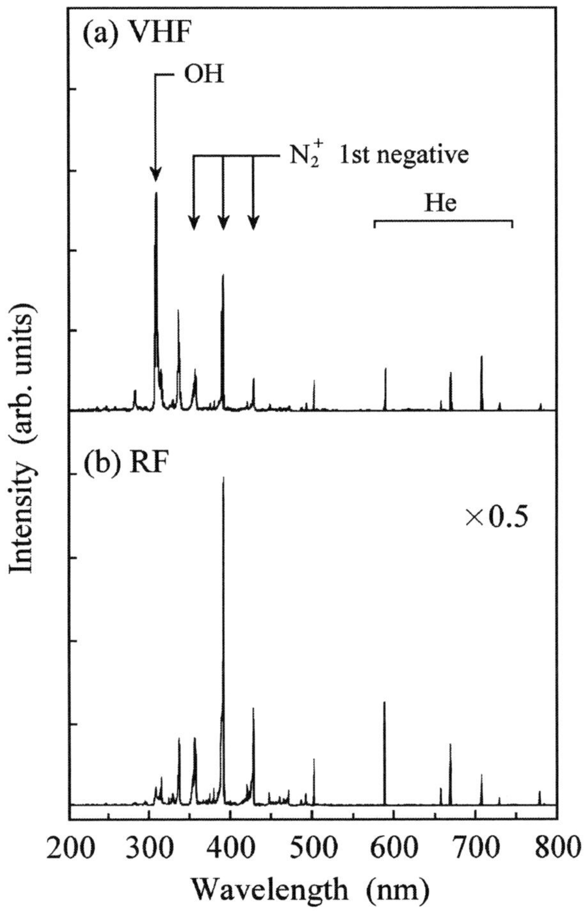

Figure 7. Optical emission spectra of the atmospheric pressure He plasma generated by VHF (a) and RF (b) excitations using the metal electrodes (without Al_2O_3 coating). The input power was fixed at 80 W.

Figure 7 shows the optical emission spectra of the atmospheric pressure He plasma generated by VHF (a) and RF (b) excitations using the metal electrodes (without Al_2O_3 coating). The power into the matching network is fixed at 80 W. In Fig. 7(a), emission lines from the first negative system of N_2^+ ($B_2\Sigma_u^+ \rightarrow X_2\Sigma_g^+$) at 427.8, 391.4 and 358.2 nm are seen, together with He peaks at 728.1, 706.5, 667.8, 587.6 and 388.9 nm. Stronger emission line at 308.9 nm is attributed to the transition in ($A_2\Sigma^+ \rightarrow X_2\Pi$) system of OH molecules, originating from the presence of water vapor in the atmosphere. The emission from the N_2^+ first negative system is caused by the Penning collisions of N_2 molecules with the metastable He (2^3S) atoms and He_2^+ ions, while the observed emission due to the OH ($A_2\Sigma^+ \rightarrow X_2\Pi$) system originates from this same Penning collision process, in addition, that with He^+ ions [14]. For the spectrum of RF plasma in Fig. 7(b), all the emission lines from the first negative system of N_2^+ and He atoms increase in intensity, while that from OH molecules drastically decreases. On the other hand, figure 8 shows the optical emission spectra of the atmospheric-pressure He plasma generated by VHF excitation using the metal (a) and Al_2O_3-coated (b) electrodes. The spectrum in Fig. 8(a) is the same one as that in Fig. 7(a). By insulating the electrode surfaces, the emission intensities decrease over the whole wavelength range.

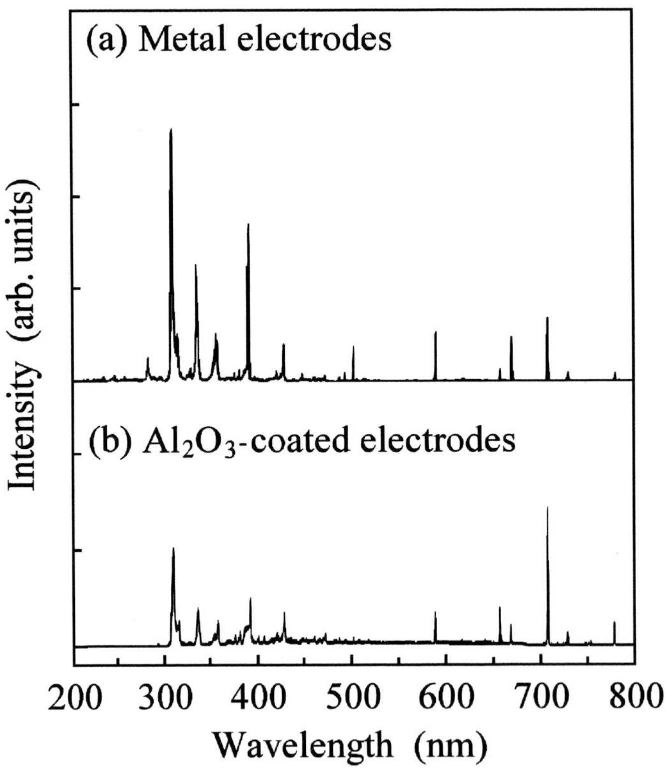

Figure 8. Optical emission spectra of the atmospheric-pressure He plasma generated by VHF excitation using the metal (a) and Al_2O_3-coated (b) electrodes. The input power is fixed at 80 W.

As described later, the plasma impedance depends on both excitation frequency and surface material of electrodes. Since plasma with a high impedance requires higher voltage to dissipate the same electric power as for low-impedance plasma, the high-impedance plasma circulate more current through the stray impedances, dissipating a greater fraction of the power outside the plasma. This means that it is difficult to discuss the results in Figs. 7 and 8 quantitatively when the input power to the matching network is held constant. However, the influence of excitation frequency on the properties of atmospheric-pressure plasma can be argued qualitatively as follows. From Fig. 7, although the density of metastable He atoms is considered to be lower in the VHF plasma than that in the RF plasma, the presence of strong emission from OH molecules at 308.9 nm implies that the VHF plasma has a considerable amount of He^+ ions, and thus, has higher electron temperature than the RF plasma. Since the electrode surfaces are not insulated in these experiments, secondary electrons can readily be emitted from the surfaces. However, it can be said from Fig. 2 that the VHF excitation enables the trapping of ionic species and even electrons in the gap region effectively. Thus, the density of electrons should be lower in the VHF plasma than in the RF plasma. These indicate that the ratio of the number of high-energy electrons to the total number of electrons is relatively larger in VHF plasma than in RF plasma when consuming the similar amount of energy in the plasma, supporting the speculation mentioned above. Taking into account these considerations, it is conceived that the impedance of the VHF plasma is higher than that of the RF plasma. When the Al_2O_3-coated electrodes are used, it is conjectured that the plasma impedance is more increased. Thus, the difference between the observed emission intensities in Fig. 8 may mainly be caused by the difference in the power dissipation inside the plasma. The influences of the excitation frequency and the surface material of the electrode on the plasma impedance are discussed in the following section.

3.3. Plasma Impedance

In general, for the investigation of the behavior of charged particles in plasma, Langmuir (electrostatic) probes have been widely used, where a probe electrode must be inserted into the plasma and biased with respect to one of the electrodes. In the case of the present study, however, the gap size between the electrodes is so narrow that Langmuir probes can hardly be used to characterize the atmospheric-pressure plasma without causing any disturbance on the plasma generation. Thus, we performed impedance analysis, which is generally recognized as a convenient and important diagnostic technique for plasma characterization. Combination of the measurements of the current and voltage waveforms applied to the electrode and the calculation using the electrical equivalent circuit on the basis of the apparatus geometry (Fig. 6) enabled us to derive the impedance Z of the plasma-filled capacitor. To quantitatively discuss the influences of the excitation frequency and the surface material of the electrode on the plasma impedance, we used the calculated power dissipated in the actual plasma, instead of the power input to the matching network.

Figure 9 shows the real (a) and imaginary (b) impedances of the atmospheric-pressure He plasma as a function of calculated power density, in which the VHF and RF excitations are compared using the metal electrodes. While, figure 10 shows the real (a) and imaginary (b) impedances of the plasma generated by VHF excitation, in which the effect of Al_2O_3 coating of the electrode surface is indicated. In Figs. 9 and 10, both real and imaginary impedances tend to decrease as the power is increased, indicating that the plasma becomes less resistive and less capacitive for the larger power. Comparing the VHF and RF plasmas in Fig. 9, the VHF plasma is more resistive and more capacitive than the RF plasma. This means that the emission of secondary electron from the electrode surfaces is more suppressed due to the trapping of charged particles at the higher excitation frequency, supporting the discussion in Fig. 7. When the electrode surfaces are insulated with Al_2O_3, the emission of secondary electron can be completely prevented, leading to the generation of more resistive and more capacitive plasma as evidenced in Fig. 10.

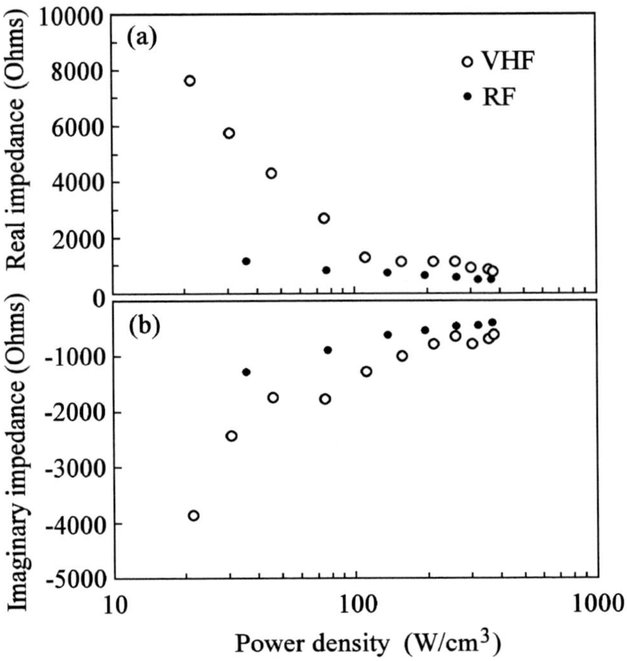

Figure 9. Real (a) and imaginary (b) impedances of the atmospheric-pressure He plasma generated using the metal electrodes as a function of calculated power density.

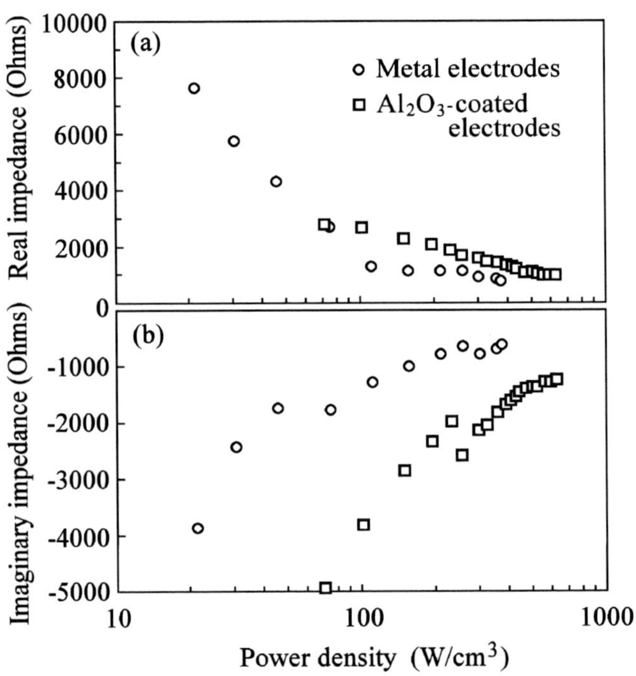

Figure 10. Real (a) and imaginary (b) impedances of the atmospheric-pressure He plasma generated by VHF excitation.

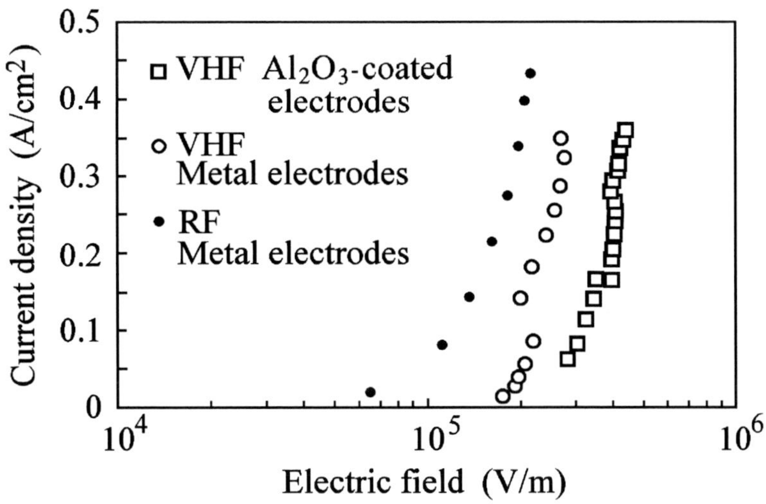

Figure 11. Current density – electric field characteristics of the atmospheric-pressure He plasma.

In the process of impedance analysis, current density–electric field (*J–E*) characteristics are easily obtained by viewing the atmospheric-pressure plasma as a frequency-dependent and lossy dielectric of relative permittivity. The results are shown in Fig. 11. Open squares and open circles in the figure indicate the data of plasmas generated by VHF excitation using the Al_2O_3-coated and the metal electrodes, respectively, and solid circles indicate the data of plasma generated by RF excitation using the metal electrodes. The *J–E* data in Fig. 11 give an approximation of the electron density, N_e, in the atmospheric-pressure plasma. The density of plasma current due to the drift of charged particles can be expressed as

$$J = Nqv_d = Nq\mu E = \sigma E, \tag{9}$$

where N, v_d and μ denote the density, drift velocity and mobility of charged particles, respectively, and σ is the electric conductivity of the plasma. Since the force which a charged particle receives from the electric field is equal to the transferred momentum per second by the drift, the equation of motion can be given as

$$qE = \nu m v_d, \tag{10}$$

where m is the mass of charged particle. Combining the equations (9) and (10), and taking into account that contribution of heavy ions to the conductivity can be neglected, σ can be written as

$$\sigma = N_e q \mu_e = N_e q v_d / E = N_e q^2 / \nu m_e. \tag{11}$$

In the case that a high frequency electric field is applied, the equation of motion corresponding to equation (10) is given by the following expression:

$$qE \exp(j\omega t) = \nu m_e v_d + m_e(dv_d/dt). \tag{12}$$

From equation (12), σ can be derived as

$$\sigma = N_e q^2(\nu - j\omega) / m_e(\nu^2 + \omega^2), \tag{13}$$

and thus, the absolute value of σ is obtained as follows:

$$|\sigma| = N_e q^2 / m_e (\nu^2 + \omega^2)^{1/2}. \tag{14}$$

Using equations (9) and (14), the values of N_e can be given on the basis of the data in Fig. 11. Since the collision frequency between electron and neutral particle, ν, usually varies with the particle velocity, the values of ν obtained from equations (4), (6), (7) and (8) for each E value in Fig. 11 have been used in the calculation. The results are shown in Fig. 12. As shown in the figure, the N_e values increase with increasing power density, but are in the range from 10^{10} to 10^{11} cm^{-3}. In the case of the high-temperature microwave plasma produced using He at atmospheric pressure by an axial injection torch, N_e ranges from 10^{14} to 10^{15} cm^{-3} for the microwave power density of the order of 10^5 W cm^{-3} [15]. Due to the extremely high power density, the gas temperature rises to approximately 3000 °C. It is also shown that the N_e value is mainly related to the incident power, not to the gas flow rate [15,16]. Assuming that N_e is proportional to the power density, the N_e values obtained in this study (Fig. 12) are considered to be reasonable for nonequilibrium plasma generated at atmospheric pressure. Although the actual gas temperature is unknown in this study, the low density of incident power (20–600 W cm^{-3}) should lead to far lower gas temperatures.

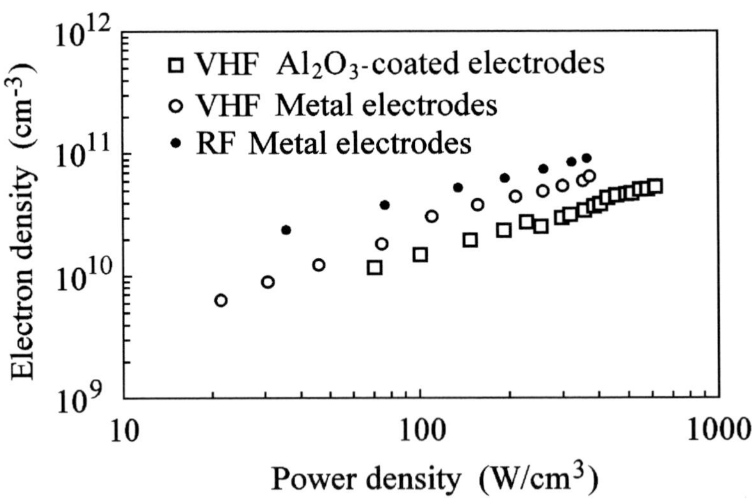

Figure 12. Electron density in the atmospheric pressure He plasma as a function of calculated power density.

In the present impedance analysis, we assume infinite and homogeneous plasma, which is made up of free electrons, immobile ions and neutral particles, and confined between the electrodes. To determine the actual characteristics of the plasma, a more complete model of the plasma including diffusion losses and the effects of sheaths should be used. Such finite-geometry effects, however, are difficult to model fully. The precise physical interpretation for the apparatus is also difficult due to the distributed nature of the stray impedances.

Although uncertainties are included in the obtained plasma properties, we can draw some conclusions from the data as follows. From the above mentioned results, the electron density is smaller in the VHF plasma than in the RF plasma. In addition, if the electrode surfaces are coated by an insulating material, the electron density is further decreased. Thus, to increase the density of active radicals in excited states, RF excitation combined with the use of metal electrodes may be the most efficient. However, in the case of depositing functional thin films it is necessary to generate homogeneous and stable plasma on a large-sized substrate. In particular, suppressing plasma gas temperature is important when substrates made of low-melting-point materials must be used. In this context, atmospheric-pressure plasma with a lower density of charged particles is desirable. Therefore, we conclude that adopting VHF excitation and using a surface-insulated electrode are essentially important for the practical applications of atmospheric-pressure plasma, as evidenced in section 5.4.

4. Atmospheric-Pressure Plasma CVD System

Utilizing the information obtained from the fundamental characterizations of atmospheric-pressure plasma, we have developed the AP-PCVD system, which is schematically shown in Fig. 13 [1,2,4]. The AP-PCVD system consists of a reaction chamber, a load lock chamber and a gas circulation system. In Fig. 13(a), a cylindrical rotary electrode with a diameter of 300 mm and a width of 100 or 200 mm is placed in the reaction chamber. The electrode surface is coated with Al_2O_3 with thickness of 0.15 mm to prevent the emission of secondary electrons from the surface. An electrode without Al_2O_3 coating is also prepared to investigate the influence of electrode surface material on the film properties. The rotary electrode is driven by a high-speed rotation motor drive at a maximum rotation speed of 5000 rpm (the velocity of the electrode surface is approximately 79 m/s). To maintain cleanliness of the process ambience, the motor drive, which contains organic lubricant, is placed outside the chamber. The shaft of the motor is introduced into the chamber with a ferrofluidic seal and connected with the shaft of the electrode using magnetic coupling. The internal volume of the AP-PCVD system is approximately 500 liters.

A substrate is loaded into the reaction chamber through the load lock chamber. We use two kinds of substrate holders to vacuum-chuck and heat the substrate. The holders can be scanned horizontally by the X-stage system. The scanning speed can be controlled in the range from 0.1 to 100 mm/s. In the case of depositing hydrogenated amorphous Si (a-Si:H) films, a substrate holder made up of a TiN-coated copper plate is used to ensure sufficient thermal conductance. The maximum heating temperature of the holder is 400 °C. On the other hand, to prepare polycrystalline Si (poly-Si) and epitaxial Si (epi-Si) films, another holder made of SiC-coated graphite is used to be able to heat the substrate up to 1000 °C. The gap size between the substrate and the electrode surface (deposition gap) can be adjusted by the stepping-motor-drive Z-stage system. Four ball-spline shafts are used as guides for the vertical motion. Therefore, the sufficient rigidity is ensured in adjusting the deposition gap at an accuracy of 10 μm. The vacuum system consists of a turbo molecular pump (1000 liter/s) and a dry roughing pump (1500 liter/min), which exhausts the reaction chamber to a pressure less than 1×10^{-4} Pa.

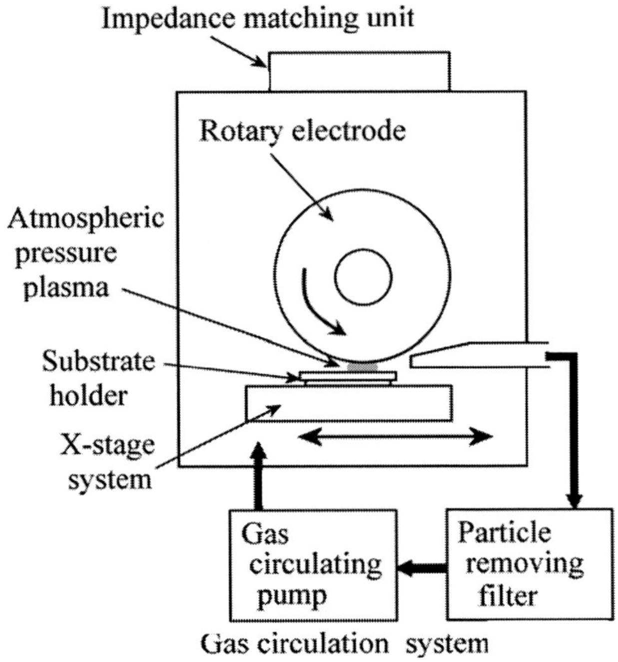

Figure 13. Schematic illustration of the AP-PCVD system.

150-MHz VHF power can be supplied to the rotary electrode through an impedance matching unit. Atmospheric-pressure plasma is confined in the gap region between the rotary electrode and a substrate. The structure of the impedance matching unit is designed to be able to minimize the skin resistance. The matching unit gives a minimum reflected power as measured by an in-line wattmeter, and thus, the AP-PCVD system plus matching unit approximates to a resistive load of 50 Ω when plasma is generated during deposition. The incident power is also measured using an in-line power meter. A Si film is deposited in a rectangular region without substrate scanning, corresponding to the plasma size. Figure 14 shows a photograph of the atmospheric-pressure plasma (side view) and a thickness profile of the a-Si:H film after 20 s of deposition. The plasma is generated in typical deposition conditions of a-Si:H films: SiH_4 concentration of 0.1%, H_2 concentration of 1%, VHF power of 500 W, electrode rotation speed of 5000 rpm, and deposition gap of 0.3 mm. The atmospheric-pressure plasma generated in the gap region stretches to some extent (plasma length). In Fig. 14(a), the plasma length is 30 mm, and the width of plasma in the direction of electrode axis (plasma width) is 100 mm, corresponding to the width of the substrate. The plasma length changes with the source gas concentration and the VHF power. The deposition rate is the highest at the position near the smallest gap, and decreases on both sides of it [Fig. 14(b)]. In this paper, the highest deposition rate near the smallest gap is mainly described. By scanning a substrate, a uniform film can be formed in the area having the width of plasma and the substrate scanning distance. The average deposition rate is calculated from the film thickness and the substrate scanning time.

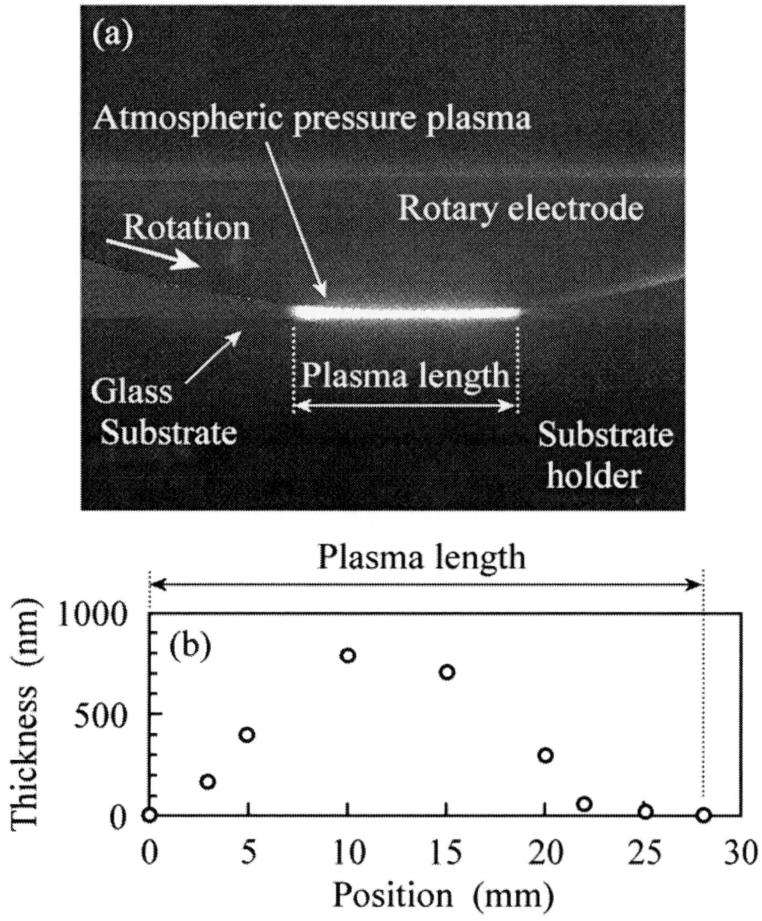

Figure 14. (a) Photograph of the atmospheric-pressure plasma confined between the rotary electrode and a glass substrate. (b) Thickness profile of the a-Si:H film after 20 s of deposition without substrate scanning. The plasma is generated in typical deposition conditions of a-Si:H films (SiH_4 = 0.1%, H_2 = 1%, electrode rotation speed = 5000rpm, and VHF power = 500 W).

The gas circulation system is connected with the reaction chamber to collect and remove particles generated by the chemical reactions in the plasma as shown in Fig. 13. The gas circulation system has a particle removing filter and a gas circulating pump. The gas circulating pump sucks the gases in the reaction chamber out through the duct behind the electrode. The front end of the duct is located between the electrode and a substrate to collect particles that float around the plasma region.

5. Silicon Thin Film Deposition by Atmospheric-Pressure Plasma CVD

5.1. Deposition and Characterization Methods

In a series of experiments, 0.7-mm-thick Corning #1737 glass plates in the size of 100 × 100 mm^2 were used for the preparation of a-Si:H films. On the other hand, thermally oxidized and bare wafers of (001) B-doped CZ-Si (ϕ 4 inch) with the resistivity of 10–20 Ωcm were respectively used to prepare poly-Si and epi-Si films. The thickness of the thermal oxide was approximately 150 nm. Before deposition, the glass substrates and thermally oxidized Si wafers were cleaned with ultra-pure water in an ultrasonic bath and with ozone water to remove organic contaminations. The bare Si wafers were cleaned by the room temperature wet cleaning process [17] shown in Table I. After a substrate was loaded into the reaction chamber, the chamber was exhausted to a pressure less than 1×10^{-4} Pa and then filled with the process gas mixtures containing He, H_2 and SiH_4 to the atmospheric pressure. He and H_2 used in this study were purified by gas purifiers to impurity levels less than 1 ppb, and the purity of SiH_4 was 99.9999% (electronic grade). To maintain the gas composition and the pressure, each process gas was supplied to the chamber through a mass flow controller, and the excessive gases in the chamber were exhausted at a constant flow rate using a variable throttle valve.

The deposition conditions of Si films are listed in Table II. The substrate temperature (T_{sub}) was monitored by a Chromel-Alumel thermocouple embedded in the substrate holder. In the case of using the substrate holder made of graphite, T_{sub} was determined by directly measuring the surface temperature of the Si wafer using an optical pyrometer, because the thermal conductivity of graphite is far smaller than that of copper. Since the optical pyrometer was unusable in a plasma atmosphere, we measured the surface temperature immediately after the plasma was turned off. The measurement error of T_{sub} deduced from the scatter of the measured values was within ± 3%. The temperatures of the substrate holder measured by the thermocouple were calibrated by the surface temperatures measured by the optical pyrometer. In addition, the temperature distribution of the substrate surface was measured by thermocouples for both substrate holders, and it was confirmed that the dispersion of surface temperature was within ± 4% even in the case that the holder was scanned with electrode rotation. Thus, we considered that the surface temperature of the substrate in the plasma area was almost constant during deposition. Experiments were conducted at various T_{sub} varying H_2 and SiH_4 concentrations and input VHF power. Optimized values were employed for the deposition gap. The efficiency in transforming the initial source gas passing through the deposition gap into useful film varied depending on the electrode rotation speed; the higher efficiency could be obtained at the slower rotation speed and the smaller deposition gap. Within the present deposition conditions, it could be estimated that nearly 10% of SiH_4 was used for Si film production. Si film was also deposited on the surface of the rotary electrode after repeated experiments. Since excessive Si deposition on the electrode might affect the Si deposition process, the electrode surface was cleaned periodically by NF_3 or H_2 plasma etching.

Table I. Cleaning procedure of Si wafers.

Process	Concentration	Time (min)
Ozonated ultrapure water	5 ppm	10
$HF/H_2O_2/H_2O$ + Surfactant	HF: 0.5 %, H_2O_2: 0.1 %	10
Ozonated ultrapure water	Surfactant: 50ppm	10
	1 ppm	
Diluted HF	0.1 %	15
Ultra pure water		10

Table II. Preparation conditions of Si films by AP-PCVD.

Material	a-Si:H, μc-Si:H	Poly-Si	Epi-Si
Base gas	He	He	He
H_2 concentration (%)	0.5 – 15	0 – 30	0 – 30
SiH_4 concentration (%)	0.01 – 1	0.001 – 0.1	0.1
Electrode rotation speed (rpm)	1700 – 5000	2000	2000
Substrate temperature (°C)	160 – 280	500	500 – 700
Process pressure (Pa)	1×10^5	1×10^5	1×10^5
Deposition gap (mm)	0.2	1	0.7
VHF power (W)	250 – 1200	500 – 2500	300 – 3000

The thickness of the a-Si:H films deposited on glass substrates was determined directly by observing the cross-sectional scanning electron microscope (SEM; Hitachi S-800) image, while that of the poly-Si and epi-Si films was calculated from the weight increase of the substrate wafer and the area of the deposited film. The morphology of the film surface was observed by SEM or atomic force microscope (AFM; Shimadzu SPM-9500). The initial photoconductivity (σ_{ph}) and dark conductivity (σ_d) of the a-Si:H films were measured at room temperature in the Ohmic region with a coplanar cell using aluminum electrodes with 1mm spacing. To know the applicability of the AP-PCVD technique to the practical use, the a-Si:H films were applied to the intrinsic layers of a-Si:H thin film solar cells. The measurements of σ_{ph} and the initial I–V characteristics of the solar cells (cell area: $1 cm^2$) were conducted under AM1.5, 100 mW/cm^2 illumination. The optical band gap (E_{opt}) of the a-Si:H films was derived together with the film thickness from reflection and transmission spectra according to the method free from the optical interference effect [18]. Fourier transform infrared (IR) absorption spectroscopy was used to investigate the hydrogen content (C_H) of the films. The measurements of IR absorption were performed with a FTIR spectrometer (Shimadzu FTIR-8600PC) in the wavenumber range of 500–4000 cm^{-1}. C_H was determined from the absorption coefficients related to Si–H (2000 cm^{-1}) and Si–H_2 (2090 cm^{-1}) stretching modes [19]. To clarify the crystallographic orientation of the poly-Si films, X-ray diffraction analysis (XRD) using Cu-K_α radiation was performed in the θ–2θ geometry. The crystallinity of the epi-Si films was investigated by reflection high-energy electron diffraction (RHEED). Transmission electron microscopy (TEM) was used to observe the cross section of the Si layer and the Si/substrate interface for both poly-Si and epi-Si films. Both RHEED and TEM observations were conducted using the same electron microscope (JEOL JEM-2000FX) operated at 200 keV with two types of attachments. The concentration of impurity atoms in the Si films was measured by the secondary ion mass spectrometer (SIMS; CAMECA IMS-5f). A Cs ion was used as the ion source for sputtering, the primary ion energy was 14.5 keV, and the beam current was 50 nA.

The cleanliness of the process ambience in the AP-PCVD system was evaluated using atmospheric-pressure ionization mass spectrometer (APIMS; Hitachi Tokyo Electronics UG-510P). By using APIMS, it is possible to quantitatively analyze ultra-low concentration of impurities contained in inert gas at atmospheric pressure. For the quantitative measurements of impurity gases in He, we conducted calibration measurements using a standard He gas (BBB Neriki Valve Co. Ltd.) that contained a known amount of analyte gases (N_2, O_2, H_2O, CH_4 and CO_2). The APIMS spectrometer was operated at 610 V discharge voltage, 5 μA discharge current and 1 *l*/min sample gas flow rate. The detection limit for H_2O was 5 ppt and that for most of the other impurities was approximately 50 ppt. When SiH_4 is contained in the analyzing gas, Si may be deposited on the discharge electrode, which deteriorates the APIMS performance. Therefore, the analysis of the process ambience was carried out by introducing only He into the chamber. While analyzing the process ambience, the gas circulation, the substrate heating, and the process gas (only He) supply were done in the same manner as that in the Si deposition process.

5.2. Deposition Characteristics of Si Films by AP-PCVD

It is generally understood that various processes are included in the plasma-enhanced Si film deposition, such as decomposition of reactive gas molecules in gas phase, hydrogen elimination, adsorption of film-forming precursors, and Si-Si bond formation at the film-growing surface. In the conventional PECVD process, the mobility of surface reactive species has often been considered as an important factor for the deposition of high quality films [20]. This means that higher quality films can be formed at a higher T_{sub} or a lower deposition rate, where the surface reactions (hydrogen elimination and Si-Si bond formation) are sufficiently enhanced. In the case of AP-PCVD process, however, the density of film-forming precursors in the plasma is much higher than that in PECVD process. Thus, the deposition rate can be remarkably increased, and hence the time for the surface reaction becomes fatally short. Therefore, we have firstly investigated the morphologies of Si films under high-rate deposition conditions.

Figure 15 shows a SEM micrograph of a cross section of the a-Si:H films deposited at 200 °C. The SiH_4 and H_2 concentrations were 0.1% and 1%, respectively, and the VHF power was 500 W. The electrode rotation speed was 5000 rpm, the deposition gap was 0.3 mm, and the gas circulation rate was 340 *l*/min. These are the same conditions as those of the film shown in Fig. 14. In these conditions, the deposition rate is approximately 40 nm/s, which is more than one order of magnitude larger than that by conventional PECVD [20]. The substrate was not scanned, and a central portion of the film was observed. As shown in the micrograph, the Si film exhibits a smooth cross-sectional morphology. Since SiH_4 molecules do not thermally decompose at 200 °C, it is suggested that the film grows solely from the film-forming precursors generated in the plasma, and that particulate contamination takes no part in the film growth. It should be noted that the atmospheric-pressure plasma supplies considerable physical and/or chemical energies to the film-growing surface, which effectively enhances the surface reactions despite the extremely high deposition rate.

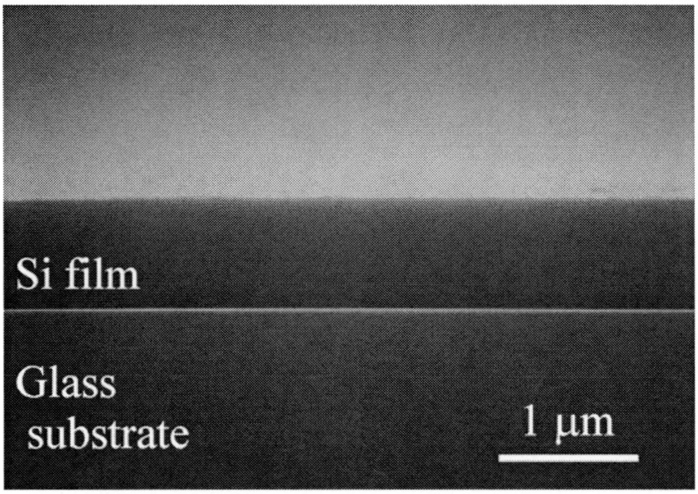

Figure 15. Cross-sectional SEM micrograph of the a-Si:H film deposited at 200 °C. The deposition conditions are the same as those of the film shown in Fig. 14. The gas circulation rate was 340 *l*/min.

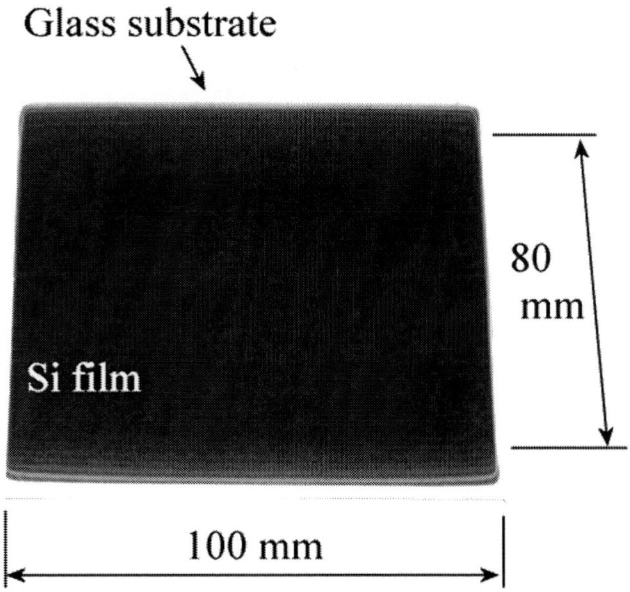

Figure 16. Photograph of the a-Si:H film formed in the same conditions as those of the film shown in Fig. 14 with substrate scanning at 1 mm/s.

Figure 16 shows a photograph of the a-Si:H film deposited in the same conditions as those in Fig. 15 with substrate scanning at 1 mm/s. The direction of the substrate scan is the same as that of the gas flow. The film width of 100 mm and the length of 80 mm correspond to the plasma width and the substrate scanning distance, respectively. The film thickness is approximately 600 nm with the variation of 5%. From this result, it is confirmed that a uniform thin film can be formed in the area with the plasma width and the substrate scanning distance. Adopting a wider electrode and a longer substrate scanning distance may enable the formation of a thin film in a larger area.

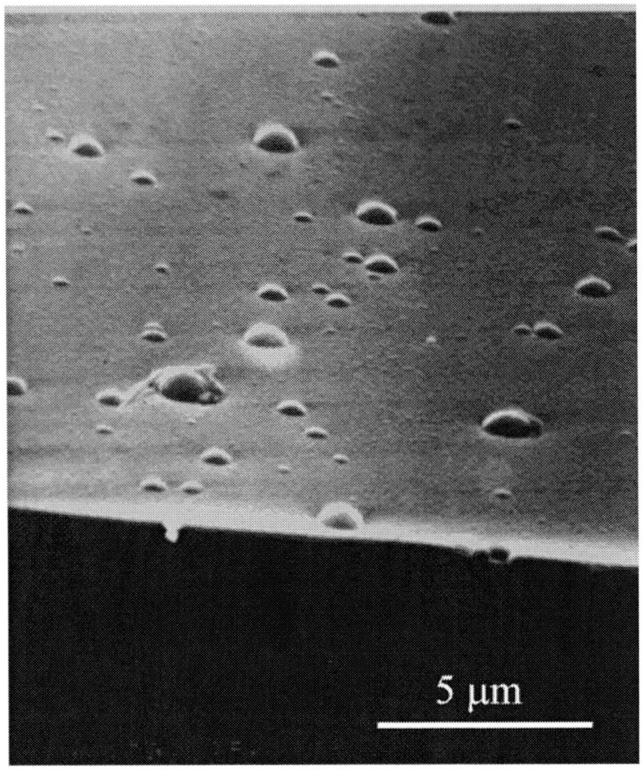

Figure 17. SEM micrograph of the surface of the a-Si:H film shown in Figure 16.

Considering the uniformity of the substrate temperature during deposition, it is expected that there is no change in the film morphology when the substrate is scanned. From SEM observations (Fig. 17), however, a large number of Si particles with sizes larger than 0.1 μm were observed on the surface of the a-Si:H film shown in Fig. 16. For the film deposited without substrate scanning (Fig. 15), no particles are observed on the film surface. These suggest that the generation of particles is occurring in the outside of the plasma region. When the substrate is scanned in the same direction as the gas flow, the film in the upstream region forms the bottom side of the resulting film, while the film in the downstream region forms the surface side. Thus, it is considered that Si particles generated near the plasma edge on the downstream side adhere to the film surface. Although the details of the formation mechanism of particles are unknown at the present stage, it is essential that the film-forming precursors (SiH_n species), which are not consumed for the film deposition, are rapidly condensed at the exit of the plasma region. Increasing plasma length by expanding the diameter of rotary electrode may lead to the enhancement of the consumption of SiH_4 molecules in the plasma and contribute to the decrease in the amount of particles.

To avoid particulate contamination of the film, it is necessary to efficiently suck the flow of gas passing through the deposition gap. In the plasma region, driven by the viscous flow around the rotating electrode, the process gas is pushed into the narrow gap. The results of the numerical simulation of gas flow have shown that a low-pressure area appears behind the minimum gap, causing complex and turbulent gas flow [21]. Thus, the increment in gas circulation rate together with the optimization of the structure of gas suction duct behind the electrode is indispensable to remove particles. Figure 18 shows a SEM micrograph of the a-Si:H film deposited with an increased gas circulation rate of 1200 *l*/min. The deposition conditions are the same as those of the film shown in Figs. 16 and 17. Comparing the micrographs in Figs. 17 and 18, it is obvious that both the size and the number density of particles markedly decrease with increasing gas circulation rate. Investigating a significant number of films deposited with various deposition parameters at gas circulation rates larger than 1000 *l*/min, we have found that condensed particles with sizes larger than 0.1 μm are not noticeable on the surface of the films deposited with substrate scanning, even when SiH_4 concentration is increased to 5%.

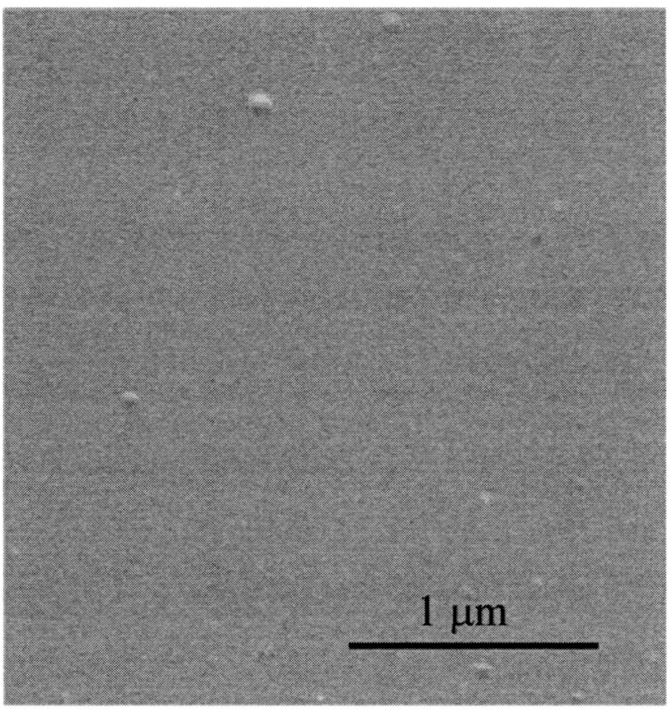

Figure 18. SEM micrograph of the surface of the a-Si:H film deposited with an increased gas circulation rate of 1200 *l*/min. The deposition conditions are the same as those of the film shown in Figs. 16 and 17.

5.3. Cleanliness of the Process Ambience

The electrical property of Si films are influenced by the incorporation of impurities, such as O, N and C. Water vapor, which is one of the origins of O impurity, is a major factor to cause deterioration of Si film quality, and thus, the concentration of H_2O in process ambience must be sufficiently reduced. For a-Si:H films, it is reported that the O incorporation more than 10^{19} cm^{-3} leads to the increase in the density of charged dangling bonds and deteriorate the film performance [22]. In the case of epi-Si growth by conventional thermal CVD, since H_2O

causes the oxide formation on the Si surface, there is a close correlation between the lowest temperature required for epi-Si growth and H_2O concentration in the atmosphere [23]. For example, H_2O concentration must be below 100 ppb for the epi-Si growth at 800 °C, below 6 ppb for 700 °C, and below 0.2 ppb for 600 °C. As for the AP-PCVD system, water vapor adsorbed on the internal surfaces of the reaction chamber and the tubes of the gas circulation system can be baked-out by heating the walls under a vacuum condition. However, the structures of the reaction chamber and the gas circulating pump are so complicated that it is very difficult to sufficiently increase the temperature of the entire surface of the AP-PCVD system only by heating the walls. This is a serious problem, because water vapor is desorbed from the internal structures and heavily contaminates the process ambience when we start the gas circulation with heating a substrate for deposition. Therefore, it is indispensable to remove water vapor in the inside of the reaction chamber and gas circulating pump for the preparation of high-quality films by AP-PCVD.

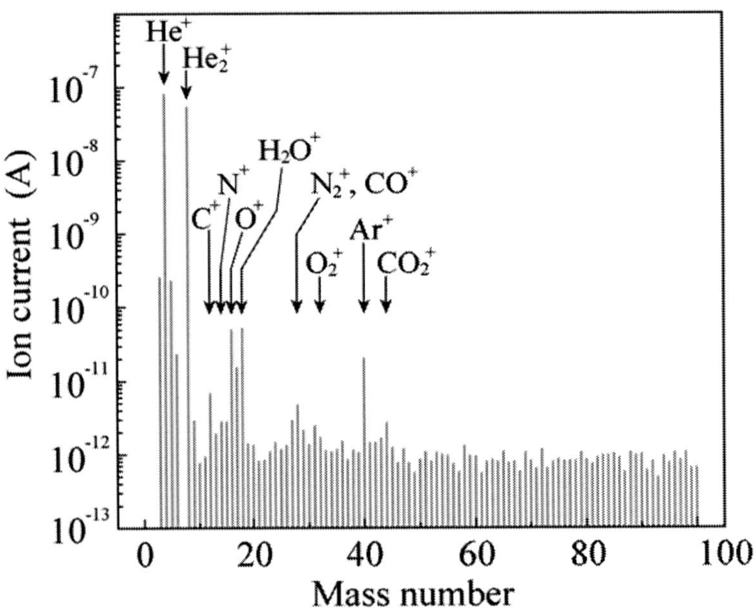

Figure 19. Mass spectrum of ultra-pure He gas used in this study measured by APIMS.

Figure 19 shows the mass spectrum of He before introducing into the chamber measured by APIMS. The horizontal axis is the mass number, and the vertical axis is the ion current proportional to the gas concentration. The main impurities in He are H_2O, N_2, O_2 and CO_2 as seen in Fig. 19. The concentration of each impurity gas is $[H_2O]$ = 0.9 ppb, $[N_2]$ = 0.6 ppb, $[O_2]$ = 0.2 ppb and $[CO_2]$ = 0.1 ppb.

After exhausting the AP-PCVD system to a pressure less than 1×10^{-4} Pa with heating the walls for 48 hours, ultrapure He was introduced, and then the process ambience was analyzed by APIMS with and without gas circulation. The gas circulating pump used in the measurements was a dry roughing pump (1200 *l*/min) generally used in conventional plasma processes. Figure 20 shows the H_2O concentrations in the process ambience measured with and without heating the walls of the AP-PCVD system at 80 °C as a function of measuring time. The substrate holder made of TiN-coated copper plates was maintained at 220 °C. The time when the gas circulation is started is indicated by arrows in the figure. In Fig. 20, the H_2O concentrations rapidly increase with time and are saturated at 598 ppb with wall heating and at 310 ppb without wall heating. The drastic increase in the H_2O concentration originates from the internal structures of the reaction chamber. When the gas circulating pump is started, further increments in the H_2O concentration to 2004 ppb with wall heating and to 610 ppb without wall heating are observed. These show that there remains a large amount of H_2O adsorbed on the surfaces of the internal structures of both the reaction chamber and the gas circulating pump, which are not removed only by heating the walls under a vacuum condition.

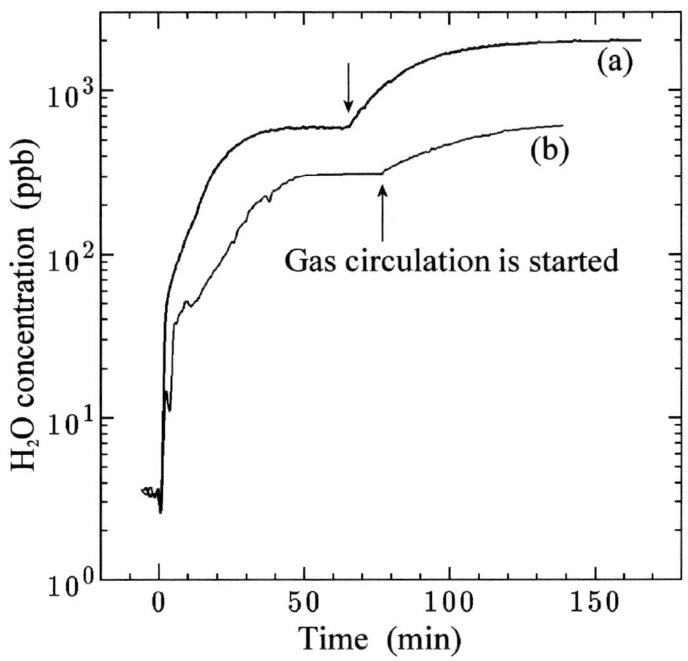

Figure 20. H_2O concentrations in the process ambience before gas circulation cleaning measured by APIMS with (a) and without (b) heating the walls of the AP-PCVD system at 80 °C as a function of measuring time.

To remove H_2O from the surfaces of the internal structures, we have utilized the reactive nature of H_2O with SiH_4. H_2O reacts with SiH_4, generating disiloxane and H_2 according to the following formula [24]:

$$2SiH_4 + H_2O \rightarrow SiH_3\text{-}O\text{-}SiH_3 + 2H_2. \quad (15)$$

In addition, a polymer siloxane derivative is simultaneously generated as written by the formula

$$SiH_4 + SiH_3\text{-}O\text{-}SiH_3 + H_2O \rightarrow SiH_3\text{-}O\text{-}SiH_2\text{-}O\text{-}SiH_3 + 2H_2, \quad (16)$$

which induces the generation of solid particles [24]. Since these reactions occur at room temperature, it is considered that H_2O, which remains adsorbed on the internal surfaces of the AP-PCVD system, can be efficiently removed by gas circulation using a high partial pressure of SiH_4 (gas circulation cleaning).

Gas circulation cleaning was conducted using a gas mixture containing He and SiH_4 at atmospheric pressure. To increase the gas temperature for the activation of reaction between H_2O and SiH_4, the walls of the AP-PCVD system and the substrate holder were heated at 80 and 220 °C, respectively. The SiH_4 concentration was 5% and the cleaning time was varied as a parameter. Particles generated by the polymerization reaction expressed in formula (16) can be trapped by the particle-removing filter of the gas circulation system. After gas circulation cleaning, the process ambience was analyzed by APIMS in the same manner as that in Fig. 20. Figure 21 shows the H_2O concentrations in the process ambience measured with and without wall heating as a function of measuring time after 70 h gas circulation cleaning. The H_2O concentrations increase with time and are saturated at 196 ppb with wall heating and at 93 ppb without wall heating. When the gas circulating pump is started, the H_2O concentrations increase to 381 ppb with wall heating and to 173 ppb without wall heating. The saturated concentrations of H_2O are apparently lower than those measured before the gas circulation cleaning shown in Fig. 20.

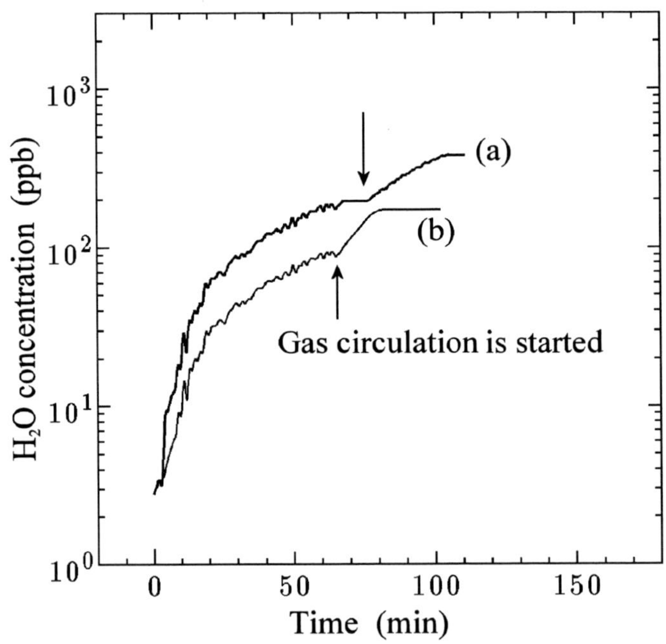

Figure 21. H_2O concentrations in the process ambience after 70 h gas circulation cleaning measured by APIMS with (a) and without (b) heating the walls of the AP-PCVD system at 80 °C as a function of measuring time.

To verify the influence of H_2O concentration in the process ambience on the film quality, we prepared a-Si:H films before and after 70 h gas circulation cleaning and measured the O content and the electrical properties of the films. The deposition conditions were as follows; the SiH_4 and H_2 concentrations were 0.1% and 1%, respectively, the VHF power was 500 W, the electrode rotation speed was 5000 rpm, the deposition gap was 0.3 mm, T_{sub} was 220 °C, and the gas circulation rate was 1200 *l*/min. The film deposition before the gas circulation cleaning was conducted with heating the walls of the AP-PCVD system at 80 °C, while the deposition after the gas circulation cleaning was performed without wall heating. In both cases, the deposition rate was almost same (approximately 40 nm/s).

Figure 22 shows the comparison between the O contents in the a-Si:H films. The H_2O concentrations in the process ambience deduced from Figs. 20 and 21 are also shown in the figure. By the gas circulation cleaning, an a-Si:H film having a sufficiently low O content of 5 $\times 10^{18}$ cm^{-3} is obtained. In addition, it is clearly observed in Fig. 22 that the O content in the film decreases closely correlating with the decrease in the H_2O concentration. This indicates that H_2O is the main factor to cause the O incorporation into the a-Si:H films. It is considered that the content of O atoms in a film can be more suppressed by the further reduction of H_2O concentration in the process ambience.

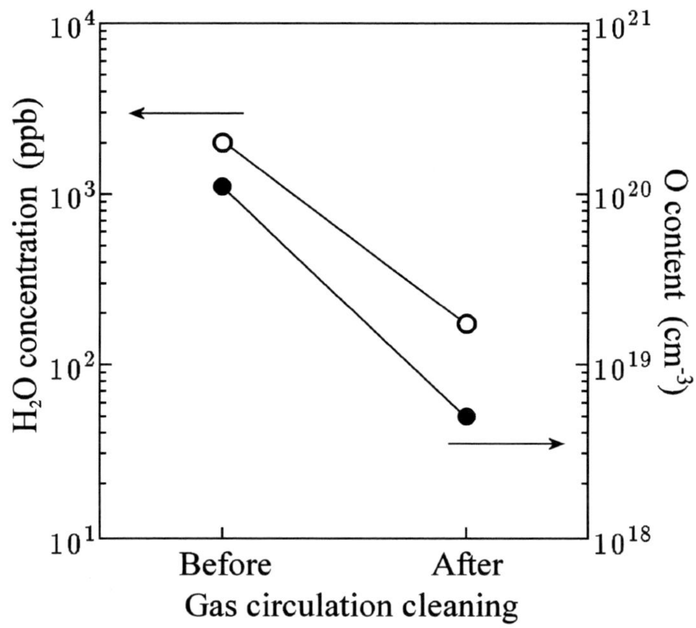

Figure 22. Correlation between H_2O concentration in the process ambience and O content in the a-Si:H film.

Figure 23 shows σ_{ph} and σ_d of the same films as those shown in Fig. 22. For the film deposited before the gas circulation cleaning, σ_{ph} and σ_d are 8.1×10^{-5} and 3.2×10^{-10} $\Omega^{-1}cm^{-1}$, respectively, and the value of photosensitivity (σ_{ph}/σ_d) is 2.5×10^5. After the gas circulation cleaning, although no remarkable change is observed in σ_{ph} (6.0×10^{-5} $\Omega^{-1}cm^{-1}$), σ_d markedly decreases to 2.4×10^{-11} $\Omega^{-1}cm^{-1}$, and thus, the photosensitivity greatly increases to 2.5×10^6. It is reported that some of O atoms incorporated in the film are in the form of positively charged threefold coordinated O (O_3^+) [25]. This plays a role of donor in an a-Si:H film and results in an upward shift of the Fermi level, causing an increase in σ_d. Thus, it is considered that the decrease in O content causes the increase in the activation energy for the electrical conduction, leading to the decrease in σ_d as shown in Fig. 23. The values of σ_{ph} and σ_d of the a-Si:H film deposited after the gas circulation cleaning are almost equivalent to those of a-Si:H films deposited by conventional PECVD [20]. Therefore, we consider that the AP-PCVD system we have developed is capable of forming a device-quality a-Si:H film at an extremely high rate of 40 nm/s.

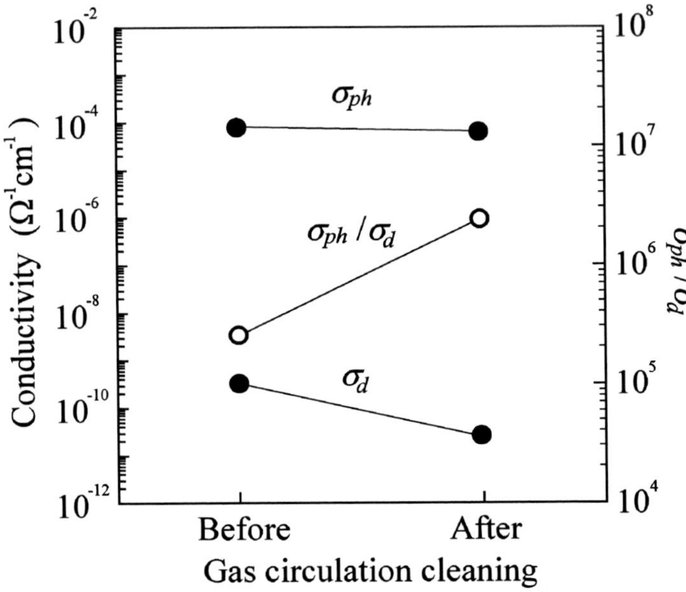

Figure 23. σ_{ph}, σ_d and σ_{ph}/σ_d of the a-Si:H films prepared in the same conditions as those of the film shown in Fig. 18, before and after gas circulation cleaning.

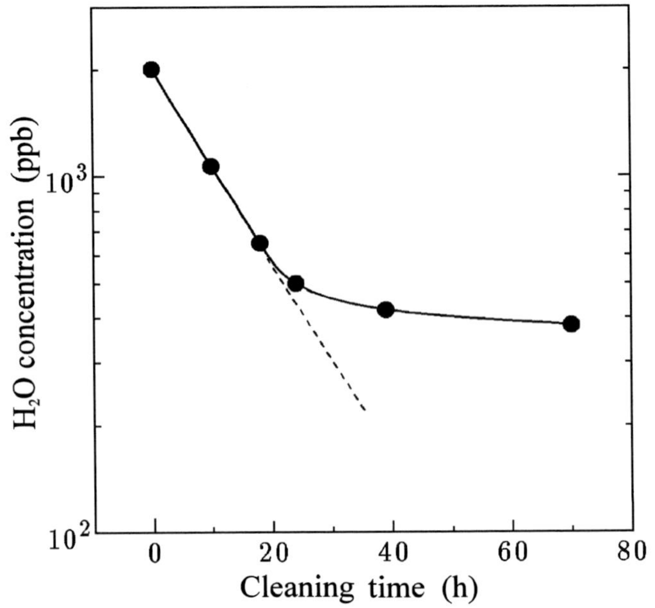

Figure 24. H_2O concentrations in the process ambience measured by APIMS with heating the walls of the AP-PCVD system at 80 °C as a function of time of gas circulation cleaning.

Figure 24 summarizes the effect of gas circulation cleaning. The H_2O concentrations at 0 and 70 h are the same as the saturated concentrations with gas circulation in Figs. 20(a) and 21(a), respectively. In Fig. 24, the H_2O concentration decreases with increasing cleaning time exponentially, and it tends to approach to a certain value for the cleaning time longer than 24 hours. We consider that increasing heating temperature of the walls of the AP-PCVD system is effective to shorten the cleaning time. However, it seems to be difficult to reduce the H_2O concentration in the process ambience further only by gas circulation cleaning. This suggests that local areas that cannot fully be purged exist in the AP-PCVD system. The most probable source of H_2O generation is the local area around the ball bearings of the motor shaft of the gas circulating pump. Therefore, the realization of a cleaner process ambience, particularly important for epi-Si growth, is considered to be difficult as far as using a dry roughing pump for the circulation of gases in the AP-PCVD process.

To improve the cleanliness of the process ambience, we made a new reaction chamber of which internal surfaces were electro-polished. Moreover, we developed an ultraclean gas circulating pump, having an active magnetic bearing (AMB) system to ensure the non-contact and lubricant-free operation. The AMB system was equipped with an axial magnetic bearing, consisting of a solid rotor disk placed between a pair of facing stator actuators, and two pairs of radial magnetic bearings. The surface materials of all internal structures of the pump were made of stainless-steel or aluminum to avoid the contamination in circulating gas. The gas circulating ability of the pump was approximately 1500 *l*/min for He. Using the new reaction chamber and the ultraclean gas circulating pump, we conducted APIMS measurements of the process ambience in the same manner as those in Figs. 20 and 21. Before measurement, H_2O adsorbed on the internal surfaces of the reaction chamber and the gas circulation system was baked-out by heating the walls to approximately 100 °C for more than 24 h under a high-vacuum condition. In addition, the local areas around the bearings of the electrode shaft in the reaction chamber were exhausted during measurement.

Figure 25 shows the mass spectrum of He in the AP-PCVD system. In the measurement, the substrate holder, made of SiC-coated graphite, was heated up to 690 °C. Comparing the spectrum in Fig. 25 with that in Fig. 19, the concentration of each impurity gas is increased to $[H_2O] = 3.1$ ppb, $[N_2] = 35$ ppb, $[O_2] = 4.5$ ppb and $[CO_2] = 2.6$ ppb. Among these impurities, H_2O may have originated from the internal surfaces of the chamber and the gas circulation system. The increases in the concentrations of O_2 and CO_2 also seem to be due to the AP-PCVD system, while it is conjectured that N_2 is generated as the reaction product during the electrode cleaning process by NF_3 plasma. It is considered that there remain diehard N-related impurities generated in the NF_3 plasma cleaning process around the electrode and substrate holder, which generate N_2 during the substrate heating. F-related impurities, such as SiF_x ($x = 1–4$), generated in the NF_3 plasma cleaning process were confirmed to be eliminated during the chamber baking process by the APIMS measurement. Additionally, it is conceived that the increase in the concentration of CO_2 is caused by the thermal reactions of H_2O and O_2 with the hydrocarbons in the ambience and on the substrate holder.

From the APIMS spectrum in Fig. 25, the H_2O concentration in the AP-PCVD system is 3.1 ppb. This enables the growth of epi-Si films at a substrate temperature higher than 680 °C by thermal CVD. In the case of AP-PCVD, since high-density atomic hydrogen generated in the atmospheric-pressure plasma removes surface oxides, further decrease in the temperature for epi-Si growth is expected. From these results, we can conclude that sufficiently clean process ambience is achieved for the deposition of high-quality Si films by AP-PCVD.

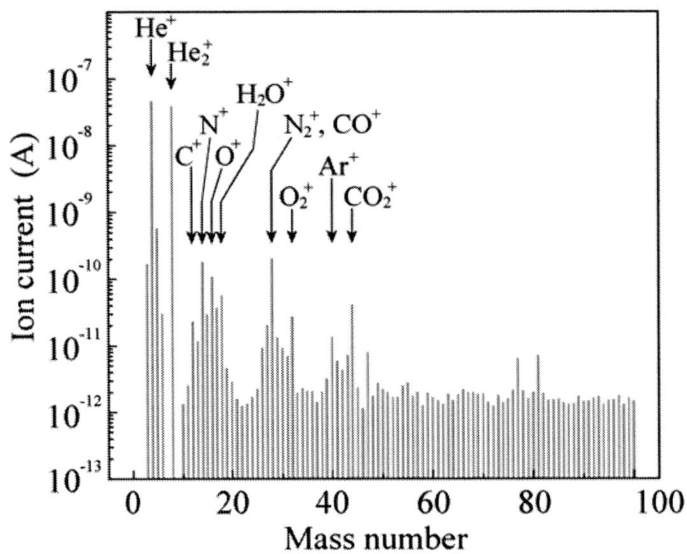

Figure 25. Mass spectrum of He gas in the AP-PCVD system measured by APIMS. The substrate holder made of SiC-coated graphite was heated at 690 °C, and an ultraclean gas circulating pump was used for the gas circulation during measurement.

5.4. Influence of Electrode Surface Material on the Film Property

To verify the influence of electrode surface material on the film deposition by AP-PCVD, we investigated a-Si:H films deposited using rotary electrodes with and without Al_2O_3 coating and evaluated σ_{ph}, σ_d and E_{opt} of the films, which are generally regarded to be a measure of the quality of a-Si:H. The deposition conditions were determined on the basis of the result shown in Fig. 23. The hydrogen dilution ratio of SiH_4 (H_2/SiH_4 ratio) was fixed at 10, and three kinds of SiH_4 concentrations (0.1, 0.3 and 0.5%) were adopted. The VHF power was 500, 800 and 1000 W for 0.1, 0.3 and 0.5% SiH_4, respectively. The other conditions were as follows; the electrode rotation speed was 5000 rpm, the deposition gap was 0.5 mm, T_{sub} was 220 °C, and the gas circulation rate was 1200 *l*/min. The deposition rates of a-Si:H for 0.1, 0.3 and 0.5% SiH_4 were approximately 60, 200 and 330 nm/s, respectively. By controlling the substrate scanning speed, a-Si:H films of 300–350 nm thickness were prepared on glass substrates with a size of 100×100 mm^2.

Figure 26 shows the deposition rate dependences of σ_{ph} and σ_d (a) and E_{opt} (b) of the a-Si:H films. Open symbols indicate the data of the films deposited using the Al_2O_3-coated electrode, while solid ones indicate those using the metal electrode. For the films deposited using the Al_2O_3-coated electrode in Fig. 26(a), although σ_d shows little variation and is as low as 10^{-11} $\Omega^{-1}cm^{-1}$, σ_{ph} monotonously decreases with increasing deposition rate. Since E_{opt} also tends to decrease with deposition rate [Fig. 26(b)], we consider that the structural relaxation becomes insufficient as the deposition rate is increased. The lack of structural relaxation can

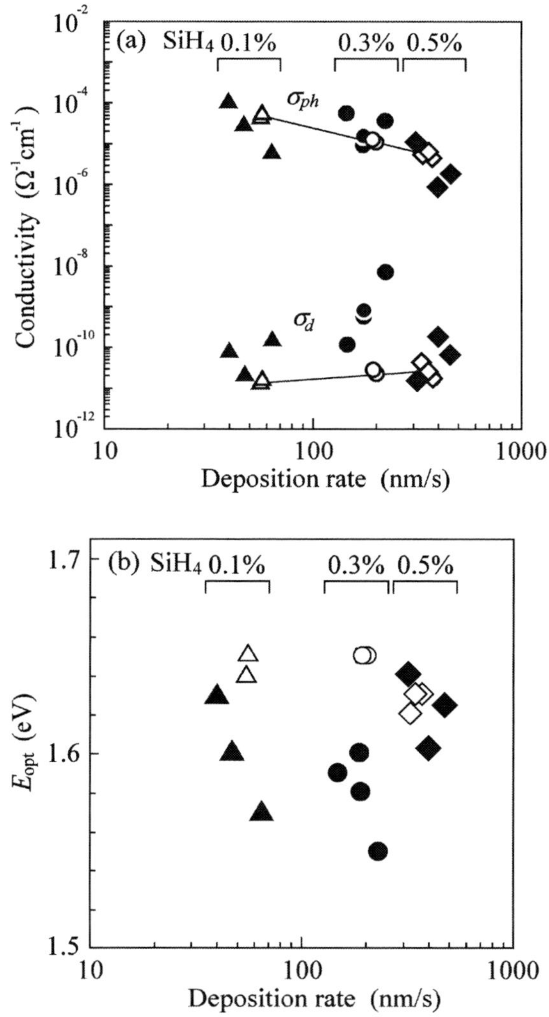

Figure 26. Deposition rate dependences of σ_{ph} and σ_d (a) and E_{opt} (b) of the films. Open symbols indicate the data of the films deposited using the Al_2O_3-coated rotary electrode, while closed ones indicate those deposited using the metal electrode.

result in an increase in the densities of the band tail states of the film and a deterioration of the film quality. For the films deposited using the metal electrode, note that relatively large values of σ_d are seen, whereas the values of σ_{ph} are comparable to those of the films deposited with the Al_2O_3-coated electrode. The marked increase in σ_d suggests that the supplied VHF power is excessive, as described in the next section. When too much VHF power is supplied, highly dissociated precursors may be generated and polymerize in gas phase, which increases the defect density of the film, causing an increase in σ_d. From the discussion in section 3.3, both electron density and gas temperature of the plasma generated with the metal electrodes are considered to be higher than those with the Al_2O_3-coated electrodes. Thus, we speculate that SiH_4 molecules are decomposed excessively in the plasma by using the metal electrode, when the same VHF power as the case of using the Al_2O_3-coated electrode is supplied. In other words, even when the metal electrode is used, the formation of a-Si:H films having equivalent quality to that of the films with the Al_2O_3-coated electrode may be possible by decreasing VHF power. However, note that a large spread of values of σ_{ph}, σ_d and E_{opt} are seen for the films deposited with the metal electrode in Fig. 26, indicating that the film deposition is not reproducible. We consider that the lack of reproducibility originates from the generation of inhomogeneous plasma. When using the metal electrode, we found that the repeated introduction of plasma damage and deposition of Si degraded the smoothness of the electrode surface. This is a fatal problem for the generation of homogeneous plasma over a wide area at a pressure as high as atmospheric pressure, because uneven electrode surface made of metal deteriorate the uniformity of field strength and the yield of the emission of secondary electrons. Indeed, we have found by carrying out numbers of deposition that both the reproducibility of deposition and the uniformity of film thickness can be significantly improved by the utilization of the Al_2O_3-coated electrode.

From the results mentioned above, we have confirmed that the utilization of a surface-insulated electrode is a promising method for the generation of stable and homogeneous plasma at atmospheric pressure, which is essentially important for the formation of high-quality films with a good reproducibility. From Fig. 26(a), however, increasing deposition rate results in a deterioration of the film property. We consider that VHF power and H_2/SiH_4 ratio are very important parameters governing the dissociation of SiH_4 molecules and the formation of Si–Si network at the film-growing surface. Therefore, the deposition parameters must be optimized to achieve the high-rate and low-temperature growths of high-quality Si films by AP-PCVD, as evidenced in the following sections.

5.5. Hydrogenated Amorphous Si

5.5.1. Deposition Rate and Properties of a-Si:H Films

To clarify whether it is possible to deposit device-quality a-Si:H films at high rates by AP-PCVD, a-Si:H films with thickness of 200 to 500 nm were prepared at various deposition conditions on glass substrates, and the electrical and optical properties of the films were studied. In a series of experiments, the deposition gap was fixed at 0.2 mm.

Figure 27 shows the VHF power dependences of σ_{ph} and σ_d (a), E_{opt} (b) and deposition rate (c) of the films prepared with the H_2 concentrations of 0.5% (H_2/SiH_4 = 1), 3% (H_2/SiH_4 = 6) and 5% (H_2/SiH_4 = 10). The SiH_4 concentration was fixed at 0.5%. The electrode rotation speed was 1700 rpm, and T_{sub} was kept constant at 220 °C. In Fig. 27(a), σ_{ph} greatly increases with VHF power and attains values greater than 10^{-5} Ω^{-1} cm^{-1} at more than 800 W, whereas σ_d monotonously increases with VHF power. Note that there is no significant relationship between H_2 concentration and the conductivity of the film. On the contrary, it can be seen in Figs. 27(b) and 27(c) that E_{opt} and deposition rate are affected not only by VHF power but also by H_2 concentration. In Fig. 27(b), E_{opt} tends to decrease with increasing VHF power, and a smaller value of E_{opt} is obtained at a higher H_2 concentration, particularly in a VHF power range larger than 800 W. In Fig. 27(c), it is seen that deposition rate tends to decrease with increasing VHF power more than 800 W, but a higher deposition rate is obtained at a higher H_2 concentration. When VHF power was larger than 1000 W, SEM observations (not shown) revealed that the film surface was very rough and that the roughness was larger under a lower H_2 concentration. It appeared that the deterioration in the surface morphology was caused by the deposition of silicon particles, which might not be removed by gas circulation during deposition. The maximum deposition rate is 200 nm/s at 700 W for 5% H_2, which corresponds to the average deposition rate of 1.0 nm/s for a substrate scanning distance of 1 m.

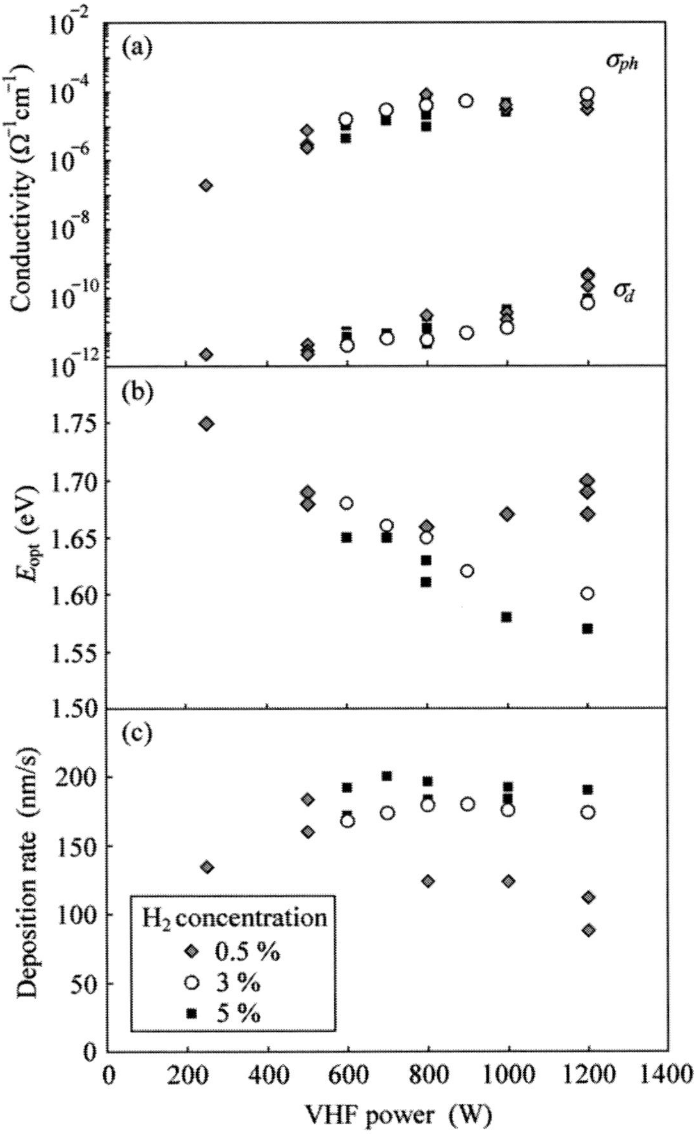

Figure 27. VHF power dependence of σ_{ph} and σ_d (a), E_{opt} (b) and deposition rate (c) of the a-Si:H films prepared with 0.5, 3 and 5% H_2. The SiH_4 concentration is 0.5%, T_{sub} is 220 °C and the electrode rotation speed is 1700 rpm. Symbols are explained in the inserts. [26]

In the case of depositing a-Si:H film in the temperature range below 350 °C by the conventional PECVD, an excessively high deposition rate generally degrades the film quality.

This is because the diffusion length of the deposition precursors on the film-growing surface primarily determines the film structure [28]. In the AP-PCVD process, however, the diffusion length of the precursors may be considerably short due to the extremely high deposition rate. It is reasonable to assume that the gas temperature gives a beneficial influence on the surface reactions, such as hydrogen elimination, diffusion of the precursors and Si-Si bond formation. Thus, it is considered from Fig. 27 that VHF power is the primarily important parameter governing the dissociation of SiH_4 molecules and enhancing the film-forming reactions in the deposition process by AP-PCVD. At low VHF power, since the E_{opt} of the film is high [Fig. 27(b)], it is considered that the hydrogen content of the film is high. This indicates that the Si-Si network structure in the film is sparse because of the insufficient dissociation of SiH_4 molecules in the plasma, resulting in the low values of σ_{ph} and σ_d [Fig. 27(a)]. On the other hand, when too much VHF power is supplied, highly dissociated precursors may polymerize in gas phase and cause the formation of silicon particles. It is conceived that the gas-phase polymerization increases the defect density of the film, causing an increase in σ_d [Fig. 27(a)]. Moreover, silicon particle formation consumes the precursors for film growth, resulting in a decrease in deposition rate as clearly seen in the case of 0.5% H_2 in Fig. 27(c). Therefore, we consider that an optimum VHF power is determined for every SiH_4 concentration. However, the obtained data implies that the hydrogen dilution of SiH_4 has considerable effects not only on suppressing gas-phase polymerization but also on forming a Si-Si network structure in the film. Indeed, in the conventional PECVD, diluting SiH_4 with H_2 is generally understood as an effective means for suppressing gas-phase polymerization and forming a denser network structure [28].

To elucidate the effect of H_2/SiH_4 ratio on the film properties, we carried out numbers of deposition varying conditions over a wide range (SiH_4 concentration: 0.1 – 1%, H_2 concentration: 0.1 – 50%, the electrode rotation speed: 500 – 5000 rpm, and T_{sub}: 160 – 280 °C). The VHF power was varied in a certain range around the optimum value for each deposition condition. Figure 28 shows σ_{ph}, σ_d (a) and E_{opt} (b) of the a-Si:H films as a function of H_2/SiH_4 ratio. The different symbol corresponds to the different SiH_4 concentration. In Fig. 28(a), values of σ_{ph} and σ_d are almost independent of H_2/SiH_4 and SiH_4 concentration. On the contrary, E_{opt} decreases with increasing H_2/SiH_4 in Fig. 28(b). Combining the results in Figs. 27 and 28, it can be noted that the lower E_{opt} is attributed to the higher value of H_2/SiH_4 within the present deposition conditions, suggesting that hydrogen content of the film decreases with increasing H_2/SiH_4.

In the atmospheric-pressure plasma, molecule-molecule interaction should play an important role in their decomposition process. Since the dissociation energies of SiH_4 and H_2 are 3.99 and 4.53 eV, respectively [29], SiH_4 molecules may easily decompose by repeated interaction with atomic hydrogen in the plasma. It is also considered that atomic hydrogen incident to the film-growing surface eliminates bonded H atoms and lowers the hydrogen coverage of the surface, leading to a more perfect Si-Si network. Actually, the amorphous to microcrystalline transition was observed for the films deposited under some deposition conditions at $H_2/SiH_4 > 50$. Additionally, it is reasonable to assume that atomic hydrogen prevents the precursors from polymerizing in gas phase through repeated interaction with them. Therefore, a higher deposition rate and a lower E_{opt} can be obtained simultaneously at a higher H_2 concentration, which is not observed in the conventional PECVD process [20]. It is noteworthy that the film growth mechanism in the AP-PCVD process is apparently different from that in the conventional PECVD process.

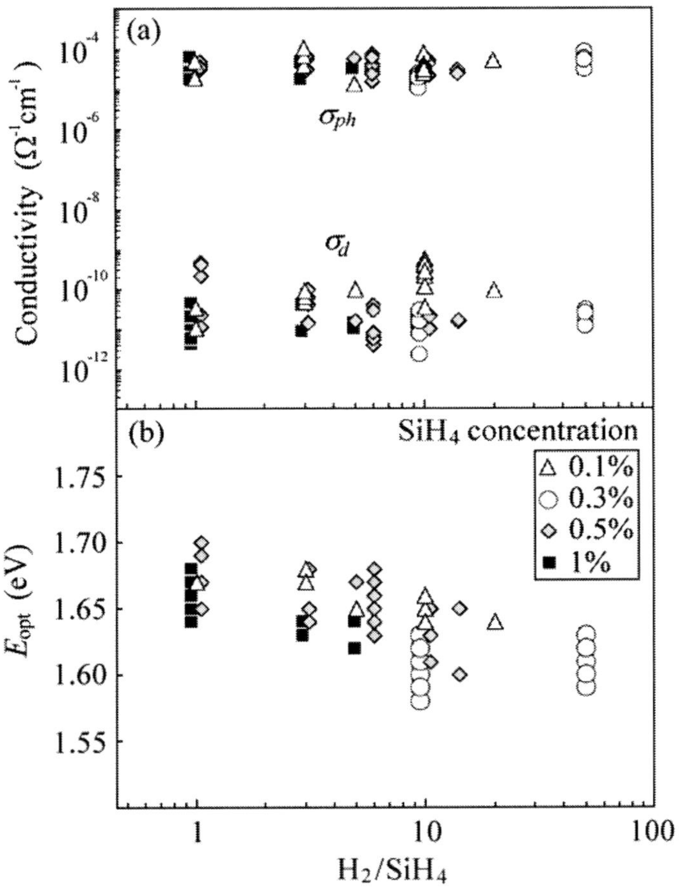

Figure 28. σ_{ph}, σ_d (a) and E_{opt} (b) of the a-Si:H films as a function of H_2/SiH_4 ratio. The deposition conditions were varied over a wide range. [27]

Figure 29 shows the deposition rate dependence of σ_{ph} and σ_d of the films described in Fig. 28. Surprisingly, σ_{ph} is almost independent of the deposition rate and is greater than 10^{-5} $\Omega^{-1}cm^{-1}$, while σ_d tends to decrease as the deposition rate increases although a relatively large spread of values are seen. The maximum deposition rate is 336 nm/s, and then the average deposition rate is calculated as 1.68 nm/s for the substrate scanning distance of 1 m. In this deposition condition, a-Si:H film of 300 nm-thick with 1 × 1 m^2 area can be deposited in approximately 180 seconds with a substrate scanning speed of 5.6 mm/s, when the rotary electrode wider than 1 m is used. Further increase in the deposition rate can be realized by adopting the higher SiH_4 concentration. Considering these results, both extremely high deposition rate and excellent film property can be simultaneously obtained by AP-PCVD, which is very attractive for mass production of a-Si:H solar cells.

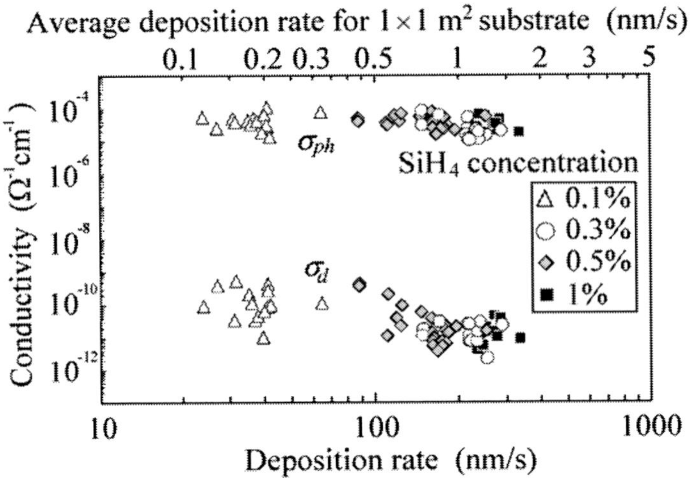

Figure 29. Deposition rate dependence of σ_{ph} and σ_d of the a-Si:H films described in Figure 28. [27]

5.5.2. Efficiency of the a-Si:H Solar Cells

To demonstrate the device applicability of the present a-Si:H films, p-i-n single junction solar cells of which i-layers were deposited by AP-PCVD were examined. The cell structure was glass substrate/textured SnO_2/p/i/n/metal electrode. Only the i-layers with thickness of 230 or 350 nm were deposited by AP-PCVD, and the p- and n-layers with thickness of approximately 50 nm were formed by conventional PECVD. Thus, the p/i and n/i interfaces of those cells were exposed to the air during the specimen transfer from the AP-PCVD to the PECVD reactor or vice versa. On the other hand, all the layers were prepared by conventional PECVD for reference cells. The deposition conditions of the i-layers were selected from those described in Figs. 28 and 29, and some additional conditions were adopted.

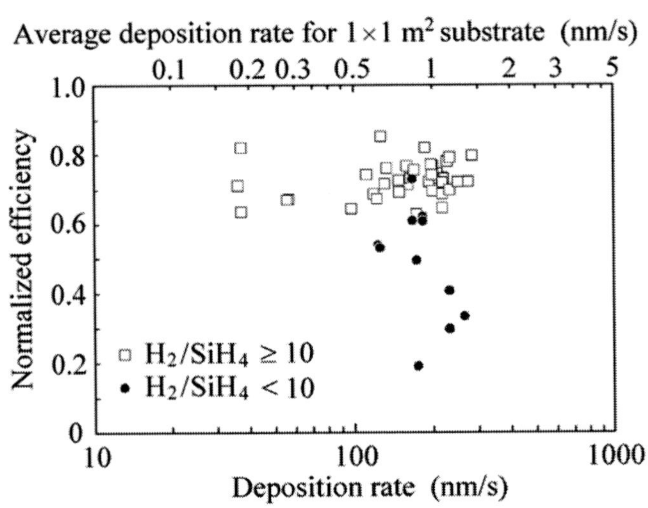

Figure 30. Relationship between the conversion efficiency of the a-Si:H solar cells normalized by that of the reference cells and the deposition rate of the i-layer. [27]

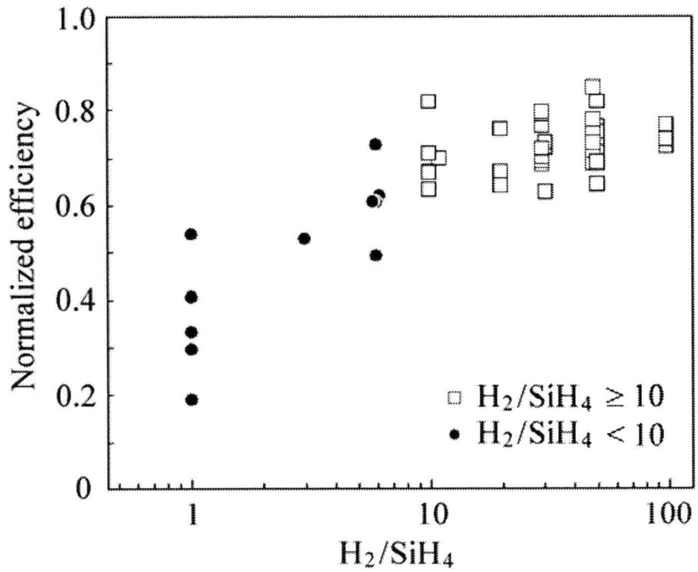

Figure 31. Normalized conversion efficiency of the a-Si:H films described in Figure 30 as a function of H_2/SiH_4 ratio. [27]

Figure 30 summarizes the relationship between the conversion efficiency of the a-Si:H solar cells normalized by that of the reference cells and the deposition rate of the i-layer. The same data are plotted as a function of H_2/SiH_4 ratio in Fig. 31. In Figs. 30 and 31, open squares and solid circles indicate the efficiencies of samples prepared at $H_2/SiH_4 \geq 10$ and $H_2/SiH_4 < 10$, respectively. As shown in the figures, the efficiencies of samples prepared at $H_2/SiH_4 > 10$ are in the range from 0.63 to 0.85, independent of both deposition rate and H_2/SiH_4 ratio. While the efficiencies of samples prepared at $H_2/SiH_4 < 10$ seem to have a correlation with the magnitude of H_2/SiH_4 and spread to smaller values. It is considered that this deterioration is partly caused by the higher E_{opt} of the i-layer at the lower H_2/SiH_4 ratio shown in Fig. 28(b). These results clarify that the deposition rate of i-layer is not the limiting factor for the efficiency and that the existence of proper amount of atomic hydrogen can improve the characteristics of i-layers even at extremely high deposition rates. At $H_2/SiH_4 \geq 10$, however, a cell having equivalent conversion efficiency to the reference cell is not obtained. Thus, the present study implies the existence of other factors that limit solar cell performance. One of these limiting factors may be the introduction of damages to the underlying SnO_2 and p-layers during the deposition of i-layers. To better understand the details of the limiting factors, further systematic studies not only on the characteristics of the high-rate deposited a-Si:H film but also on the film-forming reactions in the AP-PCVD process are necessary.

Figure 32 shows the current density–voltage characteristic of the sample that exhibits the maximum conversion efficiency in this study. The solar cell structure is also illustrated in the figure. The i-layer was prepared with 0.3% SiH_4, 15% H_2, the electrode rotation speed of 5000 rpm at T_{sub} = 220 °C. The VHF power was 800 W; then the average power density within the plasma area was 80 W/cm^2. The deposition rate of the i-layer was 128.1 nm/s, which corresponds to the average deposition rates of 0.64 nm/s for the substrate scanning distance of 1m; then the thickness of the i-layer was 350 nm at the substrate scanning speed of 1.86 mm/s. As shown in Fig. 32, an initial efficiency of 8.25% (V_{oc} = 0.93 V, I_{sc} = 13.7 mA/cm^2, F.F. = 0.65) is achieved. This is the highest efficiency ever achieved for an a-Si:H solar cell of which i-layer is fabricated at a deposition rate higher than 100 nm/s. It is considered that the cell performance can be improved by optimizing the deposition conditions of the intrinsic and the doped layers and other factors in fabricating a solar cell structure.

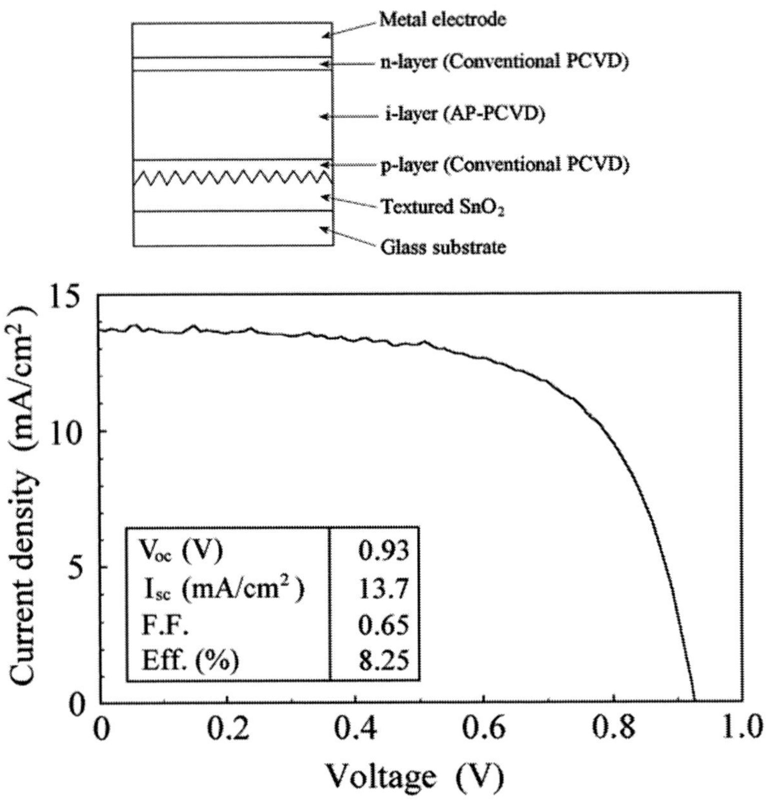

Figure 32. Current density – voltage characteristic of the a-Si:H solar cell of which i-layer is prepared at the deposition rate of 128.1 nm/s. The thickness of the i-layer is 350 nm and the cell area is 1 cm^2. [27]

5.6. Polycrystalline Si

5.6.1. Deposition Rate

The contribution of atomic hydrogen becomes more important in the case of depositing crystalline Si, as is well known for the poly-Si deposition by conventional PECVD [30,31]. Increasing H_2/SiH_4 ratio may lead to an increase in the crystalline fraction associated with a decrease in the hydrogen content in a film. Thus, we focused on investigating the influences of H_2/SiH_4 ratio on the deposition rate, structure and morphology of the poly-Si films. In addition, OES was employed to study the deposition plasma.

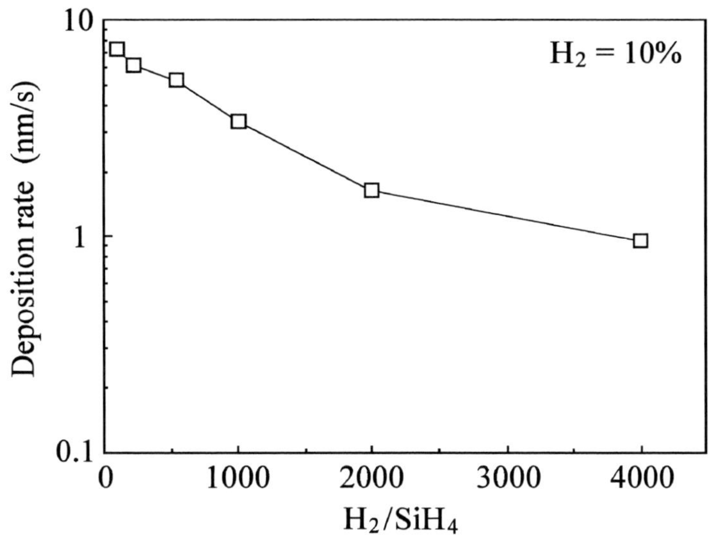

Figure 33. H_2/SiH_4 ratio dependences of deposition rate of the poly-Si films. SiH_4 concentration is varied from 0.001 to 0.1% under a constant H_2 concentration of 10%. The VHF power is 2500 W, T_{sub} is 500 °C, and the electrode rotation speed is 2000 rpm.

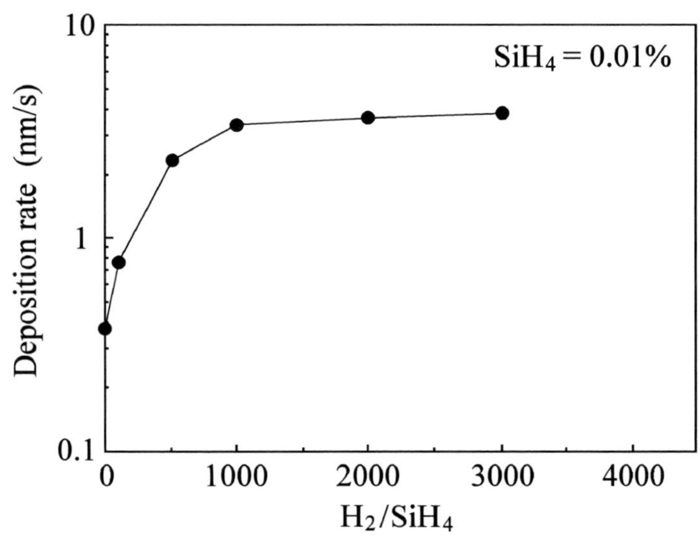

Figure 34. H_2/SiH_4 ratio dependences of deposition rate of the poly-Si films. H_2 concentration is varied from 0 to 30% at a constant SiH_4 concentration of 0.01%. The VHF power is 2500 W, T_{sub} is 500 °C, and the electrode rotation speed is 2000 rpm.

Figures 33 and 34 show the H_2/SiH_4 ratio dependences of deposition rate of the poly-Si films prepared on thermally oxidized 4-in. Si wafers with the moderately large VHF power of 2500 W at a constant T_{sub} of 500 °C. The electrode rotation speed was 2000 rpm, the deposition gap was 1 mm, and the gas circulation rate was 1000 *l*/min. SiH_4 concentration is varied from 0.001 to 0.1% under a constant H_2 concentration of 10% in Fig. 33, while H_2 concentration is varied from 0 to 30% at a constant SiH_4 concentration of 0.01% in Fig. 34.

For the films deposited with varying SiH_4 concentration in Fig. 33, the deposition rate decreases exponentially with increasing H_2/SiH_4 ratio. The maximum value of the deposition rate is 7.4 nm/s at a SiH_4 concentration of 0.1% (H_2/SiH_4 = 100). It should be noted (not shown in Fig. 33) that no film growth was observed with 0.001% SiH_4 (H_2/SiH_4 = 10000). In addition, RHEED observations revealed that a certain volume of amorphous Si phase existed in the upstream region of the deposited poly Si films at $H_2/SiH_4 < 500$ ($SiH_4 > 0.02$ %).

For the films deposited with varying H_2 concentration in Fig. 34, the deposition rate increases greatly with increasing H_2/SiH_4 ratio at $H_2/SiH_4 < 500$ ($H_2 < 5\%$), while at $H_2/SiH_4 \geq 500$, the deposition rate tends to saturate. The drastic increase in the deposition rate at $H_2/SiH_4 < 500$ suggests that the addition of H_2 to the process gas mixtures enhances the generation of atomic hydrogen in the plasma, and that atomic hydrogen has a considerable effect on dissociating SiH_4 molecules, which is also shown in the deposition of a-Si:H films in section 5.5. The saturation of the deposition rate at $H_2/SiH_4 \geq 500$ ($H_2 \geq 5\%$) presumably indicates that the H_2 concentration of 5% is sufficient to decompose all the SiH_4 molecules in the plasma. The details of the deposition process of these films in AP-PCVD are discussed later.

5.6.2. Surface and Cross-Sectional Morphologies

To elucidate the surface and cross-sectional morphologies of these poly-Si films, SEM and TEM observations were carried out. Figure 35 shows SEM images of the surface of the poly-Si films with thickness of approximately 4 μm deposited at (a) H_2/SiH_4 = 2000 and (b) H_2/SiH_4 = 200 for which the H_2 concentration is fixed at 10%. The cross-sectional SEM observations (not shown) revealed that both films had a wedge-shaped crystalline columnar structure, growing in a direction perpendicular to the substrate. However, it is clearly seen in Fig. 35 that the surface morphology is different between these films. In Fig. 35(a), the crystallites of the film deposited at the high H_2/SiH_4 ratio show anisotropic morphologies and have a maximum grain size of approximately 3 μm. In contrast, no apparent faceting growth of crystallites is observed in the film with the low H_2/SiH_4 ratio [Fig. 35(b)]. Additionally, the average grain size is approximately 1.5 μm, smaller than that of the film with the high H_2/SiH_4 ratio, and some of the enlarged grains are composed of much finer grains.

Figure 35. SEM images of the surface of the poly-Si films with thickness of approximately 4 μm deposited at (a) H_2/SiH_4 = 2000 and (b) H_2/SiH_4 = 200. The H_2 concentration is 10%. The VHF power is 2500 W, T_{sub} is 500 °C, and the electrode rotation speed is 2000 rpm.

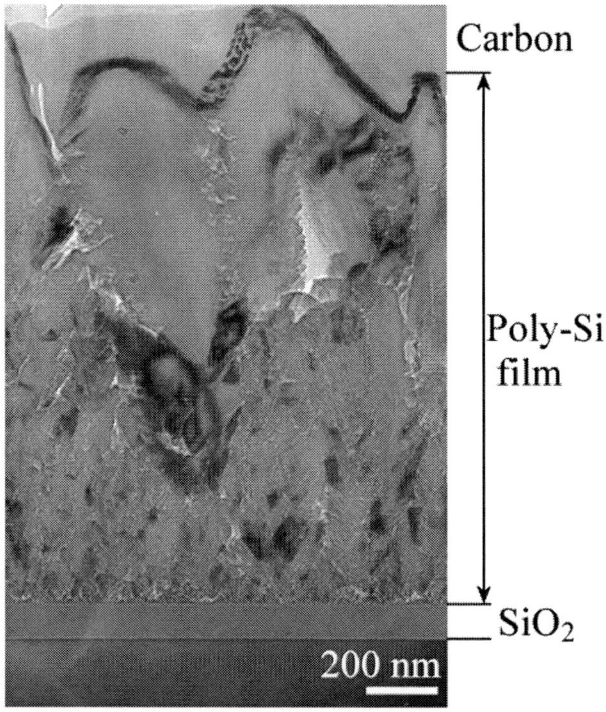

Figure 36. TEM bright field image of a cross section of the poly-Si film deposited at H_2/SiH_4 = 1000 (10% H_2 and 0.01% SiH_4).

Figure 36 shows the TEM bright field image of a cross section of the poly-Si film prepared at a high H_2/SiH_4 ratio of 1000 (10% H_2 and 0.01% SiH_4). It is clearly observed in Fig. 36 that an initial microcrystalline layer exists near the film-substrate interface, followed by wedge-shaped columnar growth of Si crystallites. An incubation layer of an amorphous phase is not observed in the image, indicating the occurrence of an initial nucleation and the growth of crystalline Si directly on the substrate. In addition, note that defects such as dislocation lines are hardly visible in the enlarged wedge-shaped grains. Therefore, it is considered that a large amount of atomic hydrogen existed in the plasma and eliminated strained Si-Si bonds at the film-growing surface, promoting the formation of a crystalline network structure of Si at the relatively low temperature of 500 °C. Although the temperature is still higher than that of conventional PECVD, we consider that the growth of good-quality crystallites in Fig. 36 is not mainly due to the temperature but rather due to the high H_2/SiH_4 ratio. This is supported by the fact that an amorphous phase exists in the upstream region of the poly Si films deposited at $H_2/SiH_4 < 500$ even at T_{sub} = 500 °C, as described earlier.

5.6.3. Structure of the Poly-Si Films

The crystal structure of the poly-Si films was investigated by XRD. Figure 37 shows a typical XRD spectrum of the film obtained in this study. Three peaks are seen at 2θ = 28.4°, 47.2° and 56°, corresponding respectively to the Si(111), Si(220) and Si(311) planes, respectively. The diffraction peaks from the Si(400) plane are barely visible. Figure 38 shows the H_2/SiH_4 ratio dependence of the relative intensity of X-ray diffraction peaks determined from dividing the peak intensity for each plane by the sum of the intensities of the three peaks. The films in Fig. 38(a) are deposited varying H_2 concentration from 0 to 30% at a constant SiH_4 concentration of 0.01%, while the films in Fig. 38(b) are deposited varying SiH_4 concentration from 0.0025 to 0.1% at a constant H_2 concentration of 10%.

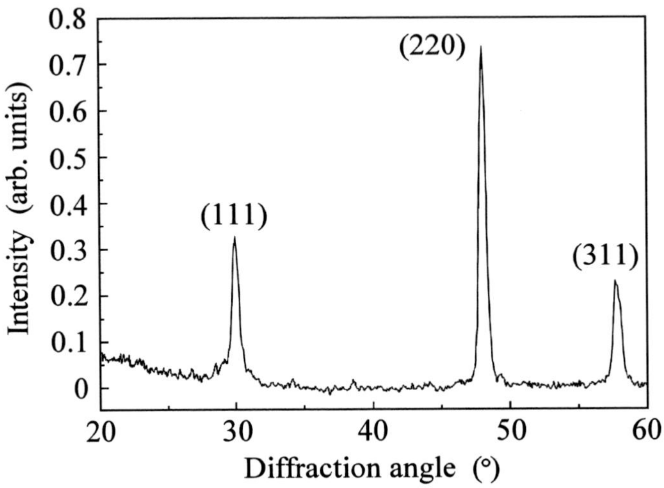

Figure 37. X-ray diffraction spectrum of the poly-Si film deposited at H_2/SiH_4 = 1000 (10% H_2 and 0.01% SiH_4).

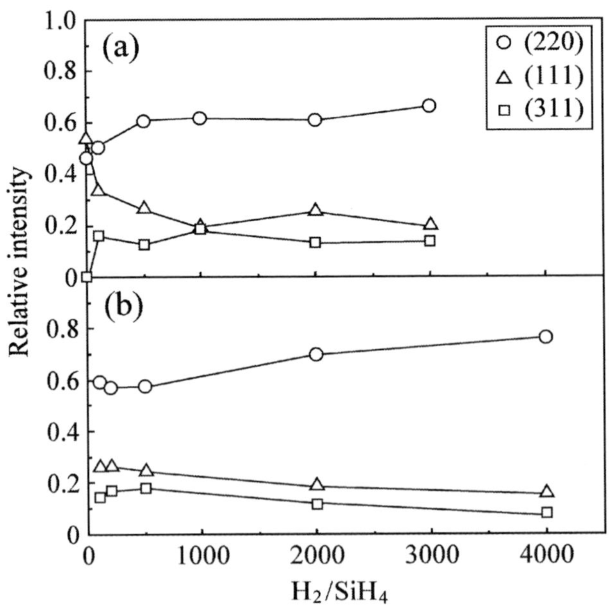

Figure 38. H_2/SiH_4 ratio dependence of the relative intensity of X-ray diffraction peaks for the poly-Si films. The films are deposited varying H_2 concentration from 0 to 30% at a constant SiH_4 concentration of 0.01% (a), and varying SiH_4 concentration from 0.0025 to 0.1% at a constant H_2 concentration of 10% (b). The VHF power is 2500 W, T_{sub} is 500 °C, and the electrode rotation speed is 2000 rpm.

As the H_2/SiH_4 ratio is increased for the constant SiH_4 concentration in Fig. 38(a), the relative intensity of (220) peak increases from 0.45 to 0.60, while that of (111) decreases from 0.54 to 0.20, and the both intensities tend to saturate over $H_2/SiH_4 = 1000$. The dependence of the relative intensity of (220) peak on the H_2/SiH_4 ratio has a similar tendency to that of the deposition rate in Fig. 34. The relative intensity of the (311) peak is almost constant at around 0.15, except for the film at $H_2/SiH_4 = 0$. Since the relative intensities of the diffraction peaks for (111), (220) and (311) planes of poly-Si having random orientation are 0.54, 0.30 and 0.16, respectively [32], it can be said that the present poly-Si films have a preferential orientation of (220). In Fig. 38(b), a different trend in dependence on the H_2/SiH_4 ratio is observed for the films prepared at the constant H_2 concentration. Although the films also exhibit the <110> preferred orientation, no appreciable saturation of the relative intensities is seen even at $H_2/SiH_4 \geq 1000$. In this case, the dependence of the relative intensity of (220) peak on the H_2/SiH_4 ratio has the opposite tendency to that of the deposition rate in Fig. 33. From these facts, it can be said that the H_2/SiH_4 ratio primarily governs the deposition process of poly-Si films, indicating that atomic hydrogen significantly influences not only the deposition rate but also the crystallographic orientation of the poly-Si films.

The deposition process of poly-Si films by AP-PCVD can be discussed as follows. Atomic hydrogen etches out the less chemically stable structure more easily in competition with film deposition. Thus, the occurrence of preferential orientation is qualitatively explained by comparing the stability and the growth rate of crystalline Si for each crystal plane. The surface free energies for the {110}, {100}, and {111} planes are at the ratio of 1.51 : 2.13 : 1.23 [33], so the {110} and {111} surfaces have similar stability against etching by atomic hydrogen. On the other hand, the growth rates of crystalline Si in the <110>, <100>, and <111> directions are at the ratio of 25 : 75 : 3 [34]. From these values, although the crystallite with an orientation of {100} grows faster than that with other orientations, it is etched much more easily at the same time, resulting in the absence of a diffraction peak from the Si(400) plane in Fig. 37. Comparing the values for the {110} plane with those for the {111} plane, it is reasonable to suggest that the <110> preferred orientation becomes dominant under the existence of a high density of atomic hydrogen. Therefore, we can conclude that a large amount of atomic hydrogen existed in the atmospheric-pressure plasma within the present deposition conditions, and the supply of atomic hydrogen to the film-growing surface is essential to form Si films with good crystallinity at low temperatures.

5.6.4. Diagnostics for the Atmospheric-pressure plasma

To confirm the discussion mentioned above, OES measurements for the atmospheric-pressure plasma was performed to investigate the emission intensities from atomic hydrogen.

Figure 39 shows the OES spectra of the atmospheric-pressure plasma for the H_2 concentrations of 5, 20 and 30% at a constant SiH_4 concentration of 0.01%. In Fig. 39, emission lines from atomic hydrogen (H_α) at 656.3 nm and SiH radical at 412.8 nm are seen together with He peaks at 728.1, 706.5, 667.8, 587.6, and 388.9 nm. The broad emission band in the 300 – 500 nm range is from H_2 ($^3\Sigma$) molecules. With increasing H_2 concentration, all the emission lines drastically decreases in intensity and the emission from H_2 molecule becomes dominant. Note that the relative intensities of the He emission lines vary with increasing H_2 concentration. For example, although the intensity of the He emission line at

706.5nm ($3^3S_1 - 2^3P$) is as strong as that at 388.9nm ($3^3P - 2^3S$) at 5% H_2, the former is not distinguishable and the latter is still existent at 30% H_2. This indicates that the electron energy distribution function changes with H_2 concentration.

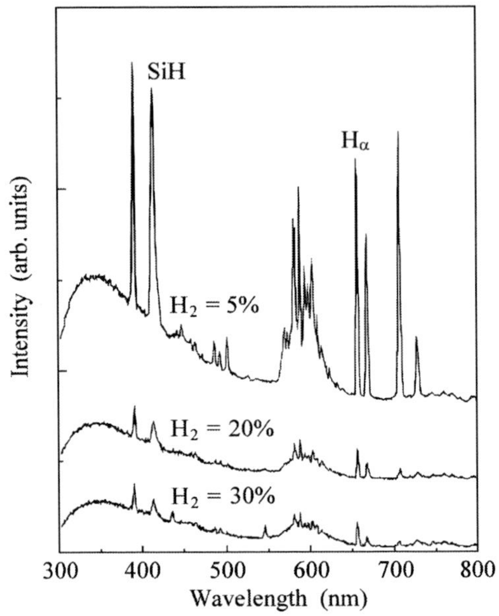

Figure 39. OES spectra of the atmospheric-pressure plasma generated varying H_2 concentration at a constant SiH_4 concentration of 0.01%. The VHF power is 2500 W.

At the present stage, it is not evidenced from the OES spectra whether the higher concentration of H_2 actually leads to the higher density of atomic hydrogen in the plasma. Therefore, further diagnostic study on the atmospheric-pressure plasma is necessary. However, it can be mentioned that atomic hydrogen is the key factor that governs the deposition process of crystalline Si at low temperatures by AP-PCVD.

5.7. Epitaxial Si

5.7.1. Effects of H_2/SiH_4 Ratio on Si Epitaxial Growth

Using (001) B-doped CZ-Si wafers as substrates, we investigated the optimum conditions for the growth of epi-Si films. From the viewpoint of the enhancement of Si growth in the <100> direction and the prevention of excessive etching by atomic hydrogen, it is considered that smaller H_2/SiH_4 ratio is appropriate for Si epitaxy by AP-PCVD. Experiments were conducted at various substrate temperatures from 500 to 800 °C and with VHF powers from 300 to 3000 W. Optimized values were employed for the electrode rotation speed (2000 rpm) and the deposition gap (0.7 mm). The Si films deposited in the appropriate conditions exhibited smooth surface morphologies in the central region, which were evidenced to be the growths of epi-Si films as described later. On the contrary, the film surfaces at the film edges on both upstream and downstream sides were very rough, which appeared to be the depositions of poly-Si films (data not shown). This is partly due to the existence of turbulent gas flow in front of and behind the rotary electrode [21]. Therefore, crystallinity in the central part (10 × 10 mm^2) of the film grown without substrate scanning was studied.

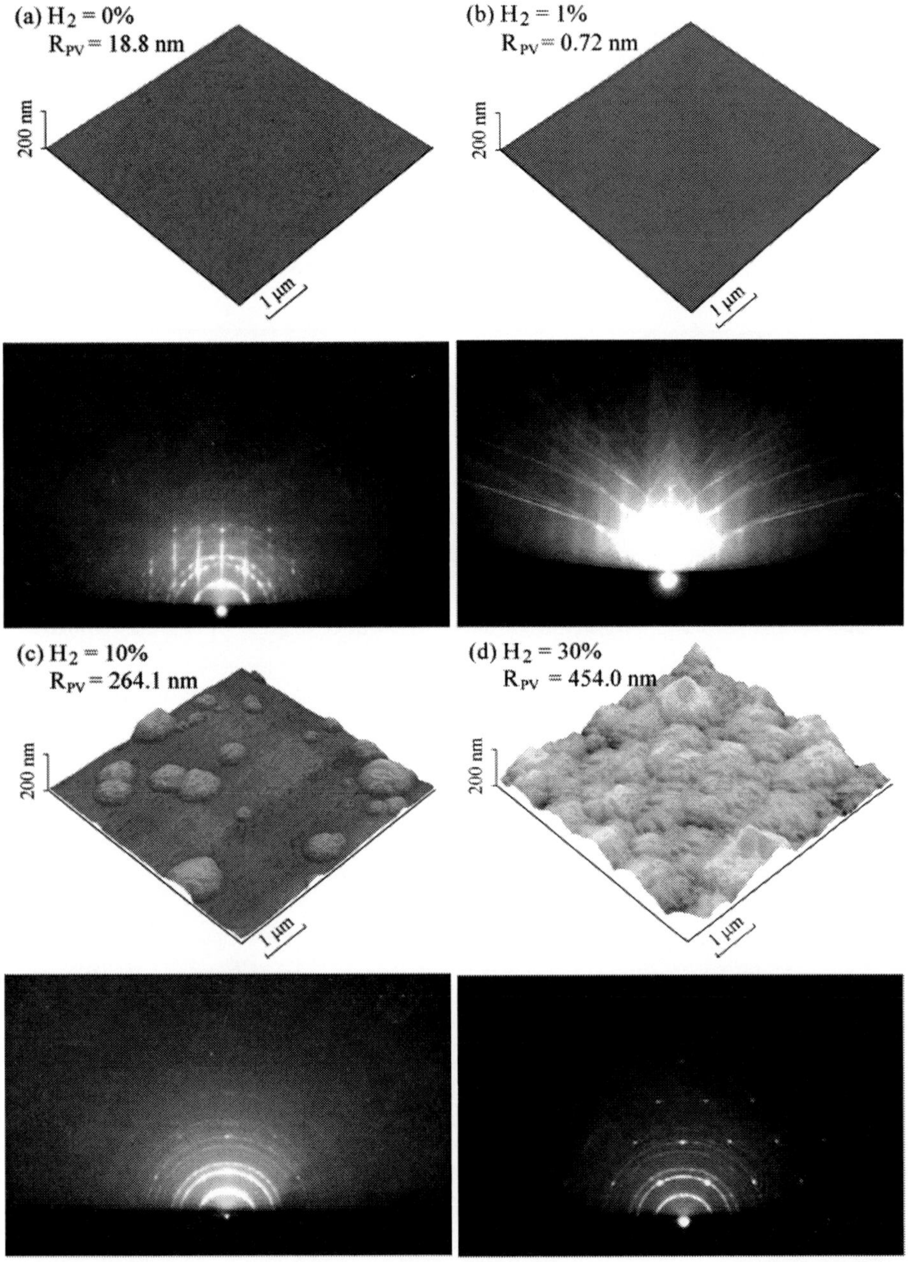

Figure 40. AFM images with corresponding RHEED patterns of the Si films deposited by AP-PCVD at 700 °C with a H_2 concentration of (a) 0%, (b) 1%, (c) 10%, and (d) 30%. The SiH_4 concentration was constant at 0.1%, and the VHF power was 1100 W. Scanned area for each AFM image is 5 × 5 μm^2.

Figure 40 shows AFM and corresponding RHEED images of deposited Si films at 700 °C as a function of H_2 concentration. The VHF power was 1100 W and the SiH_4 concentration was 0.1% in He atmosphere. When the H_2 concentration is 1% [Fig. 40(b)], single crystal growth is confirmed from the RHEED pattern. Surface of the film is very smooth with the peak to valley roughness (R_{PV}) of 0.72 nm, which is smaller than that of the CZ-Si wafer surface (R_{PV} = 1.40 nm). When no H_2 gas is added to the plasma gas, the surface of the Si film becomes rough (R_{PV} = 18.8 nm) and some polycrystalline rings are observed in the RHEED pattern [Fig. 40(a)]. Possibly, the residual H_2O molecules in the AP-PCVD chamber (20 – 30 ppb) in the routine experiments degraded the cleanliness of the Si surface during the epitaxial growth process, resulting in the roughness of the Si film surface in Fig. 40(a). On the other hand, since the film surface grown with 1% H_2 is very smooth (R_{PV} = 0.72 nm), the addition of 1% H_2 in He may be effective to maintain the cleanliness of the Si surface required for epitaxial growth before and/or during the growth process by the present AP-PCVD technique. When the H_2 concentration is increased, Si films become polycrystalline and the surface roughness increases as shown in Figs. 40(c) and 40(d). It is considered that one of the reasons of the degradation of crystallinity is the excessive etching of film-growing surface by atomic hydrogen during growth. However, an IR absorption peak at 2000 cm^{-1} associated with Si-H stretching vibration mode was detected in the poly-Si film in Figs. 40(c) and 40(d) (data not shown). This suggests that the supplied VHF power is dissipated in the H_2 molecule vibration excitation, and the energy for dissociating SiH_4 molecules and enhancing epitaxial growth reaction becomes insufficient. The insufficient plasma energy per each SiH_4 molecule may cause the incorporation of large amount of SiH_x species in the Si film and makes the film polycrystalline. Investigating numbers of deposition varying SiH_4 and H_2 concentrations, we found that VHF power should be optimized for each combination of SiH_4 and H_2 concentrations. In addition, H_2/SiH_4 ratio at around 10 was appropriate for Si epitaxy by AP-PCVD.

5.7.2. Effects of Substrate Temperature and VHF Power on Si Epitaxial Growth

Based on the results obtained in the previous section, we have studied the effects of T_{sub} and VHF power on Si epitaxial growth with a fixed H_2/SiH_4 ratio at 10 ([H_2]: 1 %, [SiH_4]: 0.1 %). Figure 41 shows a "crystallinity map" of the Si films determined from the RHEED patterns as functions of T_{sub} and VHF power. The open circles, solid squares and solid triangles indicate the growths of epi-Si, poly-Si and amorphous Si, respectively. From Fig. 41, amorphous or polycrystalline Si films grow in the lower temperature and the lower VHF power region. It is obvious that even if T_{sub} is lowered, epi-Si can be grown by increasing VHF power. This presumably indicates a beneficial influence of the gas temperature on structural relaxation of the film deposited at higher rates. When the VHF power is too high (2200 W at 700 °C), however, the film becomes polycrystalline. Therefore, even at high enough temperatures for Si epitaxial growth by AP-PCVD, there exists an appropriate range for VHF power.

From the results shown in Fig. 41, it is hypothesized that the increase in the following four physical parameters, which is caused by the increased VHF power or T_{sub}, contributes to the epi-Si growth in the AP-PCVD process.

i) The degree of decomposition and activation of SiH_4 molecules in the plasma (the form and the state of film-forming precursor).

ii) The gas temperature of the plasma (the kinetic energy of He atom).
iii) The desorption rate of the bonded hydrogen atom from the film-growing surface (the density of absorption site for Si atoms).
iv) The diffusion length of film-forming precursor at the film-growing surface.

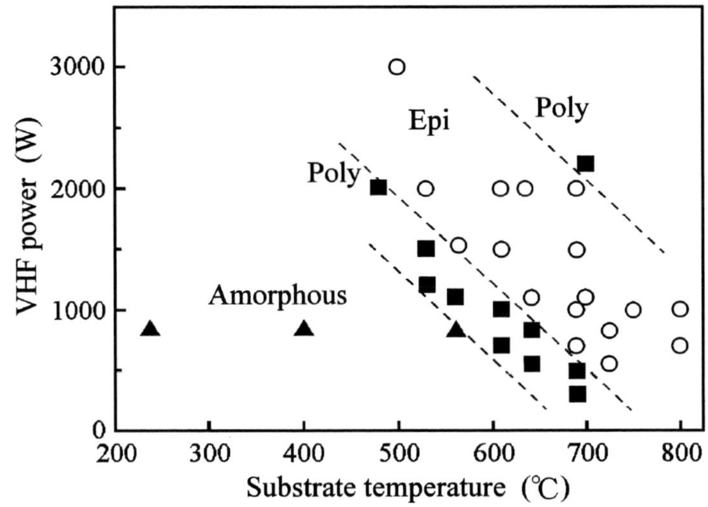

Figure 41. "Crystallinity map" of the Si films deposited by AP-PCVD as functions of T_{sub} and VHF power. The Crystallinity was determined by RHEED observations.

The increase in VHF power is related to all four of these physical parameters, while T_{sub} is mainly related to iii) and iv). Generally, epi-Si can be easily grown at high temperatures. The decrease in T_{sub} causes a decrease in iii) and iv), and is connected with the inhibition of epitaxial growth of Si. Thus, it is important to raise the gas temperature of the plasma by increasing VHF power, which supplies enough energy to enhance iii) and iv). Therefore, the increase in VHF power is essential for the further reduction of temperature for epitaxial Si growth, which is consistent with the result seen in Fig. 41. However, it should be mentioned that there exists an optimum VHF power above which another physical process (polymerization of film-forming precursors) becomes the main parameter for determining the film crystallinity. When too much VHF power is applied, SiH_4 molecules are excessively decomposed, and Si particles are generated, resulting in degradation of the crystallinity and the surface flatness. In this case, the RHEED image changes from a streaky to a spotty pattern (data not shown).

Figure 42 shows TEM images of cross sections of the epi-Si films grown at 800 °C with the VHF power of 700 W (a) and 1000 W (b), and at 500 °C with 3000 W (c). The epi-Si film grown at 800 °C with 700 W [Fig. 42(a)] contains many dislocation lines, and the interface between the film and the Si substrate is rough. Dislocation lines in the right hand side of the photograph are generated from the imperfections at the interface. Therefore, it is necessary to clean the substrate surface by the higher density of atomic hydrogen in the plasma during the growth process, which can be achieved by the larger VHF power. By increasing VHF power to 1000 W, the dislocation density in the film is lowered [Fig. 42(b)]. However, since a residual contrast is observed at the interface, further increase in VHF power may be needed to grow defect-free epi-Si films at 800 °C. For the film obtained with the highest VHF power of 3000 W at the lowest temperature (500 °C) in the present study [Fig. 42(c)], no defect or no interface contrast is seen in the epi-Si film (and also in the other TEM images taken under the different diffraction conditions) other than thickness fringes. Although epitaxial layer is not discernible from the substrate Si crystal in Fig. 42(c), the presence of protecting film of polycrystalline Al guarantees the existence of the epi-Si layer. Defect-free epi-Si growth at 500 °C was also confirmed by the observations of four TEM samples made from the different portions approximately 5 mm apart with each other, where no defect or no interface contrast was observed. AFM observation of the film revealed that the surface is very smooth with a R_{PV} roughness of 0.92 nm (data not shown). Therefore, we conclude that defect-free epi-Si growth at 500 °C is achieved by increasing VHF power to 3000 W in the optimum range for Si epitaxial growth conditions by AP-PCVD.

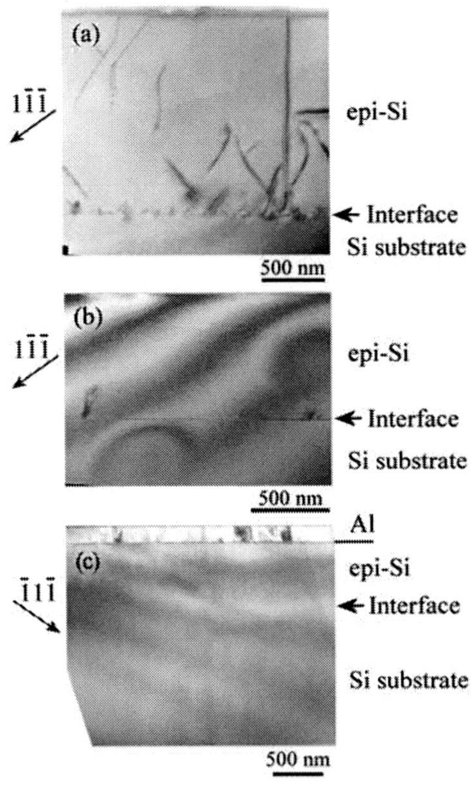

Figure 42. Cross-sectional TEM micrographs of the Si films deposited by AP-PCVD at 800 °C with the VHF power of 700 W (a) and 1000 W (b), and at 500 °C with 3000 W (c). The electrode rotation speed was 2000 rpm.

5.7.3. Temperature Dependence of Si Growth Rate

Thickness of the Si film varies along the plasma length when the substrate is not scanned. Therefore, the average growth rate of the Si film was determined from the weight increase of the substrate wafer and the area of the deposited film. Figure 43 shows a T_{sub} dependence of the average growth rate of Si films grown by AP-PCVD with 0.1% SiH_4 and 1% H_2 at the

VHF power of 1000 W. A dependence of the Si growth rate by thermal CVD with the same SiH_4 concentration is also plotted in Fig. 43 for comparison. In the thermal CVD process, the gap between the substrate and the electrode (without VHF power) was set at 28 mm not to cool the substrate by the electrode rotation. The electrode rotation speed was set at 1000 rpm to generate a moderate gas flow above the substrate.

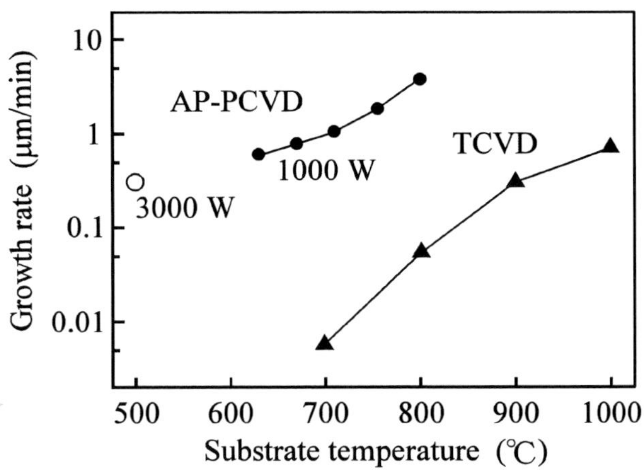

Figure 43. Tsub dependence of the average Si growth rate by AP-PCVD and thermal CVD (TCVD). The H_2 and SiH_4 concentrations were constant at 1% and 0.1%, respectively.

The Si growth rate by thermal CVD increases with increasing temperature and tends to saturate in the high temperature region where the Si growth is limited by the transport processes of the reacting molecules. It is seen from Fig. 43 that the variation of the average Si growth rate by AP-PCVD is smaller than that of thermal CVD. This indicates that the plasma energy is dominant for enhancing the CVD growth of Si rather than the thermal energy in the low temperature region.

Comparing the average Si growth rates at the same temperature, the growth rate by AP-PCVD is approximately 180 and 70 times higher than that of the thermal CVD at 700 and 800 °C, respectively. The growth rates by AP-PCVD are higher than the saturation value by thermal CVD. This is attributed to the fact that the gas supply to the deposition gap (0.7 mm) induced by the rotation (2000 rpm) of the electrode in AP-PCVD process is much larger than that by the gas flow in the present thermal CVD process. The average growth rate of the defect-free epitaxial Si by AP-PCVD at 500°C with 3000 W [the sample shown in Fig. 42(c)] is also plotted in Fig. 43. The growth rate is approximately 0.25 μm/min, which is in the acceptable range for epi-Si wafer production process [35] and is as high as that of the thermal CVD at 900 °C.

5.7.4. Impurities in the epi-Si Films

Finally, SIMS measurements were employed to clarify the contents of impurity atoms in the epitaxial Si films deposited in this study. Oxygen and carbon were mainly detected, and their contents were 4×10^{16} cm^{-3} and 1×10^{16} cm^{-3}, respectively. In the previous reports on the epitaxial Si films deposited by conventional thermal CVD, oxygen and carbon content is [O]

$\leq 2\times10^{17}$ cm^{-3} and [C] $\leq 8\times10^{16}$ cm^{-3}, respectively [36,37]. On the other hand, oxygen and carbon contents in the films fabricated by the low-pressure plasma CVD method vary widely depending on the authors, namely [O] = $1\times10^{17} - 4\times10^{20}$ cm^{-3} and [C] = $8\times10^{16} - 1\times10^{18}$ cm^{-3} [38–41]. Oxygen and carbon concentrations in the Si films deposited by AP-PCVD in the present study are both below those of the epitaxial Si films deposited by thermal CVD.

6. Conclusion

The main positive contribution of this article has been to demonstrate the high potentials of the AP-PCVD technique we have developed for depositing functional thin films with high growth rates at low temperatures. The advantageous aspects of the AP-PCVD technique are as follows. (1) It is possible to generate stable glow plasma at atmospheric pressure by the employment of a 150 MHz VHF power and a cylindrical rotary electrode. (2) Deposition rate is significantly increased by virtue of a high partial pressure of source gas and the efficient and homogeneous supply of a process gas mixture by the viscous drag effect of electrode rotation. (3) Incorporation of damages and impurities into films are suppressed due to the low kinetic energy of ions and the confinement of plasma-generating area. Throughout the course of this article, we have explored elemental technologies for the AP-PCVD system and deposition conditions for the formation of high-quality Si films by AP-PCVD. We can extract a conclusion that adopting VHF excitation and using a surface-insulated electrode are essentially important for the generation of stable atmospheric-pressure plasma with a good reproducibility. We also find that the gas temperature of the atmospheric-pressure plasma has a considerable effect on the structural relaxation of a film and that H_2/SiH_4 ratio in the process gas mixture is the critical parameter that governs both the gas-phase reactions and the growth of crystalline Si. At the present stage, we have fundamentally achieved high-rate depositions of high-quality a-Si:H, poly-Si and epi-Si films at low temperatures. Future study will focus on expanding the deposition area by generating more homogeneous and uniform plasma for the industrial applications of AP-PCVD, and on defining the actual internal parameters of atmospheric-pressure plasma, such as gas temperature, rotational and vibrational temperatures of molecules, electron density and electron temperature. The atmospheric-pressure plasma used in this study may have an intermediate nature between the conventional low-pressure plasma and the high-temperature plasma used for nuclear fusion. Full understandings of the physical and chemical properties of the atmospheric-pressure plasma await more detailed experimental and theoretical investigations.

Acknowledgements

This work was carried out at the Ultra Clean Room of Department of Precision Science and Technology and the Ultra Clean Facility of Research Center for Ultra-precision Science and Technology, Osaka University. This work was partially supported by a Grant-in-Aid for the 21st Century COE Program from the Ministry of Education, Culture, Sports, Science and Technology of Japan. The authors are grateful to Professors Emeriti Y. Mori and K. Yoshii of Osaka University for their stimulating suggestions and discussions, and Professors K. Endo and K. Yamauchi, Drs. K. Yamamura and Y. Sano of Osaka University for their helpful discussions. They also wish to thank Professor H. Mori, Dr. T. Sakata and E. Taguchi of the Research Center for Ultra-High Voltage Electron Microscopy, Osaka University for their kind support in the TEM observations, and Drs. A. Nakaue and K. Hayashi of Kobe Steel, Ltd. for their useful discussions. Technical assistance of A. Takeuchi of Osaka University is also greatly appreciated.

References

[1] Mori Y., Yoshii K., Kakiuchi H. and Yasutake K. *Rev. Sci. Instrum.* 2000, 71, 3173–3177.
[2] Kakiuchi H., Matsumoto M., Ebata Y., Ohmi H., Yasutake K., Yoshii K. and Mori Y. *J. Non-Cryst. Solids* 2005, 351, 741–747.
[3] Kakiuchi H., Ohmi H., Kuwahara Y., Matsumoto M., Ebata Y., Yasutake K., Yoshii K. and Mori Y. *Jpn. J. Appl. Phys.* 2006, 45, 3587–3591.
[4] Mori Y., Yoshii K., Yasutake K., Kakiuchi H., Ohmi H. and Wada K. *Thin Solid Films* 2003, 444, 138–145.
[5] Yasutake K., Kakiuchi H., Ohmi H., Yoshii K. and Mori Y. *Appl. Phys. A* 2005, 81, 1139–1144.
[6] Yasutake K., Ohmi H., Kakiuchi H., Wakamiya T. and Watanabe H. *Jpn. J. Appl. Phys.* 2006, 45, 3592–3597.
[7] Ohmi H., Kakiuchi H., Yasutake K., Nakahama Y., Ebata Y., Yoshii K. and Mori Y. *Jpn. J. Appl. Phys.* 2006, 45, 3581–3586.
[8] Mori Y., Kakiuchi H., Yoshii K., Yasutake K. and Ohmi H. *J. Phys. D: Appl. Phys.* 2003, 36, 3057–3063.
[9] Kakiuchi H., Ohmi H., Aketa M., Yasutake K., Yoshii K. and Mori Y. *Thin Solid Films* 2006, 496, 259–265.
[10] Kakiuchi H., Nakahama Y., Ohmi H., Yasutake K., Yoshii K. and Mori Y. *Thin Solid Films* 2005, 479, 17–23.
[11] Matsuda A., Kaga T., Tanaka H. and Tanaka K. *Jpn. J. Appl. Phys.* 1984, 23, L567.
[12] Finger F., Kroll U., Viret V., Shah A., Tang X. M., Howling A. and Hollenstein Ch. *J. Appl. Phys.* 1992, 71, 5665.
[13] Druyvesteyn M. J. and Penning F. M. *Rev. Mod. Phys.* 1940, 12, 87–174.
[14] Collins C. B. and Robertson W. W. *J. Chem. Phys.* 1964, 40, 701–712.
[15] Alvarez R., Quintero M. C. and Rodero A. *Spectrochim. Acta Part B* 2004, 59, 709–721.
[16] Rodero A., Quintero M. C., Sola A. and Gamero A. *Spectrochim. Acta Part B* 1996, 51, 467–479.
[17] Ohmi T. *J. Electrochem. Soc.* 1996, 143, 2957.
[18] Hishikawa Y., Nakamura N., Tsuda S., Nakano S., Kishi Y. and Kuwano Y. *Jpn. J. Appl. Phys.* 1991, 30, 1008.
[19] Stutzmann M. In *Properties of Amorphous Silicon and its Alloys*; Searle T.; Ed.; emis DATAREVIEWS SERIES No. 19; INSPEC: London, 1998; pp 56–60.
[20] Hishikawa Y., Tsuda S., Wakisaka K. and Kuwano Y. *J. Appl. Phys.* 1993, 73, 4227.
[21] Nakano M., Yasutake K., Kakiuchi H., Yamauchi Y., Yoshii K., Kataoka T. and Mori Y. *Proc. COE Int. Symp. Ultraprecision Science and Technology – Creation of Perfect Surface* 2001, 221–226.

[22] Shimizu T., Matsumoto M., Yoshita M. and Iwami M. *J. Non-Cryst. Solids* 1991, 137&138, 391–394.
[23] Ghidini G. and Smith F. W. *J. Electrochem. Soc.* 1984, 131, 2924.
[24] Ohki A., Ohmi T., Date J. and Kijima T. *J. Electrochem. Soc.* 1998, 145, 3560–3569.
[25] Shimizu T., Kidoh H., Morimoto A. and Kumeda M. *Jpn. J. Appl. Phys.* 1989, 28, 586.
[26] Reprint from *Jpn. J. Appl. Phys.* 2006, 45, 3587–3591, Kakiuchi H., Ohmi H., Kuwahara Y., Matsumoto M., Ebata Y., Yasutake K., Yoshii K. and Mori Y., High-Rate Deposition of Intrinsic Amorphous Silicon Layers for Solar Cells Using Very High Frequency Plasma at Atmospheric Pressure, Copyright (2006), with permission from The Japan Society of Applied Physics.
[27] Reprint from *J. Non-Cryst. Solids* 2005, 351, 741–747, Kakiuchi H., Matsumoto M., Ebata Y., Ohmi H., Yasutake K., Yoshii K. and Mori Y. Characterization of intrinsic amorphous silicon layers for solar cells prepared at extremely high rates by atmospheric-pressure plasma chemical vapor deposition, Copyright (2006), with permission from Elsevier.
[28] Matsuda A., Takai M., Nishimoto T. and Kondo M. *Sol. Energy Mater. Sol. Cells* 2003, 78, 3.
[29] Lide D. R.; Ed.; *CRC Handbook of Chemistry and Physics* 82nd ed.; CRC Press, 2001–2002; pp 9-50 – 9-64.
[30] Matsuda A. *J. Non-Cryst. Solids* 1983, 59–60, 767.
[31] Knights J. C. and Lujan R. A. *Appl. Phys. Lett.* 1979, 35, 244.
[32] Yoon S. Y., Park S. J., Kim K. H., Jang J. and Kim C. O. *J. Appl. Phys.* 2000, 87, 609.
[33] Bisaro R., Magarino J., Proust N. and Zellma K. *J. Appl. Phys.* 1986, 59, 1167.
[34] Rai-Choudhury P. and Schroder D. K. *J. Electrochem. Soc.* 1973, 120, 664.
[35] Sze S. M. *VLSI Technology*; McGraw-Hill: New York, 1988, p 65.
[36] Meyerson B.S. *Appl. Phys. Lett.* 1986, 48, 797.
[37] Geddo M., Pivac B., Borghesi A., Stella A. and Pedrotti M. *Appl. Phys. Lett.* 1991, 58, 370.
[38] Chen C. -H., Wan C. -M., Yew T. -R., Shieh M. -D. and Kung C. -Y. *Appl. Phys. Lett.* 1993, 62, 3126.
[39] Comfort J. H. and Reif R., *Appl. Phys. Lett.* 1987, 51, 2016.
[40] Ohi S., Burger W. R. and Reif R. *Appl. Phys. Lett.* 1988, 53, 891.
[41] Comfort J. H. and Reif R. *J. Electrochem. Soc.* 1989, 136, 2398.

In: Materials Science Research Trends
Editor: Lawrence V. Olivante, pp. 251-271
ISBN: 978-1-60021-654-1

Chapter 5

Overview of B-Al_5FeSi Phase in Al-Si Alloys

M. Mahta[1], M. Emamy[1], X. Cao[2,3] and J. Campbell[3]

[1] School of Metallurgy and Materials, University of Tehran, Tehran, Iran

[2] Aerospace Manufacturing Technology Center, Institute for Aerospace Research, National Research Council Canada, 5145 Decelles Avenue, Montreal, Quebec, Canada, H3T 2B2

[3] Department of Metallurgy and Materials, University of Birmingham, B15 2TT, UK

Abstract

In aluminum alloys one of the most pervasive and important impurity elements is iron, stemming from the impurities in bauxite ores and the contamination of ferrous metals such as melting tools. Since iron has a very low solid solubility in aluminum (max. 0.05%), almost all iron in aluminum alloys is present in the form of second intermetallic phases. One of the most common Fe-rich intermetallics that form in cast and wrought aluminum alloys upon solidification is the β-Al_5FeSi phase. This phase has long been thought to be brittle and responsible for the inferior mechanical properties (in particular ductility) of aluminum cast alloys. The commonly accepted method to ameliorate the harmful influence of iron is the addition of one or more corrective elements. Such additions generally convert the β-Fe platelets into α-Fe dendrites. Various studies have been carried out by researchers on the modification of β-Al_5FeSi intermetallics in aluminum alloys using Mn, Cr, Co, Mg, Sr, Li and Be. The relative effectiveness of these elements is collected and compared in the present review. The mechanisms for the action of the chemical modifiers are critically reviewed particularly in the light of the modern theory of their nucleation on oxide films present in aluminum melts, probably in large populations. The new insights into the Fe-rich phase in aluminum alloys will aid in better understanding the role of iron in aluminum alloys.

Introduction

Aluminum alloy castings are extensively used to manufacture a large number of components for automotive and aerospace industries. However, the mechanical properties and casting quality of these materials are strongly determined by the microstructures of the alloys. It is common to add a number of different alloying elements to improve mechanical properties and simultaneously suppress the impact of some unwanted impurities, particularly iron because of its serious effect on mechanical properties.

The presence of Fe, even in small amounts, degrades the mechanical properties of aluminum alloys such as strength, fatigue, fracture toughness, and especially the ductility, since it forms β-Al_5FeSi intermetallic phase in a plate-like morphology (needle-like in 2D cross-section). Figure 1 exhibits some β-Fe intermetallics in A413 alloy with high concentration of Fe and without any modifiers. Because of its importance, the phenomenon has been reviewed many times before [1-10].

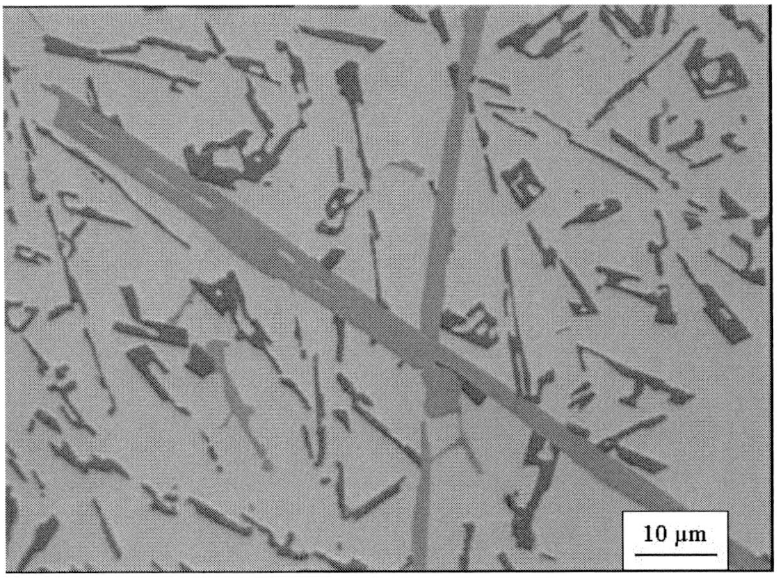

Figure 1. Microstructure of A413 alloy with 1.1 wt% Fe showing β-Fe platelets surrounded by unmodified eutectic silicon particles.

The liquid solubility limits for Fe in aluminum alloys is 1.87 wt% at 655 °C but its solid solubility is only 0.052 wt% at 655°C and is less than 0.01 wt% at 427°C [11,12]. Therefore, during solidification and cooling, there is a strong driving force for the formation of Fe-rich intermetallic compounds containing Al and other alloying elements. Typical morphologies of Fe-rich intermetallics include plate-like β-Al_5FeSi (β-Fe), and dendritic α-Al_8Fe_2Si, i.e. α-Fe, often reported as α-$Al_{12}Fe_3Si_2$ with a probable range of existence of 30-35% Fe and 6-12% Si [13]. In the presence of Mn, the α-Fe phase will precipitate as α-$Al_{15}(FeMn)_3Si_2$. Recently Kral et al. [14,15] has revealed that the Chinese script α-Fe phase is probably cubic $Al_{19}(FeMn)_5Si_2$. The crystal lattice of β-Fe is generally taken to be monoclinic [13,16,17], or $Al_3(FeMn)Si_2$ tetragonal [14,15] (or pseudo-tetragonal) [18,19], or perhaps a multi-layer composite including an orthorhombic phase [20]. Its growth is limited mainly to two dimensions, forming large plates. In contrast, the crystal structure of the α-Fe is cubic (bcc) [13-15,21]. Thus its more uniform surface energy and structure allows it to grow more freely in three dimensions, forming a variety of morphologies.

The work by Kral et al. [14] showed that the α-Fe and β-Fe phases may not clearly exhibit the dendritic (Chinese script) or plate shapes, respectively, and therefore cannot always be identified by their morphologies. Mondolfo [13] reported that Al_5FeSi phase may appear in either Chinese script or platelet form. Even so, acknowledging these reservations, intermetallic phases are generally identified based upon their morphology and/or energy dispersive X-ray spectra (EDS) [14,15,19-21], and there seems little doubt that this usually leads to correct identification. Crystallization of various morphologies of Fe-rich intermetallics is shown in Figure 2. The β-Fe has a well-defined stoichiometry and compared with the α-Fe phase, it dissolves smaller amounts of other corrective elements such as Mn, Cr and Co [22]. In addition, the final morphology of intermetallics in aluminum foundry alloys is greatly affected by casting conditions, particularly the solidification and cooling rates, and alloy chemistry.

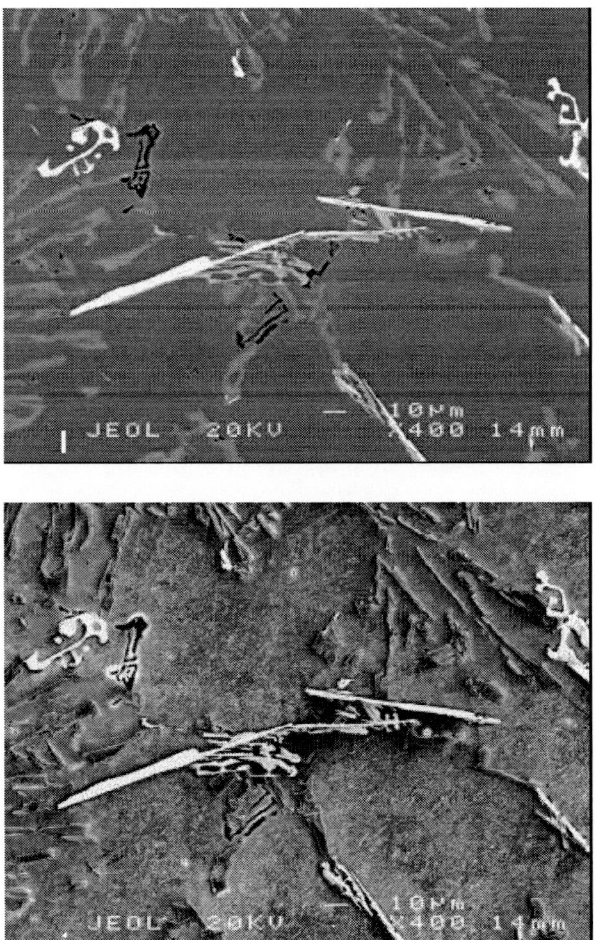

Figure 2. Typical α-Fe, β-Fe and π-Fe phases in Al-11.5Si-0.4Mg cast alloy containing Fe and Mn

In the present review work, the damage and modification mechanisms of β-Fe phase are critically reviewed, and the effectiveness of the various chemical elements as modifiers is compared.

Damage Mechanism

The powerful deleterious effect of the presence of β-Al_5FeSi has generally been attributed to its stress raising potential as a result of its plate-like morphology, and its apparently brittle nature. In addition, Iwahori and coworkers [23] reported that the presence of the plate-like intermetallics considerably increases cracking tendencies and shrinkage cavities during solidification, because of the blockage of the interdendritic channels, thus hindering the flow of liquid metal to feed solidification shrinkage at a late stage of solidification [24,25]. The presence of interdendritic shrinkage cavities adjacent to β-Fe particles has all the appearance of compelling evidence for this mechanism. Even so, there are some authors [26-28] who claim that the presence of porosity and shrinkage defects are not directly related to β-Fe intermetallic particles, but propose that the phenomenon may be due to the effect of β-Fe platelets on the nucleation and growth of eutectic silicon. A further mechanism is proposed here that harmonizes the observations on which both of these approaches are based, introducing the concept of the growth of β-Fe on double oxide film (bifilm) substrates as will be discussed below.

Recent researches have shown that doubled-over oxide films (bifilms) form substrates for the nucleation and growth of iron-rich particles [29-34]. Oxide films become incorporated into melts by an entrainment process [35-39]. This is an enfolding mechanism whereby the oxide skin of the melt becomes incorporated into the bulk liquid. Folded oxide films in melts have two sides: the dry unbonded inner surfaces and their wetted exterior surfaces. The wetted outer interfaces of oxide films appear to act as preferred substrates for the nucleation and growth of some Fe-rich phases. It is expected that there will be good atomic contact between the wetted side of the film and Fe-rich phases or aluminum matrix [29]. However, the gaps between the dry sides constitute the cracks that are commonly observed in association with β-Fe particles. Such cracks are particularly straight for β-Fe (Figure 3), but more irregular for α-Fe (Figure 4). Both α-Fe and β-Fe phases can grow either on the one side or on both sides of the doubled-over film. Thus the cracks automatically form along either the side or the centre of the precipitate. Figure 5 shows a longitudinal crack associated with a β-Fe particle showing an apparent decohesion of the matrix along its upper part (formation of the precipitate only on one side) and along a more central axis in the lower part (formation of the precipitate on both sides of the doubled-over film).

When the melt is first poured, the internal turbulence is sufficiently strong to ensure that the entrained bifilm is repeatedly folded into a small volume, in which form, as a highly convoluted and compact, if somewhat untidy, defect, it is relatively harmless. However, the growth of the monoclinic β-Fe crystal forces a progressive straightening-out action on its substrate, causing the substrate (the bifilm) to become a significant planar crack, thus degrading the properties of the alloy [40]. The reported brittleness of the β-Fe particles is a natural, but perhaps erroneous conclusion drawn from their necessarily cracked appearance as a result of the presence of bifilms. The bifilm is not always observable as a crack in the β-Fe particles; the bifilm will only open to reveal its presence if gas or shrinkage problems operate; otherwise, it will remain closed and practically invisible. Its behavior as a crack, however, is not expected to be significantly impaired. The straightening of the bifilm from relatively harmless compact features into extensive planar cracks by the growth of β-Fe appears to be the mechanism whereby Fe reduces the ductility of Al-Si alloys [35]. In contrast, the α-Fe phase also nucleates and grows on convoluted and compact bifilms, but the bifilm is not straightened; its form remains compact and so relatively harmless because the higher crystal symmetry of the α-Fe phase does not strongly influence its growth morphology, allowing the crystal to be more influenced by the irregular shape of its substrate.

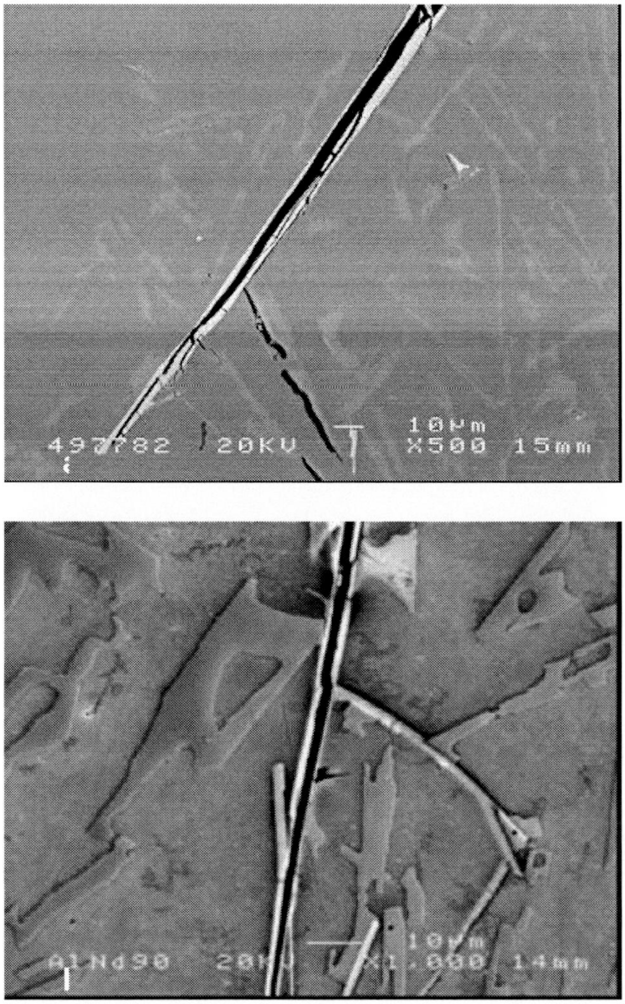

Figure 3. β-Fe platelet and its longitudinal crack in Al-11.5Si-0.4Mg cast alloy containing Fe and Mn

The second type of damage to the properties of castings linked with the presence of β-Fe particles is the apparent association of the particles with porosity, as though the particles have nucleated pores, or obstructed feeding. It is proposed here that the association with the pore occurs if the β-Fe particle happens to have formed on only one side of the bifilm as shown in Figure 6. The remaining oxide film on the opposite side of the bifilm is diaphanously thin, and therefore easily pulled away from its unbonded opposite half, and sucked deep into the dendrite mesh. By this mechanism a shrinkage pore or a gas pore is created, depending on conditions of either poor feeding reducing the external pressure, or high gas content in solution in the melt increasing the internal pressure in the bifilm [34-41].

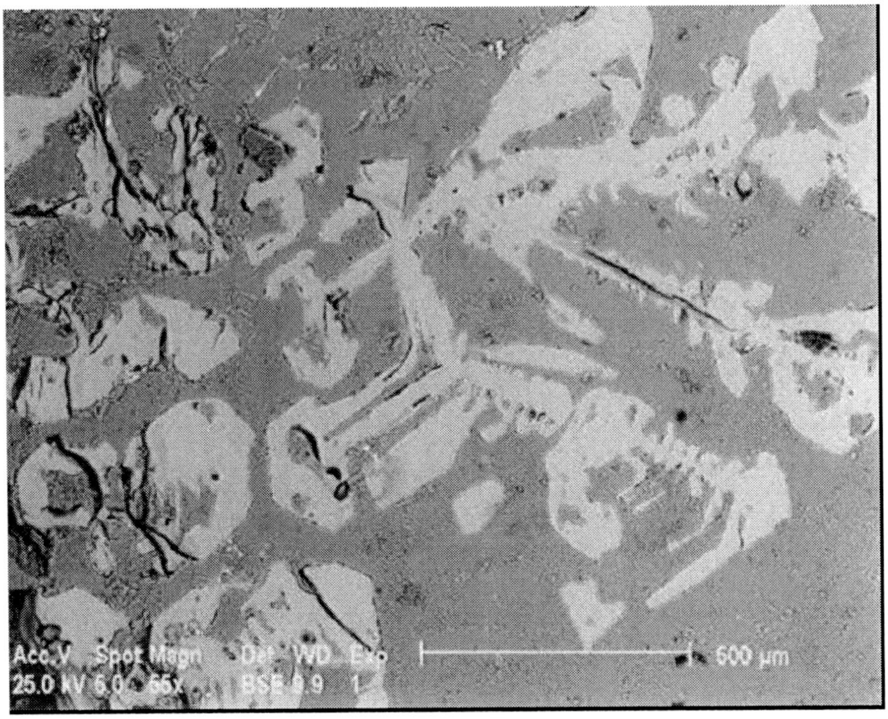

Figure 4. SEM secondary electron image of α–Fe crystals showing a mixture of irregular cracks leading to the formation of irregular particles and more regular dendrites associated with planar central cracks [41].

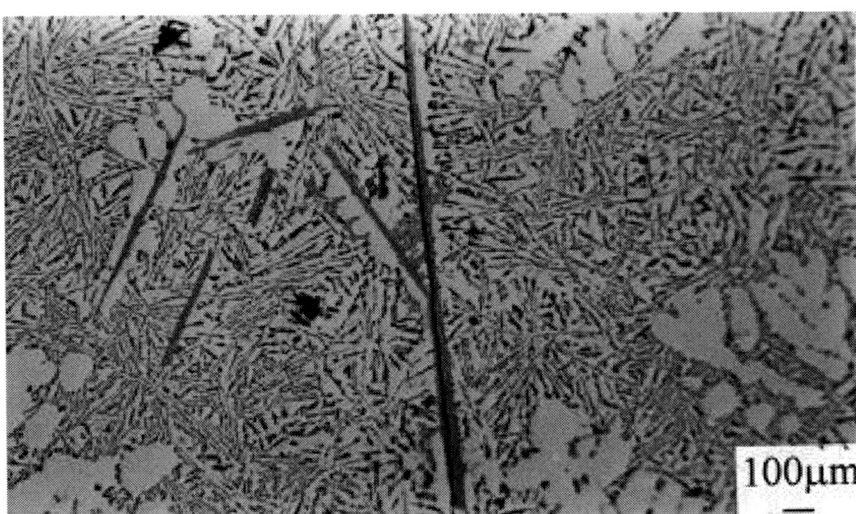

Figure 5. β-Fe platelet and its longitudinal crack in Al-11.5Si-0.4Mg cast alloy containing Fe and Mn

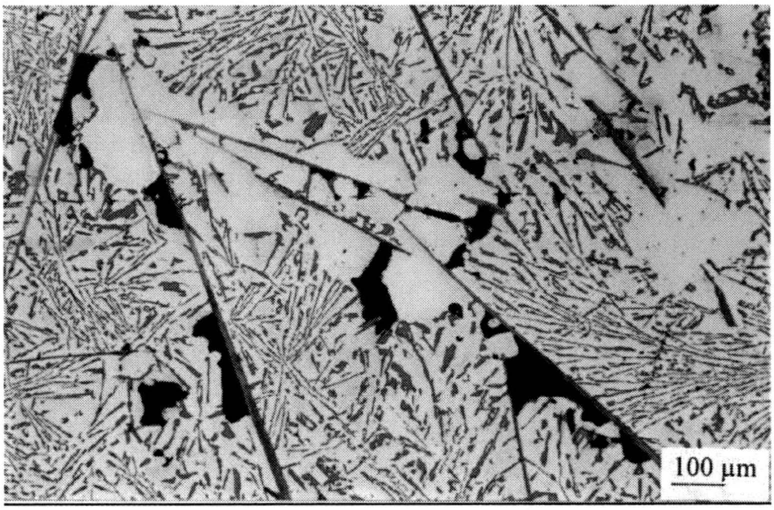

Figure 6. Porosities associated with β-Fe platelets in Al-11.5Si-0.4Mg cast alloy

Modification Mechanism

Following the bifilm hypothesis, in the presence of any modifying action, the formation of α-Fe particles is thought to be the consequence of its nucleation and growth on bifilms [29], but as a result of its high crystal symmetry, it can take up practically any growth form without undue difficulty. Thus chemical modification of β-Fe to α-Fe is advantageous because of the fixing of bifilms in their convoluted state.

Another point that is useful to clarify, is the concept of "fragmentation". This term has been used to describe the appearance of the conversion of β-Fe plates into smaller α-Fe particles. It seems that there has been an unstated assumption that the production of smaller α-Fe particles arises by some kind of mechanical fracturing process, implied by the apparently brittle nature of iron-rich phases. It seems more likely that the apparent fragmentation of plate-like β-Fe to α-Fe occurs because of (i) the dendritic form of the α-Fe appears in cross sections as separated particles, and (ii) the presence of the bifilm cracks in the particles [41,42]. (The bifilm cracks only become clearly visible if the alloy has a high gas content, or suffers some reduction of pressure due to shrinkage, that expands the residual air layer in the centre of the bifilms as shown in Figure 4).

The crystal structure of α-alumina (corundum) is hexagonal. The γ-alumina, spinel and magnesia all have cubic structures [29]. The calculation of the disregistry between the α-Fe phase with some typical oxides such as γ-alumina, α-alumina, spinel and magnesia showed that such oxides may be good substrates for the nucleation and growth of α-Fe phase [29]. To explain the effect of superheat temperature on Fe-intermetallics in Al-Si alloys, however, it has been reported that at low melt superheat temperature, the γ aluminum oxide is stabilized and acts as nucleation substrates to β-Fe platelets. Transformation of γ-alumna to α-alumina with increasing superheat temperature decreases the potential to nucleate β-Fe platelets but provides effective nuclei for the α-Fe phase [43,44].

The addition of chemical neutralizers may stabilise the α-Fe compared to the β-Fe. This effect results from the thermodynamics, i.e. the relative free energies of the phases, as would be apparent on an equilibrium phase diagram. Such considerations have been foremost in the arguments put forward by some authors and are undoubtedly a factor of major importance [45-50]. Even so, the use of equilibrium reactions to explain the changes from β-Fe to α-Fe requires some caution. Most work on the establishment of the currently accepted equilibrium diagrams (for instance the Al-Si-Fe-Mn systems) has been carried out on material that was, unwittingly, of unknown and variable bifilm content. Thus the kinetics of the reactions may have interfered with the accuracy of these early results. Thus, before placing too much faith in classical sources, for instance [13], some of the phase diagrams may require review.

As a further detail considering the equilibrium phase diagram approach, during work on the common industrial alloy A413 containing appreciable amounts of Fe and with the presence of sufficient quantity of a modifier such as Cr, α-script particles were formed probably via a peritectic reaction (L + α-Fe dendrite → α-Al + α-Fe script) [42]. Even so, following the logic presented in this work, this reaction would have taken place on the wetted outer surface of the bifilms in suspension in the melt.

A serious effect of reliance on the use of the equilibrium diagram is that non-equilibrium conditions are normally dominant in solidification. For instance, it seems possible to envisage that Fe-rich intermetallics would not precipitate at all in the absence of suitable substrates such as oxide bifilms [40]. Therefore, if the liquid metal were very clean, it is predicted that the iron would remain in supersaturated solid solution. In this form it is expected that it could be of significant benefit to the alloy, aiding solid solution strengthening, or possibly being manipulated to act as a precipitation-hardening constituent. If this is true, the role of the bifilm in aiding the approach to equilibrium is seen to be central, and its meticulous control leading to its absence can be envisaged to lead to a new generation of metallurgical alloys of low cost but high performance.

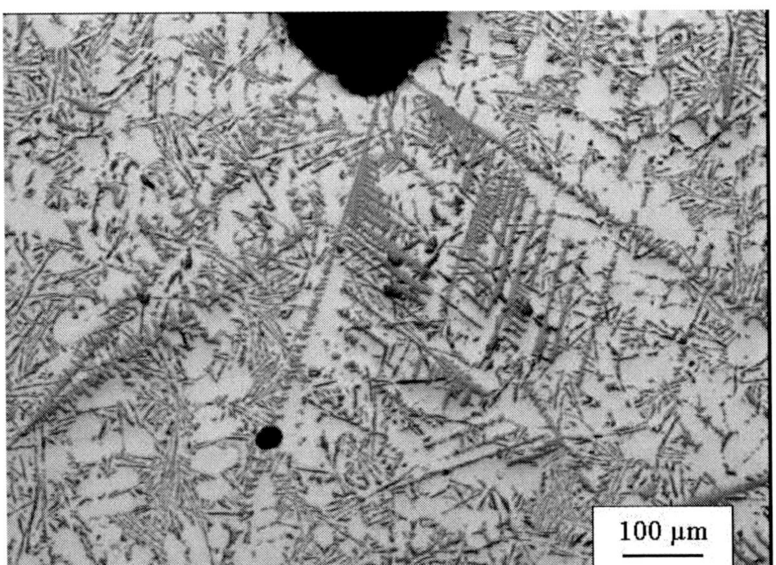

Figure 7. α-Fe dendrites associated with porosity, sedimented near the bottom of the mould in A413 alloy modified with Cr.

Finally, one of the approaches to remove suspended oxide crack defects has suggested the use of the precipitation and sedimentation of primary intermetallic compounds from the liquid metal [30-34]. For this purpose, the molten metal must be cooled to a temperature as low as possible (below the effective liquidus for the precipitation of the primary intermetallic compound, but, of course, above the general liquidus temperature for the precipitation of the α-Al dendrites). In these conditions the primary intermetallics precipitate on the oxide films and mutually sink under their combined weight to the bottom of the holding furnace or crucible [30-34]. Figure 7 illustrates this effect, showing the sedimentation of α-Fe dendrite intermetallic ($Al_{15}(FeCr)_{3.3}Si_{2.2}$) in the bottom of a vertical cylinder casting.

Modifying Elements

Manganese

Mn addition to aluminum alloys is the most commonly used modifying treatment. The effect is probably governed by equilibrium diagram type of considerations [46-49].

Its use in industrial grade 319 aluminum alloy containing 1 wt% Fe with a Fe:Mn ratio of 1.5 causes the crystallization of α-Fe phase, at low and high cooling rates (0.1 and 10 K/s). At 10 K/s cooling rate, α-$Al_{15}(FeMn)_3Si_2$ is formed in interdendritic regions but at even higher solidification rates (20 K/s) almost 50% of the Fe-rich intermetallics crystallizes in the plate-like morphology (β-Fe phase), so, optimum Fe:Mn ratio for the formation of α-Fe phase is greatly related to cooling rate [43]. Another report of work using a relatively high cooling rate of 10 K/s [51] indicated that the optimum Mn concentration is estimated to be 0.5 wt% for 319 alloy with 1.0 wt% Fe and 0.3 wt% for 413 alloy with 1.2 wt% Fe + 0.1 wt% Cr respectively [51]. There is also a report claiming that addition of more than 0.9 wt% Mn to A413 alloy with 2.5 wt% Fe (below the normally accepted Fe:Mn ratio of 2 [4, 52]) leads to the precipitation of “numerous and compacted” α-Fe phase instead of β-Fe plates [41]. Additional investigation reported that for at least partial neutralization of Fe, the ratio Fe:Mn must be less than 2.0, but even higher Mn additions (lower ratios) may not be sufficient [4]. Thus although there is a consensus that Mn is helpful in substituting α-Fe for β-Fe, it seems that the definition of an ‘optimum’ or ‘safe’ ratio for Mn in relation to Fe remains unclear at this time.

Chromium

The main benefit of adding Cr to Al-Mg, Al-Mg-Si and Al-Mg-Zn group alloys is the formation of fine dispersed phases (dispersoids) that prevent grain growth and recrystallization during hot working or heat treatment.

As a common addition to neutralize the effects of Fe, the action of Cr is likely to be analogous to that of Mn, in that the equilibrium distribution of phases is probably altered to promote the formation of α-Fe [42]. The addition of Cr causes the formation of very coarse complex compounds with other impurities such as Fe in those alloys commonly used in the pressure-die-casting industry [11,24,53,54]. The formation, growth and sedimentation of these multicomponent particles can lead to a depletion of Fe concentration, generally accepted to be beneficial to casting quality [5]. Gustafsson et al. [55] claimed that additions of Cr to 356 aluminum alloy have a strong effect on the morphology of Fe-rich intermetallic compounds and can cause the precipitation of α-$Al_{13}(Fe,Cr)_4Si_4$ Chinese-script.

Figure 8 shows various types of α-Fe and β-Fe phases in A413 alloys in the presence of Cr. The recent work by some of the present authors shows that in A413 alloy with an Fe:Cr ratio of 3.0 is required to obtain the crystallization of Fe intermetallics mainly as α-Fe Chinese-script form or branched phases with chemical composition of $Al_{15}(FeCr)_{3.7}Si_{2.3}$ [42]. The combined addition of Mn and Cr (1.0 wt% Mn + 0.1 wt% Cr to counter 1.5 wt% Fe) was found very effective to enhance the tensile properties of 319 automobile alloy [56]. In that study, Fe-rich intermetallics are precipitated completely in a star-like form [56] but higher concentration of Cr is not recommended because of the formation of branched phases and their impact on fluidity and castability [25]. Unfortunately, it was found [42] that presence of Cr also appeared to promote the precipitation of primary silicon (Figure 9).

Cobalt

Co is not a common addition to aluminum alloys [11]. It has an effect similar to Mn and Cr in the neutralization of iron, but a higher concentration of Co is needed to suppress the formation of β-Fe [4,42]. It was reported by the present authors that an optimum Fe:Co ratio of approximately 1.0 is required in commercial A413 alloy to cause the precipitation of Fe-rich intermetallics mainly as α-Fe dendrites with the composition of $Al_{15}(Fe,Co)_4Si_{2.2}$ instead of β-Fe platelets [42]. Analogously to the action of Cr, a potentially unwelcome consequence of the Co addition was the appearance of primary Si particles [42].

Magnesium

Mg is commonly used for strengthening purposes and hardness development in heat-treated Al-Si alloys [11] by the formation of the extremely fine theta phase [50,57]. In Al-Si based alloys, Mg is also known to have a minor effect on refining the silicon eutectic [58].

According to Samuel et al [58], addition of Mg, in amounts up to 0.5 wt% to the 310 alloy at very slow cooling rates (close to equilibrium) leads to the precipitation of rounded black Mg_2Si particles along the sides of the eutectic Si particles. A large proportion of the pervasive β-Al_5FeSi phase was modified into Chinese-script phase with a composition of $Al_8Mg_3FeSi_6$. It was also reported by the same authors that at higher concentrations of Mg (> 1.0 wt%) β-plates are very rarely found, but modification of eutectic Si does not occur due to the consumption of Mg in the precipitation of Mg_2Si and $Al_8Mg_3FeSi_6$ phases. Addition of Mg to A319.1 and A319.2 alloys (with 750°C pouring temperature in graphite moulds) transforms a large number of β-Al_5FeSi plates into the compacted Chinese-script phase [22]. The interesting result of the last reference is the equivalency in the effect of a 1.2 wt% Mg addition and a combination of 0.5 wt% Mg with 0.03 wt% Sr in reducing the volume fraction of β-Fe platelets in 319 alloy [22]. Addition of Mg to the 1000-series wrought alloys promotes the formation of α-Al_8Fe_2Si phase in dendritic form [22]. However, some confusion remains because of the contradictory results of Awano and Shimizu [3]. These authors found that the shape of Fe-rich compounds could not easily be altered from plate-like to Chinese-script morphology with the addition of Mg.

The conflicting findings for the effect of Mg seem likely to be the result of the effect of Mg on bifilms in suspension in the melt. For instance it is known that the iron-rich phases

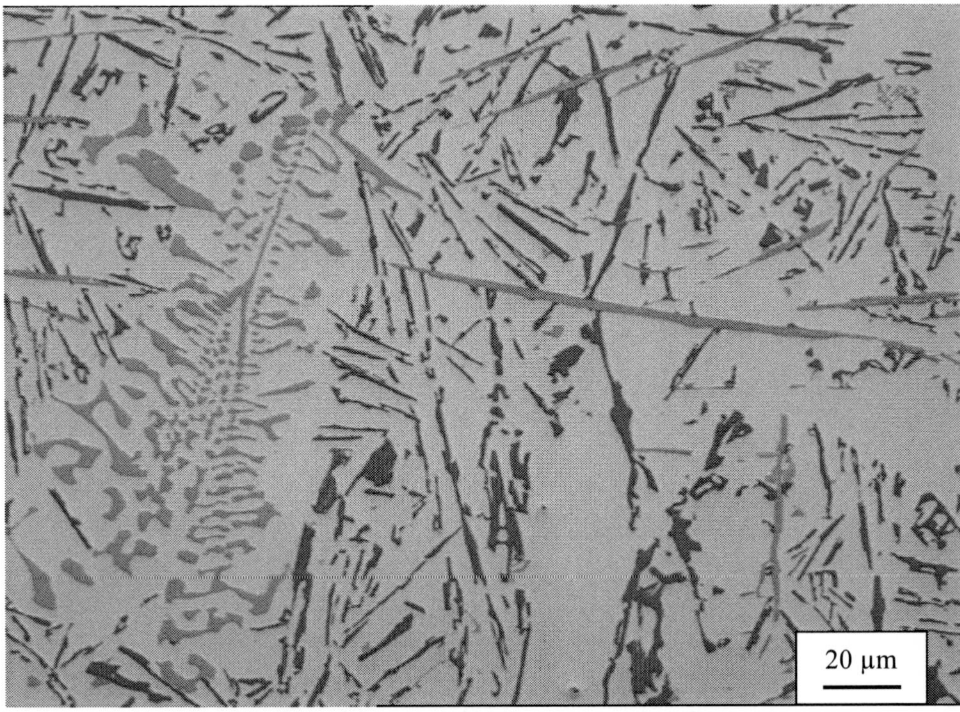

(a)

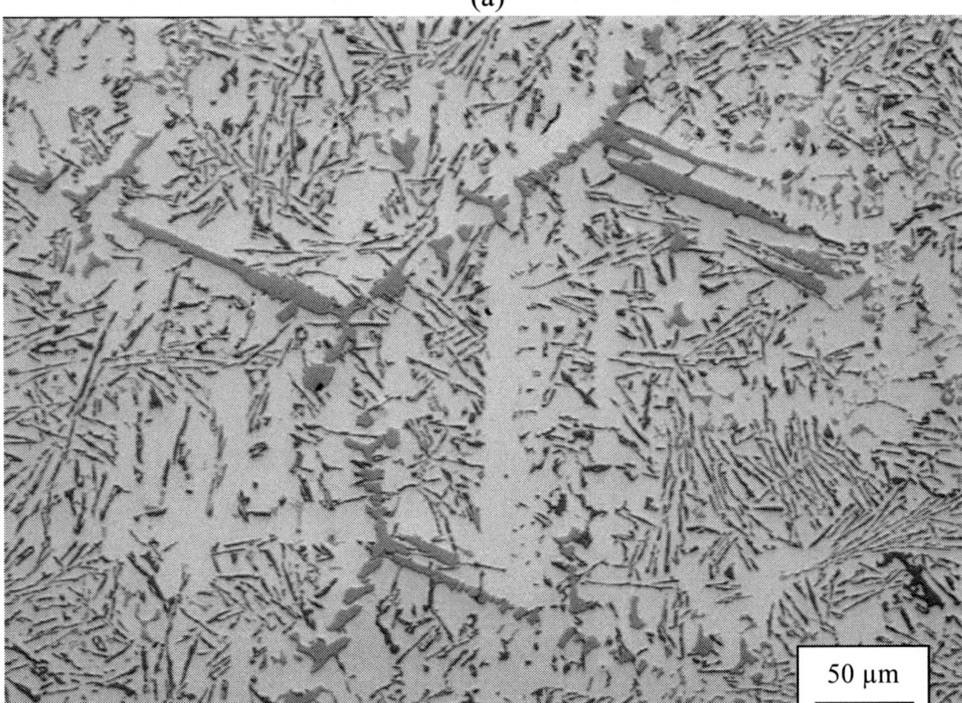

(b)

Figure 8. Various types of α-Fe phase in A413 alloy in the presence of Cr showing: (a) α-Fe Chinese-script (together with some β–Fe platelets) and (b) α-Fe dendrites.

precipitate on oxide films [29]. In addition, there is strong evidence that Si may also precipitate on oxide films [29,31], but that these favoured substrates are deactivated by modifiers such as Na and Sr [59]. By reasonable extrapolation, therefore, it seems probable that Mg, being an alkaline earth metal like Sr, will also deactivate oxide films as substrates for both Si and Fe-rich precipitates. However, being somewhat less effective than Sr will require higher concentration to give an effect equivalent to that of Sr. In addition, the variable nature of experimental results would be expected, since populations of oxide substrates will be highly variable from melt to melt; an effect so far overlooked by investigators, and in any case not easily controlled or measured [60].\

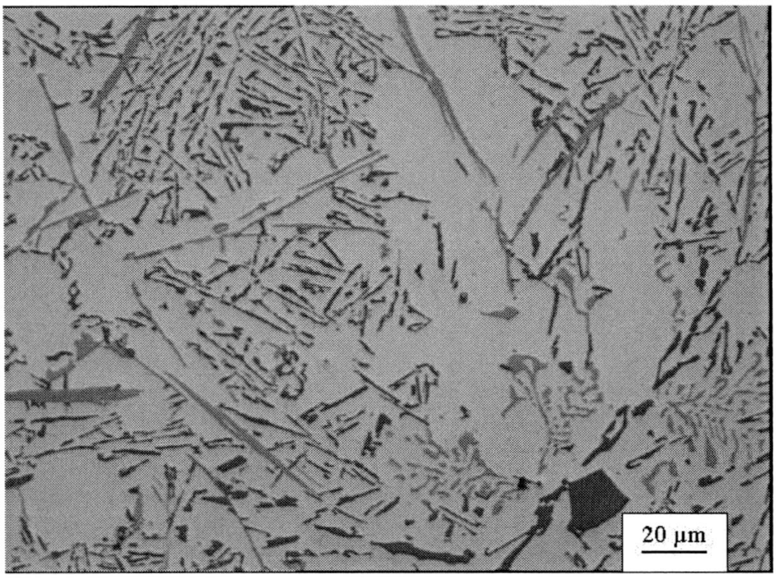

Figure 9. Numerous β-Fe plates in a partial modified Al-Si eutectic, containing both Chinese-script α-Fe particles and primary Si crystal (black) in A413 alloy modified with Cr.

Strontium

Modification of eutectic silicon is routinely carried out, using Sr, to transform the flake morphology into a fibrous form [61,62]. Mechanical properties such as ductility, UTS, hardness and machinability are generally increased appreciably only at low concentration of Sr (0.008-0.04 wt%) [11]. The effect seems to be associated, once again, with the deactivation of the oxide as a favored substrate for Si, so that Si no longer nucleates in the liquid ahead of the solidification front, thus giving a more planar front. The conversion of a ragged to a planar front has the effect of reducing the area of the front, but since the rate of extraction of heat from the casting remains unchanged, the rate of advance of the front is greatly increased, as though it were chilled, so that the eutectic is refined. This clever idea proposed in 1981 by Flood and Hunt [63] is given additional support more recently [35]. The reduced nucleation ahead of the front, and the increased rate of its advance is naturally associated with an observed increase in the undercooling at the front. The consequent reduction in Si spacing will also reduce the size of other second phases such as Fe-rich intermetallics. Most of the effect of Si (and by implication, also of Na and other Al-Si eutectic modifiers) is assumed to be explained by this mechanism. Thus to reduce the detrimental effect of Fe in both wrought and foundry aluminum alloys, Sr can be added as a powerful modifying element. The formation of β-Fe phase can be reduced and even suppressed with the use of Sr modification [64,65].

However, it is also well established that the addition of Sr increases the amount of hydrogen porosity in castings especially with slow cooling rates [11,35,66-70]. This is a complicating effect of Sr that to some extent counters its benefit with respect to the refinement of Si and Fe-rich phases. Those foundries that enjoy good melting and metal transfer technology generally experience relatively good freedom from oxide bifilms [59]. As a natural consequence they do not therefore receive any significant benefit of the suppression of the prior nucleation of Si and/or Fe phases, because there are few films on which nucleation can occur. Thus such foundries receive practically no benefit from Sr, but suffer the disadvantage of additional porosity, probably from the precipitation of the additional hydrogen on their population of old bifilms. Thus these few foundries have suffered a degradation of the mechanical properties of their alloys when attempting to modify with Sr, and such foundries have therefore abandoned the use of Sr modification [59].

There are, however, many reports of the beneficial effects of Sr in reducing the proportion of β-Fe plates and replacing these with α-Fe Chinese script. These reports are at least partly explained by the action of Sr to suppress the precipitation of β-Fe on bifilms, so that bifilms are not straightened by the subsequent growth of the β-Fe phase, and properties are not therefore impaired. The Chinese script form of α-Fe is clearly a dendritic growth form, requiring, in an extreme case, only a single nucleus (which might be an oxide) from which to start. Being relatively free of internal defects such precipitates are relatively strong, resisting failure by cracking or apparent decohesion. Some of the experimental findings are listed below.

The addition of 0.01 to 0.05 wt%Sr to 6069 wrought aluminum alloy has been reported to transform a large proportion of β-Fe plates, into the compact α-Fe Chinese-script [71]. Another report claimed the same effect of Sr in modification of 1XXX and 6XXX DC alloys [72]. It was also claimed that at approximately 0.05 wt% Sr, all of the intermetallic phases in Al-Cu-Mg-Zn wrought alloy are modified [73]. If the bifilm concept is also in operation in this instance, we may surmise that the Sr may also be deactivating the oxide bifilms, making them less favorable for precipitation of the AlCuMg-rich compounds. The Sr addition of 0.06 wt% changes the morphology of Fe-rich intermetallics from plate-like to Chinese-script form [74].

Pennors et al. [75] indicated that Sr is effective in causing the modification and control of the β-Al_5FeSi phase to a large extent in Al-6 wt%Si-3.5 wt% Cu (319) alloy, when the Fe content is 0.5 and 1.0 wt%. It was found that the optimum range of Sr in minimizing the β-plate lengths is approximately 400-600 ppm (irrespective of cooling rate) and higher levels of Sr cause an "overmodification" effect [75]. Effectiveness of Sr addition in the reduction of the total amount of intermetallics during solidification of an Al-12.2 wt% Si alloy has been demonstrated [76]. The maximum reduction in the size, number and vol% of β-Fe phase and alteration of morphology from plate-like to dendritic form was observed when adding Sr in a range of 0.04-0.06 wt% in A413 and 413P cast alloys in both sand and permanent mold castings [77]. These studies have been supported by observations carried out by Samuel et al. [73], where the authors have attributed the shortening of β-plates to the poisoning of nucleation sites by Sr, in agreement with the proposition that Sr deactivates the bifilm substrates for β-Fe. They found the effect was accelerated with increasing Sr content to approximately 300 ppm (the optimum level) in commercial 319 alloy end-chilled castings [78]. In another work by these authors [57,58], the optimum range of Sr was concluded between 0.02 and 0.04 wt% for the elimination of more than two thirds of the β-Fe plates and modification of the Mg_2Si particles in 319 alloy. The presence of Sr (300 ppm) leads to precipitation of a large part of the Fe-rich phases in the form of coarse pre-dendritic particles situated within the α-aluminum dendrites in A380.1 alloy with 1 wt% Fe [79]. Addition of Sr together with Mn was found to be effective in the conversion of β-Fe phase into α-Fe dendrites [41]. The precipitates of α-Fe are likely to be precipitated together with the bifilm substrate in the melting furnace prior to the melt being poured. Thus the poured melt is likely to be relatively free from both Fe-rich precipitates and bifilms, and so improved in properties.

The claimed optimum addition levels of Sr reported above at 300-600 ppm, contrast with other foundry operations that have found 50 ppm is ideal, and yet others that have found that 0 ppm is best. These conflicting results seem likely to be consistent with the varying bifilm contents of melts. As melts become cleaner, the increased rarity of substrates means that less area of oxide is required to be deactivated by Sr [59].

Finally, it should be noted that a quite different mechanism for the beneficial effects of Sr in reducing the numbers of β-Fe platelets has been proposed based on the assumption that Sr strengthens the oxide film on the melt [41]. This would have the effect, in some pouring operations, of holding back the oxide layer, reducing the area that enters the melt, and reducing the shredding of oxide film that does, by chance, enter the melt. It is clear that such a mechanism, if true, would be very much dependent on the detailed geometry of the pouring operation. There seems good circumstantial evidence that such an effect has been observed in the case of Be additions to Al melts [80].

Lithium

Presence of Li greatly increases the oxidation rate of molten aluminum [81,82] and changes the surface characteristics of wrought products [11]. Li increases the hydrogen solubility in aluminum melts [83] and is one of the effective modifiers of the eutectic Si phase in Al-Si alloys [84]. Thus, by analogy with the action of Sr, it is strongly suspected that Li may be expected to have a refining action on Fe-rich phases by the deactivation of favored oxide substrates and the consequential speeding up of the freezing front. This conclusion is given weight by the observation that Li addition increases the undercooling during the solidification of the eutectic [84].

The effect of adding Li to Al-6.5 wt%Si-3.5 wt%Cu-1 wt%Fe cast alloy cooled at 4.2 K/s has been reported by Ashtari et al. [84]. They indicated that the length of the β-plates decreased from 37.0 to 14.5 μm by the addition of 0.33 wt% Li. Shortening of the β-plates by Li is accompanied by the formation of AlLiSi phase, which is undesirable for mechanical properties [84], so only a low addition of Li is recommended. This understandable conclusion might be challenged if the AlLiSi phase only precipitates on bifilms in suspension in the liquid, as seems to be the case for many other intermetallics [35]. If so, we might predict that higher Li additions would create no problems if the melt were especially clean.

Beryllium

Be is used in aluminum alloys containing Mg to reduce oxidation losses from melts at elevated temperatures but only a few parts per million are required for this purpose.

Significantly higher levels of Be are required for alloying purposes. The addition of Be may be helpful in fluidity of automotive aluminum alloys [11]. According to Villeneuve et al. [56] Be additions in amounts of approximately 0.08 wt% to variants of Al-6 wt%Si-3.5 wt%Cu-0.1 wt%Mn (A319.2) with 1.4 wt% Fe caused a large proportion of β-Fe platelets to be replaced with α-Chinese-script phase identified as Al_8Fe_2SiBe. This phase was formed within the α-aluminum dendrites but a limited number of fine β-plates still remained in the microstructure. To enhance the effectiveness of Be an additional corrective element may be necessary. Exceptional improvement in ductility and strength was obtained by the addition of Be together with 200 ppm Sr, but the increase of Sr level beyond this limit led to precipitation of large β-Fe platelets [56]. Be up to 0.4 wt% can cause the formation of compact $Al_4Fe_2Be_5$ particles [3]. It was also noted that an approximately spheroidal form of Fe-rich intermetallics can be formed in aluminum containing 6 to 10 wt% Si by addition of 0.05 to 0.5 wt% Be [4]. The morphology of intermetallic compounds in Al-7 wt%Si-0.3 wt%Mg with 0.2-1.0 wt% Fe alloy in the presence of Be, transform from β-Fe plates into α-Fe Chinese-script form with the chemical formulae once again Al_8Fe_2SiBe [85,86]. The combined effect of 0.08wt%Be and 0.02wt%Sr in A380.1 alloy containing 1.4 wt% Fe is reported to be equivalent to the addition of 1 wt% Mn (Fe:Mn ratio of 1.4)[79].

The beneficial effect of Be in reducing the amount of β-Fe phase in Al melts may also be importantly influenced from a completely different mechanism. It is well known in semi-continuous DC casting of wrought alloys that attempts to eliminate Be from certain alloys has caused great problems with castability. It is thought that the Be greatly strengthens the oxide film on the surface of the liquid metal, to such an extent that the oxide formed in pouring and transfer operations 'hangs on' to the lip of the pouring vessel, and so does not find its way into the casting. Thus Be-containing castings are cleaner, containing fewer oxides entering the melt at the pouring stage of casting. Thus the precipitation of β-Fe is discouraged by reduced area of favorable substrate [80].

It must be mentioned that in practice the use of Be is currently limited, and will become more so, because of toxicity problems.

Counter-Modifying Elements

Recent work has found that Sr at low levels of addition (in contrast to its beneficial effects at higher levels [41]) used typically for the modification of the eutectic silicon in Al-Si alloys, is antagonistic towards the beneficial action of Mn in modifying β-Fe to α-Fe [41]. This negative action might also be expected to extend to other silicon modifiers such as Na, and possibly to other β-Fe modifiers such as Cr, Co, etc.

Some such effect is predictable if the action of Sr to modify the Si eutectic is a consequence of the deactivation of bifilms by Sr, making them unfavourable sites for the formation of Si [35]. Such deactivation might not unreasonably be expected to affect the effectiveness of the bifilms as sites for the formation of Fe-rich intermetallics as is observed [41].

Clearly, this effect very much complicates the overall action of Sr, since Sr has other beneficial and detrimental effects. A proper evaluation remains to be established. In particular, of course, the further complication of such an attempt at evaluation will be a challenge, because of the effects of bifilms will require to be controlled, and such control will not easily be attained.

Conclusion

1) The detrimental effect of Fe in Al alloys arises because β-Fe particles straighten bifilms, and thus convert relatively harmless convoluted cracks in suspension in the liquid to large straight cracks.
2) The prior cooling of the melt to cause the sedimentation of primary α-Fe intermetallics is beneficial to casting quality because (i) Fe remaining in the alloy is reduced, and (ii) oxide bifilms are simultaneously removed.
3) The reduction of the detrimental effect of Fe using the chemical modification of Mn, Cr, Sr or Be is found to be effective, and to be preferred to Co, Mg and Li.
4) There appear to be several different mechanisms for the different chemical modifiers.

    i) Mn and Cr appear to stabilise the α-Fe phase. Such action is in principle predictable from the equilibrium phase diagram.
    ii) Sr (and possibly other silicon eutectic modifiers) poisons the oxide film substrates for eutectic Si, deactivating and therefore effectively removing the favoured substrates, and simultaneously straightening and speeding the freezing front, giving β-Fe less time to grow.
    iii) Both Be and Sr may act by strengthening the oxide film so that it is held back during pouring, not entering the melt to provide substrates for the formation of β-Fe.

References

[1] Villeneuve, C., & Samuel, F. H. (1999). Fragmentation and dissolution of β-Al_5Fesi phase during solution heat treatment of Al-13wt%Si-Fe alloys. *International Journal of Cast Metals Research,* **12**, 145-160.
[2] Mascré, C. (1955). Influence du fer et du manganèse sur les alliages du type de l'A-S13 (Alpax). *Fonderie,* **108**, 4330-4336.
[3] Awano, Y., & Shimizu, Y. (1990). Non-equilibrium crystallization of AlFeSi compound in melt-superheated Al-Si alloy castings. *AFS Transactions,* **98**, 889-895.
[4] Crepeau, P. N. (1995). Effect of iron in Al-Si casting alloys: a critical review. *AFS Transactions,* **103**, 361-366.
[5] Wang, L., Apelian, D., & Makhlouf, M. (1997). Tensile properties of aluminum die-casting alloys. *19th International Die-Casting Congress,* November, Minneapolis, Minnesota.

[6] Murali, S., Raman, K. S., & Murthy, K. S. S. (1994). Morphological studies on β-$FeSiAl_5$ phase in Al-7-Si-0.3Mg alloy with trace additions of Be, Mn, Cr, and Co. *Materials Characterization,* **33** (2), 99-112.

[7] Westbrook, J. H., & Fleischer, R. L. (2000). *Structural Applications of Intermetallic Compounds.* Chichester, UK: John Wiley & Sons Ltd.

[8] Liu, Y. L., Kang, S. B., & Kim, H. W. (1999). The complex microstructures in an as-cast Al-Mg-Si alloy. *Materials Letters,* **41**, 267-272.

[9] Espinoza-Cuadra, J., Garíca-García, G., & Mancha-Molinar, H. (2005). Influence of defects on strength of industrial aluminum alloy Al-Si 319. *Materials & Design,* in press.

[10] Sina, H., Emamy, M., Saremi, M., Keyvani, A., Mahta, M., & Campbell, J. (2006). The influence of Ti and Zr on electrochemical properties of aluminum sacrificial anodes. *Materials Science and Engineering A,* **431** (1-2), 263-276.

[11] Hatch, J. E. (1984). *Aluminum: Properties and Physical Metallurgy.* Metals Park, OH: ASM International.

[12] Wang, L., Apelian, D., & Makhlouf, M. M. (1999). Iron-bearing compounds in Al-Si diecasting alloys: morphology and conditions under which they form. *AFS Transactions,* **146**, 231-238.

[13] Mondolfo, L. F. (1976). *Aluminium Alloys; Structure and Properties.* Oxford, UK: Butterworth & Co. Ltd.

[14] Kral, M. V., McIntyre, H. R., & Smillie, M. J. (2004). Identification of intermetallic phases in a eutectic Al–Si casting alloy using electron backscatter diffraction pattern analysis. *Scripta Materialia,* **51** (3), 215-219.

[15] Kral, M. V. (2005). A crystallographic identification of intermetallic phases in Al–Si alloys. *Materials Letters,* **59** (18), 2271-2276.

[16] Barlock, J. G., & Mondolfo, L. F. (1975). Structure of Some Aluminum-Iron-Magnesium-Manganese-Silicon Alloys. *Zeitschrift für Metallkunde,* **66**, 605-611.

[17] Tanihata, H., Sugawara, T., Matsuda, K., & Ikeno, S. (1999). Effect of casting and homogenizing treatment conditions on the formation Al-Fe-Si intermetallic compounds in 6063 Al-Mg-Si alloys. *Journal of Materials Science,* **34** (6), 1205-1210.

[18] Brandes, E. A., & Brook, G. B. (1992). *Smithells Metals Reference Book* (7th edition). Oxford, UK: Butterworth & Co. Ltd.

[19] Hansen, V., Hauback, B., Sundberg, M., Römming, Chr., & Gjönnes, J. (1998). β-$Al_{4.5}FeSi$: a combined synchrotron powder diffraction, electron diffraction, high-resolution electron microscopy and single-crystal x-ray diffraction study of a faulted structure. *Acta Crystallografica,* **B54**, 351-357.

[20] Zheng, J. G., Vincent, R., & Steeds, J. W. (2000). Crystal structure of an orthorhombic phase in beta-(Al-Fe-Si) precipitates determined by convergent-beam electron diffraction. *Philosophical Magazine A,* **80** (2), 493-500.

[21] Zheng, J. G., Vincent, R., & Steeds, J. W. (1999). Crystal structure determination of an Al-Fe-Si-Be phase by convergent-beam electron diffraction. *Philosophical Magazine A,* **79** (11), 2725-2733.

[22] Samuel, A. M. & Samuel, F. H. (1997). Modification of iron intermetallics by magnesium and strontium in Al-Si alloys. *International Journal of Cast Metals Research,* **10**, 147-157.

[23] Iwahori, H., Takamiya, H., Yonekura, K., Yamamoto, Y., & Nakamura, M. (1988). Influence of iron and manganese on feedability of AC2B aluminum alloy. *Imono,* **60** (9), 590-595.

[24] Shabestari, S. G., & Gruzleski, J. E. (1995). Gravity segregation of complex intermetallic compounds in liquid aluminum-silicon alloys. *Metallurgical and Materials Transactions A,* **26A**, 999-1006.

[25] Serák, J., & Vojktéch, D. (2002). Influence of (AlSIFeMnCr) intermetallic phases on the casting properties of AlSi9Cu2FeMnCr alloys. *ALUMINIUM,* **78**, 384-387.

[26] Taylor, J. A., Schaffer, G. B., & StJOHN, D. H. (1999). The role of iron in the formation of porosity in Al-Si-Cu-based casting alloys: Part I. Initial experimental observations. *Metallurgical and Materials Transactions A,* **30A**, 1643-1650.

[27] Taylor, J. A., Schaffer, G. B., & StJOHN, D. H. (1999). The role of iron in the formation of porosity in Al-Si-Cu-based casting alloys: Part II. A phase-diagram approach. *Metallurgical and Materials Transactions A,* **30A**, 1651-1655.

[28] Taylor, J. A., Schaffer, G. B., & StJOHN, D. H. (1999). The role of iron in the formation of porosity in Al-Si-Cu-based casting alloys: Part III. A microstructural model. *Metallurgical and Materials Transactions A,* **30A**, 1657-1662.

[29] Cao, X., & Campbell, J. (2003). The nucleation of Fe-rich phases on oxide films in Al-11.5Si-0.4Mg cast alloys. *Metallurgical and Materials Transactions A,* **34A**, 1409-1420.

[30] Cao, X., & Campbell, J. (2000). Effect of precipitation of primary intermetallic compounds on tensile properties of cast Al-11.5Si-0.4Mg alloy. *AFS Transactions,* **108**, 391-400.

[31] Cao, X., & Campbell, J. (2004). The solidification characteristics of Fe-rich intermetallics in Al-11.5Si-0.4Mg cast alloys. *Metallurgical and Materials Transactions A,* **35A**, 1425-1435.

[32] Cao, X., & Campbell, J. (2001). Convection-free precipitation of primary iron-rich phase in liquid Al 11.5Si 0.4Mg alloy. *AFS Transactions,* **109**, 501-515.

[33] Cao, X., Saunders, N., & Campbell, J. (2004). Effect of iron and manganese contents on convection-free precipitation and sedimentation of primary α-Al(FeMn)Si phase in liquid Al-11.5Si-0.4Mg alloy. *Journal of Materials Science,* **39** (7), 2303-2314.

[34] Cao, X., & Campbell, J. (2004). Effect of precipitation and sedimentation of primary α-Fe phase on liquid metal quality of cast Al-11.1Si-0.4Mg alloy. *International Journal of Cast Metals Research,* **17** (1), 1-11.

[35] Campbell, J. (2003). *Castings* (2nd Edition). Oxford, UK: Butterworth-Heinemann.

[36] Fox, S., & Campbell, J. (2000). Visualisation of oxide film defects during solidification of aluminium alloys. *Scripta Materialia,* **43**, 881-886.

[37] Yang, X., Huang, X., Dai, X., Campbell, J., & Tatler, J. (2004). Numerical modeling of entrainment of oxide film defects in filling of aluminium alloy castings. *International Journal of Cast Metals Research,* **17** (6), 321-331.

[38] Campbell, J. (2006). Entrainment defects. *Materials Science and Technology,* **22** (2) 127-145.

[39] Dai, X., Yang, X., Campbell, J., & Wood, J. (2004). Influence of oxide film defects generated in filling on mechanical strength of aluminium alloy castings. *Materials Science and Technology,* **20** (4), 505-513.

[40] Cao, X., & Campbell, J. (2006). Morphology of β-Al_5FeSi phase in Al-Si-Mg cast alloys. *Materials Transactions, JIM,* **47** (5), 1303-1312.

[41] Shabestari, S. G., Mahmudi, M., Emamy, M., & Campbell, J. (2002). Effect of Mn and Sr on intermetallics in Fe-rich eutectic Al-Si alloy. *International Journal of Cast Metals Research,* **15** (1), 17-24.

[42] Mahta, M., Emamy, M., Daman, A., Keyvani, A., & Campbell, J. (2005). Precipitation of Fe rich intermetallics in Cr- and Co-modified A413 alloy. *International Journal of Cast Metals Research,* **18** (2), 73-79.

[43] Narayanan, L. A., Samuel, F. H., & Gruzleski, J. E. (1994). Crystallization behavior of iron-containing intermetallic compounds in 319 aluminum alloy. *Metallurgical and Materials Transactions A,* **25A**, 1761-1773.

[44] Ahmad, R., & Marshall, R. I. (2003). Effect of superheating on iron-rich plate-type compounds in aluminium-silicon alloys. *International Journal of Cast Metals Research,* **15**, 497-504.

[45] Taylor, J. A., StJohn, D. H., Barresi, J., & Couper, M. J. (2000). Influence of Mg content on the microstructure and solid solution chemistry of Al-7%Si-Mg casting alloys during solution treatment. *Materials Science Forum*, **33**, 277-282.

[46] Taylor, J. A., Dahle, A., StJohn, D. H., & Schneider, W. (2003). The role of manganese and iron in porosity formation in aluminum - silicon foundry alloys. The 42nd Annual Conference of Metallurgists of CIM (COM 2003); Vancouver, Canada, 24-27 Aug. 2003. pp. 483-497.

[47] Phillips, H. W. L., & Varley, P. C. (1943). The Constitution of Alloys of Aluminum with Manganese, Silicon, and Iron. III. - The Ternary System: Aluminum-Silicon-Iron. IV. - The Quaternary System: Aluminum-Manganese-Silicon-Iron. *Journal of the Institute of Metals*, **69**, 317-350.

[48] Warmuzek, M., Gazda, A., Sieniawski, J., & Mrówka, G. (2003). Processes of the formation of the Fe (Mn)-bearing intermetallic phases in the Al-Fe-(Mn)-Si alloys. *Advances in Materials Science,* **4** (2), 81-91.

[49] Shabestari, S. G. (2004). The effect of iron and manganese on the formation of intermetallic compounds in aluminum-silicon alloys. *Materials Science and Engineering A,* **383** (2), 289-298.

[50] Backerud, L., Chai, G., & Tamminen, J. (1990). *Solidification Characteristics of Aluminum Alloys: Vol.* **2**. *Foundry Alloys.* Des Plaines, IL: AFS/Skanaluminium.

[51] Lakshmanan, A. N., Shabestari, S. G., & Gruzleski, J. E. (1995). Microstructure control of iron intermetallics in AL-Si casting alloys. *Zeitschrift für Metallkunde,* **86**, 457-464.

[52] Mackay, R. I., & Gruzleski, J. E. (1997). Quantification of iron in aluminium-silicon foundry alloys via thermal analysis. *International Journal of Cast Metals Research,* **10**, 131-145.

[53] Jorstad, J. L. (1986). Understanding sludge. *Die Casting Engineer*, Nov/Dec, 30-36. Yener, Y., *Convective Heat transfer*, 2nd ed., pp 302, 311-373, CRC Press Inc., Boca Raton, Florada (1995)

[54] Osame, Y., Toyoda, K., Tsumura, Y., Suzuki, M., Furuya, S., & Nagayama, K. (1986). Sludge deposition in molten aluminum die casting alloys in the holding furnace. *Light Metals,* **36** (12), 813-818.

[55] Gustafsson, G., Thorvaldsson, T., & Dunlop, G. L. (1986). Influence of Fe and Cr on the microstructure of cast Al-Si-Mg alloys. *Metallurgical Transactions A,* **17A**, 45-52.

[56] Villeneuve, C., Samuel, A. M., Samuel, F. H., Doty, H. W., & Valtierra, S. (2001). Role of trace elements in enhancing the performance of 319 aluminum alloys. *AFS Transactions,* **23**, 1-14.

[57] Samuel, A. M., Gauthier, J., & Samuel, F. H. (1996). Microstructural aspects of the dissolution and melting of Al_2Cu phase in Al-Si alloys during solution heat treatment. *Metallurgical and Materials Transactions A,* **27A**, 1785-1798.

[58] Samuel, F. H., Ouellet, P., Samuel, A. M., & Doty, H. W. (1998). Effect of Mg and Sr additions on the formation of intermetallics in Al-6 wt pct Si-3.5 wt pct Cu-(0.45) to (0.8) wt pct Fe 319-type alloys. *Metallurgical and Materials Transactions A,* **29A**, 2871-2884.

[59] Campbell, J. (2005). Shaped Casting: The John Campbell Symposium. P. Crepeau, & M. Tiryakioglu. San Francisco, CA: TMS.

[60] Dispinar, D., & Campbell, J. (2005). Shaped Casting: The John Campbell Symposium. P. Crepeau, & M. Tiryakioglu. San Francisco, CA: TMS.

[61] García-Hinojosa, J. A., González, C. R., González, G. M., & Houbaert, Y. (2003). Structure and properties of Al-7Si-Ni and Al-7Si-Cu cast alloys nonmodified and modified with Sr. *Journal of Materials Processing Technology,* 143-144, 306-310.

[62] Kulunk, B., Shabestari, S. G., Gruzleski, J. E. & Zuliani, D. J. (1996). Beneficial effects of strontium on A380 alloy. *AFS Transactions,* **170**, 1189-1193.

[63] Flood, S. C., & Hunt, J. D. (1981). Modification of Al-Si eutectic alloys with Na. *Metal Science,* **15** (7), 287-294.

[64] Sigworth, G. K., Shivkumar, S., & Apelian, D. (1989). The Influence of Molten Metal Processing on Mechanical Properties of Cast Al-Si-Mg Alloys. *AFS Transactions,* **97**, 811-824.

[65] Samuel, F. H., Samuel, A. M., Doty, H. W., & Valtierra, S. (2001). Decomposition of Fe-Intermetallics in Sr-Modified Cast 6XXX Type Aluminum Alloys for Automotive Skin. *Metallurgical and Materials Transactions A,* **32A**, 2061-2075.

[66] Fuoco, R., Goldenstein, H., & Gruzleski, J. E. (1994). Evaluation of effect of modification-induced eutectic undercooling on microporosity formation in 356 al alloy. *AFS Transactions,* **102**, 297-306.

[67] Emadi, D., & Gruzleski, J. E. (1994). Effects of casting and melt variables on porosity in directionally-solidified Al-Si alloys. *AFS Transactions,* **102**, 307-312.

[68] Emadi, D., Gruzleski, J. E., & Toguri, J. M. (1993). The effect of Na and Sr modification on surface tension and volumetric shrinkage of A356 Alloy and their influence on porosity formation. *Metallurgical Transactions B,* **24B**, 1055.

[69] Anson, J. P., Stucky, M., & Gruzleski, J. E. (2000). Effect of strontium modification on the nucleation and growth of microporosity during the solidification of aluminum-7%silicon foundry alloy. *AFS Transactions,* **108**, 419-426.

[70] Cao, X., & Campbell, J. (2004). Effect of Sr on primary α-Fe phase in liquid Al-11.5Si-0.4Mg cast alloy. *Materials Science and Technology,* **20** (4), 514-520.

[71] Morris, L. R., & Miners, F. B. (1975). Aluminium alloys: Aluminium alloy containing strontium and/or calcium - to enhance alpha phase formation of metal impurities. US Patent No. 3,926,690.

[72] Mulazimoglu, M. H., Zaluska, A., Gruzleski, J. E., & Paray, F. (1996). Electron microscope study of Al-Fe-Si intermetallics in 6201 aluminum alloy. *Metallurgical and Materials Transactions A*, **27A**, 929-936.

[73] Vernam, W. D., & Lifka, B. W. (1987). Aluminum base alloys of the A1-Cu-Mg-Zn type: Aluminium alloy prods. – containing copper, magnesium, zinc and strontium. US Patent No. 4,711,762.

[74] Mulazimoglu, M. H., Paray, F., Stephen, G., Kulunk, B., & Gruzleski, J. E. (1994). Modification of Intermetallic Phases by Strontium in Aluminum Wrought Alloys. *Light Metals*, Annual Meeting in San Francisco, California, February 27 - March 3. 1047-1055.

[75] Pennors, A., Samuel, A. M., Samuel, F. H., & Doty, H. W. (1998). Precipitation of β-Al_5FeSi iron intermetallic in Al-6%Si-3.5%Cu (319) type alloys: role of Sr and P. *AFS Transactions,* **106**, 251-264.

[76] Haque, M. M., & Kondic, V. (1985). Light Metals, Science and Technology, Proceedings of an International Symposium, Trans. Tech. Publications, Swiss, 159-170.

[77] Shabestari, S. G., & Gruzleski, J. E. (1995). Modification of iron intermetallics by strontium in 413 aluminum alloys. *AFS Transactions,* **103**, 285-293.

[78] Samuel, A. M., Samuel, F. H., & Doty, H. W. (1996). Observations on the formation of β-Al_5FeSi phase in 319 type Al-Si alloys. *Journal of Materials Science,* **31**, 5529-5539.

[79] Samuel, F. H., Pucella, G., Villeneuve, C., Samuel, A. M., Doty H. W., & Valtierra, S. (1999). Microstructural observations on Fe-intermetallics in unmodified and Sr-modified Al-Si-Cu (A380.1) die casting alloy. *International Journal of Cast Metals Research,* **12**, 197-210.

[80] J. Campbell, Verbal communication from Alcan UK personnel, 2004.

[81] Prasad, N. H., & Balasubramaniam, R. (1997). Influence of laser surface treatment on the oxidation behaviour of an Al-Li-Cu alloy. *Journal of Materials Processing Technology,* **68** (2), 117-120.

[82] Jayaram, V. (1996). The role of volatile solute elements in directed melt oxidation of Al-alloys. *Journal of Materials Science,* **31**, 4591-4598.

[83] Saikawa, S., Nakai, K., Sugiura, Y., & Kamio, A. (1999). Effect of hydrogen gas content on generation of porosity in Al-Li casting alloys. *Materials Transactions, JIM,* **40** (1), 57-63.

[84] Ashtari, P., Tezuka, H., & Sato, T. (2004). Influence of Li addition on intermetallic compound morphologies in AL-Si-Cu cast alloys. *Scripta Materialia,* **51**, 43-46.

[85] Murali, S., Raman, K. S., & Murthy, K. S. S. (1994). Effect of trace additions on the mechanical properties and fracture toughness of Fe-containing Al-7Si-0.3Mg alloy. *Cast Metals,* **6** (4), 189-198.

[86] Murali, S., Raman, K. S., & Murthy, K. S. S. (1995). The formation of β-$FeSiAl_5$ and Be-Fe phases in Al-7%Si-0.3%Mg alloy containing Be. *Materials Science and Engineering A,* **190**, 165-172.

In: Materials Science Research Trends
Editor: Lawrence V. Olivante, pp. 273-292
ISBN 978-1-60021-654-1

Chapter 6

SUPERSELECTION RULES INDUCED BY INFRARED DIVERGENCE

Joachim Kupsch*
Fachbereich Physik, TU Kaiserslautern, D-67653 Kaiserslautern, Germany

Abstract

Superselection rules induced by the interaction with a mass zero Boson field are investigated for a class of exactly soluble Hamiltonian models. The calculations apply as well to discrete as to continuous superselection rules. The initial state (reference state) of the Boson field is either a normal state or a KMS state. The superselection sectors emerge if and only if the Boson field is infrared divergent, i. e. the bare photon number diverges and the ground state of the Boson field disappears in the continuum. The time scale of the decoherence depends on the strength of the infrared contributions of the interaction and on properties of the initial state of the Boson system. These results are first derived for a Hamiltonian with conservation laws. But in the most general case the Hamiltonian includes an additional scattering potential, and the only conserved quantity is the energy of the total system. The superselection sectors are stable against the perturbation by the scattering processes.

1. Introduction

Superselection rules are the basis for the emergence of classical physics within quantum theory. But despite of the great progress in understanding superselection rules, see e.g. [Wig95], quantum mechanics and quantum field theory do not provide enough exact superselection rules to infer the classical probability of "facts" from quantum theory. This problem is most often discussed in the context of measurement of quantum mechanical objects. In an important paper about the process of measurement Hepp [Hep72] has presented a class of models for which the dynamics induces superselection sectors. Hepp starts with a very large algebra of observables – essentially all observables with the exception of the "observables at infinity" which constitute an a priory set of superselection rules – and the

*E-mail address: kupsch@physik.uni-kl.de

superselection sectors emerge in the weak operator convergence. But it has soon been realized that the algebra of observables, which is relevant for the understanding of the process of measurement [Emc72b] [Ara80] and, more generally for the understanding of the classical appearance of the world [Zur82] [JZ85] [JZK$^+$03] can be severely restricted. Then strong or even uniform operator convergence is possible.

A system, which is weakly coupled to an environment, which has a Hamiltonian with a continuous spectrum, usually decays into its ground state, if the environment is in a normal state; or the system approaches a canonical ensemble, if the environment is in a state with positive temperature. More interesting decoherence effects may occur on an intermediate time scale, or in systems, for which the decay or the thermalization are prevented by conservation laws. To emphasize effects on an intermediate time scale one can use a strong coupling between system and environment. This method has some similarity to the singular coupling method of the Markov approximation, which also scales the dynamics at an intermediate time period to large times. The basic model, which we discuss, has therefore the following properties: existence of conservation laws and strong coupling. Thereby strong coupling means that the spectral properties of the Hamiltonian are modified by the interaction term.

In this paper, which is an extension of [Kup00b], we investigate the emergence of superselection rules for a system, which is coupled to a mass zero Boson field. The dynamics of the total system is always generated by a semibounded Hamiltonian. The restriction to the Boson sector corresponds to a van Hove model [vH52]. As the main result we prove for a class of such models:

– The superselection rules are induced by the infrared contributions of the Boson field.
– The superselection sectors are stable for $t \to \infty$ if and only if the Boson field is infrared divergent.

The infrared divergence of the van Hove model has been studied by Schroer [Sch63] more than forty years ago. The Boson field is still defined on the Fock space, but the ground state of the Boson field disappears in the continuum. In the usual discussions of decoherence this type of infrared divergence corresponds to the ohmic or subohmic case [LCD$^+$87]. As additional result we prove that the induced superselection sectors are stable against perturbation by scattering processes.

The paper is organized as follows. In Sect. 2. we give a short introduction to the dynamics of subsystems and to superselection rules induced by the environment. The calculations are preformed in the Schrödinger picture, which allows also non-factorized initial states. We prove that the off-diagonal matrix elements of the reduced statistical operator can be suppressed in trace norm for discrete and for continuous superselection rules. In Sect. 3. we investigate a class of Hamiltonian models with the environment given by a mass zero Boson field, and the interplay between infrared divergence and induced superselection rules is derived. The resulting superselection sectors do not depend on the initial state; they finally emerge for all initial states of the total system. But to have superselection sectors, which are effective on a short time scale, the reference state of the environment has to satisfy some "smoothness" conditions.

In Sect. 3.4. we admit a KMS state of positive temperature as reference state of the Boson system. Again the same superselection sectors emerge, even on a shorter time scale. In the final Sect. 4. we prove that the induced superselection sectors are stable against additional scattering processes. Some technical details for the Sects. 2. and 3. are given in the Appendices A and B.

2. Induced Superselection Rules

2.1. General Considerations

We start with a few mathematical notations. Let $\mathcal{H}$ be a separable Hilbert space, then the following spaces of linear operators are used.
$\mathcal{B}(\mathcal{H})$: The linear space of all bounded operators A with the operator norm $\|A\|$.
$\mathcal{T}(\mathcal{H})$: The linear space of all nuclear operators A with the trace norm $\|A\|_1 = \mathrm{tr}\sqrt{A^+A}$.
$\mathcal{S}(\mathcal{H})$: The set of all positive nuclear operators W with a normalized trace, $\mathrm{tr}\, W = 1$.
If A is a closed (unbounded) linear operator, then $\mathcal{D}(A) \subset \mathcal{H}$ denotes the domain of definition of this operator.

With the exception of Sect. 3.4., where also KMS states are admitted for the environment, we assume standard quantum mechanics where any state of a quantum system is represented by a statistical operator $W \in \mathcal{S}(\mathcal{H})$; the rank one projection operators thereby correspond to the pure states. Without additional knowledge about the structure of the system we have to assume that the set of all states corresponds to $\mathcal{S}(\mathcal{H})$, and the operator algebra of all (bounded) observables coincides with $\mathcal{B}(\mathcal{H})$.

In the following we consider an *open system*, i.e. a system S which interacts with an environment E, such that the total system $S \times E$ satisfies the usual Hamiltonian dynamics. The Hilbert space $\mathcal{H}_{S\times E}$ of the total system is the tensor space $\mathcal{H}_S \otimes \mathcal{H}_E$ of the Hilbert spaces for S and for E. Let $W \in \mathcal{S}(\mathcal{H}_{S\times E})$ be the state of the total system and $A \in \mathcal{B}(\mathcal{H}_S)$ be an observable of the system S, then the expectation $\mathrm{tr}_{S\times E} W(A \otimes I_E)$ satisfies the identity $\mathrm{tr}_{S\times E}\,(A \otimes I_E)W = \mathrm{tr}_S\, A\rho$ with the reduced statistical operator $\rho_S = \mathrm{tr}_E W \in \mathcal{S}(\mathcal{H}_S)$. Here the symbols tr_S, tr_E and $\mathrm{tr}_{S\times E}$ denote the (partial) traces with respect to the Hilbert spaces $\mathcal{H}_S$, $\mathcal{H}_E$ or $\mathcal{H}_{S\times E}$, respectively. We shall refer to $\rho_S = \mathrm{tr}_E W$ as the *state* of the system S. As indicated above we consider the usual Hamiltonian dynamics for the total system, i.e. $W \to W(t) = U(t)WU^+(t) \in \mathcal{S}(\mathcal{H}_{S\times E})$ with the unitary group $U(t) = \exp(-iH_{S\times E}t)$ generated by the total Hamiltonian $H_{S\times E}$. Except for the trivial case that S and E do not interact, the dynamics of the reduced statistical operator $\rho_S(t) = \mathrm{tr}_E\, U(t)WU^+(t)$ does no longer follow a group law; and it is exactly this dynamics which can produce induced superselection sectors.

In order to define a linear dynamics $\rho_S \to \rho_S(t)$ for the state of the system S we have to assume that the initial state factorizes as

$$W = \rho_S \otimes \rho_E, \tag{1}$$

see [KSS01]. Here $\rho_S \in \mathcal{S}(\mathcal{H}_S)$ is the initial state of the system and $\rho_E \in \mathcal{S}(\mathcal{H}_E)$ is the reference state of the environment. The dynamics of the reduced statistical operator $\rho_S(t)$ then follows as

$$\rho_S \in \mathcal{S}(\mathcal{H}_S) \to \rho_S(t) = \Phi_t(\rho_S) := \mathrm{tr}_E U(t)\,(\rho_S \otimes \rho_E)\, U^\dagger(t) \in \mathcal{S}(\mathcal{H}_S). \tag{2}$$

The reduced dynamics $\Phi_t(\rho_S)$ can be extended to a continuous linear mapping $\rho_S \in \mathcal{T}(\mathcal{H}_S) \to \Phi_t(\rho_S) \in \mathcal{T}(\mathcal{H}_S)$ with the obvious properties

$$\|\Phi_t(\rho_S)\|_1 \leq \|\rho_S\|_1 \,, \ \mathrm{tr}_S\Phi_t(\rho_S) = \mathrm{tr}_S\rho_S, \ \Phi_t(\rho_S) \geq 0 \text{ if } \rho_S \geq 0. \tag{3}$$

Here $\|\cdot\|_1$ is the trace norm of operators on $\mathcal{H}_S$.

Discrete and continuous superselection rules are characterized by a self-adjoint superselection operator $F = \int_{\mathbf{R}} \lambda P(d\lambda)$. The projection operators $P(\Delta)$ of the spectral resolution of this operator are defined for all intervals $\Delta = [a, b)$ of the real line and satisfy

$$\begin{array}{l} P(\Delta^1 \cup \Delta^2) = P(\Delta^1) + P(\Delta^2) \quad \text{if} \quad \Delta^1 \cap \Delta^2 = \emptyset \\ P(\Delta^1)P(\Delta^2) = P(\Delta^1 \cap \Delta^2), \ P(\emptyset) = 0, \ P(\mathbf{R}) = 1. \end{array} \tag{4}$$

The mapping $\Delta \to P(\Delta)$ can be extended to the σ-algebra of the real line $\mathcal{B}(\mathbf{R})$ generated by open sets. In the most commonly discussed case of discrete superselection rules the function $x \in \mathbf{R} \to P((-\infty, x))$ is a step function. In the case of continuous superselection rules, in which we are mainly interested in, the function $x \in \mathbf{R} \to P((-\infty, x))$ is strongly continuous, and the projection operators $P((a, b)) = P([a, b)) = P([a, b])$ coincide.

The dynamics of the total system $S \times E$ induces superselection rules into the system S, if there exists a family of projection operators $\{P_S(\Delta) \mid \Delta \subset \mathbf{R}\}$ on the Hilbert space $\mathcal{H}_S$, which satisfies the rules (4), such that the off-diagonal parts $P_S(\Delta^1)\Phi_t(\rho_S)P_S(\Delta^2)$ of the statistical operators of the system S are dynamically suppressed, i.e. $P_S(\Delta^1)\Phi_t(\rho_S)P_S(\Delta^2) \to 0$ if $t \to \infty$ and $\mathrm{dist}(\Delta^1, \Delta^2) > 0$. In the subsequent sections we derive superselection rules, for which the off-diagonal parts of the statistical operator even vanish in trace norm

$$\left\| P_S(\Delta^1)\Phi_t(\rho_S)P_S(\Delta^2) \right\|_1 \to 0 \quad \text{if } t \to \infty \tag{5}$$

for all initial states $\rho_S \in \mathcal{S}(\mathcal{H}_S)$ and all separated intervals Δ^1 and Δ^2. This statement, which does not specify the time scale of the decoherence process, can be used as definition of induced superselection rules. But to have superselection rules, which contribute to the emergence of classical properties, the decrease of (5) has to be sufficiently fast. We shall come back to that problem later.

2.2. Models

For all models we are investigating, the total Hamiltonian is defined on the tensor space $\mathcal{H}_{S\times E} = \mathcal{H}_S \otimes \mathcal{H}_E$ as

$$\begin{aligned} H_{S\times E} &= H_S \otimes I_E + I_S \otimes H_E + F \otimes G \\ &= \left(H_S - \frac{1}{2}F^2\right) \otimes I_E + \frac{1}{2}(F \otimes I_E + I_S \otimes G)^2 + I_S \otimes \left(H_E - \frac{1}{2}G^2\right) \end{aligned} \tag{6}$$

where H_S is the positive Hamiltonian of S, H_E is the positive Hamiltonian of E, and $F \otimes G$ is the interaction potential between S and E with operators F on $\mathcal{H}_S$ and G on $\mathcal{H}_E$. To guarantee that $H_{S\times E}$ is self-adjoint and semibounded we assume

1) The operators F and F^2 (G and G^2) are essentially self-adjoint on the domain of H_S (H_E). The operators $H_S - \frac{1}{2}F^2$ and $H_E - \frac{1}{2}G^2$ are semibounded.

Since $F^2 \otimes I_E \pm 2F \otimes G + I_S \otimes G^2$ are positive operators, the operator $F \otimes G$ is $(H_S \otimes I_E + I_S \otimes H_E)$-bounded with relative bound one, and Wüst's theorem, see e.g. Theorem X.14 in [RS75], implies that $H_{S\times E}$ is essentially self-adjoint on the domain of $H_S \otimes I_E + I_S \otimes H_E$. Moreover $H_{S\times E}$ is obviously semibounded.

To derive induced superselection rules we need the rather severe restriction

2) The operators H_S and F commute strongly, i.e. their spectral projections commute.

This assumption implies that F is a conserved quantity of the dynamics generated by the Hamiltonian (6). The operator F has a spectral representation

$$F = \int_{\mathbf{R}} \lambda P_S(d\lambda) \tag{7}$$

with a family (4) of projection operators $P_S(\Delta)$ indexed by measurable subsets $\Delta \subset \mathbf{R}$. We shall see below that exactly the projection operators of this spectral representation determine the induced superselection sectors.

As a consequence of assumption 2) we have $[H_S, P_S(\Delta)] = 0$ for all intervals $\Delta \subset \mathbf{R}$. The Hamiltonian (6) has therefore the form $H_{S\times E} = H_S \otimes I_E + \int_{\mathbf{R}} P_S(d\lambda) \otimes (H_E + \lambda G)$. The operator $|G| = \sqrt{G^2}$ has the upper bound $|G| \leq aG^2 + (4a)^{-1}I$ with an arbitrarily small constant $a > 0$. Since G^2 is H_E-bounded with relative bound 2, the operator G is H_E-bounded with an arbitrarily small bound. The Kato-Rellich theorem, see e.g. [RS75], implies that the operators $H_E + \lambda G$ are self-adjoint on the domain of H_E for all $\lambda \in \mathbf{R}$. The unitary evolution $U(t) := \exp(-iH_{S\times E}t)$ of the total system can therefore be written as $U(t) = (U_S(t) \otimes I_E) \int dP_S(\lambda) \otimes \exp\left(-i\left(H_E + \lambda G\right)t\right)$, where

$$U_S(t) = \exp\left(-iH_S t\right) \tag{8}$$

is the unitary evolution of the system S. The evolution (2) of an initial state $\rho_S \in \mathcal{S}(\mathcal{H}_S)$ follows as

$$\Phi_t(\rho_S) = U_S(t)\left(\int_{\mathbf{R}\times\mathbf{R}} \chi\left(\alpha, \beta; t\right) P_S(d\alpha)\, \rho_S\, P_S(d\beta)\right) U_S^+(t) \tag{9}$$

with the trace

$$\chi(\alpha, \beta; t) = \mathrm{tr}_E\left(\mathrm{e}^{i(H_E+\alpha G)t}\mathrm{e}^{-i(H_E+\beta G)t}\rho_E\right). \tag{10}$$

For the models investigated in Sect. 3. this trace factorizes into

$$\chi(\alpha, \beta; t) = \mathrm{e}^{i\vartheta(\alpha,t)}\chi_0(\alpha - \beta; t)\mathrm{e}^{-i\vartheta(\beta,t)} \tag{11}$$

where $\vartheta(\alpha, t)$ is a real phase. The function $\chi_0(\lambda; t) = \overline{\chi_0(-\lambda; t)}$ and its derivative can be estimated by

$$|\chi_0(\lambda; t)| \leq \phi\left(\lambda^2\zeta(t)\right), \quad \int_\delta^\infty \left|\frac{\partial}{\partial\lambda}\chi_0(\lambda; t)\right| d\lambda \leq \phi\left(\delta^2\zeta(t)\right) \text{ for all } \delta \geq 0. \tag{12}$$

Thereby $\phi(s)$ is a positive non increasing function which vanishes for $s \to \infty$ such that $\int^{\infty} \phi(s)ds < \infty$, and the time dependent positive function $\zeta(t)$ increases to infinity if $t \to \infty$. The factorization (10) implies that the operator (9) is the product $\Phi_t(\rho_S) = U_S(t)U_\vartheta(t) \left(\int_{\mathbf{R}\times\mathbf{R}} \chi_1 (\alpha - \beta; t)\, P_S(d\alpha)\, \rho_S\, P_S(d\beta)\right) U_\vartheta^+(t)U_S^+(t)$ with the unitary operator $U_\vartheta(t) = \int \exp\left(i\vartheta(\alpha, t)\right) P_S(d\alpha)$. As the projection operators $P_S(\Delta)$ commute with the unitary operators $U_S(t)$ and $U_\vartheta(t)$, the trace norm of $P_S(\Delta^1)\Phi_t(\rho_S)P_S(\Delta^2)$ is given by

$$\left\|P_S(\Delta^1)\Phi_t(\rho_S)P_S(\Delta^2)\right\|_1 = \left\|\int_{\Delta^1\times\Delta^2} \chi_0 \left(\alpha - \beta; t\right) P_S(d\alpha)\, \rho_S\, P_S(d\beta)\right\|_1 . \tag{13}$$

The phase function does not contribute to this norm. In Appendix A we prove that the estimate (12) is sufficient to derive the upper bound

$$\left\|P_S(\Delta^1)\Phi_t(\rho_S)P_S(\Delta^2)\right\|_1 \leq \phi\left(\delta^2\zeta(t)\right) \tag{14}$$

for arbitrary intervals Δ^1 and Δ^2 with a distance $\delta \geq 0$. This estimate is uniform for all initial states $\rho_S \in \mathcal{S}(\mathcal{H}_S)$. The arguments of Appendix A are applicable to superselection operators (7) F with an arbitrary spectrum. For operators $F = \sum \lambda_n P_n^S$ with a discrete spectrum, which has no accumulation point, uniform norm estimates can be derived with simpler methods, see [Kup00a] or Sect. 7.6 of [JZK$^+$03].

The norm (14) vanishes for all intervals with a positive distance $\mathrm{dist}(\Delta^1, \Delta^2) = \delta > 0$ on a time scale, which depends on the functions $\zeta(t)$ and $\phi(s)$. The function $\zeta(t)$ is mainly determined by the Hamiltonian, whereas $\phi(s)$ depends strongly on the reference state of the environment.

Remark 1 *A simple class of explicitly soluble models, which yield estimates similar to (14), can be obtained under the additional assumptions*
– the operator G has an absolutely continuous spectrum,
– the Hamiltonian H_E and the potential G commute strongly.
Models of this type have been investigated (for operators F with a discrete spectrum) by Araki [Ara80] and by Zurek [Zur82], see also Sect. 7.6 of [JZK$^+$03] and [Kup00a]. With these additional assumptions the trace (10) simplifies to $\chi(\alpha, \beta; t) = \mathrm{tr}_E\left(\mathrm{e}^{i(\alpha-\beta)Gt}\rho_E\right)$. Let $G = \int_{\mathbf{R}} \lambda P_E(d\lambda)$ be the spectral representation of the operator G. Then the measure $d\mu(\lambda) := \mathrm{tr}_E\left(P_E(d\lambda)\,\rho_E\right)$ is absolutely continuous with respect to the Lebesgue measure for any $\rho_E \in \mathcal{S}(\mathcal{H}_E)$, and the function $\chi(t) := \mathrm{tr}_E\left(\mathrm{e}^{iGt}\rho_E\right) = \int_{\mathbf{R}} \mathrm{e}^{i\lambda t}\, d\mu(\lambda)$ vanishes for $t \to \infty$. Under suitable restrictions on the reference state the measure $d\mu(\lambda) = \mathrm{tr}_E\left(P_E(d\lambda)\,\rho_E\right)$ has a smooth density, and we can derive a fast decrease of the Fourier transform $\chi(t)$ and its derivatives. That implies upper bounds similar to (12) and a fast decrease of (10) $\chi(\alpha, \beta; t)$ for $\alpha \neq \beta$.

Remark 2 *Instead of the dynamics (2) in the Schrödinger picture we can use the Heisenberg dynamics*

$$A \in \mathcal{B}(\mathcal{H}_S) \to \Psi_t(A) = \Psi_t(A) := \mathrm{tr}_E U^\dagger(t)\left(A \otimes I_E\right) U(t)\rho_E \in \mathcal{B}(\mathcal{H}_S) \tag{15}$$

to investigate induced superselection rules. As the estimate (14) is uniform with respect to the initial state $\rho_S \in \mathcal{S}(\mathcal{H}_S)$*, the duality relation* $\mathrm{tr}_S\left(P_S(\Delta^1)\Phi_t(\rho_S)P_S(\Delta^2)A\right) = \mathrm{tr}_S\,\rho_S\Psi_t\left(P_S(\Delta^2)AP_S(\Delta^1)\right)$ *leads to a criterion for induced superselection rules in the Heisenberg picture:*

$$\lim_{t\to\infty}\left\|\Psi_t\left(P_S(\Delta^1)AP_S(\Delta^2)\right)\right\| = 0 \tag{16}$$

for all observables $A \in \mathcal{B}(\mathcal{H}_S)$ *and for all intervals* Δ^1 *and* Δ^2 *with a distance* $\mathrm{dist}(\Delta^1, \Delta^2) > 0$. *In the case of models with the Hamiltonian (6) which satisfy Assumption 2), the condition (16) is equivalent to a more transparent condition. For these models the full dynamics* $U(t) = \exp(-iH_{S\times E}t)$ *commutes with* $P_S(\Delta)\otimes I_E$*, and the Heisenberg dynamics (15) satisfies the identities* $P_S(\Delta)\Psi_t(A) = \Psi_t(P_S(\Delta)A)$ *and* $\Psi_t(A)P_S(\Delta) = \Psi_t(AP_S(\Delta))$. *These identities and (16) imply that the off-diagonal parts of* $\Psi_t(A)$ *have to vanish for all observables* $A \in \mathcal{B}(\mathcal{H}_S)$ *and for all disjoint intervals with a non-vanishing distance at large* t

$$\lim_{t\to\infty}\left\|P_S(\Delta^1)\Psi_t\left(A\right)P_S(\Delta^2)\right\| = 0. \tag{17}$$

This criterion resembles the definition of the exact superselection rules: $P_S(\Delta^1)\,A\,P_S(\Delta^2) = 0$ *for all* $\Delta^1 \cap \Delta^2 = \emptyset$*, see e.g. [Jau60] or [Wig95]. The criterion (17) has been used in [Kup00b] to derive induced superselection rules for the model of Sect. 3..*

3. The Interaction with a Boson Field

3.1. The Hamiltonian

We choose a system S which satisfies the constraints 1) and 2), and the environment E is given by a Boson field. As specific example we may consider a spin system with Hilbert space $\mathcal{H}_S = \mathbf{C}^2$ and Hamiltonian $H_S = \alpha\sigma_3$ and $F = \beta\sigma_3$ where $\alpha \geq 0$ and β are real constants and σ_3 is the Pauli spin matrix. A more interesting example is a particle on the real line with velocity coupling. The Hilbert space of the particle is $\mathcal{H}_S = \mathcal{L}^2(\mathbf{R})$. The Hamiltonian and the interaction potential of the particle are

$$H_S = \frac{1}{2}P^2 \text{ and } F = P \tag{18}$$

where $P = -i\,d/dx$ is the momentum operator of the particle. The identity $H_S - \frac{1}{2}F^2 = 0$ guarantees the positivity of the first term in (6).

As Hilbert space $\mathcal{H}_E$ we choose the Fock space of symmetric tensors $\mathcal{F}(\mathcal{H}_1)$ based on the one particle Hilbert space $\mathcal{H}_1$. The Hamiltonian is generated by a one-particle Hamilton operator M on $\mathcal{H}_1$ with the following properties

(i) M is a positive operator with an absolutely continuous spectrum,

(ii) M has an unbounded inverse M^{-1}.

The spectrum of M is (a subset of) $\mathbf{R}_+$, which – as a consequence of the second assumption – includes zero. The Hamiltonian of the free field is then the derivation $H_E = d\Gamma(M)$ generated by M, see Appendix B. As explicit example we may take $\mathcal{H}_1 = \mathcal{L}^2(\mathbf{R}^n)$ with inner product $(f \mid g) = \int_{\mathbf{R}^n}\overline{f(k)}g(k)d^nk$. The one-particle Hamilton operator can be chosen as $(Mf)(k) := \varepsilon(k)f(k)$ with the positive energy function $\varepsilon(k) = c|k|$, $c > 0$, $k \in \mathbf{R}^n$. Let $a_k^\#$, $k \in \mathbf{R}^n$, denote the distributional creation/annihilation operators, such that $a^+(f) = \int a_k^+ f(k)d^nk$ and $a(f) = \int a_k \overline{f(k)}d^nk$ are the creation/annihilation operators of the vector $f \in \mathcal{H}_1$, normalized to $[a(f), a^+(g)] = (f \mid g)$. The Hamiltonian $H_E = d\Gamma(M)$ coincides with $H_E = \int \varepsilon(k)a_k^+ a_k d^nk$. The interaction potential G is chosen as the self-adjoint field operator

$$G = \Phi(h) := a^+(h) + a(h) \tag{19}$$

where the vector $h \in \mathcal{H}_1$ satisfies the additional constraint

$$2\left\|M^{-\frac{1}{2}}h\right\| \leq 1. \tag{20}$$

This constraint guarantees that $H_E - \frac{1}{2}\Phi^2(h)$ is bounded from below, and the Hamiltonian (6) is a well defined semibounded operator on $\mathcal{H}_S \otimes \mathcal{F}(\mathcal{H}_1)$, see Appendix B. In the sequel we always assume that (20) is satisfied.

To derive induced superselection sectors we have to estimate the time dependence of the traces (10) $\chi(\alpha, \beta; t) = \mathrm{tr}_E U_{\alpha\beta}(t)\rho_E$ where ρ_E is the reference state of the Boson field, and the unitary operators $U_{\alpha\beta}(t)$ are given by

$$U_{\alpha\beta}(t) := \exp(iH_\alpha t)\exp(-iH_\beta t), \text{ with } H_\alpha = H_E + \alpha\Phi(h),\ \alpha, \beta \in \mathbf{R}. \tag{21}$$

The Hamiltonians H_α are Hamiltonians of the van Hove model [vH52]. Details for the following statements are given in the Appendix B. The Hamiltonian $H_E + \Phi(h)$ is defined on the Fock space $\mathcal{F}(\mathcal{H}_1)$ as semibounded self-adjoint operator if $h \in \mathcal{H}_1$ is in the domain of $M^{-\frac{1}{2}}$, $h \in \mathcal{D}(M^{-\frac{1}{2}})$. But this Hamiltonian has a ground state only if the low energy contributions of h are not too strong, more precisely, if

$$h \in \mathcal{D}(M^{-1}) \tag{22}$$

is satisfied. Under this more restrictive condition the Hamiltonian has another important property: $H_E + \Phi(h)$ is unitarily equivalent to the free Hamiltonian H_E

$$H_E + \Phi(h) = T^+(M^{-1}h)H_E T(M^{-1}h) - \left\|M^{-\frac{1}{2}}h\right\|^2. \tag{23}$$

Thereby the intertwining operators are the unitary Weyl operators $T(f) = \exp\left(a^+(f) - a(f)\right)$ defined for $f \in \mathcal{H}_1$.

3.2. Coherent States as Reference State

For the further calculations we first choose as reference state a coherent state. Let $f \in \mathcal{H}_1 \to \exp f = 1_{vac} + f + \frac{1}{2}f \circ f + .. \in \mathcal{F}(\mathcal{H}_1)$ be the convergent exponential series of the symmetric tensor algebra of the Fock space. Thereby $1_{vac} \in \mathcal{F}(\mathcal{H}_1)$ is the vacuum vector. Then $T(f)1_{vac} = \exp\left(f - \frac{1}{2}\|f\|^2\right)$ is a normalized exponential vector or coherent state. The reference state ρ_E is the projection operator $\omega(f)$ onto this vector, i.e. $\omega(f) = T(f)P_{vac}T^+(f)$ where P_{vac} is the projection operator onto the vacuum. The basic identity which characterizes the coherent states is the expectation of the Weyl operators

$$\mathrm{tr}_E T(h)\omega(f) = \exp\left(-\frac{1}{2}\|h\|^2\right)\exp\left(2i\,\mathrm{Im}\,(f \mid h)\right) \tag{24}$$

Under the assumption (22) the trace (10) is calculated in Appendix B using (23) and properties of the Weyl operators. The result is

$$\mathrm{tr}_E U_{\alpha\beta}(t)\omega(f) = \exp\left(-(\alpha-\beta)^2 \zeta(t)\right) \exp\left(i\left(\vartheta(\alpha,t) - \vartheta(\beta,t)\right)\right) \tag{25}$$

with

$$\zeta(t) = \frac{1}{2}\left\|\left(I - \exp(iMt)\right) M^{-1} h\right\|^2; \tag{26}$$

the phase function $\vartheta(\alpha, t)$ is given in (57). This result implies an estimate (12) of the trace where $\phi(s)$ is the exponential

$$\phi(s) = \exp(-s) \tag{27}$$

and $\zeta(t)$ is the function (26).

In this first step the identities (25) and (26) have been derived assuming (22). But under this restriction the function (26) is almost periodic. It may grow to large numbers, but it cannot diverge to infinity. Hence the traces (25) do not vanish for $t \to \infty$. One can achieve a strong decrease which persists for some finite time interval; but inevitably, recurrences exist.

To derive induced superselection rules one has to violate the condition (22). If $h \in \mathcal{D}(M^{-\frac{1}{2}}) \setminus \mathcal{D}(M^{-1})$ we prove in Appendix B that the identities (25) and (26) are still valid. Then an evaluation of (26) implies that $\zeta(t)$ diverges for $t \to \infty$, and superselection rules follow from (14). The time scale of the decoherence depends only on the vector h in the interaction potential (19), and (26) can increase like $\log t$ or also like t^α with some $\alpha \in (0,1)$. The assumption $h \notin \mathcal{D}(M^{-1})$ is therefore necessary and sufficient for the emergence of superselection rules, which persist for $t \to \infty$. Exactly under this condition the Boson field is known to be infrared divergent. It is still defined on the Fock space, but the bare Boson number diverges and its ground state disappears in the continuum, see [Sch63] [AH00].

3.3. Arbitrary Normal States as Initial State

The results of Sect. 3.2. can be easily extended to reference states which are superpositions of a finite number of exponential vectors, see Appendix B. Estimates like (14) remain valid with an additional numerical factor, which increases with the number of exponential vectors involved. The linear span $\mathcal{L}\{\exp f \mid f \in \mathcal{H}_1\}$ of the exponential vectors is a dense linear subset of the Fock space $\mathcal{H}_E = \mathcal{F}(\mathcal{H}_1)$, and the convex linear span of all projection operators onto these vectors is a dense subset $\mathcal{S}_{coh} \subset \mathcal{S}(\mathcal{H}_E)$ of all states of the Boson system. We finally obtain for all reference states $\rho_E \in \mathcal{S}_{coh}$ an estimate like (14)

$$\left\|P_S(\Delta^1)\Phi_t(\rho_S)P_S(\Delta^2)\right\|_1 \leq c(\rho_E)\phi\left((1-\varepsilon)\delta^2\zeta(t)\right), \tag{28}$$

where ζ and ϕ are again the functions (26) and (27), but with some small $\varepsilon > 0$ and an additional numerical factor $c(\rho_E)$ which depends on the reference state. If $h \in \mathcal{D}(M^{-\frac{1}{2}}) \setminus \mathcal{D}(M^{-1})$ this estimate implies for all $\rho_E \in \mathcal{S}_{coh}$

$$\lim_{t\to\infty} \left\|P_S(\Delta^1)\Phi_t(\rho_S)P_S(\Delta^2)\right\|_1 = 0 \tag{29}$$

if the intervals Δ^1 and Δ^2 are separated by a distance $\delta > 0$. Due to the factor $c(\rho_E)$ the emergence of the superselection sectors $\{P_S(\Delta)\mathcal{H}_S\}$ is not uniform with respect to ρ_E; but for suitably restricted subsets of reference states a fast suppression of the off-diagonal matrix elements of $\Phi_t(\rho_S)$ can be achieved.

So far we have assumed that the initial state factorizes. The Schrödinger picture allows to start from the more general initial states

$$W = \sum_{\mu=1}^{N} c_\mu \, \rho_{S\mu} \otimes \rho_{E\mu} \tag{30}$$

with $\rho_{S\mu} \in \mathcal{S}(\mathcal{H}_S)$, $\rho_{E\mu} \in \mathcal{S}_{coh}$ and real (positive and negative) numbers c_μ, which satisfy $\sum_\mu c_\mu = \operatorname{tr} W = 1$. Thereby N is an arbitrary finite number. The set of states (30) is dense in $\mathcal{S}(\mathcal{H}_{S+E})$ and will be denoted by $\mathcal{S}_{fin}(\mathcal{H}_{S+E})$. The reduced dynamics for such an initial state

$$\rho_S(t) = \widehat{\Phi}_t(W) := \operatorname{tr}_E U(t) W U^\dagger(t) \tag{31}$$

decomposes into $\rho_S(t) = \sum_\mu c_\mu \Phi_t^\mu(\rho_{S\mu})$, where $\Phi_t^\mu(\,.\,)$ is the reduced dynamics (2) with reference state $\rho_{E\mu}$. For all contributions $\Phi_t^\mu(\rho_{S\mu})$ estimates of the type (28) are valid. Hence $\left\| P_S(\Delta^1)\widehat{\Phi}_t(W)P_S(\Delta^2)\right\|_1 \leq \sum_\mu |c_\mu| \left\| P_S(\Delta^1)\Phi_t^\mu(\rho_{S\mu})P_S(\Delta^2)\right\|_1$ implies

$$\lim_{t\to\infty} \left\| P_S(\Delta^1)\widehat{\Phi}_t(W)P_S(\Delta^2)\right\|_1 = 0 \tag{32}$$

for $W \in \mathcal{S}_{fin}(\mathcal{H}_{S+E})$ and all separated intervals.

By a continuity argument on the mapping (31) we finally derive that the superselection sectors $\{P_S(\Delta)\mathcal{H}_S \mid \Delta \subset \mathbf{R}\}$ emerge for all initial states $W \in \mathcal{S}(\mathcal{H}_{S+E})$ of the total system.

Theorem 1 *If the interaction is determined by a vector $h \in \mathcal{D}(M^{-\frac{1}{2}})\backslash\mathcal{D}(M^{-1})$ with norm restriction (20), then (32) is true for all initial states $W \in \mathcal{S}(\mathcal{H}_{S+E})$ and all intervals with distance* $\operatorname{dist}(\Delta^1, \Delta^2) > 0$.

Proof. The mapping (31) can be extended to a linear mapping $W \in \mathcal{T}(\mathcal{H}_{S+E}) \to \widehat{\Phi}_t(W) \in \mathcal{T}(\mathcal{H}_S)$ which is continuous with respect to the trace norms of these spaces

$$\left\| \widehat{\Phi}_t(W) \right\|_1 \leq \| W \|_1 . \tag{33}$$

Given some $\varepsilon > 0$ and a state $W \in \mathcal{S}(\mathcal{H}_{S+E})$, then we can find a $W_1 \in \mathcal{S}_{fin}(\mathcal{H}_{S+E})$ such that $\|W - W_1\|_1 < \varepsilon$. The limit (32) implies: for intervals Δ^1 and Δ^2 with a non-vanishing distance there is a time $T(\varepsilon) < \infty$ such that
$\left\| P_S(\Delta^1)\widehat{\Phi}_t(W_1)P_S(\Delta^2)\right\|_1 < \varepsilon$ if $t > T(\varepsilon)$. By linearity of $\widehat{\Phi}_t$ we have
$\widehat{\Phi}_t(W) = \widehat{\Phi}_t(W_1) + \widehat{\Phi}_t(W - W_1)$, and we derive the upper bound

$$\begin{aligned} &\left\| P_S(\Delta^1)\widehat{\Phi}_t(W)P_S(\Delta^2)\right\|_1 \\ &\leq \left\| P_S(\Delta^1)\widehat{\Phi}_t(W_1)P_S(\Delta^2)\right\|_1 + \left\| P_S(\Delta^1)\widehat{\Phi}_t(W - W_1)P_S(\Delta^2)\right\|_1 \\ &\leq \left\| P_S(\Delta^1)\widehat{\Phi}_t(W_1)P_S(\Delta^2)\right\|_1 + \|W - W_1\|_1 < 2\varepsilon \end{aligned} \tag{34}$$

As ε can be chosen arbitrarily small, the Theorem follows. ■

3.4. KMS States as Reference States

The considerations presented so far can be extended to an environment with positive temperature $\beta^{-1} > 0$. That means the Boson field is in a KMS state[1], which is uniquely characterized by the following expectation of the Weyl operators $T(h)$

$$\langle T(h)\rangle_\beta = \exp\left(-\left(h \mid \left((e^{\beta M} - I)^{-1} + \frac{1}{2}\right) h\right)\right). \tag{35}$$

The calculations for a KMS state correspond to the calculations for coherent states. Only the expectation (24) has to be substituted by the expectation (35). As $(\exp(\beta M) - I)^{-1}$ is a positive operator we have

$$\langle T(h)\rangle_\beta < \exp\left(-\frac{1}{2}\|h\|^2\right) = |\mathrm{tr}_E T(h)\omega(f)|. \tag{36}$$

Hence in an environment of temperature $\beta^{-1} > 0$ the superselection sectors are induced on shorter time scale than for coherent states, see Appendix B.

4. Scattering Processes

In this final section we investigate the stability of the induced superselection sectors against additional scattering processes. We restrict the initial state of the total system to a normal state $W \in \mathcal{S}(\mathcal{H}_{S+E})$ to apply standard scattering theory.

The Hamiltonian (6) is generalized to

$$H = H_{S\times E} + V, \tag{37}$$

where V is a scattering potential on $\mathcal{H}_{S\times E} = \mathcal{H}_S \otimes \mathcal{H}_E$. There are no constraints on the commutators $[H_{S\times E}, V]$ or $[F \otimes I_E, V]$, and in general the dynamics has no conservation law except energy conservation. The restriction to scattering potentials means that the wave operator $\Omega = \lim_{t\to\infty} U^+(t)U_0(t)$ with $U(t) = \exp(-itH)$ and $U_0(t) = \exp(-itH_{S\times E})$ exists as strong limit. To simplify the arguments we assume that there are no bound states and that the wave operator is unitary on $\mathcal{H}_{S\times E}$. Then the time evolution $U(t) = \exp(-itH)$ behaves asymptotically like $U_0(t)\Omega^+$ with $U_0(t) = \exp(-itH_{S\times E})$. More precisely, the existence of wave operators implies

$$\lim_{t\to\infty} \left\|U(t)WU^+(t) - U_0(t)\Omega^+W\Omega U_0^+(t)\right\|_1 = 0 \tag{38}$$

for all $W \in \mathcal{S}(\mathcal{H}_{S+E})$. As Ω is unitary we have $\Omega^+W\Omega \in \mathcal{S}(\mathcal{H}_{S\times E})$ for all $W \in \mathcal{S}(\mathcal{H}_{S+E})$. Let us denote the reduced dynamics with the full Hamiltonian (37) by

$$\rho_S(t) = \widehat{\Phi}_t(W) = \mathrm{tr}_E\left(U(t)WU^+(t)\right), \tag{39}$$

[1]The KMS states of an environment which has a Hamiltonian with a continuous spectrum cannot be represented by a statistical operator in $\mathcal{S}(\mathcal{H}_E)$. In such a case the algebra of observables has to be restricted to the Weyl algebra, which is strictly smaller than $\mathcal{B}(\mathcal{H}_E)$, and the KMS states are positive linear functionals on that algebra.

and the reduced dynamics with the Hamiltonian (6) by $\widehat{\Phi}_t^0(W) = \mathrm{tr}_E\left(U_0(t)WU_0^+(t)\right)$. The linearity of the trace implies

$$\widehat{\Phi}_t(W) = \widehat{\Phi}_t^0(\Omega^+ W\Omega) + \mathrm{tr}_E\left(U(t)WU^+(t) - U_0(t)\Omega^+ W\Omega U_0^+(t)\right). \quad (40)$$

The off-diagonal contributions of the reduced dynamics (39) with scattering can therefore be estimated by

$$\begin{aligned}\left\|P_S(\Delta^1)\rho_S(t)P_S(\Delta^2)\right\|_1 \leq & \left\|P_S(\Delta^1)\,\widehat{\Phi}_t^0(\Omega^+ W\Omega)\,P_S(\Delta^2)\right\|_1 \\ & + \left\|U(t)WU^+(t) - U_0(t)\Omega^+ W\Omega U_0^+(t)\right\|_1. \quad (41)\end{aligned}$$

As $\Omega^+ W\Omega \in \mathcal{S}(\mathcal{H}_{S\times E})$ the first term vanishes for $t \to \infty$ under the conditions of Theorem 1, and the second term vanishes for $t \to \infty$ as a consequence of (38). Hence we have derived

Theorem 2 *If the interaction is determined by a vector* $h \in \mathcal{D}(M^{-\frac{1}{2}})\backslash\mathcal{D}(M^{-1})$ *with norm restriction (20), then*

$$\lim_{t\to\infty}\left\|P_S(\Delta^1)\rho_S(t)P_S(\Delta^2)\right\|_1 = 0 \quad (42)$$

follows for the reduced dynamics (39) with the Hamiltonian (37) for all initial states $W \in \mathcal{S}(\mathcal{H}_{S+E})$ *and all intervals with distance* $\mathrm{dist}\,(\Delta^1, \Delta^2) > 0$.

Therefore scattering processes do not destroy or modify the induced superselection sectors $\{P_S(\Delta)\mathcal{H}_S.\mid \Delta \subset \mathbf{R}\}$, but the time scale of there emergence increases. Estimates on the time scale require a more detailed investigation of the scattering process, which is not given here.

5. Conclusion

We have investigated a class of systems, which are coupled to a mass zero Boson field. These models exhibit the following properties:

- The Boson field induces superselection rules into the system, if and only if the field is infrared divergent. Thereby infrared divergence means that the bare Boson number diverges and the Boson vacuum disappears in the continuum, but the Hamiltonian remains bounded from below.
- The superselection sectors are fully determined by the Hamiltonian, they finally emerge for all normal initial states of the total system, – including non-product states – and for KMS states as reference states of the Boson system.
- The time scale of the decoherence depends on the interaction and on the initial state. There are restrictions on the reference state of the Boson field to obtain superselection rules, which are effective within a short time.
- The superselection sectors persist, if additional scattering processes take place. In this case the total system may have no conservation law except energy conservation.

These results underline the known importance of low frequency excitations of the environment for the process of decoherence [LCD+87] [Del03].

A Estimates of Operators

Let $P : \Delta = [a, b) \subset \mathbf{R} \to \mathcal{L}(\mathcal{H})$ be a family of orthogonal projectors in $\mathcal{H}$ with the properties (4). The mapping $P(\Delta)$ can be extended to a σ-additive measure on the Borel algebra $\mathcal{B}(\mathbf{R})$ generated by open subsets of the real line $\mathbf{R}$. The operators $P([a, b))$ are naturally left continuous in both variables a and b. In what follows we investigate some integrals of bounded operator-valued functions with respect to P.

Lemma 1 *Let $f : \mathbf{R} \to \mathcal{T}(\mathcal{H})$ be a differentiable function with a Bochner integrable derivative $f'(x) \in \mathcal{T}(\mathcal{H})$. Then for any interval $\Delta = [a, b) \subset \mathbf{R}$ the following identity holds*

$$\begin{aligned} \textstyle\int_a^b P(dx)f(x) &= P([a, b))f(b) - \textstyle\int_a^b P([a, x))f'(x)dx \\ &= P([a, b))f(a) + \textstyle\int_a^b P([x, b))f'(x)dx, \end{aligned} \tag{43}$$

and the norm of this integral has the upper bound

$$\left\| \int_\Delta P(dx)f(x) \right\| \leq \min\left(\|f(a)\|, \|f(b)\| \right) + \int_\Delta \|f'(x)\|_1 \, dx. \tag{44}$$

Proof. The identities (43) are just the integration by parts formula of the Stieltjes integral, see e.g. [BW83] Sect. 5.1. The norm estimate (44) is then a consequence of $\|P(\Delta)\| \leq 1$ and the rule $\|AB\|_1 \leq \|A\| \, \|B\|_1$ for the trace norm. ■

The same type of identities and norm estimates can be derived for integrals $\int_\Delta f(x)P(dx) = \left(\int_\Delta P(dx)f^+(x)\right)^+$ with a reversed order of the operators. An immediate consequence of Lemma 1 is the

Corollary 2 *Let $f : \mathbf{R} \to \mathcal{T}(\mathcal{H})$ be a function with a Bochner integrable derivative $f'(x) \in \mathcal{T}(\mathcal{H})$. If $\|f(x)\|_1$ vanishes for $x \to \pm\infty$ the identities*

$$\begin{aligned} \textstyle\int_a^\infty P(dx)f(x) &= -\textstyle\int_a^\infty P([a, x))f'(x)dx \quad \text{and} \\ \textstyle\int_{-\infty}^b P(dx)f(x) &= \textstyle\int_{-\infty}^b P([x, b))f'(x)dx \end{aligned} \tag{45}$$

hold for all $a, b \in \mathbf{R}$ and the estimate

$$\left\| \int_\Delta P(dx)f(x) \right\|_1 \leq \int_\Delta \|f'(x)\|_1 \, dx. \tag{46}$$

follows for the infinite intervals $\Delta = [a, \infty)$ and $(-\infty, b)$.

We now consider operators

$$S_\varphi = \int_{\mathbf{R}\times\mathbf{R}} \varphi(x, y)P(dx)SP(dy) \tag{47}$$

where $S \in \mathcal{T}(\mathcal{H})$ and $\varphi : \mathbf{R} \times \mathbf{R} \to \mathbf{C}$ is a differentiable function. We obviously have $P(\Delta^1)S_\varphi P(\Delta^2) = \int_{\Delta^1\times\Delta^2} \varphi(x, y)P(dx)SP(dy)$. First let us notice that

$$P(\Delta^1)S_\varphi P(\Delta^2) = \int_{\Delta^1} P(dx)S \int_{\Delta^2} \varphi(x, y)P(dy) = \int_{\Delta^1} P(dx)SA(x), \tag{48}$$

where the function $A(x)$ is defined by $A(x) = \int_{\Delta^2} \varphi(x,y)P(dy) \in \mathcal{L}(H)$. Its derivative is $A'(x) = \int_{\Delta^2} \varphi_1(x,y)P(dy)$ with $\varphi_1(x,y) = \frac{\partial}{\partial x}\varphi(x,y)$. The operator norm of this derivative has the upper bound $\|A'(x)\| \leq \sup_{y\in\Delta^2} |\varphi_1(x,y)|$. Then (48) can be estimated by Corollary 2. We formulate the final result for operators (46) S_φ with a function $\varphi(x,y) = \chi(x-y)$ which depends only on the difference $x - y$.

Theorem 3 *Let $\chi : x \in \mathbf{R} \to \mathbf{C}$ be a differentiable complex-valued function with $\chi(x) \to 0$ if $|x| \to \infty$ and $|\chi'(x)| \leq \phi(|x|)$, where $\phi(s)$ is non-increasing for $s \geq 0$ with a bounded integral $\int_0^\infty \phi(x)dx < \infty$. Then for any nuclear operator S the operator S_φ with $\varphi(x,y) = \chi(x-y)$ is again nuclear, and for the disjoint intervals $\Delta^1 = (-\infty, b_1)$ and $\Delta^2 = [a_2, \infty)$ with $\delta = a_2 - b_1 \geq 0$ the following estimate holds*

$$\left\| P(\Delta^1) S_\varphi P(\Delta^2) \right\|_1 \leq \|S\|_1 \int_\delta^\infty \phi(x)dx. \tag{49}$$

B The van Hove Model

B1. The Hamiltonian

Let $F \circ G$ denote the symmetric tensor product of the Fock space $\mathcal{F}(\mathcal{H}_1)$ with vacuum 1_{vac}. For all $f \in \mathcal{H}_1$ the exponential vectors $\exp f = 1_{vac} + f + \frac{1}{2} f \circ f + \ldots$ converge within $\mathcal{F}(\mathcal{H}_1)$, the inner product being $(\exp f \mid \exp g) = \exp(f \mid g)$. Coherent vectors (states) are the normalized exponential vectors $\exp\left(f - \frac{1}{2}\|f\|^2\right)$. The linear span of all exponential vectors $\{\exp f \mid f \in \mathcal{H}_1\}$ is dense in $\mathcal{F}(\mathcal{H}_1)$. The creation operators $a^+(f)$ are uniquely determined by $a^+(f)\exp g = f \circ \exp g =$
$\frac{\partial}{\partial\lambda}\exp(f + \lambda g) \mid_{\lambda=0}$ with $f, g \in \mathcal{H}_1$ and the annihilation operators are given by $a(g)\exp f = (g \mid f)\exp f$. These operators satisfy the standard commutation relations $[a(f), a^+(g)] = (f \mid g)$. If M is a operator on $\mathcal{H}_1$ then $\Gamma(M)$ is uniquely defined as operator on $\mathcal{F}(\mathcal{H}_1)$ by $\Gamma(M)\exp f := \exp(Mf)$, and the derivation $d\Gamma(M)$ is defined by $d\Gamma(M)\exp f := (Mf) \circ \exp f$.

For arbitrary elements $g \in \mathcal{H}_1$ the unitary Weyl operators are defined on the set of exponential vectors by $T(g)\exp f = \exp\left(-(g \mid f) - \frac{1}{2}\|g\|^2\right)\exp(f+g)$. This definition is equivalent to $T(g) = \exp(a^+(g) - a(g))$. The Weyl operators are characterized by the properties

$$\begin{aligned} T(g_1)T(g_2) &= T(g_1+g_2)\exp(-i\,\mathrm{Im}\,(g_1 \mid g_2)) \\ (1_{vac} \mid T(g)\,1_{vac}) &= \exp\left(-\tfrac{1}{2}\|g\|^2\right). \end{aligned} \tag{50}$$

The matrix element of $T(h)$ between coherent vectors $\exp\left(f - \frac{1}{2}\|f\|^2\right) = T(f)1_{vac}$ follows from these relations as

$$(1_{vac} \mid T^+(g)T(h)T(f)\,1_{vac}) = \exp\left(-\frac{1}{2}\|h+f-g\|^2 + i\,\mathrm{Im}\{(g \mid f) + (f+g \mid h)\}\right). \tag{51}$$

For a free field the time evolution on the Fock space is given by $U(t) = \exp(-iH_E t) = \Gamma(V(t))$ with $V(t) := \exp(-iMt)$. For exponential vectors we obtain $U(t)\exp f = \exp(V(t)f)$. From these equations the dynamics of the Weyl operators follows as

$$U^+(t)T(g)U(t) = T\left(V^+(t)\,g\right). \tag{52}$$

For fixed $h \in \mathcal{H}_1$ the unitary operators $T^+(h)U(t)T(h)$, $t \in \mathbf{R}$, form a one parameter group which acts on exponential vectors as
$T^+(h)U(t)T(h) \exp f = \exp\left((h \mid V(t)(f+h) - f) - \|h\|^2\right) \exp\left(V(t)(f+h) - h\right)$.
For $h \in \mathcal{H}_1$ with $Mh \in \mathcal{H}_1$ the generator of this group is easily identified with
$T^+(h)H_ET(h) = H_E + \Phi(Mh) + (h \mid Mh)$, where $\Phi(.)$ is the field operator. This identity was first derived by Cook [Coo61] by quite different methods. If h satisfies $M^{-1}h \in \mathcal{H}_1$ we obtain

$$T^+(M^{-1}h)H_ET(M^{-1}h) - \left\|M^{-\frac{1}{2}}h\right\|^2 = H_E + \Phi(h) \tag{53}$$

which is the Hamiltonian of the van Hove model [vH52], see also, [Ber66] p.166ff, [Emc72a] and [AH00].

For all $h \in \mathcal{H}_E$ with $M^{-\frac{1}{2}}h \in \mathcal{H}_E$ the field operator $\Phi(h)$ satisfies the estimate

$$\|\Phi(h)\psi\| \leq 2\left\|M^{-\frac{1}{2}}h\right\| \left\|\sqrt{H_E}\psi\right\| + \|h\| \, \|\psi\|, \tag{54}$$

where $\psi \in \mathcal{F}(\mathcal{H}_1)$ is an arbitrary vector in the domain of H_E, see e.g. eq. (2.3) of [AH97]. As consequences we obtain the following Lemma for the Hamiltonian of the van Hove model, see [Sch63] and [AH00].

Lemma 3 *The operators $H_E + \lambda\Phi(h)$, $\lambda \in \mathbf{R}$, are self-adjoint on the domain of H_E if $h \in \mathcal{D}(M^{-\frac{1}{2}})$.*

Proof. From (54) and the numerical inequality $\sqrt{x} \leq ax + (4a)^{-1}$, valid for $x \geq 0$ and $a > 0$, we obtain a bound $\|\Phi(h)\psi\| \leq c_1 \|H_E\psi\| + c_2 \|\psi\|$ with positive numbers $c_1, c_2 > 0$ where c_1 can be chosen arbitrarily small. Then the Kato-Rellich Theorem yields the first statement. ■

A further consequence is

Lemma 4 *The operator $H_E - \frac{1}{2}\Phi^2(h)$ has the lower bound $H_E - \frac{1}{2}\Phi^2(h) \geq -\|h\|^2$, if $h \in \mathcal{H}_1$ and $\left\|M^{-\frac{1}{2}}h\right\| \leq 2^{-1}$.*

Proof. From (54) we obtain
$\|\Phi(h)\psi\|^2 \leq 4\left\|M^{-\frac{1}{2}}h\right\|^2 (\psi \mid H_E\psi) + 4\left\|M^{-\frac{1}{2}}h\right\| \|h\| \left\|\sqrt{H_E}\psi\right\| \|\psi\| + \|h\|^2 \|\psi\|^2$
$\leq 8\left\|M^{-\frac{1}{2}}h\right\|^2 (\psi \mid H_E\psi) + 2\|h\|^2 \|\psi\|^2$. Hence the operator inequalities
$0 \leq \frac{1}{2}\Phi^2(h) \leq 4\left\|M^{-\frac{1}{2}}h\right\|^2 H_E + \|h\|^2 I_E$ hold, and Lemma 4 follows. ■

Therefore the total Hamiltonian (6) is semibounded, and the unitary operators $U_\lambda(t) = \exp\left(-i(H_E + \lambda\Phi(h))t\right)$ are well defined if (20) is satisfied.

B2. Evaluation of the Traces

In a first step we evaluate the expectation value of (21) $U_{\alpha\beta}(t) = U_\alpha(-t)U_\beta(t)$ for a coherent state (= normalized exponential vector) $\exp\left(f - \frac{1}{2}\|f\|^2\right) = T(f)\,1_{vac}$ under the additional constraint $h \in \mathcal{D}(M^{-1})$. This assumption allows to use the identity (53) which reduces all calculations to the Weyl relations and the vacuum expectation (51). The extension to the general case, which violates $h \in \mathcal{D}(M^{-1})$, can then be performed by a continuity argument.

If $M^{-1}h \in \mathcal{H}_1$ the identity (53) implies
$U_\lambda(t) = T(-\lambda M^{-1}h)U_0(t)T(\lambda M^{-1}h)\exp\left(i\lambda^2 \left(h \mid M^{-1}h\right) t\right)$. Then
$U_{\alpha\beta}(t) = U_\alpha(-t)U_\beta(t)$ can be calculated with the help of (50) and (52) as

$$\begin{aligned} U_{\alpha\beta}(t) &= T\left((\alpha-\beta)\left(V^+(t) - I\right) M^{-1}h\right) \exp\left(-i\eta(t)\right), \\ \eta(t) &= \left(\alpha^2 - \beta^2\right)\left\{\left(h \mid M^{-1}h\right) t + \left(M^{-1}h \mid M^{-1}\sin(Mt)h\right)\right\}. \end{aligned} \tag{55}$$

The matrix element of $U_{\alpha\beta}(t)$ between the coherent states $T(f)\,1_{vac}$ and $T(g)\,1_{vac}$ is then evaluated with the help of (51)

$$\begin{aligned} \left(1_{vac} \mid T^+(g)U_{\alpha\beta}(t)T(f)\,1_{vac}\right) = \exp\left(-\tfrac{1}{2}(\alpha-\beta)^2 \left\|\left(V^+(t)-I\right)M^{-1}h + f - g\right\|^2\right) \\ \times \exp\left(i\,\mathrm{Im}(g \mid f)\right) \times \exp\left(i\left(\vartheta(\alpha,t) - \vartheta(\beta,t)\right)\right) \end{aligned} \tag{56}$$

with the phase function

$$\vartheta(\alpha,t) = \alpha\,\mathrm{Im}\left(f + g \mid \left(I - V^+(t)\right)M^{-1}h\right) - \alpha^2\left(M^{-1}h \mid ht + M^{-1}\sin(Mt)h\right). \tag{57}$$

For $g = f$ the identity (56) leads to the trace (25).

So far we have assumed $h \in \mathcal{D}(M^{-1})$. Then the norm $\left\|\left(V^+(t) - I\right)M^{-1}h\right\| = \left\|\left(I - \exp(iMt)\right)M^{-1}h\right\|$ is an almost periodic function of t, and induced superselection rule can emerge only in an approximate sense on an intermediate time scale. But $\left(V^+(t) - I\right)M^{-1}h$ is a vector in $\mathcal{H}_1$ also under the weaker condition $h \in \mathcal{D}(M^{-\frac{1}{2}}) \supset \mathcal{D}(M^{-1})$. Moreover from Lemma 3 we know that the van Hove Hamiltonians $H_E + \lambda\Phi(h)$ and the groups $U_\lambda(t)$ are defined under this weaker condition. In the next step we shall use a continuity argument to prove that (56) is indeed still valid for vectors $h \in \mathcal{D}(M^{-\frac{1}{2}})$ without knowing whether $h \in \mathcal{D}(M^{-1})$ or not. Then we derive the essential statement that the norm $\left\|\left(V^+(t) - I\right)M^{-1}h\right\|$ diverges for $t \to \infty$ if $h \notin \mathcal{D}(M^{-1})$. As this behaviour is possible under the condition (20), which guarantees the existence of a semibounded Hamiltonian (6), stable superselection sectors emerge if h is chosen such that $h \in \mathcal{D}(M^{-\frac{1}{2}}) \setminus \mathcal{D}(M^{-1})$ with the additional constraint (20).

For the proof of this statement we introduce the norm

$$|||h||| := \|h\| + \left\|M^{-\frac{1}{2}}h\right\|. \tag{58}$$

Let $h_n \in \mathcal{H}_1$, $n = 1, 2, \ldots$, be a sequence of real vectors which converges in this topology to a vector h, then we know from (54) and the proof of Lemma 3 that there exist two null sequences of positive numbers c_{1n} and c_{2n} such that

$$\left\|\left(\Phi(h_n) - \Phi(h)\right)\psi\right\| \le c_{1n}\left\|\left(H_E + \Phi(h)\right)\psi\right\| + c_{2n}\|\psi\|.$$

Hence the operators $H_E + \Phi(h_n)$ converge strongly to $H_E + \Phi(h)$ and the groups $U(h_n;t) = \exp\left(-i\left(H_E + \Phi(h_n)\right)t\right)$ converge strongly to the group
$U(h;t) = \exp\left(-i\left(H_E + \Phi(h)\right)t\right)$, uniformly in any finite interval $0 \le t \le s < \infty$; see e.g. Theorem 4.4 on p. 82 of [Mas72], or Theorem 3.17 of [Dav80]. The operators

$U_{\alpha\beta,n}(t) := \exp\left(i\left(H_E + \alpha\Phi(h_n)\right)t\right)\exp\left(-i\left(H_E + \beta\Phi(h_n)\right)t\right)$ converge therefore in the weak operator topology to $U_{\alpha\beta}(t)$. For $n = 1, 2, ..$ we can calculate the corresponding traces $\mathrm{tr}_E U_{\alpha\beta,n}(t)\omega(f)$ with the result (26), where h has to be substituted by h_n. Since (26) is continuous in the variable h in the topology (58) the limit for $n \to \infty$ is again given by (26).

To prove the divergence of $\left\|\left(V^+(t) - I\right)M^{-1}h\right\|$ for $t \to \infty$ we introduce the spectral resolution $P_M(d\lambda)$ of the one-particle Hamilton operator M. The energy distribution of the vector $h \in \mathcal{H}_1$ is given by the measure $d\sigma_h(\lambda) = (h \mid P_M(d\lambda)h)$. The exponent (26) is then the integral

$$\zeta(t) = \frac{1}{2}\left\|\left(I - \exp\left(Mt\right)\right)M^{-1}h\right\|^2 = 2\int_{\mathbf{R}_+} \lambda^{-2}\sin^2\frac{\lambda t}{2}\,d\sigma_h(\lambda). \tag{59}$$

This integral is well defined for all $h \in \mathcal{H}_1$, and $\zeta(t)$ is a differentiable function for $t \in \mathbf{R}$. The requirement $h \in \mathcal{D}(M^{-\frac{1}{2}}) \setminus \mathcal{D}(M^{-1})$ is equivalent to the conditions $\int_0^\infty \lambda^{-1}d\sigma_h(\lambda) < \infty$ and

$$\int_\varepsilon^\infty \lambda^{-2}\,d\sigma_h(\lambda) \nearrow \infty \quad \text{if } \varepsilon \to +0. \tag{60}$$

Lemma 5 *If $h \notin \mathcal{D}(M^{-1})$, i.e. (60), the integral (59) diverges for $t \to \infty$.*

Proof. Since the operator M has an absolutely continuous spectrum, the measure $d\sigma_h(\lambda)$ is absolutely continuous with respect to the Lebesgue measure $d\lambda$ on $\mathbf{R}_+$. Consequently, the measure $\lambda^{-2}\,d\sigma_h(\lambda)$ is absolutely continuous with respect to the Lebesgue measure on any interval (ε, ∞) with $\varepsilon > 0$. The identity $\sin^2\frac{\lambda t}{2} = \frac{1}{2}\left(1 - \cos\lambda t\right)$ and the Lebesgue Lemma therefore imply
$\lim_{t\to\infty}\int_\varepsilon^\infty \lambda^{-2}\sin^2\frac{\lambda t}{2}\,d\sigma_h(\lambda) = \frac{1}{2}\int_\varepsilon^\infty \lambda^{-2}\,d\sigma_h(\lambda)$. Given a number $\Lambda > 0$ the assumption (60) yields the existence of an $\varepsilon > 0$ such that

$$\lim_{t\to\infty}\int_\varepsilon^\infty \lambda^{-2}\sin^2\frac{\lambda t}{2}\,d\sigma_h(\lambda) = \frac{1}{2}\int_\varepsilon^\infty \lambda^{-2}\,d\sigma_h(\lambda) > \Lambda. \tag{61}$$

From the inequality $\int_{\mathbf{R}_+}\lambda^{-2}\sin^2\frac{\lambda t}{2}\,d\sigma_h(\lambda) \geq \int_\varepsilon^\infty \lambda^{-2}\sin^2\frac{\lambda t}{2}\,d\sigma_h(\lambda)$ we then obtain $\int_0^\infty \lambda^{-2}\sin^2\frac{\lambda t}{2}\,d\sigma_h(\lambda) > \Lambda$ for sufficiently large t. Since the number Λ can be arbitrarily large the integral (59) diverges for $t \to \infty$. ■

If $d\sigma_h(\lambda)$ satisfies additional regularity conditions, we can obtain more precise statements. A powerlike behaviour $d\sigma_h(\lambda) \cong c\cdot\lambda^{2\mu}d\lambda$, $c > 0$, near $\lambda = +0$ is compatible with the requirement $h \in \mathcal{D}(M^{-\frac{1}{2}}) \setminus \mathcal{D}(M^{-1})$ if $0 < \mu \leq \frac{1}{2}$. For the ohmic case $d\sigma_h(\lambda) \cong c\cdot\lambda d\lambda$ we obtain

$$\begin{aligned}\zeta(t) &= 2\int_0^\infty \lambda^{-2}\sin^2\frac{\lambda t}{2}\,d\sigma_h(\lambda) = \int_0^\infty \lambda^{-2}\left(1 - \cos\lambda t\right)d\sigma_h(\lambda)\\ &\simeq c\int_0^t s^{-1}\left(1 - \cos s\right)ds \simeq c\log t \quad \text{for } t \to \infty;\end{aligned}$$

and the subohmic case $d\sigma_h(\lambda) \cong c \cdot \lambda^{2\mu} d\lambda$ with $0 < \mu < \frac{1}{2}$ implies a powerlike divergence

$$\begin{aligned} \zeta(t) &= \int_0^\infty \lambda^{-2} (1 - \cos \lambda t)\, d\sigma_h(\lambda) \\ &\simeq c t^{1-2\mu} \int_0^t s^{-2+2\mu} (1 - \cos s)\, ds \sim t^{1-2\mu} \quad \text{for } t \to \infty. \end{aligned}$$

So far the reference state ω has been a coherent state. But the results remain true if we take as reference state ω the projection onto a vector $\psi = \sum_{n=1}^{N} c_n \exp f_n$, $f_n \in \mathcal{H}_1$, which is a finite linear combination of exponential vectors. In that case the trace (10) is a sum of terms (56) with f and g given by the vectors f_n. The exponent $\zeta_{\alpha\beta}[f, g](t) := \frac{1}{2} \left\| (\alpha - \beta) (V^+(t) - I) M^{-1} h + f - g \right\|^2$ in (56) diverges for $\alpha \neq \beta$ under the same condition as (59) does. Moreover, the asymptotic behaviour of $\zeta_{\alpha\beta}[f, g](t)$ is dominated by the asymptotics of $(\alpha - \beta)^2 \zeta(t)$. Hence uniform estimates like (14) remain valid, but the function $\phi(\delta^2 \zeta(t))$ must be substituted by $c \cdot \phi((1 - \varepsilon)\delta^2 \zeta(t))$ with a small $\varepsilon > 0$, and a constant $c \geq 1$, which depends on the coefficients c_n and on the norms $\| f_m - f_n \|$. This factor increases with the number N of the exponential vectors.

In the case of a KMS state of temperature $\beta^{-1} > 0$ the calculations essentially follow the calculations for coherent states. The expectation of $U_{\mu\nu}(t)$ is calculated using (35). The result

$$\langle U_{\mu\nu}(t) \rangle_\beta = \exp\left(-(\mu - \nu)^2 \zeta_\beta(t) \right) \exp\left(i \left(\vartheta(\mu, t) - \vartheta(\nu, t) \right) \right) \tag{62}$$

has the same structure as (25) with the temperature dependent function

$$\begin{aligned} \zeta_\beta(t) &= \left(\left(I - e^{Mt} \right) M^{-1} h \mid \left((e^{\beta M} - I)^{-1} + \frac{1}{2} \right) \left(I - e^{Mt} \right) M^{-1} h \right) \\ &\geq \frac{1}{2} \left\| (I - \exp(Mt)) M^{-1} h \right\|^2 = \zeta(t), \end{aligned} \tag{63}$$

and the phase function $\vartheta(\mu, t) = -\mu^2 \left(M^{-1} h \mid ht + M^{-1} \sin(Mt) h \right)$, which originates from (55). The inequality (63) implies that for $h \in \mathcal{D}(M^{-\frac{1}{2}}) \setminus \mathcal{D}(M^{-1})$ superselection sectors are induced on a shorter time scale than for coherent states.

As a final remark we indicate a modification of the model, which does not use the absolute continuity of the spectrum of M. But we still need a dominating low energy contribution in the interaction. More precisely, we assume that $\sigma_h(\lambda) \equiv \int_0^\lambda d\sigma_h(\alpha)$ behaves at low energies like

$$\lambda^{-2} \sigma_h(\lambda) \nearrow \infty \text{ if } \lambda \to +0. \tag{64}$$

Then we can derive the divergence of (59) by the inequalities $\zeta(t) \geq 4 \int_0^{\frac{\pi}{t}} \lambda^{-2} \sin^2 \frac{\lambda t}{2}\, d\sigma_h(\lambda) \geq \frac{4}{\pi^2} t^2 \int_0^{\frac{\pi}{t}} d\sigma_h(\lambda) = \frac{4}{\pi^2} t^2 \sigma_h(\frac{\pi}{t})$ using $\sin x \geq \frac{2}{\pi} x$ if $0 \leq x \leq \frac{\pi}{2}$. For measures $d\sigma_h(\lambda) \sim \lambda^{2\mu} d\lambda$ the assumption (64) is more restrictive than (60) – it excludes $d\sigma_h(\lambda) \sim \lambda d\lambda$ which satisfies the conditions of Lemma 2. But (64) is also meaningful for point measures $d\sigma_h(\lambda)$, and M may be an operator with a pure point spectrum. The Boson field can therefore be substituted by an infinite family of harmonic oscillators, which have zero as accumulation point of their frequencies. Such an example has been discussed – also for KMS states – by Primas [Pri00].

References

[AH97] A. Arai and M. Hirokawa. On the existence and uniqueness of ground states of a generalized spin-boson model. *J. Funct. Anal.*, **151**:455–503, 1997.

[AH00] A. Arai and M. Hirokawa. Ground states of a general class of quantum field Hamiltonians. *Rev. Math. Phys.*, **12**:1085–1135, 2000.

[Ara80] H. Araki. A remark on Machida-Namiki theory of measurement. *Prog. Theor. Phys.*, **64**:719–730, 1980.

[Ber66] F. A. Berezin. *The Method of Second Quantization*. Academic Press, New York, 1966.

[BW83] H. Baumgärtel and M. Wollenberg. *Mathematical Sattering Theory*. Birkhäuser, Basel, 1983.

[Coo61] J. M. Cook. Asymptotic properties of a Boson field with given source. *J. Math. Phys.*, **2**:33–45, 1961.

[Dav80] E. B. Davies. *One-Parameter Semigroups*. Academic Press, London, 1980.

[Del03] G. Dell'Antonio. On decoherence. *J. Math. Phys.*, **44**:4939–4956, 2003.

[Emc72a] G. G. Emch. *Algebraic Methods in Statistical Mechanics and Quantum Field Theory*. Wiley-Interscience, New York, 1972.

[Emc72b] G. G. Emch. On quantum measurement processes. *Helv. Phys. Acta*, **45**:1049–1056, 1972.

[Hep72] K. Hepp. Quantum theory of measurement and macroscopic observables. *Helv. Phys. Acta*, **45**:236–248, 1972.

[Jau60] J. M. Jauch. Systems of observables in quantum mechanics. *Helv. Physica Acta*, **33**:711–726, 1960.

[JZ85] E. Joos and H. D. Zeh. The emergence of classical properties through interaction with the environment. *Z. Phys.*, **B59**:223–243, 1985.

[JZK$^+$03] E. Joos, H. D. Zeh, C. Kiefer, D. Giulini, J. Kupsch, and I. O. Stamatescu. *Decoherence and the Appearance of a Classical World in Quantum Theory*. Springer, Berlin, 2nd edition, 2003.

[KSS01] J. Kupsch, O. G. Smolyanov, and N. A. Sidorova. States of quantum systems and their liftings. *J. Math. Phys.*, **42**:1026–1037, 2001.

[Kup00a] J. Kupsch. Mathematical aspects of decoherence. In Ph. Blanchard, D. Giulini, E. Joos, C. Kiefer, and I. O. Stamatescu, editors, *Decoherence: Theoretical, Experimental, and Conceptual Problems*, volume 538 of *Lecture Notes in Physics*, pages 125–136, Berlin, 2000. Springer. Proceedings of a ZiF Workshop Bielefeld 10. – 14. Nov. 1998.

[Kup00b] J. Kupsch. The role of infrared divergence for decoherence. *J. Math. Phys.*, **41**(9):5945–5953, 2000.

[LCD+87] A. J. Leggett, S. Chekravarty, A. T. Dorsey, M. P. A. Fischer, A. Garg, and W. Zwerger. Dynamics of the dissipative two state system. *Rev. Mod. Phys.*, **59**:1–85, 1987.

[Mas72] V. P. Maslov. *Théorie des Perturbations et Méthodes Asymptotiques*. Études Mathematiques. Dunod, Paris, 1972.

[Pri00] H. Primas. Asymptotically disjoint quantum states. In Ph. Blanchard, D. Giulini, E. Joos, C. Kiefer, and I.-O. Stamatescu, editors, *Decoherence: Theoretical, Experimental, and Conceptual Problems*, volume 538 of *Lecture Notes in Physics*, pages 161–178, Berlin, 2000. Springer. Proceedings of a ZiF Workshop Bielefeld 10. – 14. Nov. 1998.

[RS75] M. Reed and B. Simon. *Methods of Modern Mathematical Physics II: Fourier Analysis, Self-Adjointness*. Academic Press, New York, 1975.

[Sch63] B. Schroer. Infrateilchen in der Quantenfeldtheorie. *Fortschr. Physik*, **11**:1–32, 1963.

[vH52] L. van Hove. Les difficultés de divergences pour un modèle particulier de champ quantifié. *Physica*, **18**:145–159, 1952.

[Wig95] A. S. Wightman. Superselection rules; old and new. *Nuovo Cimento*, **110B**:751–769, 1995.

[Zur82] W. H. Zurek. Environment induced superselection rules. *Phys. Rev.*, **D26**:1862–1880, 1982.

In: Materials Science Research Trends
Editor: Lawrence V. Olivante, pp. 293-320
ISBN: 978-1-60021-654-1

Chapter 7

MICROSTRUCTURE EVOLUTION AND ELECTRONIC TRANSPORT IN ULTRA THIN AL FILMS

Niraj Joshi, A. K. Debnath, D. K. Aswal*, S. K. Gupta and J. V. Yakhmi

Technical Physics and Prototype Engineering Division, Modular Laboratory, Bhabha Atomic Research Center, Mumbai 400085, India

Abstract

The microstructure evolution of ultra-thin Al films deposited on Si and SiO_2 substrates using molecular beam epitaxy (MBE) and, the effect of microstructure on electronic properties has been studied. First, we present a literature review on the "microstructure formation phenomena" and "structure zone model" for metallic films and, various existing theoretical models to explain electronic transport in these films. We present a systematic study on the evolution of microstructure in ultra-thin Al films on Si as a function of: (i) Film thickness: film thickness is varied between 10 and 200 nm, while keeping deposition temperature to a fix value; (ii) Deposition temperature: films are insitu deposited at different temperature between 25 and 600°C, while keeping thickness fixed; (iii) Post-annealing: annealing the room temperature deposited at higher temperature under UHV conditions.

The results reveal that in-situ deposited films grow in a columnar structure, forming a random 2D network of islands. The low temperature electrical transport in these films could not be accounted by the existing theoretical models. We have found that the charge conduction is governed by 2D variable range hopping mechanism. The coalescence of columnar Al islands is found to take place at a critical thickness, and this thickness is found to anomalously increase with increasing deposition temperature and we have proposed an explanation for this phenomenon. Post-annealing of films leads to the normal and abnormal growth, owing to the grain boundary migration. On SiO_2 substrates, the Al film picks up oxygen during in-situ deposition at elevated temperature as well as during post-annealing process, leading to the formation of Al_2O_3 at the grain boundaries.

---

* E-mail address: dkaswal@yahoo.com

1. Why Ultra-thin Films of Al?

Al thin films are currently being used as interconnects in very large scale integration (VLSI) technology for making computer chips. In the next decade, as predicted by the Semiconductor Industry Association (SIA) the minimum width of interconnection wiring in semiconductor metallization will decrease to approximately 50nm, which is comparable to the mean free path of the conduction electrons in Al at room temperature [1]. Ultra-thin (thickness <50 nm) Al films are being considered to be a potential candidate as a buffer layer for Ag or Cu metallization in futuristic ultra large scale integration (ULSI) or giga scale integration (GSI) technologies [2-11] and, as metallic counter-electrodes in molecule-based nanoelectronics [12,13]. Moreover, the studies on ultra-thin Al films assume significance from the viewpoint of the impending nanotechnology where the feature size of the devices would be in the nanometer range.

Most of the previous works have been carried on polycrystalline Al films, which were deposited on SiO_2 or glass at ambient temperature [14,15]. The microstructure evolution and electrical transport in these films have been well studied. In the present studies, the interest in ultra-thin Al films has been two fold:

(i) How does the microstructure of ultra-thin Al films evolve on technologically important Si substrates?

(ii) Are existing theories sufficient to explain the electron transport in ultra-thin Al films?

In order to address above two issues, we first review the literature on "structure formation phenomena" in thin metallic films and, various theoretical models describing electrical transport in these films. Then, we describe the deposition of ultra-thin Al films on (111) Si or SiO_2 substrates by molecular beam epitaxy with varying thickness and/or at different deposition temperatures. The microstructure evolution has been investigated using qualitative and quantitative atomic force microscopy (AFM) analyses, x-ray photoelectron spectroscopy (XPS) and x-ray diffraction (XRD) measurements. The temperature dependence of electrical resistivity has been measured using four-probe method and, the data have been analyzed using different models.

2. Microstructure Forming Phenomena

It is well established that films can grow on a substrate by three mechanisms namely, the layer-by-layer or Frank-van der Merwe mode, the island or Volmer-Weber mode and the layer-by-layer plus island or Stranski-Krastanov growth mode [1617]. Which of these mechanisms will occur in a particular case depends on the relative surface energy of substrate to film, that is, $(\gamma_s-\gamma_f)/\gamma_f$, where γ_s and γ_f are free surface energy per unit area of the substrate and the film, respectively, and the lattice mismatch between substrate and film i.e. $(a_s-a_f)/a_f$.

(i) If $\gamma_s>\gamma_f$ and the lattice mismatch is zero then layer-by-layer growth (Frank-van der Merwe mode) can be achieved under proper conditions.

(ii) If the film has a high surface energy per unit area compared to the substrate i.e. $\gamma_s<\gamma_f$, the film grow in the form of clusters to minimize overall surface energy leading to the Volmer-Weber mode. With increased lattice mismatch, island growth takes place even if $\gamma_s>\gamma_f$.

(iii) In the intermediate Stranski-Krastanov mode, a few monolayers grow before clusters are formed. At a later stage of growth, since the exposed surface and islands are the same materials, γ_s becomes equal to γ_f.

In the present case the surface energy of the Al and Si are 1.25 and 1.15 J/m^2 respectively [18,19], while their lattice constants are 0.405 nm and 0.543 nm, respectively. A negative surface energy ratio and a significant lattice mismatch indicate that Al film will grow on Si via island or Volmer-Weber mechanism, and this in fact has been observed in the literature. Now we make a literature review on how the island formation of metallic films evolves using fundamental growth processes controlling the microstructure evolution, and then explain the well established "structure zone model".

2.1 Fundamental Growth Processes

The evolution of the microstructure/morphology of thin films via island or Volmer-Weber mechanism is a very complex phenomenon and exhibits different features at different stages of film growth, as schematically shown in Fig. 1. It is well known that the growth of thin films proceeds through different fundamental growth processes of microstructure evolution, that is, nucleation, island (crystal) growth and grain growth via coalescence of islands. In addition, the process-induced phenomena, such as, effect of impurities etc can play an active role during the structure evolution. These phenomena depend on elementary atomic processes, which are influenced by various parameters, such as, impinging flux, deposition temperature, structural conditions of the substrates [20-25], etc.

The fundamental microstructure forming processes are summarized in the following.

(a) The nucleation

Nucleation of individual islands takes place on the substrate surface at the very first stage of the condensation, known as *primary nucleation* or at later stage on the bare substrate surface developing upon liquid like coalescence called *secondary nucleation.* A peculiar case of nucleation, known as *repeated nucleation,* shows up on the surface of a growing crystal when its growth is blocked by a surface-covering layer of an impurity phase. The primary nucleation and the film growth start simultaneously on the whole substrate area. However, the secondary and the repeated nucleation initiate a local growth in the later stages of the film formation. It is important to note that on amorphous substrates the nuclei are randomly oriented, while on single crystalline substrates the nuclei can have a preferred orientation. The details on the kinetics of nucleation can be found in literature [20-25].

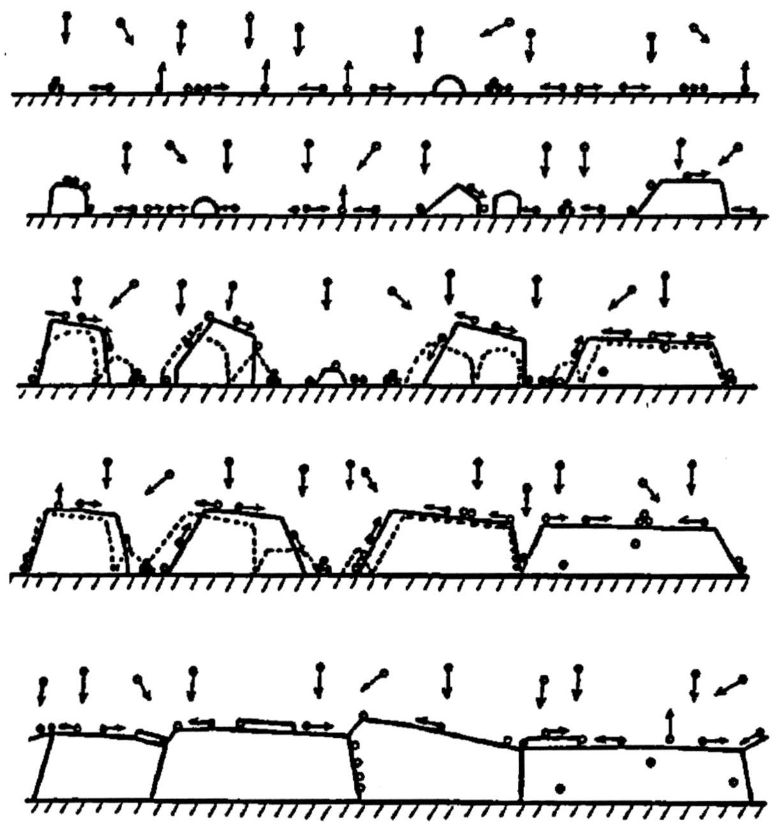

Figure 1. Schematic diagram showing fundamental growth processes: nucleation, island growth, impingement and coalescence of islands, grain coarsening and formation of polycrystalline islands and channels, development of a continuous structure and film growth.

(b) The island (crystal) growth

Island growth phenomenon incorporates the depositing material into the condensed phase. Crystals growing from the nuclei are either randomly oriented or textured depending upon the nuclei. The coalescence, that is, islands touching each other, is known as grain coarsening and, results in the development of a continuous film. The intersection lines of the island side faces and the substrate can be active or passive for monolayer nucleation and its growth on the side faces. In the active case, the movement of the monolayer growth steps proceeds from the intersection line to the top of the crystal, while in the passive case, the movement of the growth steps proceeds in the opposite direction i.e. to the direction of the intersection line. In the presence of impurities, the direction of the movement of the growth steps will be important in determining the location of the developing second phase.

In the case of polycrystalline structure, a growth competition can start among the neighboring crystals having different orientations. The faster growing crystals will grow over the slower growing ones, developing V-shaped crystal forms. This competition is terminated when only crystals exhibiting the same type of crystal faces proceed to the free surface. The consequence of this competitive growth is the development of a changing morphology and texture along the film thickness. In this case, a small-grained structure corresponding mainly to the nucleation density of random orientation exists in the substrate-near part of the film. It is followed by a part of V-shaped grains accompanied by an increase of the volume of preferentially oriented crystals. This process concludes later in the development of a columnar structure with a nearly unique crystal orientation. The intersection lines of grain boundaries with the free surface can be active (for high purity case) or passive (in case of contaminated grain boundaries) for monolayer nucleation.

(c) Grain growth

During structure evolution of a film, two types of grain growth take place: (i) dispersed small islands form a large new island during the coalescence stage and (ii) normal and abnormal grain growth by grain boundary migration.

The coalescence of the contacting grains, beside the increase of grain size, also results in a change of crystal orientations due to lowering of the free energy of the developing crystals. In the case of complete coalescence, a single crystal island on the substrate is formed.

Abnormal grain growth takes place in the structure by grain boundary migration and the direction of this migration is governed by the minimization of the substrate–film interface and free surface energy. During abnormal grain growth, the grain size distribution is bimodal. When the abnormal grain growth is completed the film again has a monomodal grain size distribution.

(d) Process-induced segregation of impurity species

Impurity species impinging onto the film surface can be adsorbed and segregated on the growing crystal faces or dissolved in the crystal lattice. Experimentally it has been found that impurities can act as *inhibitor* or *promoter* of the grain growth.

2.2. Structure Zone Model

Considering the basic microstructure forming phenomena as discussed above a basic structure zone model has been constructed in the literature. This model interprets the structure evolution at various temperatures. Fig. 2 illustrates the structure evolution of a metallic film of a particular thickness as a function of temperature T_s/T_m (T_s and T_m are respectively, substrate temperature and melting point of the metal) [26]. One can identify four different structure zones, which are described below.

(a) Zone-I

The *zone I* belongs to the temperature interval $0 < T_s/T_m < 0.2$ where neither the bulk diffusion nor the self surface diffusion has a remarkable value. The film is composed of fibers of small diameter, 1–10 nm, determined by the nucleation density and statistical fluctuations. The crystalline fibers grow out of the primary nuclei and proceed to the top of the film. The fibers are often collected into bundles. This is a rather homogeneous structure along the thickness of the film with increasing diameter of fibers by increasing T_s/T_m.

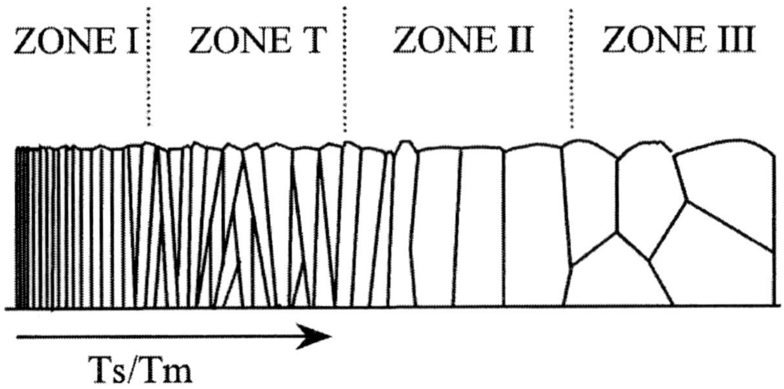

Figure 2. Main characteristics of structure zones used in the literature.

(b) Zone-T

In *zone T*, the structure is inhomogeneous along the film thickness. This zone belongs generally to the temperature interval $0.2 < T_s/T_m < 0.4$ in which the surface diffusion is remarkable, but the grain boundary migration is strongly limited. In the lower part of this temperature interval, grain boundary migration is expected to be very slow. Therefore, the lateral size of competing crystals at the substrate is determined by the nucleation density. At higher temperatures more and more grain boundaries become mobile resulting in the lateral growth of grains situated on the substrate. Due to this process, a weak preferred orientation develops strengthening gradually with temperature. The development of V-shaped crystals is a result of the competition taking place among the differently oriented neighboring crystals.

(c) Zone-II

This zone is characteristic for high substrate temperatures $T_s/T_m > 0.4$. In this temperature interval, the effect of grain boundary migration becomes decisive. The first structure of randomly oriented small grains is dissolved gradually by the coalescence and grain coarsening. This strong restructuration is controlled by the minimization of the interface and surface energy and, develops the *restructuration growth texture*. Being the grain boundaries mobile, the minimization of the grain boundary energy can also take place, resulting in grain boundaries perpendicular to the film plane. The film is composed of columnar crystals with similar orientation. The lateral size of the grains increases with increasing temperature.

(d) Zone-III

In *zone III*, the structure is characterized by globular three-dimensional network of grains, which is a direct indication that the crystal growth has been blocked periodically. If no impurities are present, this kind of structure is generally observed in the high substrate temperature range. However, this kind of structure can exist at any substrate temperature in the presence of inhibitors, which results in grains of different sizes, depending upon the nature of the impurity.

3. Theoretical Models for Electrical Resistivity

The room temperature electrical resistivity of metallic thin films is in general higher than that of the bulk. The various mechanisms contributing to an enhanced resistivity of films are (i) electron scattering from various defects [27-30], such as, impurities, vacancies and dislocations, (ii) external size effect [31] and (iii) internal size effects [32]. These effects are briefly discussed below.

(a) Electron scattering

The valance electrons in a metal are regarded as free to move through the lattice. The number of free electrons (*n*) in a metal is of the same order as the number of atoms. The precise number, however, depends on the detailed configuration of the energy bands of the metal and need not be a simple multiple or submultiple of the number of atoms [27-30].

The variation of *n* with temperature is negligible, since at ordinary temperatures the free electrons form a highly degenerate Fermi-Dirac gas. An electron can move freely through a perfect and rigid crystal lattice and there is no resistance. In a pure metal a finite mean free path (λ) is caused by the thermal vibrations of the lattice, which is large compared with the inter-atomic distance and is increased by lowering the temperature. The mean free path does not increase indefinitely as the temperature is lowered, however at very low temperatures it tends to a constant 'residual' value, which is determined by static lattice imperfections, such as, the presence of impurity atoms.

The following theoretical formula has been deduced for the electrical resistivity of a metal [27-30], which was subject to many simplifying assumptions concerning the interaction between the electrons and the lattice vibrations:

$$\rho(T) = \frac{mv}{nq^2\lambda_r} + \left(\frac{m}{2}\right)^2 \frac{9\pi h^2 C^2}{8n\Delta q^2 Mk\Theta\zeta^{3/2}} \left(\frac{T}{\Theta}\right)^5 \int_0^{\Theta/T} \frac{z^5 dz}{(e^z - 1)(1 - e^{-z})} \qquad (1)$$

where h is Planck's constant, q is the charge of an electron, and v is the velocity of an electron at the surface of the Fermi distribution, k is Boltzmann's constant, ξ is the Fermi energy level ($\zeta = 1/2mv^2$), Θ is the Debye temperature, m and M are the masses of electron and atom, respectively, Δ is the volume of the unit cell, and C is a constant which determines the interaction between the electrons and the lattice. According to this formula, the 'ideal' and 'residual' resistances are additive, and the ideal resistance is proportional to T at high and to T^5 at very low temperatures.

Thin films are more susceptible to the formation of various defects, such as, impurities, vacancies and dislocations. Thus, enhanced electron scattering at these defects leads to a higher electrical resistivity of films as compared to the bulk.

(b) External size effect:

The size effect theory of electrical resistivity for metallic thin films was developed by Fuschs and extended by Sondheimer (known as FS theory) [31]. According to this theory, as thickness (t) of the film becomes smaller than the mean free path (λ) of the electrons, the electrons are scattered at the external surface, which results in an increase in the film resistivity.

(c) Internal size effect

Mayadas et al [32] have shown that the electron scattering at the grain boundaries enhances the sample resistivity and termed it as internal size effect.

(d) Resistivity of metallic films

Mozrzymas and Warkusz [33] considered all the above three factors and derived following theoretical relation for the metallic thin films:

$$\rho = \rho_b + \rho_d + \rho_t \tag{2}$$

In this equation,

ρ_b is the bulk value of resistivity and takes into account of impurities, vacancies and dislocations;

ρ_d is the resistivity that depends on the grain diameter (*d*) via relation:

$$\rho_D = \frac{3\lambda R \rho_b}{2d(1-R)} \tag{3}$$

where λ is the electron mean free path, R is the grain boundary scattering coefficient; and

ρ_t is the resistivity term that depends on the film thickness (t) via

$$\rho_t = \frac{3\lambda(1-p)\rho_b}{8t} \tag{4}$$

where p is the fractions of electrons specularly scattered at the external surface.

It can be seen from equation (2) that all the contributions of resistivities are additive, which in accordance with Matthiessen's rule and allows us to make the following inferences.

The film resistivity increases with decreasing grain size, and this is physically understandable as at low grain size the grain boundary scattering is enhanced.

For a given grain diameter the resistivity of film increases with lowering film thickness.

4. Morphology Evolution of Ultra-thin Al films

4.1. Film Deposition

Thin films of Al of varying thickness at different temperatures were grown on Si (111) using MBE (RIBER EVA 32 E) system. The preparation/cleaning process adopted for Si (111) substrates (Wacker-Chemitronic GmBH, Germany) is as follows. First, Si (111) wafer was degreased by trichloroethylene and cleaned ultrasonically using acetone. The cleaned substrates were etched in dilute HF acid (4%) and rinsed in de-ionized water and finally in acetone. The chemically cleaned substrates were loaded in the Introduction chamber of the MBE system and baked at 500°C for 30 min under a vacuum (~10^{-8} torr). After baking the substrates were transferred to the Analysis chamber and the surface conditions of the substrates were examined by XPS. A typical Si-2p spectrum for a chemically cleaned and baked Si substrate is shown in Fig.3 Presence of only a single peak at ~99 eV indicates that only pure Si exists on the surface (for SiO_2 a peak at ~103 eV is expected to be present) [34,35]. The AFM image of the Si substrate is shown in Fig. 4. The average roughness of the surface was found to be <0.2 nm, indicating atomically flat surface.

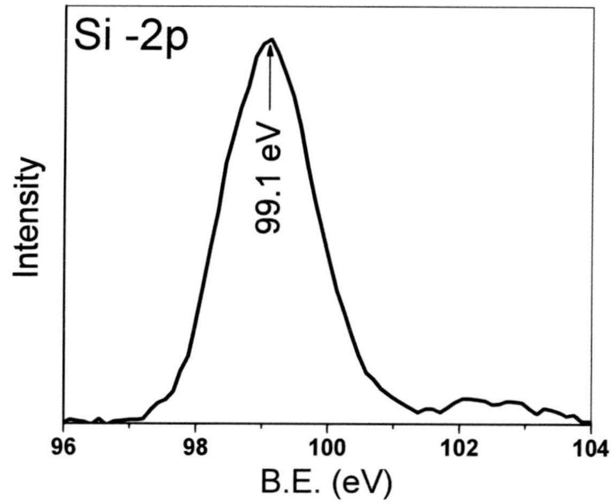

Figure 3. Si-2p spectrum of a bare Si substrate (after chemical cleaning and vacuum baking, as described in the text).

After baking and surface analysis using XPS, the Si substrates were transferred to the Growth chamber for the Al film deposition. The base vacuum in the Growth chamber during deposition was better than 10^{-9} torr. The Al was evaporated using an effusion cell loaded with 99.999% pure Al. For measurement of deposition rate, pressure (p) of material in substrate position was measured using flux gauge and flux of the material was calculated using the relation

$$F = 3.52 \times 10^{22} \frac{p}{\sqrt{MT}} \text{ molecules/cm}^2\text{.sec} \quad (5)$$

where p is in Torr, M is the molecular weight of the elements and T is the temperature of the effusion cell. The deposition rate was found to be ~0.15 Å/sec. The thickness (t) of the films was controlled by controlling the deposition time. We have also confirmed the deposition rates by measuring thickness of a 300 nm films using a Dektek profilometer. The error between the two measurements was found within 10%. After deposition, Al films were transferred to the Analysis chamber for in-situ XPS measurements.

Figure 4. AFM image of a bare Si substrate (immediate (after chemical cleaning and vacuum baking, as described in the text).

In order to understand the microstructure evolution in Al thin film following sets of experiments were conducted.

(i) **Effect of thickness:** Films of varying thickness between 10 and 500nm were deposited at a fixed substrate temperature (250°C or 500°C).

(ii) **Effect of substrate temperature**: Films of a fixed thickness (10 nm or 40 nm) were deposited at different temperatures between 25 -500°C.

(iii) **Effect of post annealing**: Films of a fixed thickness (10 or 40nm) deposited at different temperatures (250 or 500°C) and annealed at different temperatures.

4.2. Effect of Thickness

(a) AFM and SEM Studies

The AFM images depicting the morphology of Al thin films of varying thickness deposited at a fixed substrate temperature of 250°C and 500°C are shown in Figs. 6 and 7, respectively. Qualitatively, from these images it is evident that Al films, as expected, grow via island or Volmer-Weber growth mechanism on Si substrates. At low thickness the grains have spike-like features with small average grain size. As the thickness increases, the morphology takes columnar shape. A coalescence of the grains takes place at a particular thickness, which depends on the deposition temperature viz 60 nm for a deposition temperature of 250°C and 150 nm for 500°C.

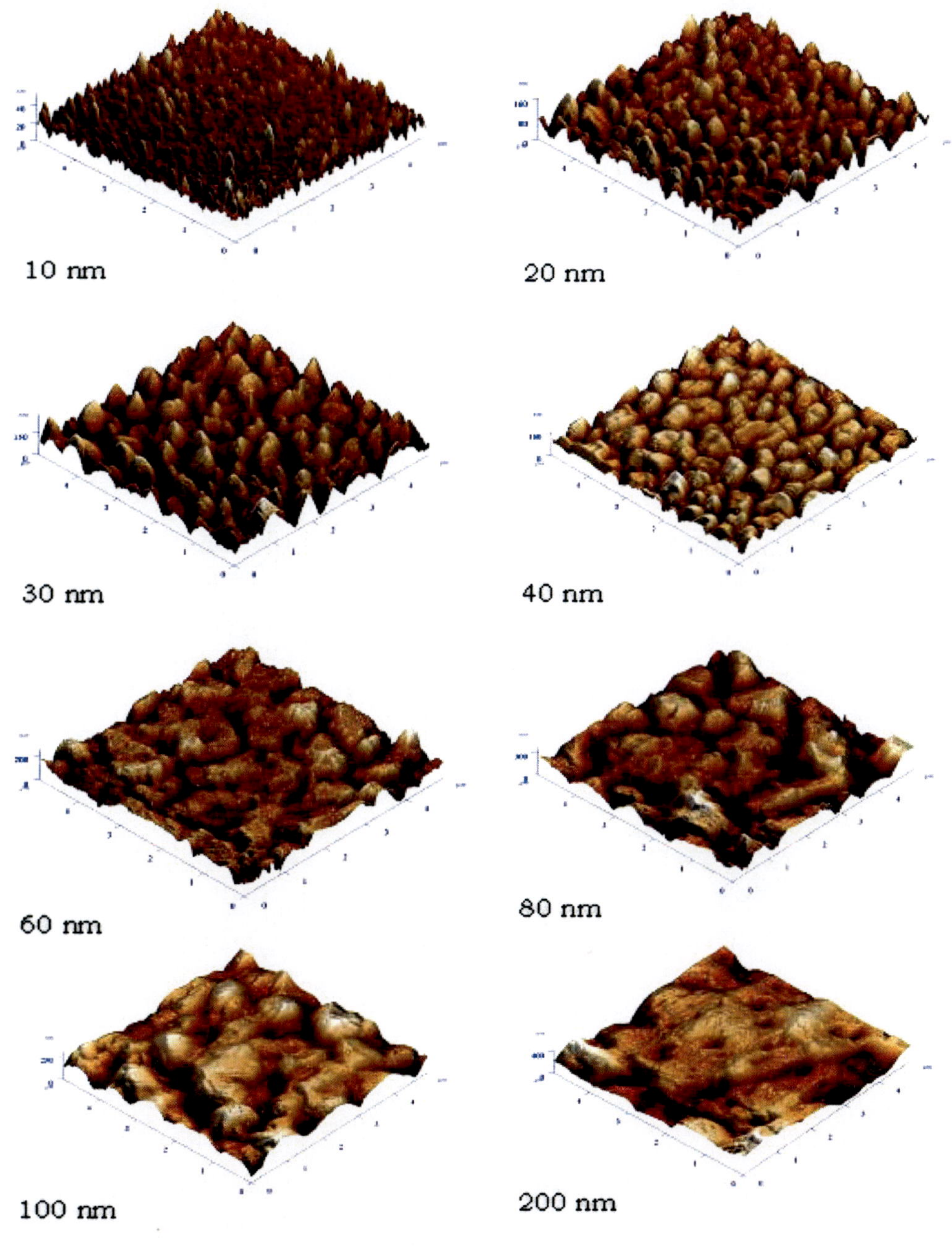

Figure 5. AFM images of Al films with varying thickness deposited at 250°C.

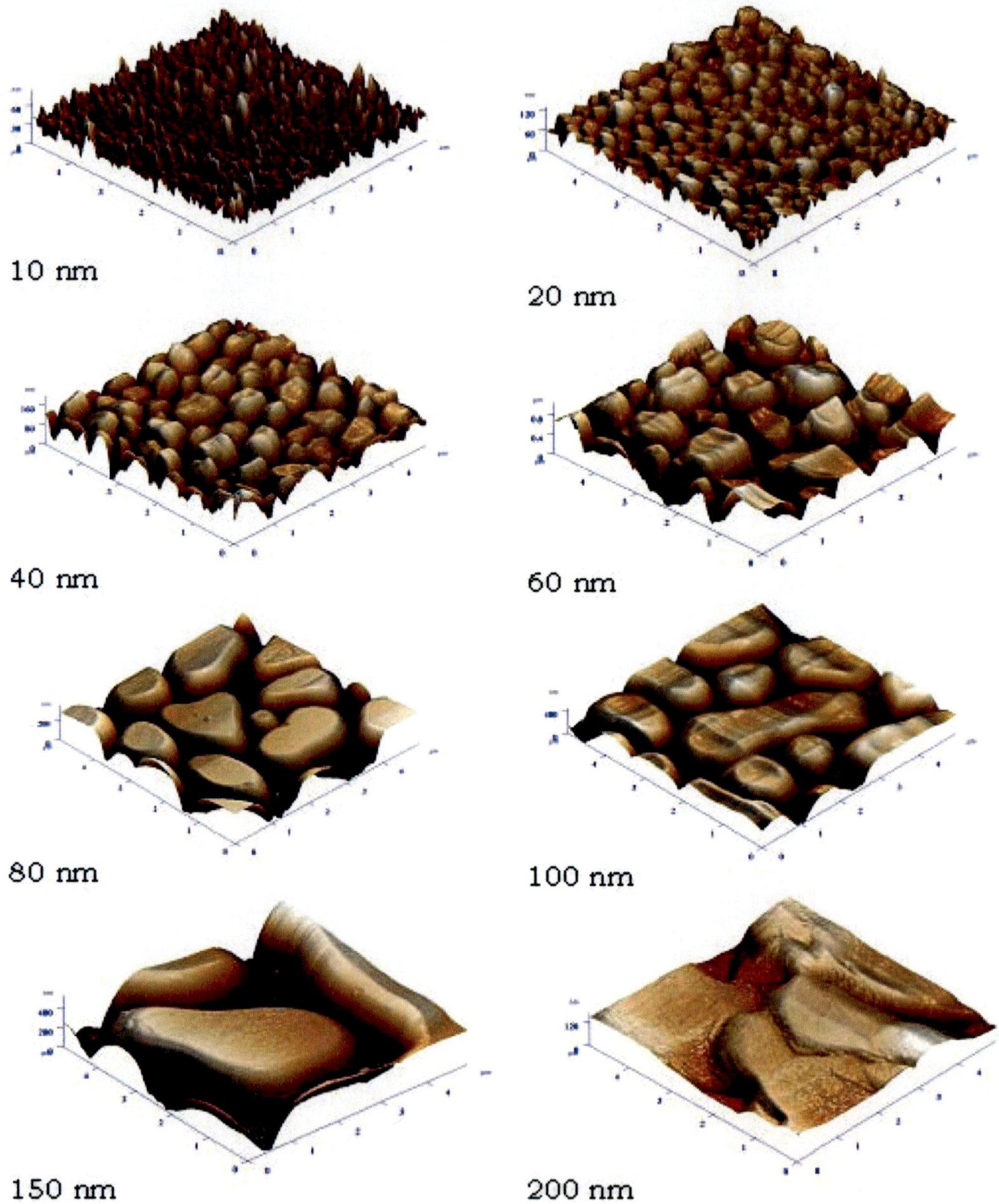

Figure 6. AFM images of Al films with varying thickness deposited at 500°C.

The SEM images of Al films of higher thickness deposited at 500°C, as shown in Fig. 7, is in agreement with the AFM data that coalescence in these films occur around 150 nm. The advantage of AFM technique is that one can quantify various parameters of the grains. These include the average grain size (d), average grain height (h), the angle (Φ) between the side face of the grain and the substrate plane, and roughness or step patterns on the surface of the grains. The determination of these parameters is demonstrated in Fig. 8. First, one records a height profile across a line in a 2D AFM image, and then from this profile h, d and Φ can be determined. Φ is determined by measuring the slope of the line corresponding to the side face. Such measurements are done over a large number of grains to obtain the average values. From these parameters one can infer the mechanism of grain growth in a film.

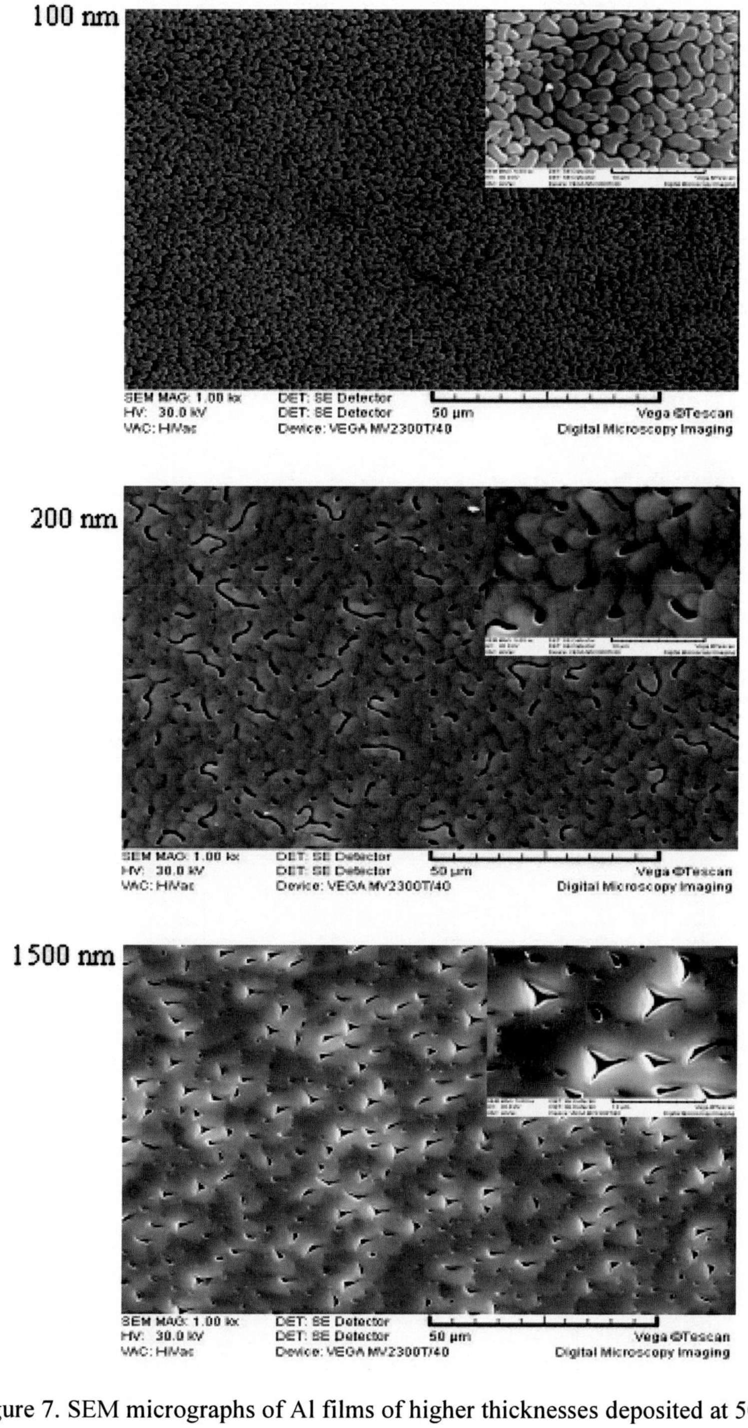

Figure 7. SEM micrographs of Al films of higher thicknesses deposited at 500°C.

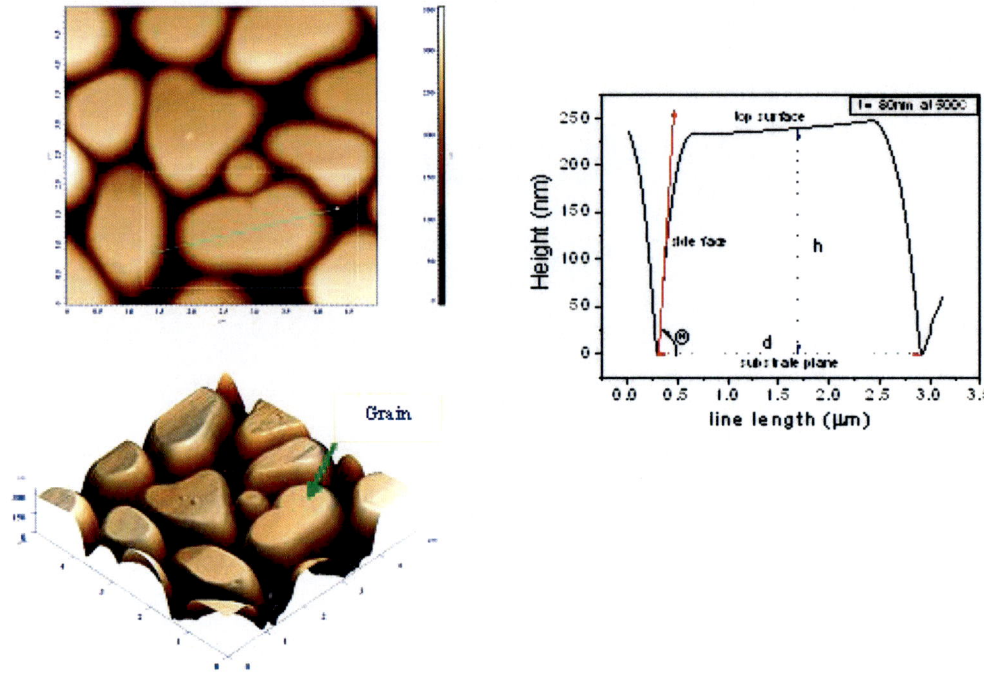

Figure 8. A typical example demonstrating quantification of an AFM image. The line-height profile across a grain yields various parameters (as discussed in the text).

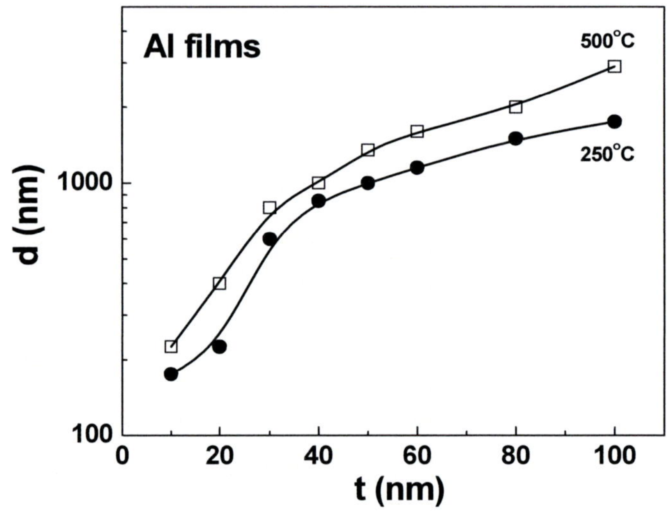

Figure 9. Variation of average grain size of Al with thickness at deposition temperatures of 250°C and 500°C.

The dependence of average grain size (*d*) on thickness (*t*) for film grown at 250 and 500°C is plotted in Fig. 9. For a substrate temperature of 250°C, *d* increases monotonically up to a thickness of 100 nm and above that the films become continuous across the substrates. The values of *d* are usually 20-25 times larger than *t*. Almost similar trend is observed for of the substrate temperature of 500°C, except that the grains are slightly larger in size. This is expected as at a high substrate temperature, the grain boundary migration becomes faster. Interestingly however, a complete coalescence is observed at a much higher thickness i.e. ~150 nm. This phenomenon is quite unusual as at high temperatures the coalescence is anticipated to be at lower thickness, owing to a higher surface diffusion of the atoms. We will address this issue later.

The variation of average grain height to thickness (*h/t*) ratio for films deposited at 250°C is shown in Fig. 10. An *h/t* >1 implies that the film grows in columnar morphology. Also, the angle (Φ) between the side face of the grain and the substrate plane were found to be ~89°, and was nearly independent of the thickness. These results indicate that the grains grow columnar almost perpendicular to the substrate surface, which is also apparent from 3D AFM images presented in Fig. 6. The *h/t* ratio becomes less than 1 for $t \geq 60$ nm, which indicate that the coalescence of grains begins for a thickness of 60 nm. The *h/t* ratio for film thickness of 100 nm is only 0.046 indicating that all the grains are fused and the resultant film is continuous.

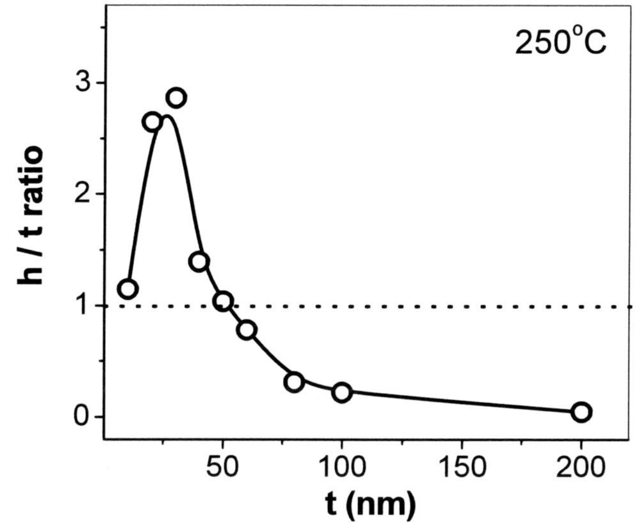

Figure 10. Variation of grain height to thickness (h/t) ratio with thickness.

In order to gain insight how the film grow as the thickness increases, we have taken a height profile at the top surface of a particular grain of 80nm thick film deposited at 500°C. The height profile shown in Fig. 11, clearly indicate a step-and-terrace structure with 2-3 nm of step height and 70-80 nm of terrace widths. This indicates layer-by-layer growth mode of films at higher thickness, and this expected as the Al film is growing on Al surface i.e. homoepitaxy.

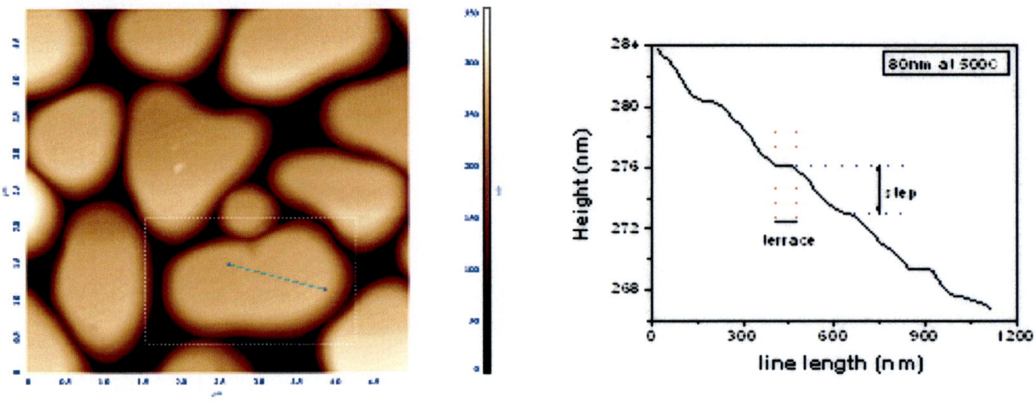

Figure 11. Height profile across a line (right graph) on the top surface of a grain (left image).

(b) XPS studies

The chemical state of the films was examined using in-situ XPS. The core level Al-2p and Si-2p XPS spectra recorded for Al thin films of different thickness are shown in Fig. 12. It is seen that Al is mainly in the metallic state (corresponding to a peak at 72.4 eV, however presence of a faint peak at 74.8 eV reveals formation of oxide, which indicates strong affinity of Al towards the residual oxygen present in the growth chamber [34,35]. In addition, a peak corresponding to Si is observed for *t* range of 10-60 nm, though the intensity monotonically decreases and becomes zero for 60 nm film. The observation of Si peak is quite interesting considering the facts that (i) Si and Al do not form compounds and are rather immiscible at this temperature [36,37], and (ii) XPS is a surface sensitive technique with a sampling depth of 3 nm [34]. Since the minimum thickness of our films is 10 nm, therefore the XPS signals of Si are attributed to originate from the inter-island regions of Al films. The Si peak vanishes for $t \geq 60$ nm, and this is expected as Al islands coalesce. Thus, the XPS data provides an additional support to the AFM analyses on thickness dependent columnar growth and coalescence of Al films.

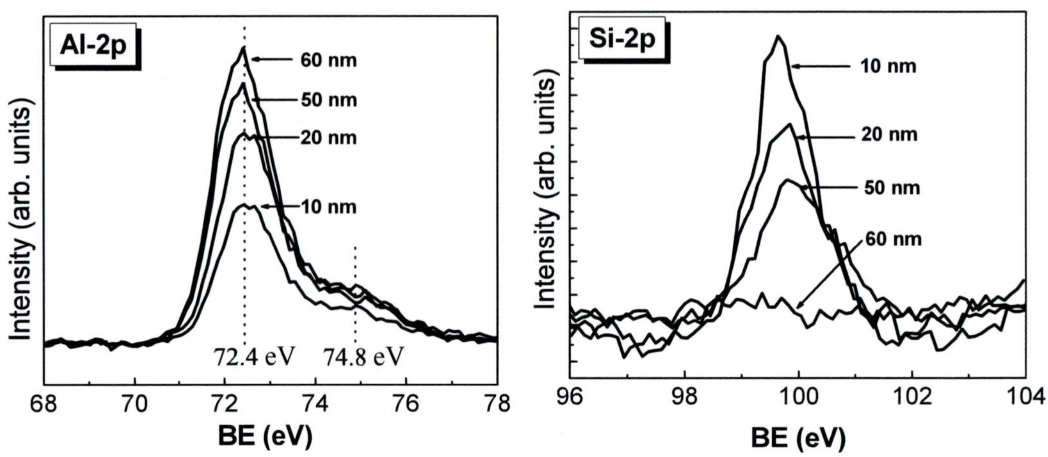

Figure 12. Al-2p and Si-2p spectra of Al films of different thickness.

(c) XRD studies

In order to determine the orientation of the columnar grains of Al films, XRD patterns were recorded, which are plotted in Fig. 13. The rocking curve of an Al(111) peaks revealed a full-width at half-maximum (FWHM) of ~0.4°. These results indicate that, irrespective of the thickness, Al films grow epitaxially on Si with (111) orientation. This result is understandable because (i) Al has fcc crystal structure and for this structure (111) face has the lowest surface energy and (ii) the Si substrate itself has (111) orientation.

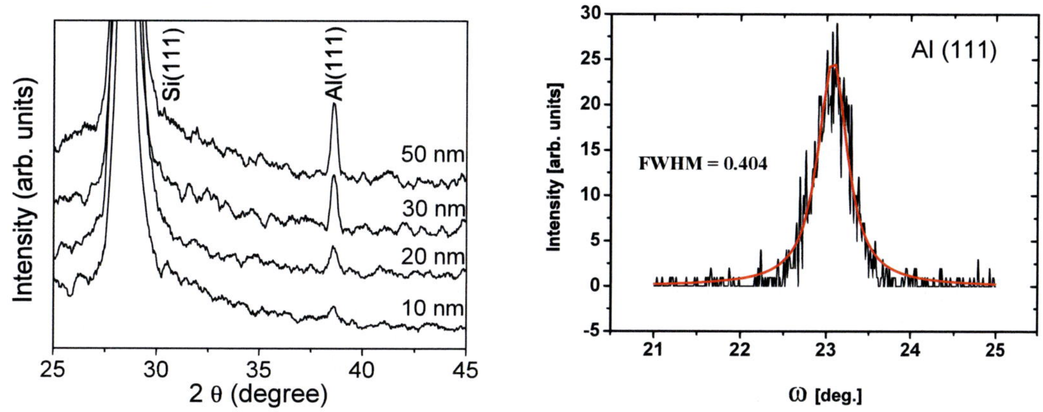

Figure 13. XRD patterns recorded for Al films having different thickness (left), and rocking curve of Al(111) peak of a 40 nm film.

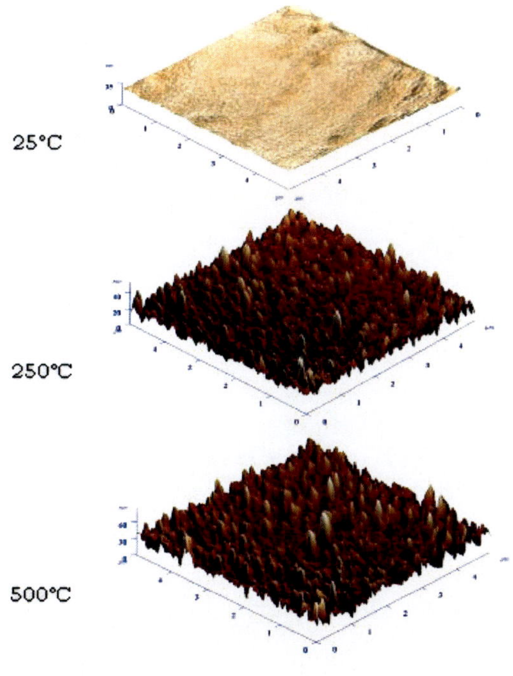

Figure 14. AFM images of 10nm Al film deposited at different temperatures.

4.2. Effect of Substrate Temperature

In order to investigate the effect of substrate temperature on the microstructure evolution, films of a fixed thickness (10 nm or 40 nm) were deposited at different temperatures between 25 -500C. The AFM images of these films are shown in Fig. 14 and 15, respectively. It is seen that at room temperature the film has a featureless morphology, indicating an amorphous like nature. As the deposition temperature increases, the grains initially grow in the spike-shape followed by well defined columnar-shape at high temperatures. At higher thickness the sharpness of the grain boundaries enhances.

Figure 15. AFM images for 40 nm thick film deposited at different temperatures.

From the AFM images, the average grain size (*d*) and height to thickness ratio (*h/t*) have been computed for films deposited at different temperatures, and are plotted in Fig. 16. It is seen that the *d* increases monotonically with increasing temperature; while *h/t* first increases rapidly upto 300°C and thereafter increases slowly. These results indicate that the island growth becomes more prominent as the temperature increases. The observed microstructure evolution with temperature is in agreement with the "structure zone model" discussed earlier.

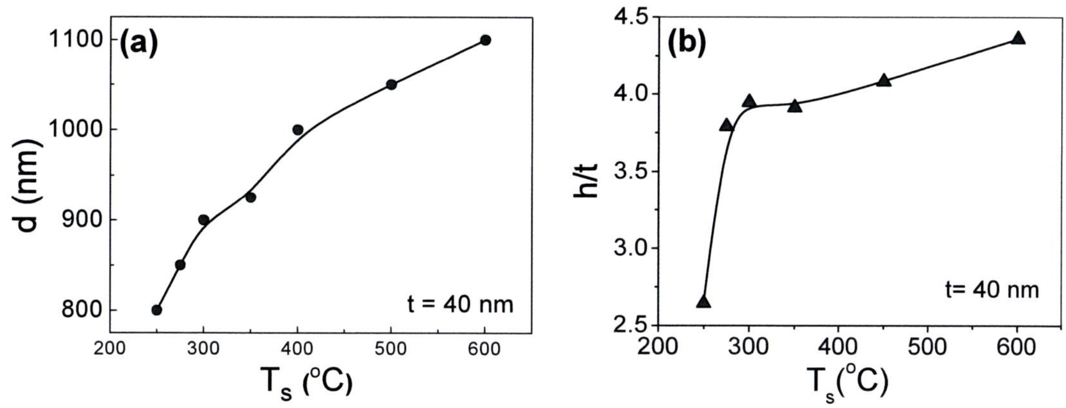

Figure 16. Variation of the average grain size (d) and height to thickness ratio (h/t) with deposition temperature for a 40 nm film.

The additional parameter that may influence the morphology of films at high substrate temperature is the interface reaction between film and substrate. The phase binary phase diagram of Al and Si indicates formation of Al-Si at high temperatures (>300°C). In order to confirm the existence of an Al-Si alloy phase at the interface, a film deposited at 600°C was mechanically scraped out and XPS of the interface was recorded. The obtained spectrum of Si-2p is shown in Fig. 17. It is seen that Si-2p has three components attributed to Si (~99 eV), Al-Si (102 eV) and SiO_2 (~104 eV). The intensity of the peak corresponding to Al-Si alloy [38-40] was calculated to be ~24%. Thus, it is postulated that formation of Al-Si alloy at interface has a significant role in governing the morphology of the films at high temperatures. This also explains the uncharacteristic coalescence behavior observed in the previous Section, i.e. higher is the deposition temperature larger is the film thickness required for complete coalescence.

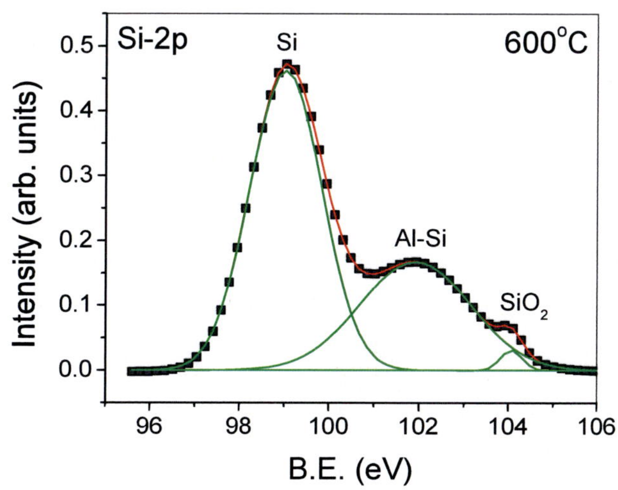

Figure 17. Si-2p spectrum of an interface between Si and Al film deposited at 600°C.

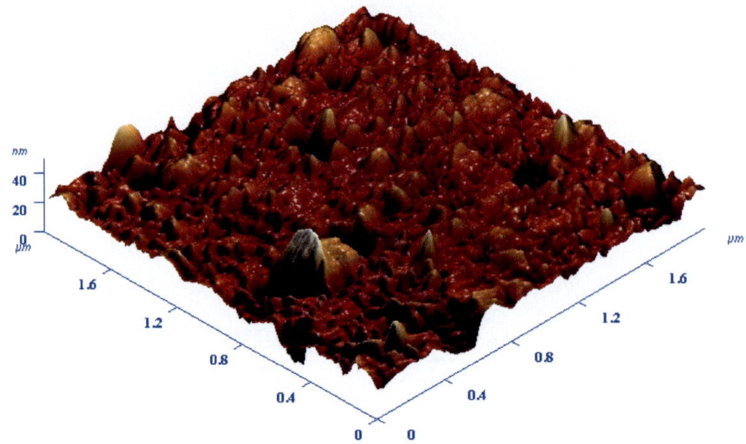

Figure 18. AFM image of a 10 nm Al film post-annealed at 250°C for 30 min under a base vacuum of (~10^{-9} torr).

4.3 Effect of Post Annealing

In order to investigate how the microstructure evolves in Al films on post-annealing, a 10 nm Al film deposited at room temperature was annealed at 250°C for 30 min under a base vacuum of (~10^{-9} torr). An AFM image of the post-annealed film is shown in Fig. 18, which is very different than the columnar growth observed for in-situ prepared film of same thickness and deposited at same temperature. In the case of the post-annealing, a complete coalescence has been observed. In addition, few very large size grains, due to abnormal grain growth, are also developed. This indicates that during post-annealing film-substrate interface does not play a significant role and, the grains grow laterally by normal and 'abnormal grain growth", owing to the grain boundary migration.

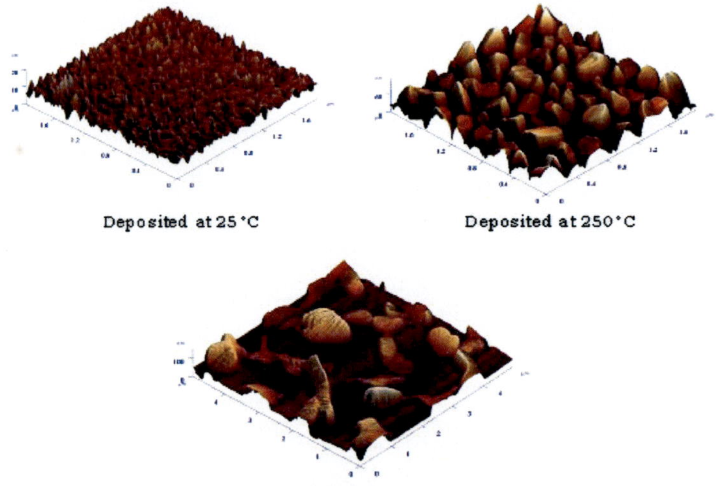

Figure 19. AFM images of Al films (10 nm) grown on SiO_2 under different conditions.

4.4. Effect of Substrate

In the preceding three Sections we have discussed the morphology evolution of Al on Si substrates as a function of thickness, deposition temperature and post-annealing temperature. In order to see how the substrate makes a difference to the mico-structure evolution of Al films, we have used SiO_2 as a substrate. It may be noted that SiO_2 has much lower surface energy compared to Al, and therefore microstructure evolution via island growth mechanism is anticipated.

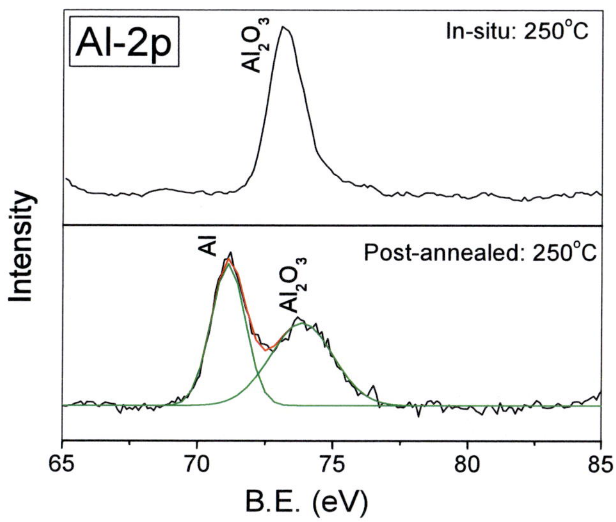

Figure 20. XPS spectra of the Al films deposited on the SiO_2 substrate.

The AFM images show that at room temperature, Al grow in fiber-like structure, which is in accordance with a larger surface energy difference between film and substrate. At higher deposition temperature (250°C), the columnar growth is observed, which is similar to that observed in Si substrate. Post-annealing at 250°C however, results in normal and abnormal growth of islands.

The XPS data, shown in Fig. 20, indicates that insitu deposition at high temperatures leads to the formation of Al_2O_3, as the Al picks up oxygen from the substrate. Post annealing results in a bit lesser formation of Al_2O_3.

5. Electrical Transport

In this section we analyze the electrical transport of the films described in the previous section.

5.1. Resistivity as a Function of Film Thickness

The room temperature value of resistivity (ρ_{RT}) of Al thin films grown at 250°C – AFM images shown in Fig. 5 - is plotted against thickness (t) in Fig. 21. The ρ_{RT} value of 10 nm films is 59.9 μΩ cm, which increases monotonically up to a value of 196.6 μΩ cm as t increases to 40 nm. The ρ_{RT} is found to decrease for $t > 40$ nm. It may be noted that ρ_{RT} drops by an order of magnitude as t increases from 50 nm to 60 nm. The bulk value of ρ_{RT} (2.59 μΩ cm) is obtained for a t value of 200nm. Almost similar behavior is observed for films grown at 500°C, except that the film resistance drops down sharply at a higher film thickness i.e. 150 nm.

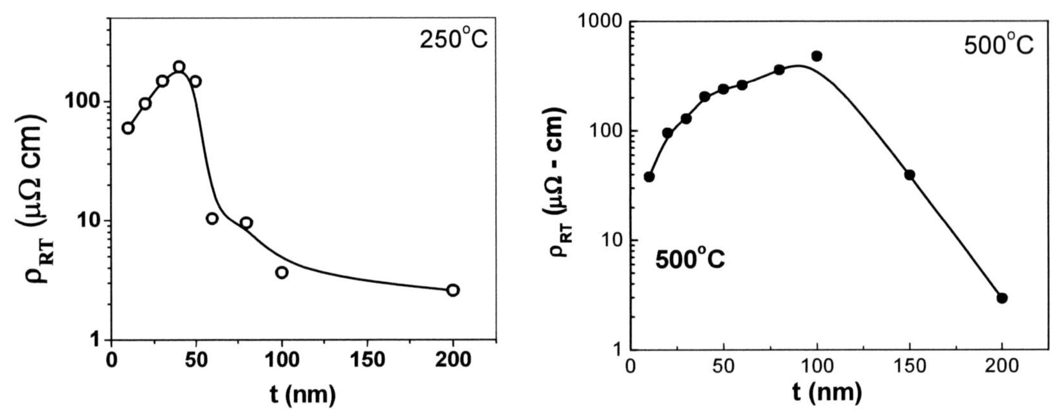

Figure 21.Variation of room temperature resistivity of Al films as a function of thickness.

We analyze the data of Fig. 21 within the framework of Mozrzymas and Warkusz model [33] described in Section 3. This model takes care of external size effect, internal size effect and electron scattering due to impurities, vacancies and dislocations. According to this theory, equations 2 to 4, the ρ should decrease with increasing film thickness (t) as well as grain size (d) [41-46]. If we consider the case of film deposited at 250°C, this theory appears to be valid for t in the range between 50 and 200 nm, as ρ decreases with increasing t. However, for t in the range 10 to 40 nm – contrary to the prediction of the theory – ρ increases with t despite of the fact that in this range both t and d increase. Almost similar condition prevails in the thickness range between 10 and 100 nm for films deposited at 500°C. Thus in the low thickness range the mechanism of electrical transport in Al films appears to be radically different, and there is a need to explore other transport mechanisms.

As we have discussed in Section 4.2, the growth morphology of films having in the low thickness range consists of isolated columnar grains. Moreover, as the XPS data of Fig. 12 suggest, the grain boundaries of such Al islands are covered with aluminum oxide. These results indicate indicates that the charge carriers are localized within the columnar grains and, therefore, can be treated as disordered materials. In highly disordered materials, electrical conduction occurs by the hopping [47] of electrons between localized sites. The hopping model qualitatively explains an increase in film resistivity with thickness as the carrier hopping would decrease owing to (i) an increase in the barrier height due the formation of aluminum oxide at the grain boundaries with thickness and (ii) the probability of carrier hopping would decrease with the number of surrounding islands, and infact the number of surrounding islands decrease with increasing thickness.

This hopping conduction results in a temperature dependence of resistivity given by [48]

$$\rho(T) = \rho_0 e^{\left(\frac{T_0}{T}\right)^x} \tag{6}$$

where T is temperature and ρ_0, T_0, and x are constants which depend on the disorder, the details of the interactions, and the dimensionality of the system. Simple activated hopping over a constant barrier results in the Arrhenius form with x=1. For noninteracting electrons, when the average hopping distance depends on temperature due to the compromise between hopping to sites which are close in energy, but farther away, Mott variable range hopping is expected, with

$$x = \frac{1}{d+1} \tag{7}$$

where d is the dimension. This implies that x has a value of 2, 3 and 4 for 1D, 2D and 3D systems, respectively.

Efros and Shklovskii (ES) showed that including Coulomb interactions between electrons results in a soft gap in the density of states at the Fermi energy, which changes the variable range hopping exponent to x=1/2 in all dimensions [48]. Hopping conduction has been investigated in a wide variety of materials, such as, doped semiconductors, semiconducting heterostructures, amorphous metals, magnetic materials, and superconductors. Both the Mott and the ES forms of variable range hopping have been observed, as well as a crossover between the two regimes [48].

The temperature dependence of ρ for films deposited at 250°C or 500°C having different thickness are shown in Fig. 22. It is evident from Fig. 22 that all the ρ-T curves for films deposited at 250°C with t in the range 10-50 nm exhibit a metal-to-semiconductor (M-I) transition around 100-120 K. Since the film morphology consists of columnar grains, it is expected that the charge transport would take place via 2D hopping mechanism. In order to verify this, the insulating part of ρ-T is plotted as ln(ρ) vs $(1/T)^{1/3}$ as shown in Fig. 23. A linear fit of data confirms the charge transport takes place via 2D VRH. From the linear fits the values of T_0 and ρ_0 have been calculated and are plotted as a function of thickness in the inset of the Fig. 23. It is seen that both T_0 and ρ_0 decrease with increasing film thickness. Similar observations are also made for films grown at 500°C.

On the other hand, for thickness ≥ 60 nm of the films deposited at 250C the M-I transition is absent. This is expected as the delocalization of charge carriers takes place owing to the coalescence of Al grains. For films deposited at 500C, the M-I transition is absent for t ≥ 150 nm, which is in agreement with the experimental observation that the coalescence of the grains occurs at this thickness only. These results therefore establish the fact that the M-I transitions in films is intimately related with the isolated columnar growth of films at lower thickness.

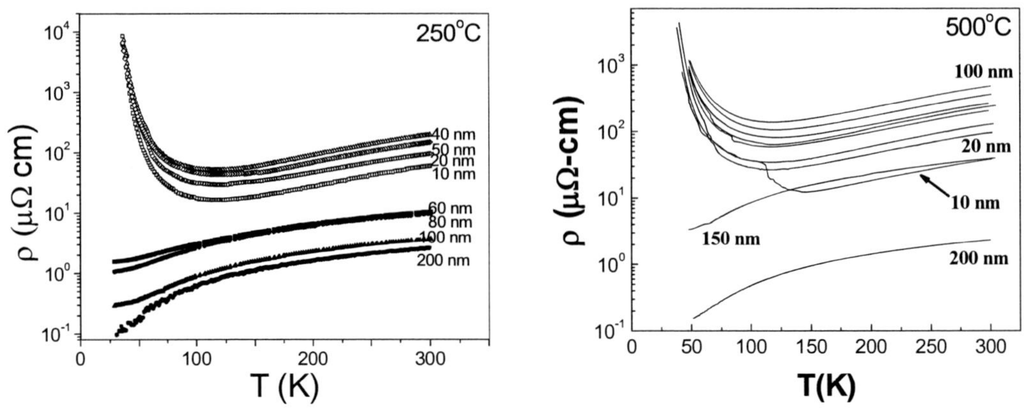

Figure 22. Temperature dependence of ρ having different thickness.

It may be noted that the M–I transition observed in the present case, as one may suspect, is not due to the interdiffusion of Si into Al. The solubility of Si in Al is about ~1% at an annealing temperature of 500°C [49-51]. If the presence of Si or the interfacial reaction is considered solely responsible for the M–I transition, then the M–I transition temperature should systematically shift to lower temperature with increasing film thickness as due to diffusion limitation less Si would be incorporated in thicker films. But it is seen that the M–I transition temperature is nearly independent of the film thickness. Further absence of M–I transition for a thickness greater or equal to that at which coalescence occurs indicates that growth morphology is responsible for the M–I transition.

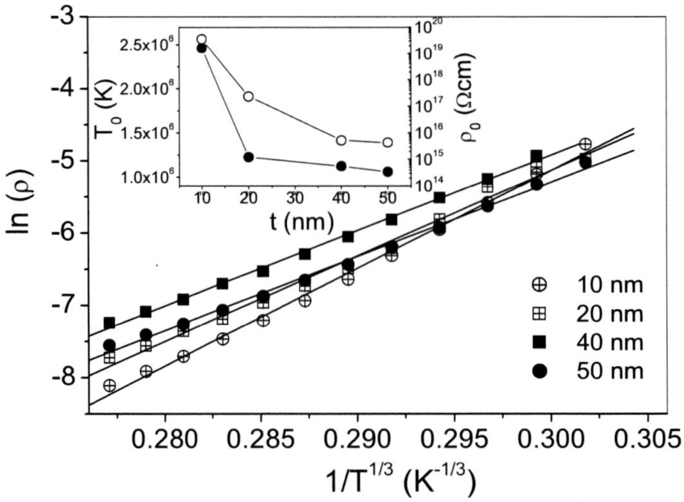

Figure 23. Insulating part of ρ-T curve of Fig. 22 (for films deposited at 250°C) is plotted as ln(ρ) vs $(1/T)^{1/3}$. The inset shows variation of T_0 and ρ_0 as a function of thickness.

5.2. Effect of Substrate, Deposition Temperature and post Annealing

The ρ-T plots of 40 nm thick films, deposited at different temperatures (AFM images shown in Fig. 15) on, are shown in Fig. 24. It is seen that except a marginal increase in the room temperature value of resistivity with increasing deposition temperature, the temperature variation of resistivity is nearly independent of the deposition temperature. The data in the insulating region exhibited a linear relationship between $ln(\rho)$ and $(1/T)^{1/3}$, indicating that 2D variable hopping conduction mechanism does not change with increasing deposition

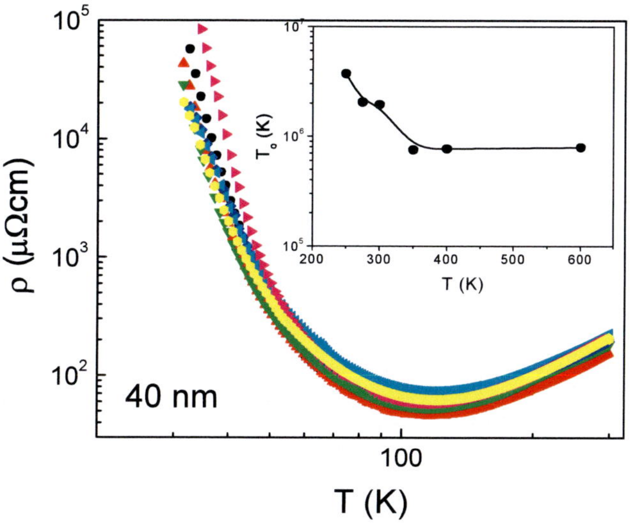

Figure 24. ρ - T plots for 40 nm thick film deposited at different temperatures. Inset shows the variation of T_0 with deposition temperature.

temperature. The computed value of T_0, as shown in the inset of Fig. 24, decreases marginally with increasing deposition temperature. These results indicate that the diffusion of Si (expected to be high at high deposition temperature) does not alter the conduction mechanism. Since the grains remained isolated even at a very high deposition temperature, the conduction is governed by 2D-VRH mechanism.

It has been demonstrated in the previous section that while in-situ deposition at higher temperatures leads to a columnar growth of grains; post annealing of a room temperature deposited film undergoes normal and abnormal growth due to grain boundary migration. The effect of such morphological changes on resistivity of the films is shown in Fig. 25. As usual, the in-situ deposition at higher temperatures leads to a temperature induced M-I transition, while post-annealing not only retains the metallic conduction down to the lowest temperature but also reduces the room temperature resistivity value. Thus connectivity of the grains is essential for metallic transport. However, as shown in the Fig. 26, the film shows different behavior when deposited on SiO_2 and post-annealed at 250°C. Despite of the grain connectivity owing to the normal and abnormal growth, the film shows temperature induced M-I transition. This is attributed to the formation of Al_2O_3 at the grain-boundaries (as supported by the XPS data), which confines the electrons. However, the 2D-VRH mechanism did not explain the low temperature transport in this case, and thus demands further investigations.

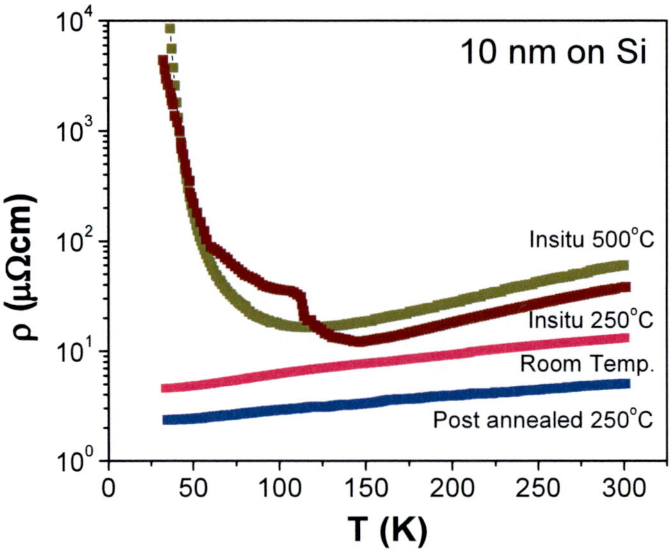

Figure 25. ρ - T plots for 10 nm thick films deposited at different temperatures, and room temperature deposited film post annealed at 250°C for 30 min.

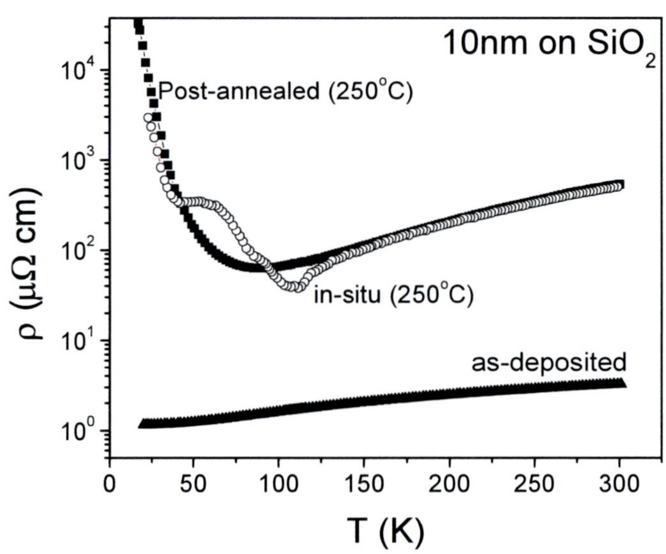

Figure 26. ρ - T plots for 10 nm thick films deposited at room temperature, in-situ at 250°C and room temperature deposited film post-annealed at 250°C for 30 min.

6. Conclusion

The microstructure evolution of ultra thin films of Al deposited on Si and SiO_2 substrates have been investigated as a function of film thickness, deposition temperature and post-annealing treatment. It has been found that in-situ growth of films leads to a columnar morphology, which is consistent with the structure zone model. The electrical transport in such films is governed by 2D variable range hopping conduction mechanism owing to the localization of charge in the columnar grains. Post-annealing treatment of the films however leads to the normal and abnormal growth of films owing to the grain boundary migration.

The results demonstrate that ultrathin films of Al may not be suitable for ULSI/GSI applications owing to the columnar grain growth. However, these films can be utilized as a buffer layer in the bilayer/multilayer schemes of the interconnect metallization, if the Al films are deposited at room temperature and post annealed at a moderate temperature (<250°C).

References

[1] The National Technology Roadmap for Semiconductors, Semiconductor Industry Association, San Jose, CA, 1994.

[2] S.P. Murarka, *Mater. Sci. Eng. R* **19**, 87 (1997) and references therein.

[3] Y. Wang, T.L. Alford, *Appl. Phys. Lett.* **74**,52 (1999).

[4] Y. Wang, T.L. Alford, J.W. Mayer, *J. Appl. Phys.* **86**,5407 (1999).

[5] D. K. Aswal, Niraj Joshi, A. K. Debnath, K. P. Muthe, S. K. Gupta, J. V. Yakhmi, *J. Appl. Phys.* **98**, 026103 (2005).

[6] Niraj Joshi, A.K. Debnath, D.K. Aswal, K.P. Muthe, M. Senthil Kumar, S.K. Gupta, J.V. Yakhmi, *Vacuum* **79**, 178–185, (2005).

[7] D.K. Aswal, Niraj Joshi, A.K. Debnath, S.K. Gupta, D. Vuillaume, J.V. Yakhmi, *Phys. Stat. Solidi (a),* **203**, 1254-1258, (2006).

[8] A.K. Debnath, Niraj Joshi, K.P. Muthe, J.C. Vyas, D.K. Aswal, S.K. Gupta, J.V. Yakhmi, *Appl. Surf. Sci.* **243**, 220, (2005).

[9] D.K. Aswal, K.P. Muthe, Niraj Joshi, A.K. Debnath, S.K. Gupta, J.V. Yakhmi, *J. Crystal Growth* **256**, 201 (2003).

[10] Niraj Joshi, A.K. Debnath, D.K. Aswal, K.P. Muthe, S.K. Gupta, J.V. Yakhmi, *Bull. Indian Vacuum Society*, **7** (2004) 3.

[11] D.K. Aswal, A.K. Debnath, Niraj Joshi, K.P. Muthe, S.K. Gupta and J.V. Yakhmi, Advances in surface treatment: research & applications (ASTRA), Ed. T.S. Sudarshan, G. Sundarajan, G.E. Totten, and S.V. Joshi, Hyderabad, 2003.p-163

[12] D. Vuillaume, *J. Nanosci. Nanotechnol.* **2,** 267 (2002).

[13] D.K. Aswal, S. Lenfant, D. Guerin, J.V. Yakhmi and D. Vuillaume, *Analytica Chimica Acta*, **568**, 84 (2006).

[14] J. W. C. de Vries, *Thin Solid Films* **167**, 25 (1988).

[15] J. Gogl, J. Vancea, and H. Hoffmann, J. Phys.: Condens. Matter 2, 1795 (1990).

[16] Thin film phenomena, KL Chopra, McGraw-Hill New York, (1969).

[17] Thin-Film Deposition: Principles and Practice, Donald L. Smith, McGraw Hill, New York (1995).

[18] C. Stamp, M. Schefier, H. Over, J. Burchhardt, M. Nielsen, D.L. Adams, W. Moritz, *Phys. Rev. B* **49**, 4959 (1994).
[19] Q Jiang, H M Lu, M Zhao, *J. Phys.: Condens. Matter*, **16**, 521 (2004).
[20] Nucleation and Growth of Thin Films, B Lewis, J. C. Anderson, Academic Press, London, 1978.
[21] C. B. Duke, ed., Surface Science: The First Thirty Years, Surf. Sci. 299/300 (1994).
[22] Introduction to Surface Physics, M. Prutton, Clarendon Press, Oxford, (1994).
[23] R. W. Vook, *Int. Metals Rev.*, **27**, 209 (1992).
[24] Solid State Theory, W. A. Harrison, Mcgraw-Hill, New York, (1970).
[25] Materials Fundamentals of Molecular Beam Epitaxy, J. Y. Tsao, Academic Press, Boston, (1993).
[26] R. Messier, A.P. Giri, and R.A. Roy, *J. Vac. Sci. Technol. A* **2**, 500 (1984).
[27] *Electrical resistance of metals*, G. T. Meaden, Plenum, New York, (1965).
[28] D. N. Langenberg, Amer. *J. Phys.*, **36**, 777, (1968).
[29] *Introduction to Solid State Physics*, Charles Kittel, Seventh edition, Wiley, (2005).
[30] *Solid State Theory: An Introduction*, Ulrich Roessler, Springer, Heidelber, Germany, (1969).
[31] E. H. Sondheimer, *Adv. Phys.* **1**, 1 (1952).
[32] F. Mayadas, M. Shatzkes, and J. F. Janak, *Appl. Phys. Lett.* **14**, 343 (1969).
[33] E. Dobierzewska-Mozrzymas and F. Warkusz, *Thin Solid Films* **43**, 267 (1977).
[34] D. Briggs, M.P. Seah, *Practical Surface Analysis by Auger and X-ray Photoelectron Spectroscopy*, Wiley, 1983.
[35] *An Introduction to Surface Analysis by XPS and AES*, John F. Watts, John Wolstenholme, John Wiley and Sons, 2003.
[36] *Materials Science of Thin Films*, Milton Ohring, Academic Press (1991).
[37] J. O. McCaldin and H. Sankur, *Appl. Phys. Lett.* **19**, 524 (1971).
[38] T. M. Klein et al., *Appl. Phys. Lett.* **75**, 4001 (1999).
[39] F. A. Ponce, *Appl. Phys. Lett.* **41**, 271 (1982).
[40] H. Iizuka, K. Yokoo, S. Ono, *Appl. Phys. Lett.* **61**, 2978 (1992).
[41] A.F. Mayadas and M. Shatzkes, *Phys. Rev., Sect. B*, **1** 1382 (1970).
[42] V.P. Nagpal and V. P. Duggal, *Thin Solid Films*, **9,** 213 (1972).
[43] S. Arajs, B. F. Oliver and J. T. Michalak, *J. Appl. Phys.*, **38** 1676 (1967).
[44] P.V. Andrews, *Phys. Lett.*, **19**, 558 (1965).
[45] P. Wissmann, *Thin Solid Films*, **5**, 329 (1970).
[46] A.F. Mayadas, R. Feder and R. Rosenberg, *J. Vac. Sci. Technol.*, **6**, 690 (1969).
[47] N. F. Mott, Metal-Insulator Transitions, Taylor & Francis, London, (1990).
[48] Y. Meir, *Phys. Rev. Lett.* **26**, 5265 (1996).
[49] R. L. Boatright and J. O. McCaldin, *J. Appl. Phys.* **47**, 2260 (1976).
[50] J. L. Murray and A. J. McAlister, *Bull. Alloy Phase Diagrams* **5**, 74 (1984).
[51] *Thin Films Interdiffusion and Reactions*, edited by J. M. Poate, K. N. Tu, and J. W. Mayer, p.380, Wiley, New York, 1978.

In: Materials Science Research Trends
Editor: Lawrence V. Olivante, pp. 321-340
ISBN: 978-1-60021-654-1

Chapter 8

THE DOUBLE IGNITION MAPS FOR COMBUSTION-SYNTHESIZING NIAL COMPOUNDS

Hung-Pin Li*
Jinwen University of Science and Technology, Taipei, Taiwan

Abstract

Combustion synthesis is a novel processing technique in which the compacted powders are first ignited by an external heating source to induce the chemical reaction inside the heated materials. Propagation of a combustion front during Ni-Al unstable combustion synthesis often extinguishes in the half way, due to the lower exothermic heat of the metallic reactions. To facilitate the combustion front to propagate completely, the reaction is always ignited again during the experimental demonstration. In this numerical study, the different second ignition positions in the combusted region, the reacting region, and the pre-heating region as well as the different second ignition times before and after the stop of the first combustion front are chosen to study the effect of the second ignition. The second ignition position and time are found to influence the subsequent temperature profiles. The stable propagation is observed as the reaction is ignited again in the reacting region. When the reaction is ignited secondly in the combusted region or the pre-heating region, part of the specimens cannot be synthesized at the theoretical combustion temperature due to low combustion temperature. In addition, the combustion temperature may be significantly enhanced for some area, and results in heterogeneous microstructure. Delay of the second ignition time is also found to increase the initial propagation velocity of the new combustion front. From the results generated in this study, the process map of double ignitions is established. The process map provides appropriate double-ignition circumstances to propagate the combustion front completely and achieve homogeneous microstructure product.

Keywords: second ignition, ignition, NiAl compound, self-propagating high-temperature synthesis (SHS), micropyretic synthesis, combustion synthesis

* E-mail address: hli@just.edu.tw / hungpin.li@msa.hinet.net. TEL: +886-932383482 FAX:+886-223813621. Hung-Pin Li, Ph.D., Professor, Dean of Academic Affairs Office, Jinwen University of Science and Technology, Hsintien, Taipei County, Taiwan

Introduction

Many exothermic non-catalytic solid-solid or solid-gas reactions, after being ignited locally, can release enough heat to sustain the self-propagating combustion front throughout the specimen without additional energy. Since the 1970's, this kind of exothermic reaction has been used in the process of synthesizing refractory compounds in the former Soviet Union. This novel technique, so-called Combustion / Micropyretic synthesis or Self-propagating High-temperature Synthesis (SHS), has been intensively studied for process implication [1-25]. This technique employs exothermic reaction processing, which circumvents difficulties associated with conventional methods of time and energy-intensive sintering processing.

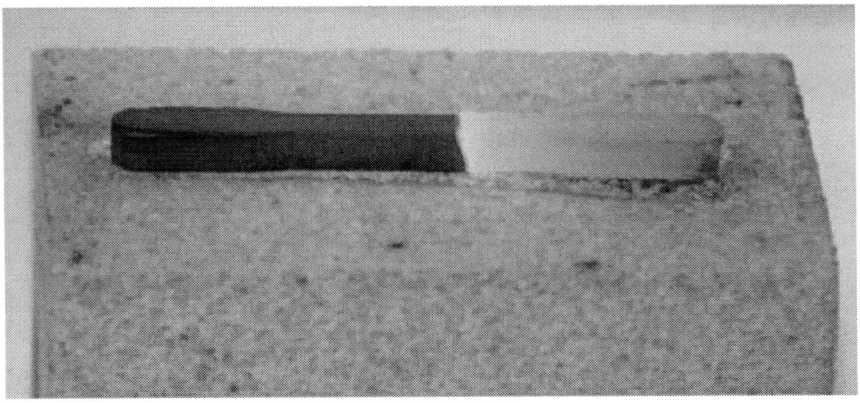

Figure 1. The combustion front propagates from right to left in the combustion synthesis of 95 wt.%(Ti+2B) + 5 wt.% Cu. [1]

Two basic combustion synthesis modes are commonly employed, namely the wave propagation mode and the thermal explosion mode. In the wave propagation mode, the compacted powders are ignited at a point by a heat source. After ignition, the heat to propagate the combustion wave is obtained from the heat released by the formation of the synthesized product, as shown in Figs. 1 [1]. The unreacted portion in front of the combustion wave is heated by this exothermic heat, undergoes synthesis, the wave propagates, thus causing further reaction and synthesis. In the thermal explosion mode, the specimen is heated in a furnace. The furnace may be kept at the ignition temperature or the specimen may be heated in the furnace at a predetermined heating rate to the ignition temperature. The combustion reaction in this mode may occur more or less simultaneously at all points in the specimen. Although the synthesized product phases obtained by both techniques are similar [2], there may be differences in the amount of residual porosity, final dimensions, and the thermal gradient during the processing. In both the modes, solid-solid reactions are most commonly encountered, sometimes solid-gas reactions are also noted as in the case of synthesis of refractory nitrides like TiN where nitrogen gas is used [3].

The advantages of combustion synthesis techniques include rapid net shape processing and clean products. When compared with conventional powder metallurgy operations, combustion synthesis not only offers shorter processing time but also excludes the requirement for high-temperature sintering. Volatile contaminants or impurities may be eliminated as the high temperature combustion wave propagates through the sample, and thus the synthesized products have the higher purity [4,5]. The steep temperature gradient also gives rise to the occurrence of metastable or non-equilibrium phases, which are not available in the conventional processing [4,5].

Combustion synthesized products have also been reported to have better mechanical and physical properties [5,6]. An example is the formation of shape-memory alloys of nickel and titanium [6]. It has been reported that those prepared by combustion synthesis, possess greater shape-recovery force than corresponding alloys produced by conventional methods [6]. On account of the high thermal gradients encountered in combustion synthesis, it has been speculated that the products of such a process may contain a high defect concentration. The presence of high levels of defects has led to expectation of higher reactivity, namely higher sinterability [7].

Several numerical and analytical models of combustion synthesis in a composite system have been well developed [8-19]. Lakshmikantha and Sekhar firstly explored the numerical model that includes the effects of dilution and porosity, and melting of each constituent of the reactants and products [13,14]. The analytical modeling of the propagation of the combustion front in solid-solid reaction systems is also reported [15]. The analytical model gives good results when compared with the experimentally determined numbers and the numerically calculated values. In addition, a dynamic modeling of the gas and solid reaction has also been carried out to illustrate the effects of various parameters on the combustion synthesis [16]. These numerical and analytical analyses provide the better understanding of the reaction sequence during combustion synthesis reactions.

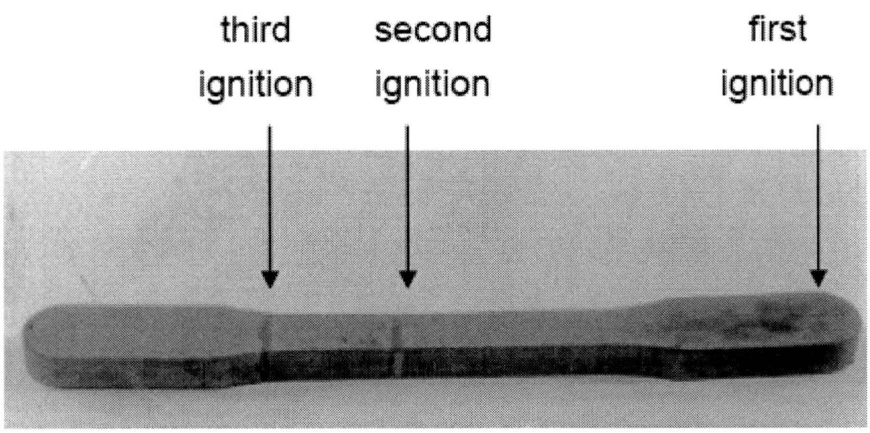

Figure 2. The combustion front propagates from right to left in the combustion synthesis of NiAl compound. Because of low exothermic heat and coarse Ni particles, the combustion front of Ni-Al reaction extinguishes in the half way. The reaction is ignited three times so as to propagate to the end completely.

One of the important quality controls of combustion synthesis processing is to create a uniform temperature profile across the sample surface in a rapid way. The temperature and fraction reacted profiles in the combustion zone are found to be strongly dependent on the process of heat dispersion [20]. Any change in the heat dispersion directly impacts the temperature and fraction reacted profiles in the combustion zone, and further affects the propagation velocity. Since the combustion synthesis of Ni and Al is a low exothermic reaction, it is noted that the combustion front is difficult to completely propagate to the end of the reaction, as shown in Fig. 2. In the experimental demonstration, the reaction is always ignited again for aiding the front to propagate completely. The second ignition time and position are found to influence the subsequent temperature profiles. The inappropriate second ignition may result in the unstable combustion reaction and further heterogeneous microstructure. Understanding the influence of the second ignition is thus important to acquire the homogeneous combustion synthesized products. In this research, the second positions in the combusted region, the reacting region, and the pre-heating region as well as the ignition time before and after the combustion front extinguished are chosen to study the effect of second ignition numerically. In addition, the effect of the delay of second ignition time is also studied. Using results from this study, a process map of the second ignition is generated and the appropriate second-ignition condition can be decided to acquire the homogeneous microstructure.

Numerical Calculation Procedure

During the passage of a combustion front in the reaction, the energy equation for transient heat conduction, including the source term, containing heat release due to the exothermic combustion reaction is given as [13,15,20]:

$$\rho C_p(\frac{\partial T}{\partial t}) = \frac{\partial}{\partial z}\left(\kappa(\frac{\partial T}{\partial z})\right) - \frac{4h(T-T_o)}{d} + \rho Q\Phi(T,\eta) \tag{1}$$

Each symbol in the equation is explained in the nomenclature section. The reaction rate, $\Phi(T,\eta)$, in Eq.(1) is given as :

$$\Phi(T,\eta) = \frac{\partial \eta}{\partial t} = K_o(1-\eta)\exp(-\frac{E}{RT}) \tag{2}$$

In this study, a numerical calculation for Eq.(1) is carried out with the assumption of the first order kinetics. In the Eq. (1), the energy required for heating the synthesized product from the initial temperature to the adiabatic combustion temperature is shown on the left-hand side. The terms on the right-hand side are the conduction heat transfer term, the surface heat loss parameter, and the heat release due to the exothermic combustion reaction, respectively. The surface heat loss is assumed radically Newtonian in this study. The previous studies [13,21] have shown that the surface heat loss is much less than the exothermic heat of the reaction, thus, the surface heat loss is taken to be zero in the numerical calculation.

The middle-difference approximation and an enthalpy-temperature method coupled with Guass-Seidel iteration procedure are used to solve the equations of the combustion synthesis problems. In the computational simulation, a one-dimensional sample of 1 cm long is divided into 1201 nodes (regions) to calculate the local temperature using an enthalpy-temperature method. The choice of 1 cm sample length is only for computational purpose, and the simulation results are applicable to practical experimental conditions. Firstly, the proper initial and boundary conditions are used to initialize the temperatures and enthalpies at all nodes. The initial conditions in the simulation are taken as follows: (1) At the ignition node, at time $t \geq 0$, the temperature is taken to be the adiabatic combustion temperature, ($T = Tc$ and $\eta = 1$). (2) At the other nodes, at time $t = 0$, the temperatures are taken to be the same as the substrate temperature, ($T = To$ and $\eta = 0$). Depending on the values of the temperature and enthalpy occurred in the reaction, the proper thermophysical / chemical parameters are considered and the limits of the reaction zone are determined for each node in the numerical calculation. At any given time, the fraction reacted and enthalpy of the current iteration are calculated from the previous fraction reacted and enthalpy of the earlier iteration. The range of the enthalpy as well as the molar ratio among each material for each node is thus determined, and the values of temperature, density, and thermal conductivity at each node can be further calculated in appropriate zone.

The various microscale events, that is, local processes such as heating of the sample, melting of lower refractory reactant, and formation of product, are included in this calculation procedure. In addition, the various thermophysical / chemical parameters, such as thermal conductivity, density, and heat capacity of the reactants and product, are assumed to be independent of temperature, but they are different in each state. The effect of melting of reactants and product is included in the calculation procedure. The porosities of the reactants and product which influence the density and thermal conductivity profiles are also considered in the calculation. The porosities of the reactants and product are both taken as 30 % in this study. The average values of these parameters vary when the reaction proceeds, depending upon the degree of reaction. Depending on the values of the temperature and extent of enthalpy released in the reaction, the proper thermophysical / chemical parameters are considered in the numerical calculation. At any given time, the reacted fraction and the enthalpy of the current iteration are calculated from the previous reacted fraction, enthalpy, and other parameters of the earlier iteration. The range of the enthalpy as well as the molar ratio among each material for each node is thus determined, and the values of temperature, density, and thermal conductivity at each node can be further calculated in the appropriate zone.

The criterion used to ascertain whether the reacted fraction (η) and the enthalpies (ϕ) at each time level converge or not, is determined from the relative error criterion, *i.e.*, for all nodes $\left|(\eta^{t+1} - \eta^t)/\eta^t\right| \leq 10^{-6}$ and $\left|(\phi^{t+1} - \phi^t)/\phi^t\right| \leq 0.001$. The superscripts $t+1$ and t denote the current and previous iterations, respectively. Once the convergence criterion for every node is met, the enthalpy and the reacted fraction of the last iteration in a time step are considered to be the corresponding final values. The calculations are normally performed 500 to 2000 times, depending upon the calculated thermal parameters to make all 1201 nodes meet the criterion for each time step. At least 600 time steps are calculated to allow the propagation of the combustion front across the 1-cm-long specimen completely. The parameter values used in the computational calculation are shown in Table I [26-28] and Table II [26,29]. In this study, the combustion temperature is defined as the highest reaction temperature during combustion synthesis and the propagation velocity is the velocity of the combustion front propagation. The pre-heating zone is calculated from the end of reaction nodes (zones) until the position where the temperature is decreased to the original substrate temperature. The detailed numerical calculation procedure is shown in the previous studies [8-12,17,21-25,30].

Table I. The thermophysical/chemical parameters for the reactants (Ni and Al) and product (NiAl) at 300 K and liquid state [26-28] .

Thermophysical/chemical parameters	Al	Ni	NiAl
Heat capacity (at 300 K) (J/(kgK))	902 [26]	445 [26]	537 [26]
Heat capacity (liquid state) (J/(kgK))	1178 [26]	735 [26]	831 [26]
Thermal conductivity (at 300 K) (J/(msK))	238 [28]	88.5 [28]	75 [27]
Thermal conductivity (liquid state) (J/(msK))	100 [28]	53 [27]	55 [27]
Density (at 300 K) (kg/m^3)	2700 [28]	8900 [28]	6050 [27]
Density (liquid state) (kg/m^3)	2385 [28]	7905 [28]	5950 [28]

Table II. The values of various parameters used in the numerical calculation [26,29].

Parameters	NiAl
Combustion Temperature (K)	1912
Activation Energy (kJ/mole)	139 [26]
Exothermic Heat (kJ/mole)	118.5 [29]
Pre-Exponential Factor (1/second)	8×10^8
Time Step (second)	0.00025

Results and Discussion

The propagation for the single ignition of Ni-Al combustion front stops before the reaction completed is shown in Fig. 3. Here the combustion front extinguishes in the half way, because of taking lower pre-exponential factor that is equivalent to using the coarse particles in the low exothermic metallic Ni-Al reaction. Small interval of time step is used in the numerical calculation of Fig. 3 to accurately find the exact time step of the extinguished front. The dark lines (1-25 time steps) and gray lines (26-34 time steps) in Fig. 3 denote the temperature profiles before and after the stop of the combustion front, respectively. Combustion front stops propagating from the twenty-sixth time step, 0.0130 seconds after the first ignition. The combustion front propagates to 0.2367 cm in this period and the specimen starts cooling. Figure 3 also illustrates that the temperature at the extinguished position of 0.2367 cm is decreased from the adiabatic combustion temperature (1912 K) to 1540 K in 0.0040 seconds (from the 26th to the 34th time step) and 3% of the specimen has been cooled down in this period. Therefore, the second ignition is required to facilitate the combustion front to propagate completely.

The different positions and times of the second ignition are chosen to study the influence on the subsequent temperature profiles, as shown in Table III. Figure 4 shows the typical temperature profiles for the second-ignition that is secondly ignited at 0.0500 seconds after the first ignition, or 15 time steps / 0.0375 seconds after the extinguish of combustion front. The specimen is ignited at 0.5833 cm far from the left end. The combustion front generated from the second ignition is noted to propagate to both sides. The propagation velocity for the new generated combustion front is noted to accelerate from 20.6 cm/s (before the second ignition) to 27.7 cm/s (after the second ignition) due to a higher substrate temperature. As the new combustion front propagates to the extinguished area, the combustion temperature is increased to ~ 2115 K, exceeding the theoretical combustion temperature (1912 K). The reaction becomes more violent and the porosity has been observed to correspondingly increase [18]. Such an unusual increase in the temperature may possibly result in the different phase changes and the non-homogeneous microstructure.

Table III. The studied positions and times of the second ignition.

Ignition positions			Ignition times	
Area I : combusted region	a	0.1875 cm	before the stop of the front	0.0125 s
	b	0.2083 cm	after the extinguish of the front	0.0150 s
	c	0.2291 cm		0.0175 s
Area II : reacting region	d	0.2500 cm		0.0200 s
Area III : pre-heating region	e	0.2708 cm		0.0225 s
	f	0.2917 cm		0.0250 s
	g	0.3125 cm		0.0275 s

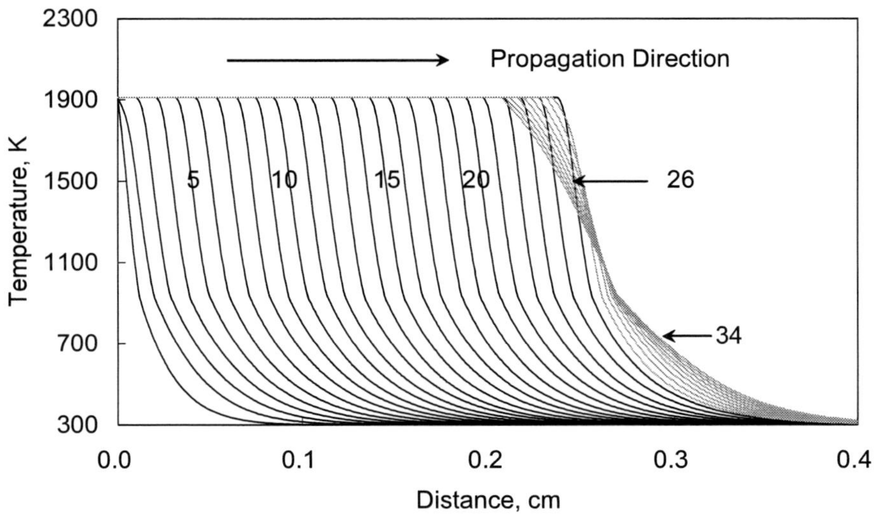

Figure 3. A plot of the combustion front temperature at various times along the specimen. The pre-exponential factor, Ko, is taken to be $8x10^7$ 1/s and the gray lines are the temperature profiles after the front stops propagating. The numbers shown in the figure are the sequences of time steps and the interval time between two consecutive time steps (profiles) is 0.0005 s.

In addition, since the second ignition point is far from the extinguished spot, the time to get the second combustion front initiating the combustion reaction in the extinguished area is delayed. Therefore, the significant cooling effect of the extinguished spot is found before the start of the new reaction and propagation. Such a decrease in the substrate temperature delays the initiation of the combustion synthesis in this area. Note from Fig. 4 that it takes 0.0100 seconds (*i.e.*, from the 20th to the 24th time step) to increase temperature of the extinguished spot up to 1900 K and "unreacted area" is observed in the local reaction. A portion of the specimen (indicated by "m") cannot be synthesized at the theoretical combustion temperature (1912 K), which is equivalent to the melting point of NiAl product. Such as a "unreacted area" causes no melting and solidification of product phase and may produce the different microstructure as compared with other region. Both un-reacting and over-reacting generate non-uniform structures and mechanical properties locally during the inappropriate second ignition.

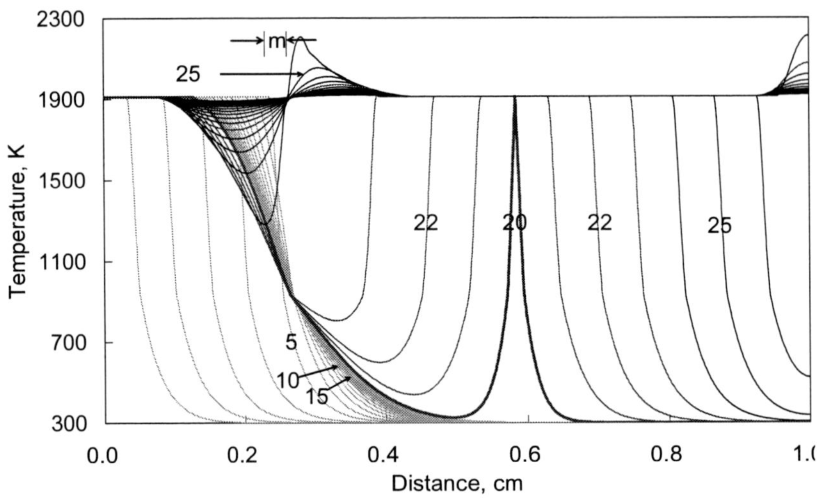

Figure 4. A plot of the combustion front temperature at various times along the specimen. The interval time between two consecutive time steps (profiles) is 0.0025 s. The reaction is ignited again at 0.5833 cm and the ignition time is 0.0500 s (the 20th time step) after the first ignition.

From the results shown in Fig. 4, it is noted that the inappropriate second ignition situation may cause parts of combustion synthesis over-reacting or under-reacting, further resulting in the heterogeneous microstructure. This study is thus aiming at understanding the effect of the second ignition condition on the propagation of the combustion front.

Figure 5 (a) shows the temperature and fraction reacted profiles at the time of 0.0125 seconds and 0.0170 seconds, respectively. Note from Fig. 3 that the profile with the time of 0.0125 seconds is the last time step sequence while the combustion front can propagate appropriately. On the other hand, the specimen has been cooled down in 0.0040 seconds while at the time of 0.0170 seconds. The propagating thermal profile becomes flatter as the specimen starts cooling. The flatter profiles imply that the energy and heat are not able to accumulate in the region in front of the combustion front. An insufficient energy cannot sustain the propagation of the combustion front; thus, propagation is noted to stop. Figure 5(a) also illustrates that each small portion in the specimen has different temperature and offers the different thermal condition for the new ignition. Therefore, the distribution of the subsequent thermal profile will have direct effect on the choice of the time and position of the second ignition. To systematically study the effect of the second ignition position and time on the thermal profiles during combustion synthesis, several different positions and times of ignition are chosen to study. It includes the different positions in the combusted region (Area

I in Fig. 5(b)), the reacting region (Area II in Fig. 5(b)), and the pre-heating region (Area III in Fig. 5(b)) as well as the various times before and after the stop of the combustion front. In this study, the combusted region is defined as the location where the reactions have reacted complete and the value of fraction reacted is equal to one. The area between the combusted region and the pre-heating region is a reacting region. The pre-heating region starts as the fraction reacted becomes zero. The relationship between the second ignition condition and the subsequent propagating thermal profiles is expected to generate. One of the objectives in this study is to choose the second ignition condition properly to sustain the propagation completely and acquire the homogeneous structure.

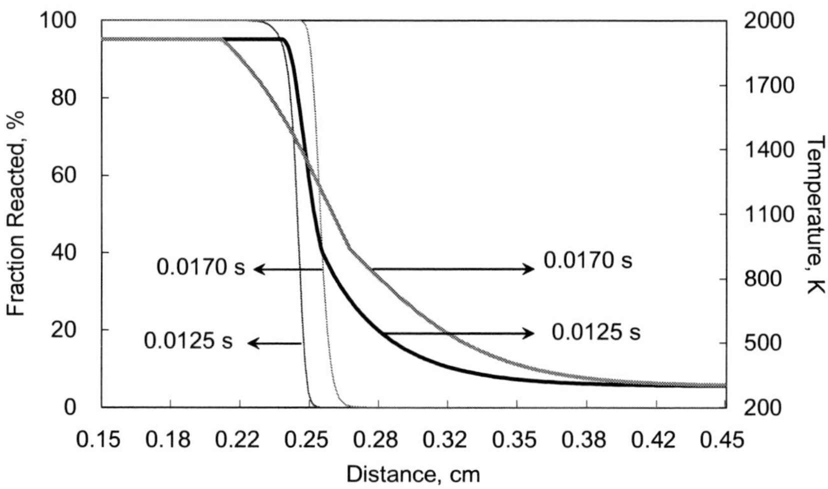

(a)

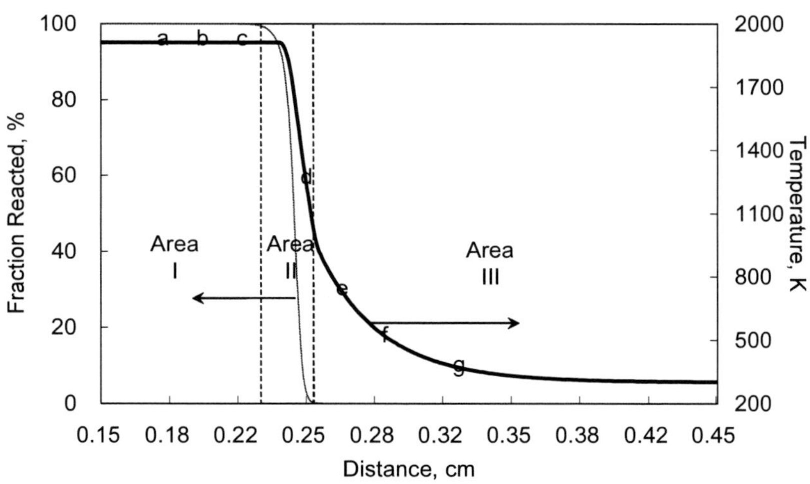

(b)

Figure 5. (a) The combustion temperature and reacted fraction profiles at the time of 0.0125 s and 0.0170 s, respectively. (b) For the profiles with the time of 0.0125 s, areas I, II, and III denote the combusted region, the reacting region, and the pre-heating region, respectively. The characters A-G are the positions of the second ignition in this study.

1. Second Ignition in Combusted Region (Area I)

Figure 6 firstly shows the temperature profiles for the reaction ignited again in the combusted region (Area I in Fig. 5(b)). As the reaction is ignited secondly in the combusted region, no new reaction is occurred and no additional exothermic heat from the second ignition is released to aid the propagation of the subsequent reaction. The tasks of the second ignition are

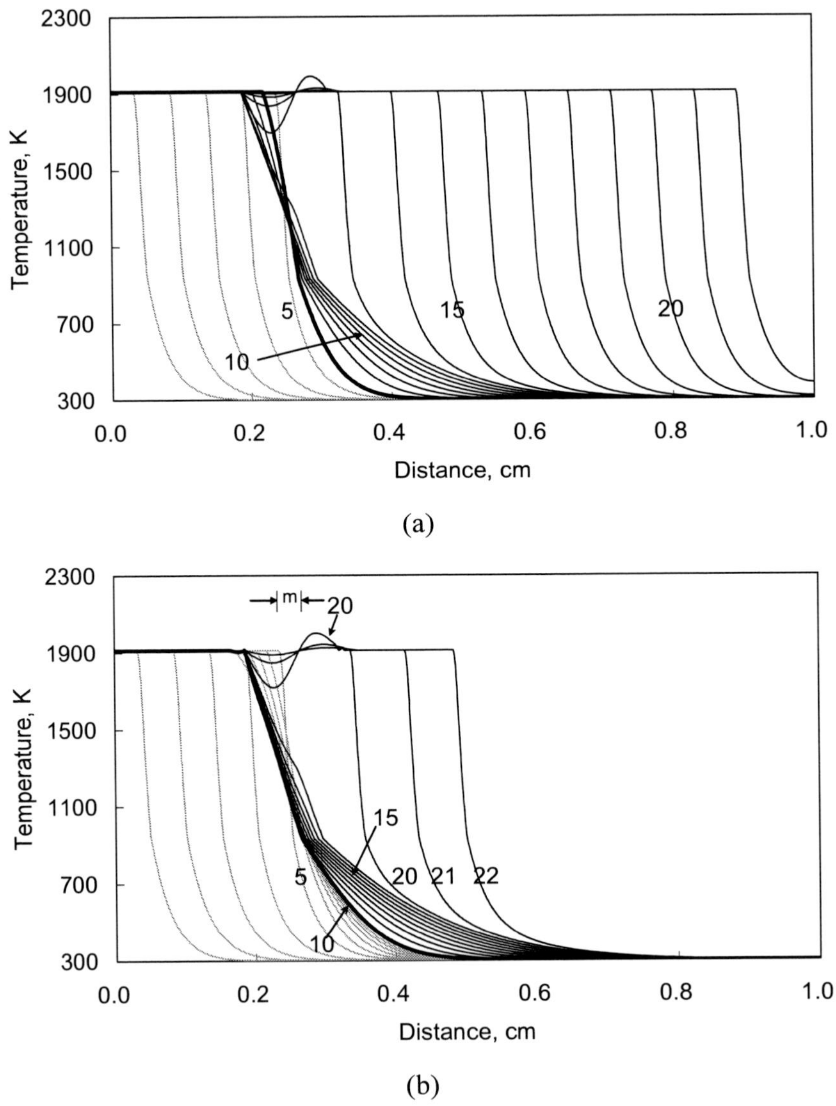

Figure 6. A plot of the combustion front temperature at various times for the reaction ignited again in the combusted region. The interval time between two consecutive time steps (profiles) is 0.0025 s and the reaction is ignited again at 0.1875 cm. The new temperature profile generated from the second ignition is denoted by the bold line. The second ignition time in (a) is 0.0150 s (the 6th time step) and in (b) is 0.0275 s (the 11th time step) after the first ignition. The figures only show the first twenty-two time steps of the combustion temperature profiles to clearly illustrate the zone "m".

transferring the heat to increase temperature to the original combustion temperature (1912 K) and to offer adequate energy for the second propagation. When the reaction is ignited at 0.0150 seconds, right after the combustion front extinguished, the area in front of the extinguished point is still kept at higher temperature. Therefore, only few heat and energy from the combusted region is required to transfer to the un-reacted zone for initiating the subsequent propagation. The transferred heat from the second ignition quickly aids the combustion front to continuously propagate. Figure 6(a) shows that it only takes 0.0175 seconds (from the 6th to the 13th time step) to propagate again. However, if the reaction is ignited at 0.0275 seconds, which is 0.0125 seconds after the stop of the first combustion front, the delay of ignition time results in the cooling of the specimen and decreases the temperature of the extinguished point. More energy is necessary in the pre-heating zone for sustaining combustion front propagation, therefore, it spends longer time (0.0225 seconds, from the 11th to the 20th time step) to initiate new propagation, as shown in Fig. 6(b). The calculated results also show that the longer delay between the first and the second ignition time decreases the initial propagation velocity of the second combustion front.

It is also noted from Fig. 6(b) that as the combustion front stops propagating, the temperature is gradually decreasing and the thermal profiles become flatter. Because part of the specimen is still kept at high temperature, the reaction is still in progress and the fraction reacted is calculated to increase gradually as the time step is increased. Thus, it is found from Fig. 5(a) that ~ 2% of the specimen has been reacted completely (fraction reacted profile "propagates" from ~0.25 cm to ~0.27 cm) in between 0.0125 and 0.0170 seconds even though reaction temperature has been gradually decreased. When the substrate temperature of the un-reacted zone is increased to threshold ignition temperature, the portions (indicated by "m" in Fig. 6(b)) that have been reacted completely after the stop of the first combustion front, cannot be further synthesized by combustion reaction and no more exothermic energy is released to heat up itself. The heat transferred from the new exothermic reaction now only heats up the reacted parts and the temperature is noted to gradually increase. The reaction re-starts from the end of zone m where no reactants are synthesized during the cooling. Therefore, Fig. 6(b) clearly shows that part of the specimen (zone m) is under-reacting and the synthesized temperature is below the theoretical combustion temperature even the reaction has been ignited again.

Figure 7 shows the plot of the temperature profiles with the reaction time for different positions in the extinguished region. The second ignition condition is the same as one in Fig. 6(b). The combustion front propagates till 0.0125 seconds and then starts cooling from the 0.0130 second. The position of 0.2367 cm indicated by bold line in Fig. 7 is the last node on which the normal propagation is obtained. As the combustion front extinguishes, the temperature gradually decreases. When the reaction is ignited again at 0.0275 seconds on the position of 0.1875 cm (around the point T-5 in Fig. 7), the temperature in the vicinity of the second ignition node quickly increases up to the theoretical combustion temperature (1912 K). However, the temperature in some areas indicated by dash lines in Fig. 7, whose position is the same as the zone "m" in Fig 6(b), is noted to increase slowly because the reaction in this portion has been reacted completely and no additional exothermic heat is released to quickly heat up itself. Figure 7 clearly shows that the reactants in zone m (gray lines) are synthesized at the lower heating rate. The temperature in zone m is also noted not to reach the theoretical combustion temperature (1912 K) that is equivalent to the melting point of NiAl product. No melting and solidification are occurred in zone m. Thus, this under-reacting portion of the specimen may possess the different reaction mechanisms, phase change, and microstructure as compared with the other region.

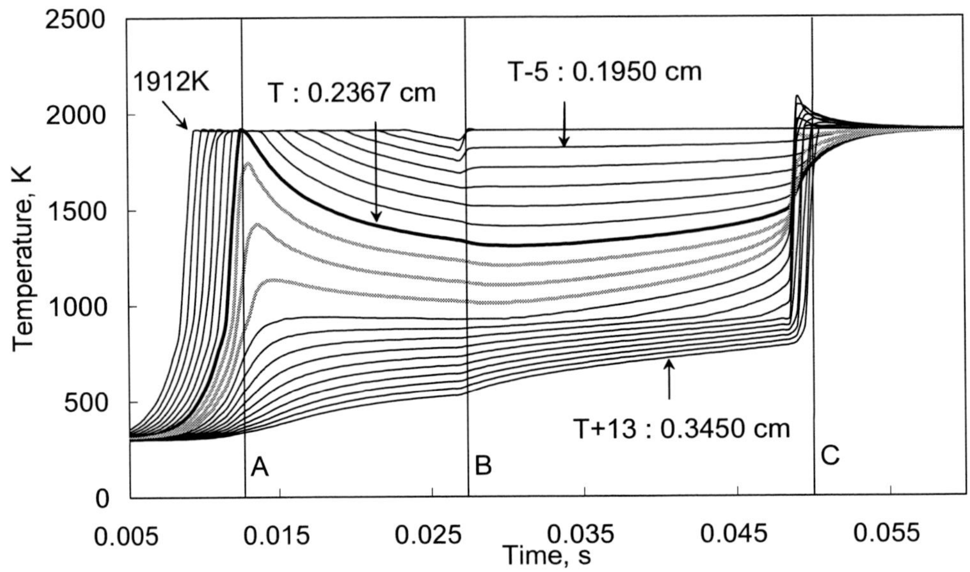

Figure 7. A plot of the time-resolved temperature at different specimen positions. The bold line whose location is 0.2367 cm far from the left end (ignition position) is the last temperature profile reaching the combustion temperature, 1912 K, before the second ignition. The distance between two consecutive lines is 0.0083 cm. The time of the second ignition is 0.0275 s and the position of the second ignition is 0.1875 cm (combusted region). Characters A, B, and C in the figure denote the stop of propagation, the second ignition time, and the 20th time step, respectively.

2. Second Ignition in Reacting Region (Area II)

Figures 8(a) and (b) illustrate the temperature profiles for the reaction that is ignited secondly in the reacting region (Area II in Fig. 5(b)). When the reaction is ignited again right after the stop of the combustion front, the second ignition instantaneously offers the heat and temperature required for propagation. The combustion front is noted to propagate again without any impediment, as shown in Fig. 8(a). Even the reaction is ignited again lately at 0.0275 seconds after the first ignition (Fig. 8(b)), the combustion front is also found to immediately propagate from the second ignition point. Examine the profiles in Fig. 8(b) carefully and it is found that all the unreacted portions in the reacting region are ignited and heated up to the adiabatic combustion temperature. Figures 8(a) and (b) both illustrate that the combustion fronts take eleven time steps (~0.0275 seconds) to propagate from the second ignition point to the end. The second ignition in the reacting region can quickly offer the energy required for combustion front propagation. This implies that the second ignition time has little influence on the propagation if the reaction is ignited in the reacting region.

The time-resolved temperature changes at different specimen positions for the second ignition in reacting region are also studied. The results show that the temperature in the reacted area and the un-reacted area both decreased as the combustion front extinguishes. As the reaction is ignited again, the temperature of the reacted area is found to increase gradually (lower heating rate) to the adiabatic combustion temperature because no new combustion reaction has occurred. However, the temperature of the un-reacted area is increased sharply to the adiabatic combustion temperature (higher heating rate), suggesting the occurrence of the combustion synthesis. It is also noted that the temperature at each position reaches the theoretical combustion temperature.

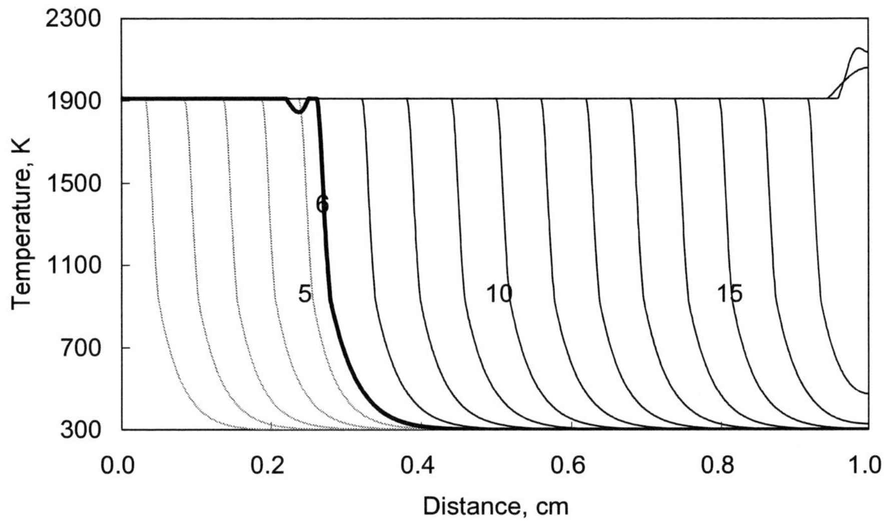

(a)

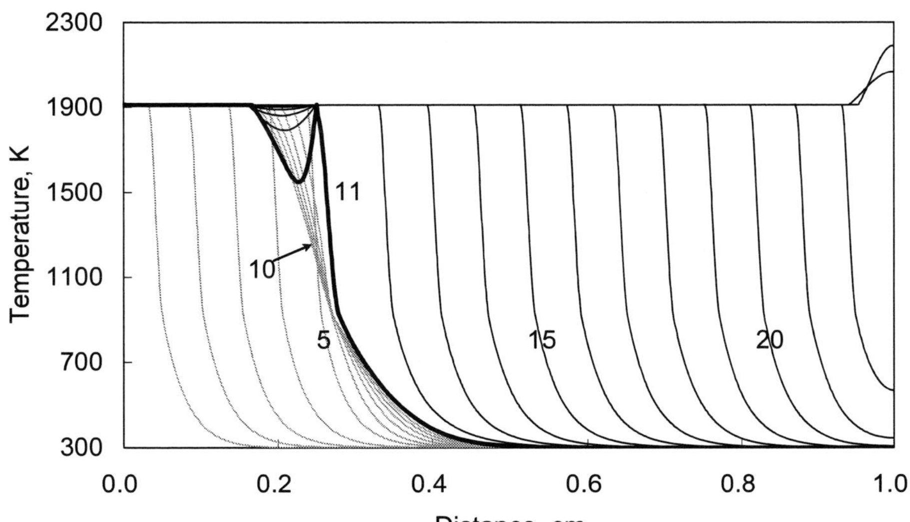

(b)

Figure 8. A plot of the combustion front temperature at various times for the reaction ignited again in the reacting region. The interval time between two consecutive time steps (profiles) is 0.0025 s and the reaction is ignited again at 0.2500 cm. The second ignition time in (a) is 0.0150 s (the 6th time step) and in (b) is 0.0275 s (the 11th time step) after the first ignition.

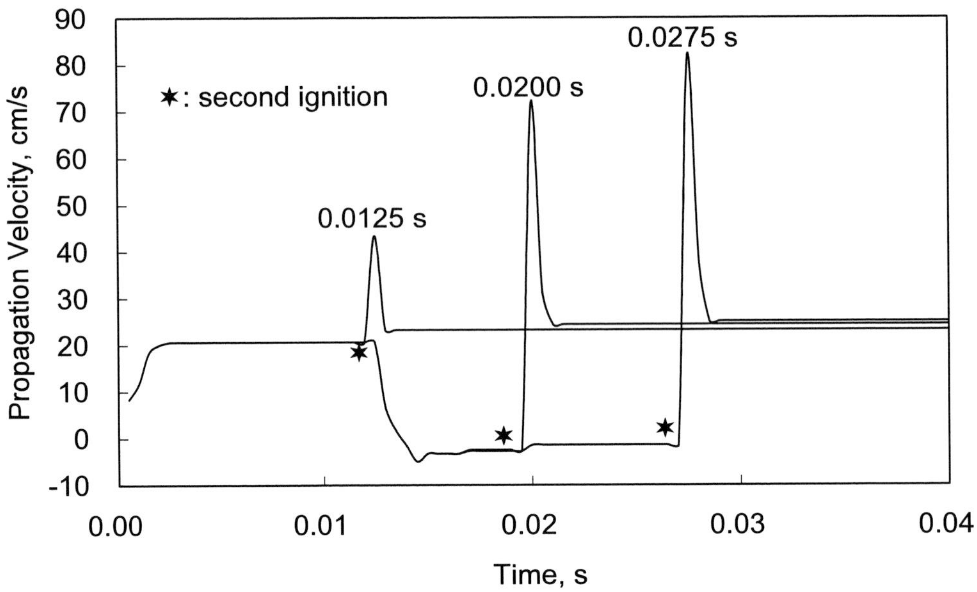

Figure 9. A plot of the propagation velocity with the reaction time for the different second ignition time. The position for the second ignition is 0.2500 cm far from the left end.

The delay of ignition in the reacting region also offers the time for transferring heat to the pre-heating region, thus increasing the substrate temperature. Therefore, the propagation velocity is expected to increase after the second ignition. The change of propagation velocity after the second ignition is shown in Figure 9, where the first ignition extinguishes at 0.0130 s, and the second ignition point is 0.2500 cm. It shows that the propagation velocity of the combustion front is increased to 20.7 cm/s after the first ignition. The combustion front propagates steady until 0.0125 seconds and the propagation velocity is gradually decreased, where the first combustion reaction extinguishes. When the reaction is secondly ignited again, the propagation velocity is significantly increased and then propagates at the steady velocity of 23.3, 24.4, and 25.1 cm/s to the end, respectively for different delay times between the first and second ignition. An extension in the second ignition time increases the propagation velocity of the new combustion front.

Figures 8(a) and (b) also illustrate that the propagations are complete and no over-reacting or under-reacting is found in the thermal profiles if the second ignition occurs in the reacting region. The calculated results also show that for a given ignition times the maximum propagation velocity is normally obtained for the reaction ignited again in the reacting region.

3. Second Ignition in Pre-Heating Region (Area III)

For the reaction ignited secondly in the pre-heating region, the new combustion front generated from the second ignition is found to propagate to the both sides. Similar to the reaction ignited secondly in the reacting region, the new combustion fronts by second ignition in Figs. 10(a) and (b) both take ~0.0275 seconds (11 time steps) to propagate to another end. In addition, the delay of the second ignition time results in an increase in the substrate temperature of the pre-heating zone, further correspondingly increases the propagation velocity of the combustion front. However, the extent of the over-reacting decreases as the second ignition time increases. When the reaction is ignited again lately in the pre-heating zone, part of the specimen in the reacted area has been cooled down. As the new combustion front arrives, the cooling specimen will consume the more heat and energy from the new combustion front to heat up itself. Therefore, the combustion temperature is decreased as the second ignition time is deferred. In addition, the under-reacting (similar to the zone m in Fig. 6(b)) is also observed as the reaction is ignited secondly in the pre-heating zone.

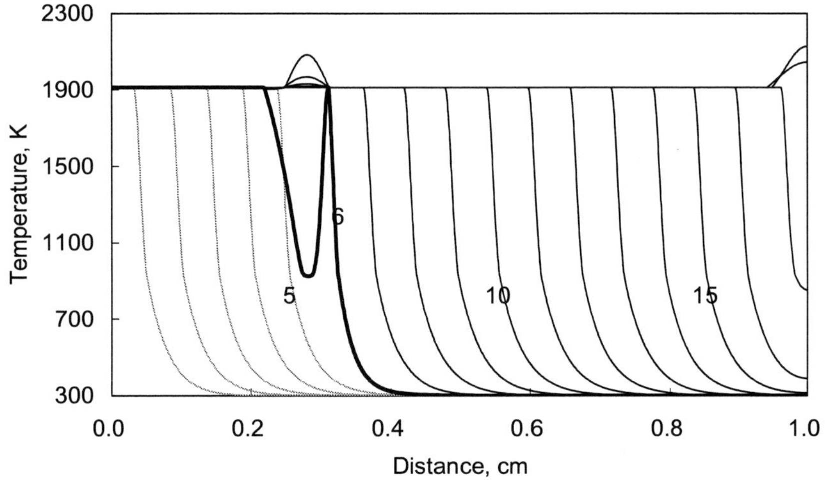

(a)

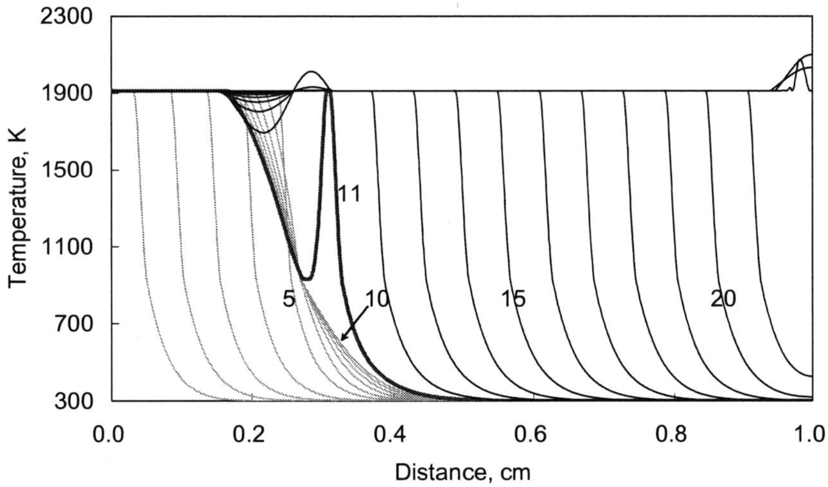

(b)

Figure 10. A plot of the combustion front temperature at various times for the reaction ignited again in the pre-heating region. The interval time between two consecutive time steps (profiles) is 0.0025 s and the reaction is ignited again at 0.3125 cm. The second ignition time in (a) is 0.0150 s (the 6th time step) and in (b) is 0.0275 s (the 11th time step) after the first ignition.

4. Ignition before the Stop of Combustion Front

Figure 11 shows the temperature profiles for the reaction ignited before the stop of the combustion front. No cooling before the second ignition is found no matter which position is ignited secondly. If the reaction is ignited behind the current combustion front (i.e., combusted region), the standard propagation profile is obtained. On the other hand, the new combustion front is formed as the reaction is ignited again in front of the combustion front (i.e., pre-heating region). A significantly increase in the temperature is found due to the accumulation of two combustion fronts.

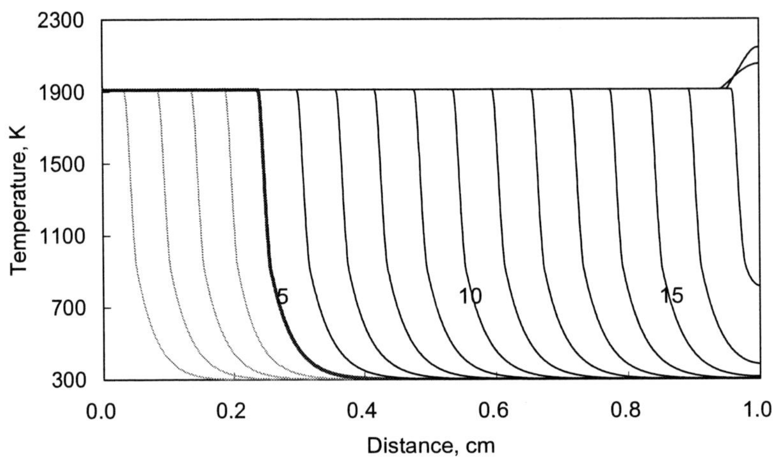

(a)

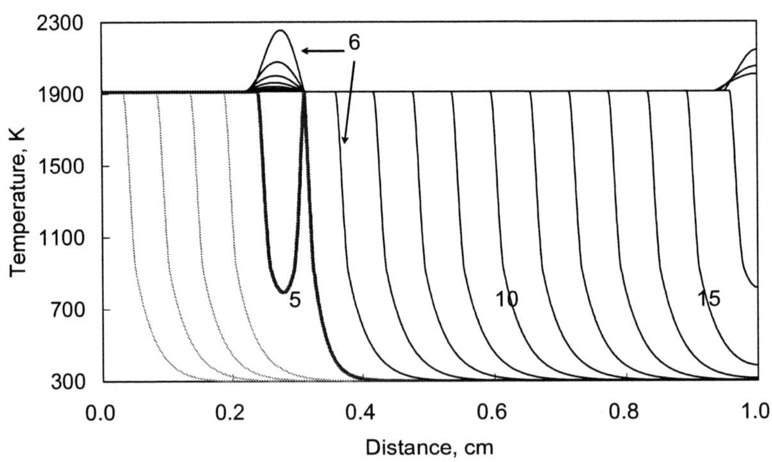

(b)

Figure 11. A plot of the combustion front temperature at various times for the reaction ignited again before the combustion front stopping. The interval time between two consecutive time steps (profiles) is 0.0025 s and the reaction is ignited again at the time of 0.0125 s (the 5th time step) after the first ignition. The second ignition position in (a) is 0.1875 cm (combusted region) and in (b) is 0.3125 cm (pre-heating region), respectively.

Process Map for the Second Ignition

To systematically study the effects of the over-reacting or under-reacting caused by the second ignition during combustion synthesis, the extents of the over-reacting or under-reacting for the different ignition conditions are calculated. Normally the combustion front

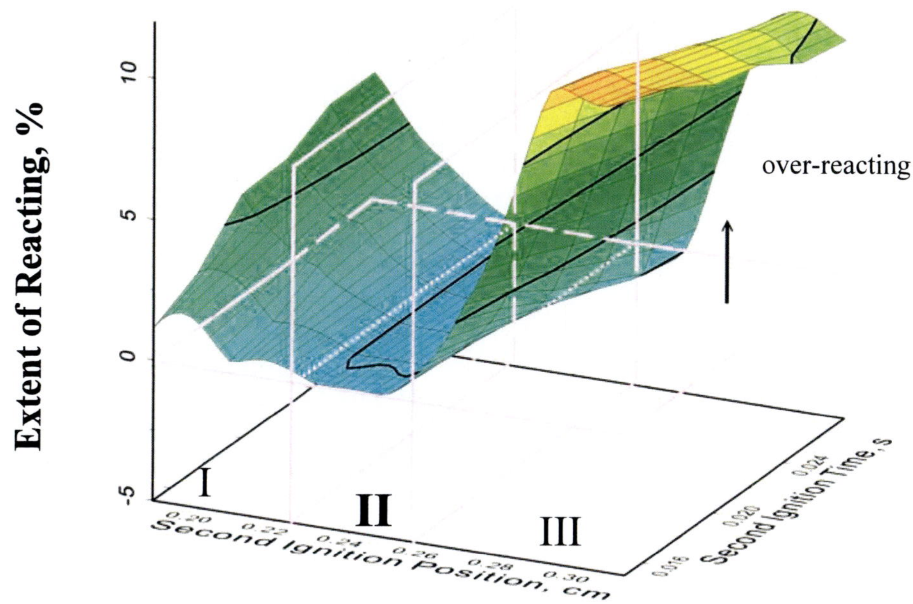

(a)

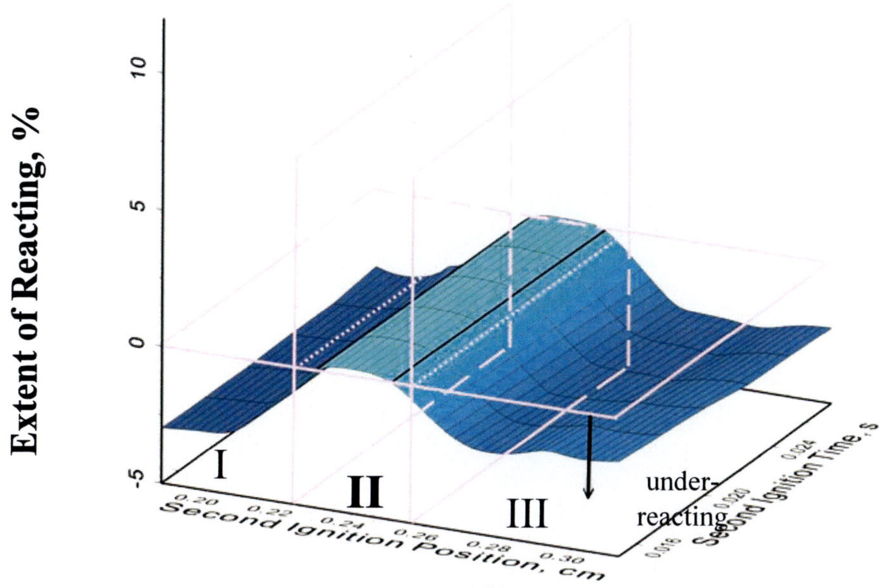

(b)

Figure 12. The process map for the extent of (a) over-reacting and (b) under-reacting. Areas I , II, and III are the combusted zone, the reacting zone, and the pre-heating in Fig. 5(b), respectively. The horizontal plane is the boundary between the over-reacting and under-reacting.

propagates at the steady values of the combustion temperature and propagation velocity to acquire the homogeneous combustion synthesized product. Any over-reacting or under-reacting may produce the heterogeneous microstructure and different mechanical properties. In this study, the extent of over-reacting (f_o) is defined as the percentage of the exceeded temperature, i.e. f_o =(combustion temperature - theoretical combustion temperature, 1912 K) / theoretical combustion temperature, and the extent of under-reacting (f_u) is the percentage of the portion that cannot be synthesized (reacted) at the theoretical combustion temperature.

The extents of the over-reacting or under-reacting for different ignition conditions are calculated and shown in Fig. 12. No over-reacting or under-reacting occurs if the reaction is ignited again in the reacting region (0.2500 cm) or the vicinity of the reacting region (0.2291 cm, in the combusted region). For the other second ignited positions, the extent of the under-reacting (absolute values) varies from the 1.67% to 3.00%. If the reaction is ignited in the combusted region, the extent of under-reacting (absolute values) is decreased as the second ignition time is prolonged or the reaction is ignited secondly near the extinguished spot (or the reacting region, 0.2500 cm). However, if the reaction is ignited in the pre-heating region, the relationship of the extent of under-reacting (absolute values) with the ignition time shows the different trend. The delay in the second ignition time increases the extent of under-reacting (absolute values). If the reaction is ignited in the combusted region, the extent of over-reacting decreases as the ignition position approaches to the extinguished spot (or the reacting region). However, the extent of over-reacting is decreased as the ignition time is prolonged as the reaction is ignited in the pre-heating region. From the knowledge of the process map, the appropriate second-ignition condition can be chosen to acquire the homogeneous microstructure. The ideal second ignition condition (no over-reacting or under-reacting occurring) is found as the reaction is ignited secondly in the reacting region or the vicinity of the combusted region.

Summary and Conclusion

The effect of the second ignition on combustion synthesizing Ni-Al compound has been numerically studied. Due to the low metallic exothermic reaction, the combustion front in the Ni-Al combustion synthesis is found to extinguish for some reactions. The reaction is ignited again to aid the front to propagate completely and achieve uniform processing. The second ignition time and position have been found to influence the subsequent temperature profiles.

The simulated results show that the heat from the second ignition is transferred to the un-reacted zone only for offering the required energy to propagate when the reaction is ignited again in the combusted region. The delay of the second ignition time or igniting far from the extinguished spot both increase the quantity of transferred heat from the extinguished spot to the pre-heating zone, correspondingly increasing the combustion temperature and resulting in the over-reacting. Meanwhile, the reaction is still in progress after the stop of the first combustion front but before the second ignition, because the high temperature aids the continuation of the combustion reaction. No over-reacting or under-reacting is observed when the reaction is ignited again in the reacting region. An extension in the delay time only increases the substrate temperature and further initial propagation velocity. When the extinguished front is ignited again in the pre-heating region, the new combustion front is formed to propagate to both sides. The reaction temperature is enhanced due to the increase in the substrate temperature and may cause the reaction over-reacting.

A process map for the second ignition is generated in this study to find the optimal condition of the second ignition position and time for facilitating combustion front propagation completely and for uniform processing. No over-reacting or under-reacting occurs if the reaction is ignited secondly in the reacting region or the vicinity of the combusted region. From the knowledge of the process map, the appropriate second-ignition condition can be chosen to acquire the homogeneous microstructure.

Nomenclature

Cp	heat capacity of product (general form), kJ/kg/K
E	activation energy, kJ/kg
Ko	pre-exponential constant, (s^{-1} for zero order reaction)
Q	heat of reaction, kJ/kg
R	gas constant, kJ/kg/K
T	temperature, K
To	initial temperature, K
z	dimensional coordinate, m
d	diameter of the specimen, m
h	surface heat transfer coefficient, $J/m^2/K/s$
t	time, s
ρ	density, kg/m^3
κ	thermal conductivity (general form), kJ/m/K/s
η	fraction reacted
$\Phi(T, \eta)$	reaction rate, 1/s

References

[1] Li, H. P., Bhaduri, S.B., and Sekhar, J.A. *Metall. Mater. Trans. A* 1993, 24*A*, 251-261.

[2] Naiborodenko, Y. S., Itin, V. I., and Savitskii, K. V. *Powder. Metall. Met. Ceram.* 1970, 7(91), 562.

[3] Munir, Z. A. and Holt, J. B., *J. Mater. Sci.* 1987, 22, 710-714.

[4] Munir, Z.A., *Am. Ceram. Bull.* 1988, 67(2), 342-349.

[5] Munir, Z.A. and Anselmi-Tamburini, U. *Mater. Sci. Reports* 1989, 3, 277-365.

[6] Booth, F. *Trans. Farad. Soc.*, 1953, 49, 272-281.

[7] Walton, J. D. Jr. and Poulos N. E. *J. Am. Ceram. Soc.* 1959, 42(1), 40-49.

[8] Li, H. P. *Modelling Sim. in Mater. Sci. Eng.* 2005, 13, 1331-1339.

[9] Li, H. P. *Modelling Sim. in Mater. Sci. Eng.* 2006, 14, 1293-1305.

[10] Li, H. P. *Mater. Sci. Eng. A* 2003, 345(1-2), 336-344.

[11] Li, H. P. *Mater. Sci. Eng. A* 2005, 392(1-2), 262-268.

[12] Li, H. P. *Mater. Sci. Eng. A* 2005, 404(1-2), 146-152.

[13] Lakshmikantha, M. G.., Bhattacharys, A., and Sekhar, J. A. *Metall. Mater. Trans. A* 1992, 23*A*, 23.
[14] Lakshmikantha, M. G. and Sekhar, J. A., *J. Mater. Sci.* 1993, 28, 6403-6408.
[15] Lakshmikantha, M. G., and Sekhar, J. A. *J. Am. Ceram. Soc.* 1994, 77(1), 202.
[16] Subramanian V, Lakshmikantha M. G. , and Sekhar J. A. *J. Mater. Res.* 1995, 10(5), 1235-1246.
[17] Li H. P. *Scripta Mater.* 2003, 50(7), 999-1002.
[18] Li, H. P., and Sekhar, J.A. *J. Mater. Res.*, 1993, 8(10), 2515-2523.
[19] Dey, G.. K., Arya A., and Sekhar J. A. *J. Mater. Res.* 2000, 15(1), 63-75.
[20] Merzhanov, A. G.. and Khaikin, B. I. *Prog. Energy Combust. Sci.* 1988, 14, 1-98.
[21] Li, H. P. *Mater. Chem. Phys.* 2003, 80(3), 758-767.
[22] Li, H. P. *Mater. Chem. Phys.* 2005, 89(1), 130-137.
[23] Li, H. P. *Acta Mater.* 2003, 51, 3213-3224.
[24] Li, H. P. *Acta Mater.* 2005, 53, 2405-2412.
[25] Li, H. P. *J. Mater. Res.* 2002 , 17(12), 3213-3221.
[26] Brain, I., Knacke, O., and Kubaschewski, O. *Thermochemical Properties of Inorganic Substances*; Springer-Verlag : New York, NY, 1973.
[27] Lide, D. R. *CRC Handbook of Chemistry and Physics* CRC : Boca Raton, FL, 1990.
[28] Brandes E. A., Brook G. B. *Smithells Metals Reference Book*; Butterworth-Heinemann Ltd. : Washington, DC, 1992.
[29] Naiborodenko, Y. S. and Itin, V. I. *Combust. Explos. Shock Waves*, 1975, 11(3), 293-300.
[30] Li, H. P. (2007). Heterogeneous Combustion Synthesis. In H. P. Glick (Eds.) *Materials Science Research Horizons* (in print) Hauppauge, NY, USA : Nova Science Publisher.

INDEX

A

B

E

F

G

H

I

J

K

L

M

N

O

P

Q

R

S

T

U

V

W

X

Y

Z